Lecture Notes in Computer Scie

Commenced Publication in 1973
Founding and Former Series Editors:
Gerhard Goos, Juris Hartmanis, and Jan van Leeuwen

Holger H. Hoos   David G. Mitchell (Eds.)

Theory and Applications of Satisfiability Testing

7th International Conference, SAT 2004
Vancouver, BC, Canada, May 10-13, 2004
Revised Selected Papers

 Springer

Volume Editors

Holger H. Hoos
University of British Columbia, Computer Science Department
2366 Main Mall, Vancouver, BC, V6T 1Z4, Canada
E-mail: hoos@cs.ubc.ca

David G. Mitchell
Simon Fraser University, School of Computing Science
Burnaby, BC, V5A 1S6, Canada
E-mail: mitchell@cs.sfu.ca

Library of Congress Control Number: 2005928808

CR Subject Classification (1998): F.4.1, I.2.3, I.2.8, I.2, F.2.2, G.1.6

ISSN	0302-9743
ISBN-10	3-540-27829-X Springer Berlin Heidelberg New York
ISBN-13	978-3-540-27829-0 Springer Berlin Heidelberg New York

Springer is a part of Springer Science+Business Media

springeronline.com

© Springer-Verlag Berlin Heidelberg 2005
Printed in Germany

Typesetting: Camera-ready by author, data conversion by Scientific Publishing Services, Chennai, India
Printed on acid-free paper   SPIN: 11527695   06/3142   5 4 3 2 1 0

Preface

The 7th International Conference on Theory and Applications of Satisfiability Testing (SAT 2004) was held 10–13 May 2004 in Vancouver, BC, Canada. The conference featured 9 technical paper sessions, 2 poster sessions, as well as the 2004 SAT Solver Competition and the 2004 QBF Solver Evaluation. It also included invited talks by Stephen A. Cook (University of Toronto) and Kenneth McMillan (Cadence Berkeley Labs). The 89 participants represented no less than 17 countries and four continents. SAT 2004 continued the series of meetings which started with the Workshops on Satisfiability held in Siena, Italy (1996), Paderborn, Germany (1998) and Renesse, The Netherlands (2000); the Workshop on Theory and Applications of Satisfiability Testing held in Boston, USA (2001); the Symposium on Theory and Applications of Satisfiability Testing held in Cincinnati, USA (2002); and the 6th International Conference on Theory and Applications of Satisfiability Testing held in Santa Margherita Ligure, Italy (2003).

The International Conference on Theory and Applications of Satisfiability Testing is the primary annual meeting for researchers studying the propositional satisfiability problem (SAT), a prominent problem in both theoretical and applied computer science. SAT lies at the heart of the most important open problem in complexity theory ($\mathcal{P}$ vs $\mathcal{NP}$) and underlies many applications in, among other examples, artificial intelligence, operations research and electronic design engineering. The primary objective of the conferences is to bring together researchers from various areas and communities, including theoretical and experimental computer science as well as many relevant application areas, to promote collaboration and the communication of new theoretical and practical results in SAT-related research and its industrial applications.

The 28 technical papers contained in this volume were selected as follows. Of the 72 technical papers submitted to the SAT 2004 conference, 30 were accepted for full presentation at the conference, and a further 18 were selected for posters and short presentations. These selections were made by the Program Committee based on a strict peer-review process, in which each submission received between two and four reviews by Program Committee members or auxiliary reviewers. Authors of accepted papers were invited to submit extended versions of those papers to this volume. From these submissions, 24 were selected for inclusion in this volume, based on another round of rigorous peer reviews. Two additional papers report on the 2004 SAT Solver Competition and the 2004 QBF Solver Evaluation. These were prepared, by invitation, by the organisers of the respective events. Furthermore, authors of the three SAT solvers which placed first in one or more categories of the 2004 SAT Solver Competition were invited to submit papers. Among these latter, one team of authors declined (because their work is presented in part in a previous publication, and in part in another paper included in this volume). These invited papers were peer-reviewed according to the same standards as the other papers in this volume.

We are very grateful to the many people who contributed to the organisation of SAT 2004, most of whom are listed on the following pages. We thank in particular Dave Tompkins for help in preparing this volume. We also thank the authors, presenters and all other attendees for making SAT 2004 a successful and memorable event.

Vancouver, Canada, 8 April 2005   Holger H. Hoos and David G. Mitchell

Organisation

Conference and Programme Co-chairs

Holger H. Hoos (University of British Columbia)
David G. Mitchell (Simon Fraser University)

Organising Committee

John Franco (University of Cincinnati)
Enrico Giunchiglia (Università di Genova)
Holger H. Hoos (University of British Columbia)
Henry Kautz (University of Washington)
Hans Kleine Büning (Universität Paderborn)
David G. Mitchell (Simon Fraser University)
Bart Selman (Cornell University)
Ewald Speckenmeyer (Universität zu Köln)
Hans van Maaren (Delft University of Technology)

Programme Committee

Dimitris Achlioptas (Microsoft Research)
Fahiem Bacchus (University of Toronto)
Paul Beame (University of Washington)
Armin Biere (ETH Zürich)
Olivier Dubois (Université Paris 6)
John Franco (University of Cincinnati)
Ian Gent (University of St Andrews)
Enrico Giunchiglia (Università di Genova)
Carla Gomes (Cornell University)
Ziyad Hanna (Intel Corporation)
Edward A. Hirsch (Steklov Institute of Mathematics at St. Petersburg)
Holger H. Hoos (University of British Columbia)
Henry Kautz (University of Washington)
Lefteris Kirousis (University of Patras)
Hans Kleine Büning (Universität Paderborn)
Daniel Le Berre (Université d'Artois)
Chu-Min Li (Université de Picardie Jules Verne)
Sharad Malik (Princeton University)
David G. Mitchell (Simon Fraser University)
Rémi Monasson (Ecole Normale Supérieure)
Bart Selman (Cornell University)

Laurent Simon (Université Paris-Sud)
Ewald Speckenmeyer (Universität zu Köln)
Allen Van Gelder (University of California at Santa Cruz)
Hans van Maaren (Delft University of Technology)
Miroslav Velev (Carnegie Mellon University)
Toby Walsh (University College Cork)

SAT Competition Organisers

Daniel Le Berre (Université d'Artois)
Laurent Simon (Université Paris-Sud)

QBF Evaluation Organisers

Daniel Le Berre (Université d'Artois)
Massimo Narizzano (Università di Genova)
Laurent Simon (Université Paris-Sud)
Armando Tacchella (Università di Genova)

Local Organisation

Olga German (PIMS, Simon Fraser University)
Valerie McRae (LCI, University of British Columbia)
Dave Tompkins (β-Lab, University of British Columbia)

Additional Referees

Andrei Bulatov
Gilles Dequen
Laure Devendeville
Yulik Feldman
Alan Frisch
Zhaohui Fu
Yong Gao
Marijn Heule
Dmitry Itsykson
Abhijit Jas
Peter Jeavons
Bernard Jurkowiak

Alexis Kaporis
Jacob Katz
Zurab Khasidashvili
Zeynep Kiziltan
Arist Kojevnikov
Boris Konev
Alexander Kulikov
Yogesh Mahajan
Marco Maratea
Alex Nadel
Massimo Narizzano
Peter Nightingale

Stefan Porschen
Steve Prestwich
Bert Randerath
Andrew Rowley
Kevin Smyth
Armando Tacchella
Daijue Tang
Dimitrios Thilikos
Michael Trick
Yinlei Yu

Sponsoring Institutions

Pacific Institute for the Mathematical Sciences (PIMS)
Intel
Intelligent Information Systems Institute (IISI) at Cornell University
CoLogNet
Simon Fraser University (SFU)
University of British Columbia (UBC)

Table of Contents

2004 SAT Solver Competition and QBF Solver Evaluation (Invited Papers)

Mapping Problems with Finite-Domain Variables to Problems with Boolean Variables*

Carlos Ansótegui and Felip Manyà

Computer Science Department,
Universitat de Lleida,
Jaume II, 69, E-25001 Lleida, Spain
{carlos, felip}@eup.udl.es

Abstract. We define a collection of mappings that transform many-valued clausal forms into satisfiability equivalent Boolean clausal forms, analyze their complexity and evaluate them empirically on a set of benchmarks with state-of-the-art SAT solvers. Our results provide empirical evidence that encoding combinatorial problems with the mappings defined here can lead to substantial performance improvements in complete SAT solvers.

1   Introduction

In the last few years, the AI community has investigated the generic problem solving approach which consists of modeling hard combinatorial problems as instances of the propositional satisfiability problem (SAT) and then solving the resulting encodings with algorithms for SAT. The success in solving SAT-encoded problems depends on both the SAT solver and the SAT encoding used. While there has been a tremendous advance in the design and implementation of SAT solvers, our understanding of SAT encodings is very limited and is yet a challenge for the AI community working on propositional reasoning.

In this paper we define a collection of mappings that transform many-valued clausal forms into satisfiability equivalent Boolean clausal forms and analyze their complexity. Given a combinatorial problem encoded as a many-valued clausal form, the mappings defined allow us to generate six different Boolean SAT encodings. We evaluated empirically the Boolean SAT encodings generated for a number of combinatorial problems (graph coloring, random binary CSPs, pigeon hole, and all interval series) using Chaff [21] and Siege_v4.[1] Our results provide empirical evidence that encoding combinatorial problems with the mappings defined here can lead to substantial performance improvements in complete

---

* Research partially supported by projects TIC2001-1577-C03-03 and TIC2003-00950 funded by the *Ministerio de Ciencia y Tecnología*. We thank Carla Gomes for allowing us to use computational resources of the Intelligent Information Systems Institute (Cornell University).

[1] Siege_v4 is publicly available at http://www.cs.sfu.ca/~loryan/personal

H.H. Hoos and D.G. Mitchell (Eds.): SAT 2004, LNCS 3542, pp. 1–15, 2005.

SAT solvers. The behaviour of different SAT encodings of graph coloring and all interval series instances on local search solvers was analyzed in [1, 23].

These results are part of a research program about many-valued satisfiability that our research group has developed during the last decade (see e.g. [2, 5, 9, 11, 18, 20]). Our research program is aimed at bridging the gap between Boolean SAT encodings and constraint satisfaction formalisms. The challenge is to combine the inherent efficiencies of Boolean SAT solvers operating on uniform encodings with the much more compact and natural representations, and more sophisticated propagation techniques of CSP formalisms.

We have used before mappings between many-valued clausal forms and Boolean clausal forms to identify new polynomially solvable many-valued SAT problems [7, 19], to known which additional deductive machinery is required to design many-valued SAT solvers from Boolean SAT solvers [7, 10], and to solve many-valued SAT encodings with Boolean SAT solvers [3, 4]. We invite the reader to consult two survey papers [8, 17] that contain a summary of our previous work.

The paper is structured as follows. In Section 2, we formally define the syntax and semantics of the many-valued clausal forms used in the paper. In Section 3, we define six mappings that transform many-valued clausal forms into satisfiability equivalent Boolean clausal forms. In Section 4, we report the empirical investigation conducted to assess the performance of those mappings.

2   Many-Valued Formulas

We first formally define the syntax and semantics of signed CNF formulas, and then present monosigned and regular CNF formulas, which are the subclasses of signed CNF formulas that are considered in this paper.

Definition 1. *A* truth value set *N is a non-empty finite set $\{i_1, i_2, \ldots, i_n\}$ where $n \in \mathbb{N}$. The cardinality of N is denoted by $|N|$. A total order $\leq$ is associated with N, which may be the empty order.*

Definition 2. *A* sign *is a set $S \subseteq N$ of truth values. A* signed literal *is an expression of the form $S\!:\!p$ where S is a sign and p is a propositional variable. The* complement *of a signed literal $S\!:\!p$, denoted by $\overline{S}\!:\!p$, is $(N \setminus S)\!:\!p$. A* signed clause *is a disjunction of signed literals. A* signed CNF formula *is a conjunction of signed clauses. The* size *of a signed clause C, denoted by $|C|$, is the total number of literals occurring in C, and the* size *of a signed CNF formula Γ, denoted by $|\Gamma|$, is the sum of the sizes of the clauses of Γ.*

Definition 3. *An* interpretation *is a mapping that assigns to every propositional variable an element of the truth value set. An interpretation I satisfies a signed literal $S\!:\!p$ iff $I(p) \in S$, satisfies a signed clause C iff it satisfies at least one of the signed literals in C, and satisfies a signed CNF formula Γ iff it satisfies all clauses in Γ. A signed CNF formula is* satisfiable *iff it is satisfied by at least one interpretation; otherwise it is* unsatisfiable.

Definition 4. *A* sign *S is* monosigned *if it either is a singleton (i.e. it contains exactly one truth value) or the complement of a singleton. A monosigned sign S is* positive *if it is identical to $\{i\} : p$, and is* negative *if it is identical to $\{\bar{i}\} : p$ for some $i \in N$. A signed literal $S : p$ is a* monosigned literal *if its sign S is monosigned. A signed clause (a signed CNF formula) is a* monosigned clause *(a monosigned CNF formula) if all its literals are monosigned.*

Definition 5. *Given a monosigned CNF formula Γ, the* domain *of a variable p occurring in Γ is $N_\Gamma(p) = \{i \in N \mid \{i\} : p$ or $\{\bar{i}\} : p$ occur in $\Gamma\}$ if $N_\Gamma(p) = N$, and $N_\Gamma(p) \cup \{j\}$, where j is any element of $N \setminus N_\Gamma(p)$, otherwise. The* Boolean signature *of Γ is $\Sigma = \{\{i\} : p \mid \{i\} : p$ or $\{\bar{i}\} : p$ occur in $\Gamma\}$.*

Definition 6. *For all $i \in N$, let $\uparrow i$ denote the sign $\{j \in N \mid j \geq i\}$, where $\leq$ is the total order associated with N, and let $\overline{\uparrow i}$ denote the complement of $\uparrow i$. A sign S is* regular *if it either is identical to $\uparrow i$ (positive) or to $\overline{\uparrow i}$ (negative) for some $i \in N$. A signed literal $S : p$ is a* regular literal *if its sign S is regular. A signed clause (a signed CNF formula) is a* regular clause *(a regular CNF formula) if all its literals are regular.*

Definition 7. *Given a regular CNF formula Γ, the* domain *of a variable p occurring in Γ is $N_\Gamma(p) = \{i \in N \mid \uparrow i : p$ or $\overline{\uparrow i} : p$ occur in $\Gamma\}$. The* Boolean signature *of Γ is $\Sigma = \{\uparrow i : p \mid \uparrow i : p$ or $\overline{\uparrow i} : p$ occur in $\Gamma\}$.*

Example 1. Suppose that $N = \{1, 2, 3, 4\}$. Then, we have that the signed clause $\{1, 2, 3\} : p_1 \vee \{4\} : p_2$ can be represented as a monosigned clause by $\{\bar{4}\} : p_1 \vee \{4\} : p_2$, and as a regular clause by $\overline{\uparrow 4} : p_1 \vee \uparrow 4 : p_2$.

The notation used in this paper is the one used in the many-valued logic community, which is the notation we used in our previous work on many-valued satisfiability. Since some readers can find hard to read that notation, we next show how to encode a graph coloring problem as a signed CNF formula.

Example 2. Given an undirected graph $G = (V, E)$, where V is the set of vertices ans E is the set of edges, the 3-colorability problem of G is encoded as a signed CNF formula as follows: for each edge $[u, v] \in E$, we define three signed binary clauses

$$(\{2, 3\} : u \vee \{2, 3\} : v) \wedge (\{1, 3\} : u \vee \{1, 3\} : v) \wedge (\{1, 2\} : u \vee \{1, 2\} : v)$$

and take as truth value set $N = \{1, 2, 3\}$.[2] The intended meaning of the previous signed clauses is that there are no two adjacent vertices with the same color.

Signed CNF formulas and their subclasses have been studied since the early 90's by the research community working on automated theorem proving in many-valued logics [6, 13, 15, 16, 22]. A few years later, Frisch and Peugniez [14] used the term non-Boolean formulas to refer to signed CNF formulas.

---

[2] These clauses are represented as monosigned clauses by $(\overline{\{1\}} : u \vee \overline{\{1\}} : v) \wedge (\overline{\{2\}} : u \vee \overline{\{2\}} : v) \wedge (\overline{\{3\}} : u \vee \overline{\{3\}} : v)$.

3   Mappings

We define a number of mappings that transform a monosigned CNF formula into a satisfiability equivalent Boolean CNF formula. In the most straightforward mappings, the derived formula consists of the input monosigned CNF formula under Boolean semantics (i.e., monosigned literals are interpreted as Boolean literals, and the notion of satisfiability is Boolean) plus a set of clauses that link many-valued interpretations with Boolean interpretations. The additional clauses ensure that exactly one of the literals of the Boolean signature of the monosigned CNF formula which correspond to a certain many-valued variable evaluates to true under Boolean semantics. We consider several cases: using only the Boolean signature of the monosigned CNF formula; extending the Boolean signature with regular literals under Boolean semantics; and extending the Boolean signature with a logarithmic number of Boolean variables for each many-valued variable (i.e., using a logarithmic encoding of the many-valued variables). In the most involved mappings, monosigned literals are replaced by their regular or logarithmic encoding in the input monosigned CNF formula, and its Boolean signature is replaced by a regular or logarithmic signature.

We analyze the complexity of the Boolean CNF formula derived by each mapping as a function of the size of the input monosigned CNF formula and the cardinality of the truth value set.

3.1   Standard Mapping (S)

The most straightforward mapping consists of dealing with the Boolean signature of the input monosigned CNF formula. In the standard (S) mapping, each positive monosigned literal of the input monosigned CNF formula is taken as a Boolean variable, and each negative monosigned literal is replaced with the negation of its complement and is taken as a negative Boolean literal; i.e., we take the input monosigned CNF formula under Boolean semantics. Moreover, we add for each many-valued variable p, a clause that states that p takes at least one value of its domain (ALO clause) and a set of clauses that state that p takes at most one value of its domain (AMO clauses). Assume that the domain of p in the input monosigned CNF formula Γ is $N_\Gamma(p) = \{i_1, \ldots, i_{m(p)}\}$. Then, the ALO clause is $\{i_1\} : p \vee \cdots \vee \{i_{m(p)}\} : p$, and the set of AMO clauses contains a clause $\neg(\{i_j\} : p) \vee \neg(\{i_k\} : p)$ for all j and k such that $1 \leq i < j \leq m(p)$.

Example 3. Let $N_\Gamma(p_1) = \{1, 2, 3\}$, let $N_\Gamma(p_2) = \{1, 2, 3, 4\}$, and let Γ be the following monosigned CNF formula:

$$\overline{\{1\}} : p_1 \wedge (\{3\} : p_1 \vee \overline{\{1\}} : p_2) \wedge (\overline{\{2\}} : p_2 \vee \overline{\{3\}} : p_2)$$

Mapping S generates the following formula:

$$\neg(\{1\} : p_1) \wedge (\{3\} : p_1 \vee \neg(\{1\} : p_2)) \wedge (\neg(\{2\} : p_2) \vee \neg(\{3\} : p_2))$$
$$\{1\} : p_1 \vee \{2\} : p_1 \vee \{3\} : p_1$$
$$\neg(\{1\} : p_1) \vee \neg(\{2\} : p_1)$$

$$\neg(\{1\}:p_1) \vee \neg(\{3\}:p_1)$$
$$\neg(\{2\}:p_1) \vee \neg(\{3\}:p_1)$$
$$\{1\}:p_2 \vee \{2\}:p_2 \vee \{3\}:p_2 \vee \{4\}:p_2$$
$$\neg(\{1\}:p_2) \vee \neg(\{2\}:p_2)$$
$$\neg(\{1\}:p_2) \vee \neg(\{3\}:p_2)$$
$$\neg(\{1\}:p_2) \vee \neg(\{4\}:p_2)$$
$$\neg(\{2\}:p_2) \vee \neg(\{3\}:p_2)$$
$$\neg(\{2\}:p_2) \vee \neg(\{4\}:p_2)$$
$$\neg(\{3\}:p_2) \vee \neg(\{4\}:p_2)$$

The size of the SAT instance generated by mapping S from a monosigned CNF formula Γ is in $\mathcal{O}(|\Gamma|\,|N|^2)$: The size of the instance generated by S is the sum of the size of Γ, denoted by $|\Gamma|$, plus the sum of the size of the ALO clauses and the size of the AMO clauses. For every many-valued variable p, there is an ALO clauses of size $|N_\Gamma(p)|$, where $|N_\Gamma(p)|$ is the size of the domain of p. If the number of distinct many-valued variables occurring in Γ is var, the size of all the ALO clauses is in $\mathcal{O}(var\,|N|)$. For every many-valued variable p, there are $\frac{|N_\Gamma(p)|(|N_\Gamma(p)|-1)}{2}$ AMO clauses of size two, and the size of all the AMO clauses is in $\mathcal{O}(var\,|N|^2)$. Therefore, the size of the instance generated by S is in $\mathcal{O}(|\Gamma| + var\,|N|^2)$. Since $|\Gamma| \geq var$, the size of the instance generated by S is in $\mathcal{O}(|\Gamma|\,|N|^2)$.

3.2   Full Logarithmic Mapping (FL)

In the full logarithmic (FL) mapping, a logarithmic encoding is used to represent a many-valued variable as a Boolean variable. To encode a many-valued variable p, using a base 2 encoding, only $\lceil \log_2 |N_\Gamma(p)| \rceil$ Boolean variables are required. For example, if p has domain $\{1,2,3,4\}$, then the monosigned literal $\{1\}:p$ is mapped to $\neg p^2 \wedge \neg p^1$, the monosigned literal $\{2\}:p$ is mapped to $\neg p^2 \wedge p^1$, the monosigned literal $\{3\}:p$ is mapped to $p^2 \wedge \neg p^1$, and the monosigned literal $\{4\}:p$ is mapped to $p^2 \wedge p^1$. If the size of the domain of p is not a power of 2, then two combinations are mapped to some monosigned literals. For example, if the domain of p is $\{1,2,3\}$, then $\{1\}:p$ is mapped to $\neg p^2$ (which subsumes $\neg p^2 \wedge p^1$ and $\neg p^2 \wedge \neg p^1$), $\{2\}:p$ is mapped to $p^2 \wedge \neg p^1$, and $\{3\}:p$ is mapped to $p^2 \wedge p^1$.[3]

Given a monosigned CNF formula Γ, the signature of mapping FL is $\Sigma = \{p^j \mid 1 \leq j \leq \lceil \log_2 |N_\Gamma(p)| \rceil, p \text{ occurs in } \Sigma\}$, each positive monosigned literal occurring in the input monosigned CNF formula is replaced with its logarithmic encoding, and each negative monosigned literal of the form $\{\bar{i}\}:p$ is replaced with the negation of the logarithmic encoding of $\{i\}:p$.

Example 4. Let $N_\Gamma(p_1) = \{1,2,3\}$, let $N_\Gamma(p_2) = \{1,2,3,4\}$, and let Γ be the following monosigned CNF formula:

---

[3] This approach of dealing with domains which are not a power of 2 is used in [14]. Another possibility is to map $\{1\}:p$ to $\neg p^2 \wedge \neg p^1$, $\{2\}:p$ to $\neg p^2 \wedge p^1$, $\{3\}:p$ to $p^2 \wedge \neg p^1$, and to add the clause $\neg p^2 \vee \neg p^1$ to exclude the combination $p^2 \wedge p^1$.

$$\overline{\{1\}} : p_1 \wedge (\{3\} : p_1 \vee \overline{\{1\}} : p_2) \wedge (\overline{\{2\}} : p_2 \vee \overline{\{3\}} : p_2)$$

Mapping FL generates the following formula:

$$p_1^2 \wedge ((p_1^2 \wedge p_1^1) \vee p_2^2 \vee p_2^1) \wedge (p_2^2 \vee \neg p_2^1 \vee \neg p_2^2 \vee p_2^1)$$

The size of the SAT instance generated by mapping FL is, in the worst case, exponentially larger than the size of the input monosigned CNF formula. The problem is that we must apply distributivity to get a clausal form when we encode positive monosigned literals. To overcome that drawback, Frisch and Peugniez [14] defined the logarithmic mapping.

3.3   Logarithmic Mapping (L)

Frisch and Peugniez [14] defined the logarithmic (L) mapping, which combines mapping S and mapping FL. Given a monosigned formula Γ, the signature of mapping L is the union of the Boolean signature and the signature of mapping FL. The Boolean CNF formula derived by mapping L is formed by Γ plus an additional set of clauses that link monosigned literals with the logarithmic encoding; this way they avoid incorporating the ALO and AMO clauses. For example, if the many-valued variable p has domain $\{1, 2, 3, 4\}$, then they add the following clauses to link the monosigned literals containing variable p with their logarithmic encoding:[4]

$$\{1\} : p \leftrightarrow \neg p^2 \wedge \neg p^1, \ \{2\} : p \leftrightarrow \neg p^2 \wedge p^1, \ \{3\} : p \leftrightarrow p^2 \wedge \neg p^1, \ \{4\} : p \leftrightarrow p^2 \wedge \neg p^1$$

Example 5. Let $N_\Gamma(p_1) = \{1, 2, 3\}$, let $N_\Gamma(p_2) = \{1, 2, 3, 4\}$, and let Γ be the following monosigned CNF formula:

$$\overline{\{1\}} : p_1 \wedge (\{3\} : p_1 \vee \overline{\{1\}} : p_2) \wedge (\overline{\{2\}} : p_2 \vee \overline{\{3\}} : p_2)$$

Mapping L generates the following formula:

$$\neg(\{1\} : p_1) \wedge (\{3\} : p_1 \vee \neg(\{1\} : p_2)) \wedge (\neg(\{2\} : p_2) \vee \neg(\{3\} : p_2))$$

$$\{1\} : p_1 \leftrightarrow \neg p_1^2 \qquad \{1\} : p_2 \leftrightarrow \neg p_2^2 \wedge \neg p_2^1$$
$$\{2\} : p_1 \leftrightarrow p_1^2 \wedge \neg p_1^1 \quad \{2\} : p_2 \leftrightarrow \neg p_2^2 \wedge p_2^1$$
$$\{3\} : p_1 \leftrightarrow p_1^2 \wedge p_1^1 \quad \{3\} : p_2 \leftrightarrow p_2^2 \wedge \neg p_2^1$$
$$\{4\} : p_2 \leftrightarrow p_2^2 \wedge p_2^1$$

Note that, with the ALO and AMO clauses, the number of clauses needed in mapping S to state that a many-valued variable takes exactly one value from its domain is in $\mathcal{O}(|N|^2)$, but with the previous transformation the number of clauses needed is in $\mathcal{O}(|N| \log_2 |N|)$. The size of the SAT instance generated by mapping L from a monosigned CNF formula Γ is in $\mathcal{O}(|\Gamma| \log_2 |N|)$.

---

[4] In the rest of the paper, we write $A \leftrightarrow B$, where A and B are propositional formulas, instead of its clausal form for the sake of readability. For instance, when we write $\{1\} : p \leftrightarrow \neg p^2 \wedge \neg p^1$, we mean $(\neg(\{1\} : p) \vee \neg p^2) \wedge (\neg(\{1\} : p) \vee \neg p^1) \wedge (\{1\} : p \vee p^2 \vee p^1)$.

3.4 Full Regular Mapping (FR)

Béjar, Hähnle and Manyà [10] defined the full regular (FR) mapping, which transforms a regular CNF formula Γ into a satisfiability equivalent Boolean CNF formula whose size is in $\mathcal{O}(|\Gamma|)$. In this section we reformulate mapping FR in the case that the input formula is a monosigned CNF formula instead of a regular CNF formula.

Given a regular CNF formula Γ, the signature of mapping FR is $\Sigma = \{\uparrow i : p \mid \uparrow i : p \text{ or } \overline{\uparrow i} : p \text{ occur in } \Gamma\}$; i.e., the Boolean signature of Γ. In mapping FR, each positive regular literal is taken as a positive Boolean literal, and each negative regular literal is taken as a negative Boolean literal. Moreover, we add, for each many-valued variable p, a set of clauses that link regular interpretations with Boolean interpretations [10]. Assume that the domain of p in the input regular CNF formula Γ is $N_\Gamma(p) = \{i_1, \ldots, i_{m(p)}\}$ and $i_1 \leq i_2 \leq \cdots \leq i_{m(p)}$ under the order $\leq$ associated with N. Then, the set of clauses added is:

$$\{\neg(\uparrow i_{(j+1)} : p) \vee \uparrow i_j : p \mid 1 \leq j < m(p)\}.$$

The variant of mapping FR for monosigned CNF formulas takes the same signature as mapping FR for regular CNF formulas. Given a monosigned CNF formula Γ and a many-valued variable p occurring in Γ whose domain is $N_\Gamma(p) = \{i_1, \ldots, i_{m(p)}\}$ and $i_1 \leq i_2 \leq \cdots \leq i_{m(p)}$ under the order $\leq$ associated with N, each positive monosigned literal occurring in the input monosigned CNF formula of the form $\{i_1\} : p$ is replaced with $\neg(\uparrow i_2 : p)$, of the form $\{i_{m(p)}\} : p$ is replaced with $\uparrow i_{m(p)} : p$, and of the form $\{i_j\} : p$, where $1 < j < m(p)$, is replaced with $\uparrow i_j : p \wedge \neg(\uparrow i_{j+1} : p)$; and each negative monosigned literal occurring in the input monosigned CNF formula of the form $\{\overline{i_1}\} : p$ is replaced with $\uparrow i_2 : p$, of the form $\{\overline{i_{m(p)}}\} : p$ is replaced with $\neg(\uparrow i_{m(p)} : p)$, and of the form $\{\overline{i_j}\} : p$, where $1 < j < m(p)$, is replaced with $\neg(\uparrow i_j : p) \vee \uparrow i_{j+1} : p$. In addition, it is added the set of clauses that link regular interpretations with Boolean interpretations as in the regular case.

Example 6. Let $N_\Gamma(p_1) = \{1, 2, 3\}$, let $N_\Gamma(p_2) = \{1, 2, 3, 4\}$, and let Γ be the following CNF formula:

$$\overline{\{1\}} : p_1 \wedge (\{3\} : p_1 \vee \overline{\{1\}} : p_2) \wedge (\overline{\{2\}} : p_2 \vee \overline{\{3\}} : p_2)$$

Mapping FR generates the following formula:

$$\uparrow 2 : p_1 \wedge (\uparrow 3 : p_1 \vee \uparrow 2 : p_2) \wedge (\neg(\uparrow 2 : p_2) \vee \uparrow 3 : p_2 \vee \neg(\uparrow 3 : p_2) \vee \uparrow 4 : p_2)$$

$$\neg(\uparrow 3 : p_1) \vee \uparrow 2 : p_1 \quad \neg(\uparrow 4 : p_2) \vee \uparrow 3 : p_2$$
$$\neg(\uparrow 2 : p_1) \vee \uparrow 1 : p_1 \quad \neg(\uparrow 3 : p_2) \vee \uparrow 2 : p_2$$
$$\neg(\uparrow 2 : p_2) \vee \uparrow 1 : p_2$$

The problem with mapping FR for monosigned CNF formulas is that the size of the derived formula can be exponential in the size of the input formula.

This is so because we must apply distributivity when mapping clauses containing positive monosigned literals. For instance, if instead of the CNF formula of Example 6, we have the CNF formula $(\{2\}:p_1 \vee \{2\}:p_2) \wedge (\{2\}:p_1 \vee \{3\}:p_2)$, we get the formula $((\uparrow 2:p_1 \wedge \neg(\uparrow 2:p_1)) \vee (\uparrow 2:p_2 \wedge \neg(\uparrow 2:p_2))) \wedge ((\uparrow 2:p_1 \wedge \neg(\uparrow 2:p_1)) \vee (\uparrow 3:p_2 \wedge \neg(\uparrow 3:p_2)))$, whose clausal form is exponential in the size of the input formula.

3.5   Regular Mapping (R)

The regular (R) mapping, which combines mapping S and mapping FR, is a new mapping whose complexity is better than the complexity of the previous mappings.

Given a monosigned CNF formula Γ, the signature of mapping R is $\Sigma = \{\{i\}:p, \uparrow i:p \mid \{i\}:p$ or $\{\overline{i}\}:p$ occur in $\Gamma\}$; i.e., the Boolean signature of Γ extended with regular signs. The Boolean CNF formula derived by mapping R is formed by (i) the clauses of Γ under Boolean semantics; (ii) the set of clauses of mapping FR that link the Boolean variables representing regular literals; and (iii) a set of clauses, for each variable p occurring in Γ, that link monosigned literals with regular literals. Assume that $N_\Gamma(p) = \{i_1, i_2, \ldots, i_{m(p)}\}$. Then, we add the following clauses

$$\{\{i_1\}:p \leftrightarrow \neg(\uparrow i_2:p)\} \cup \{\{i_j\}:p \leftrightarrow \uparrow i_j:p \wedge \neg(\uparrow i_{j+1}:p) \mid 1 < j < m(p)\} \cup$$

$$\{\{i_{m(p)}\}:p \leftrightarrow \uparrow i_{m(p)}:p\}$$

The idea of mapping R is that we maintain the input monosigned CNF formula under Boolean semantics but we use both regular and monosigned literals to link many-valued interpretations with Boolean interpretations. This way we avoid applying distributivity.

Example 7. Let $N_\Gamma(p_1) = \{1, 2, 3\}$, let $N_\Gamma(p_2) = \{1, 2, 3, 4\}$, and let Γ be the following monosigned CNF formula:

$$\overline{\{1\}}:p_1 \wedge (\{3\}:p_1 \vee \overline{\{1\}}:p_2) \wedge (\overline{\{2\}}:p_2 \vee \overline{\{3\}}:p_2)$$

Mapping R generates the following formula:

$$\neg(\{1\}:p_1) \wedge (\{3\}:p_1 \vee \neg(\{1\}:p_2)) \wedge (\neg(\{2\}:p_2) \vee \neg(\{3\}:p_2))$$

$\{1\}:p_1 \leftrightarrow \neg(\uparrow 2:p_1)$	$\{1\}:p_2 \leftrightarrow \neg(\uparrow 2:p_2)$
$\{2\}:p_1 \leftrightarrow \uparrow 2:p_1 \wedge \neg(\uparrow 3:p_1)$	$\{2\}:p_2 \leftrightarrow \uparrow 2:p_2 \wedge \neg(\uparrow 3:p_2)$
$\{3\}:p_1 \leftrightarrow \uparrow 3:p_1$	$\{3\}:p_2 \leftrightarrow \uparrow 3:p_2 \wedge \neg(\uparrow 4:p_2)$
	$\{4\}:p_2 \leftrightarrow \uparrow 4:p_2$

$\neg(\uparrow 3:p_1) \vee \uparrow 2:p_1$	$\neg(\uparrow 4:p_2) \vee \uparrow 3:p_2$
$\neg(\uparrow 2:p_1) \vee \uparrow 1:p_1$	$\neg(\uparrow 3:p_2) \vee \uparrow 2:p_2$
	$\neg(\uparrow 2:p_2) \vee \uparrow 1:p_2$

The size of the SAT instance generated by mapping R from a monosigned CNF formula Γ is in $\mathcal{O}(|\Gamma|)$.[5]

3.6   Half Regular Mapping (HR)

We now define another mapping, called half regular (HR) mapping, which is between FR and R. We defined R in order to avoid applying distributivity. To this end, R maintains the input monosigned CNF formula under Boolean semantics. Since the blowup is only due to the encoding of positive monosigned literals, HR maps negative monosigned literals as in mapping FR and positive monosigned literals as in mapping R. This way, the size of the SAT instance generated by mapping HR from a monosigned CNF formula Γ is also in $\mathcal{O}(|\Gamma|)$.

Example 8. Let $N_\Gamma(p_1) = \{1, 2, 3\}$, let $N_\Gamma(p_2) = \{1, 2, 3, 4\}$, and let Γ be the following monosigned CNF formula:

$$\overline{\{1\}} : p_1 \wedge (\{3\} : p_1 \vee \overline{\{1\}} : p_2) \wedge (\overline{\{2\}} : p_2 \vee \overline{\{3\}} : p_2)$$

Mapping HR generates the following formula:

$$\uparrow 2 : p_1 \wedge (\{3\} : p_1 \vee \uparrow 2 : p_2) \wedge (\neg(\uparrow 2 : p_2) \vee \uparrow 3 : p_2 \vee \neg(\uparrow 3 : p_2) \vee \uparrow 4 : p_2)$$

$\{1\} : p_1 \leftrightarrow \neg(\uparrow 2 : p_1)$ $\{1\} : p_2 \leftrightarrow \neg(\uparrow 2 : p_2)$
$\{2\} : p_1 \leftrightarrow \uparrow 2 : p_1 \wedge \neg(\uparrow 3 : p_1)$ $\{2\} : p_2 \leftrightarrow \uparrow 2 : p_2 \wedge \neg(\uparrow 3 : p_2)$
$\{3\} : p_1 \leftrightarrow \uparrow 3 : p_1$ $\{3\} : p_2 \leftrightarrow \uparrow 3 : p_2 \wedge \neg(\uparrow 4 : p_2)$
                                   $\{4\} : p_2 \leftrightarrow \uparrow 4 : p_2$

$\neg(\uparrow 3 : p_1) \vee \uparrow 2 : p_1$ $\neg(\uparrow 4 : p_2) \vee \uparrow 3 : p_2$
$\neg(\uparrow 2 : p_1) \vee \uparrow 1 : p_1$ $\neg(\uparrow 3 : p_2) \vee \uparrow 2 : p_2$
                                   $\neg(\uparrow 2 : p_2) \vee \uparrow 1 : p_2$

4   Experimental Investigation

We next report the experimental investigation we conducted to evaluate the performance of the mappings on a number of benchmarks: graph coloring, random binary CSPs, pigeon hole, and all interval series. All the experiments were performed with PC's Pentium III with 1.1 Ghz under Linux, and the SAT solvers used were Chaff and Siege_v4.

In the first experiment, we considered flat graph coloring problems, generated with the generator of Culberson [12]. The parameters of the generator are: number of vertices (n), number of colors (k), and edge density (p). We created

---

[5] Observe that all the added clauses have at most three literals, and the number of added clauses is in $\mathcal{O}(lit)$, where *lit* is the number of occurrences of distinct literals occurring in Γ. Since $|\Gamma| \geq lit$, the size of the instance generated by HR is in $\mathcal{O}(|\Gamma|)$.

Table 1. Experimental results for graph coloring with Chaff. Time in seconds

parameters			S			FR			HR			R			FL			L		
n	p	k	m	md	%	m	md	%	m	md	%	m	md	%	m	md	%	m	md	%
400	0.02	3	494	335	80	606	186	68	523	194	60	670	504	66	556	183	92	441	176	72
200	0.13	5	518	208	66	726	555	76	603	472	72	445	157	60	1052	1207	56	1214	1214	2
80	0.5	13	137	9	84	61	4	88	69	6.45	88	111	4.2	84	4.4	2.4	98	65	17.17	96
70	0.5	8	228	82	78	116	12	98	177	20	98	330	36	92	255	75	98	424	94	46
60	0.5	8	284	101	58	173	30	84	313	54	90	238	368	76	200	60	92	902	631	48
50	0.5	8	418	125	52	436	132	88	413	212	92	490	133	62	501	231	90	548	117	66

a sample formed by 6 sets of 50 instances; the number of variables (n) ranges from 50 to 400, the number of colors (k) ranges from 3 to 8 and the edge density (p) ranges from 0.01 to 0.5. The parameter settings were designed to sample across the phase transition following the recommendations given by Culberson.[6] Table 1 shows the experimental results obtained: for each set we give the percentage of instances solved (%) using a cutoff of 5000 seconds, and the mean (m) and median (md) time of the solved instances. The best performing mapping is FL, and then FR, HR and R; and the worst performing are L and S.

In the second experiment, we considered SAT-encoded random binary CSPs using the direct encoding [25]. We used a publicly available generator of uniform random binary CSPs[7] —designed and implemented by Frost, Bessière, Dechter and Regin— that implements the so-called model B: in the class $\langle n, d, p_1, p_2 \rangle$ with n variables of domain size d, we choose a random subset of exactly $p_1 n(n-1)/2$ constraints (rounded to the nearest integer), each with exactly $p_2 d^2$ conflicts (rounded to the nearest integer); p_1 may be thought of as the *density* of the problem and p_2 as the *tightness* of constraints. We incorporated into the generator the automatic generation of all the classes of SAT encodings, and created a representative sample of instances of the hard region of the phase transition described in [24] that could be solved within a reasonable time. The sample is formed by 9 sets of 100 instances; the number of variables ranges from 15 to 70, the domain size was selected in such a way that the instances could be solved within a reasonable time, the density was set at values greater than 0.3 in order to avoid sparse constraint problems, and the tightness was derived from the remaining parameters using the equation $p_2 = 1 - m^{\frac{-2}{p_1(n-1)}}$ in order to generate instances of the hard region of the phase transition [24]. The experimental results obtained are shown in Table 2. We used a cutoff of 2500 seconds. The first column contains the parameters given to the generator of random binary CSPs. The best performing mappings are FR and HR, and then mapping R, and the worst performing are S, FL, and L.

In the third experiment, whose results are shown in Table 3, we studied the scaling behavior of the mappings on pigeon hole instances, where the number of

---

[6] http://web.cs.ualberta.ca/~joe/Coloring/Generators/settings.html
[7] http://www.lirmm.fr/~bessiere/generator.html

Table 2. Experimental results for Random Binary CSPs with Chaff. Time in seconds

parameters $\langle n,d,p_1,p_2\rangle$	S m md %			FR m md %			HR m md %			R m md %			FL m md %			L m md %		
$\langle 15,25,\frac{80}{105},\frac{283}{625}\rangle$	23	31	100	18	21	100	20	23	100	22	26	100	117	109	100	23	28	100
$\langle 15,30,\frac{80}{105},\frac{424}{900}\rangle$	94	102	100	52	60	100	54	69	100	79	94	100	448	428	100	87	103	100
$\langle 25,15,\frac{198}{300},\frac{65}{225}\rangle$	254	236	100	77	73	100	86	80	100	229	207	100	514	502	100	1022	884	100
$\langle 25,20,\frac{198}{300},\frac{126}{400}\rangle$	329	208	56	504	470	96	477	523	96	437	397	60	415	452	34	85	82	52
$\langle 35,10,\frac{305}{595},\frac{23}{100}\rangle$	116	96	100	38	35	100	43	39	100	96	82	100	145	132	100	147	121	100
$\langle 35,15,\frac{305}{595},\frac{60}{225}\rangle$	106	88	12	564	623	44	511	479	42	229	192	16	653	653	4	155	146	14
$\langle 40,8,\frac{400}{780},\frac{12}{64}\rangle$	46	39	100	16	15	100	18	17	100	44	39	100	46	44	100	66	59	100
$\langle 45,10,\frac{415}{990},\frac{22}{100}\rangle$	587	649	78	428	386	100	451	372	100	594	619	84	646	717	88	560	520	70
$\langle 70,5,\frac{880}{2415},\frac{3}{25}\rangle$	10	8.5	100	6	5	100	7.5	6.5	100	4	8	100	9	8.5	100	21	19	100

Table 3. Experimental results for the pigeon hole problem with Chaff. Time in seconds

holes	S	FR	HR	R	FL	L
9	2.3	0.6	0.6	80.25	4	2
10	21	3	8	540	12	204
11	466	86	34	1230	172	3000
12	3040	150	220	2140	940	1114
13	> 5000	3600	872	> 5000	3890	> 5000
14	> 5000	> 5000	> 5000	> 5000	> 5000	> 5000

Table 4. Experimental results for the all interval series problem with Chaff. Time in seconds

$|v|$	S	R	HR	L
9	0.01	0	0.02	0.38
11	2.5	0.07	2.47	280
13	1066	47.51	185.58	1878
15	> 5000	527.85	> 5000	> 5000
17	> 5000	> 5000	> 5000	> 5000

holes ranges from 9 to 14. We used a cutoff of 5000 seconds. The best performing mapping is HR, and then FR and FL, and the worst performing are S, R and L.

In the fourth experiment, whose results are shown in Table 4, we studied the scaling behavior of the mappings on all interval series instances, where the size of the vector ranges from 9 to 17. We used a cutoff of 5000 seconds. The best performing mapping is R, and then HR, and the worst performing are L and S.

Table 5. Experimental results for graph coloring with Siege_v4. Time in seconds

parameters		S			FR			HR			R			FL			L		
n	p	k	m	md	%	m	md	%	m	md	%	m	md	%	m	md	%	m	md	%
400	0.02	3	468	136	96	284	46	100	520	91	98	476	94	100	411	135	96	286	58	96
200	0.13	5	32	7	100	22	10	100	25	9	100	25	5	100	2358	2220	4	2783	2600	18
50	0.5	8	13	2	100	37	23	100	46	8	100	23	3	100	63	16	100	9	2	100

Table 6. Experimental results for Random Binary CSPs with siege_v4. Time in seconds

parameters	S			FR			HR		
$\langle n, d, p_1, p_2 \rangle$	m	md	%	m	md	%	m	md	%
$\langle 25, 20, \frac{198}{300}, \frac{126}{400} \rangle$	1596	1427	90	1124	909	100	1320	919	96
$\langle 35, 15, \frac{305}{595}, \frac{60}{225} \rangle$	2907	3395	40	2367	2303	74	2457	2366	48
$\langle 45, 10, \frac{415}{990}, \frac{22}{100} \rangle$	841	630	100	402	336	100	430	355	100

parameters	R			FL			L		
$\langle n, d, p_1, p_2 \rangle$	m	md	%	m	md	%	m	md	%
$\langle 25, 20, \frac{198}{300}, \frac{126}{400} \rangle$	1004	717	100	1445	1390	20	1265	846	90
$\langle 35, 15, \frac{305}{595}, \frac{60}{225} \rangle$	2122	1880	56	> 5000	> 5000	0	3156	3539	32
$\langle 45, 10, \frac{415}{990}, \frac{22}{100} \rangle$	410	340	100	1638	1416	96	1081	845	100

Table 7. Experimental results for the pigeon hole problem with Siege_v4. Time in seconds

holes	S	FR	HR	R	FL	L
9	63	2.14	2.46	2.59	15	6.56
10	289	10	8.75	9	18	56
11	> 5000	30	51	170	49	238
12	> 5000	162	246	196	74	> 5000
13	> 5000	> 5000	533	> 5000	345	> 5000
14	> 5000	> 5000	> 5000	> 5000	1460	> 5000

Table 8. Experimental results for the all interval series problem with Siege_v4. Time in seconds

$|v|$	S	R	HR	L
9	0.06	0.04	0.01	0.03
11	0.87	1.36	0.41	2.05
13	3.96	0.75	2.98	0.01
15	59	22	127	12
17	> 5000	375	> 5000	> 5000

We can conclude that mapping S, which is commonly found in SAT repositories, is not the best option, and it is worth exploring alternative encodings. On the one hand, mappings FL and FR are the best for the first two problems but mapping HR has a very good behaviour on average. On the other hand, mapping HR has a linear complexity and does not need to apply distributivity; that fact leads to a poor performance of mappings FL and FR on some problems because of the size of the derived formula.

We believe that the good performance is due to the fact that Boolean variables of regular and logarithmic encodings capture subsets of elements of the domain which are not captured when dealing with the Boolean monosigned signature. This leads to learn shorter clauses; for example, on the hardest binary CSP and coloring instances, the learned clauses by Chaff with mapping HR are between two and three times shorter than the learned clauses by Chaff with mapping S.

Finally, in order to see if a similar behaviour is observed with other SAT solvers, we repeated the above experiments with Siege_v4. The experimental results obtained are shown in Tables 5–8. In all the experiments we used a cutoff of 5000 seconds. For random binary CSPs and graph coloring instances we only report the results of the hardest instances for Chaff. From the tables, we can conclude that mapping S is not generally the best option and it is worth to try the others mappings we have defined when solving SAT-encoded combinatorial problems with Siege_v4. For the graph coloring instances we have tested we observe that FL is not as good as it was for Chaff, and we do not see many differences among the other encodings. For the random binary CSPs instances, we observe a behaviour similar to Chaff: mappings FR, HR and R allow us to solve more instances with our cutoff. For the pigeon hole instances, the best mapping is FL, but mapping HR, which is the best mapping for Chaff, also scales nicely. For the all interval series instances mapping R is, like in Chaff, the best option.

References

1. T. Alsinet, R. Béjar, A. Cabiscol, C. Fernández, and F. Manyà. Minimal and redundant SAT encodings for the all-interval-series problem. In *Proceedings of the Catalan Conference on Artificial Intelligence, CCIA 2002, Castellón, Spain*, pages 139–144. Springer LNCS 2504, 2002.

2. C. Ansótegui, R. Béjar, A. Cabiscol, C. M. Li, and F. Manyà. Resolution methods for many-valued CNF formulas. In *Fifth International Symposium on the Theory and Applications of Satisfiability Testing, SAT-2002, Cincinnati, USA*, pages 156–163, 2002.

3. C. Ansótegui, J. Larrubia, C. M. Li, and F. Manyà. Mv-Satz: A SAT solver for many-valued clausal forms. In *4th International Conference Journées de L'Informatique Messine, JIM-2003, Metz, France*, 2003.

4. C. Ansótegui, J. Larrubia, and F. Manyà. Boosting Chaff's performance by incorporating CSP heuristics. In *9th International Conference on Principles and Practice of Constraint Programming, CP-2003, Kinsale, Ireland*, pages 96–107. Springer LNCS 2833, 2003.

5. C. Ansótegui, F. Manyà, R. Béjar, and C. Gomes. Solving many-valued SAT encodings with local search. In *Proceedings of the Workshop on Probabilistics Approaches in Search, 18th National Conference on Artificial Intelligence, AAAI-2002, Edmonton, Canada, 2002*, 2002.

6. M. Baaz and C. G. Fermüller. Resolution-based theorem proving for many-valued logics. *Journal of Symbolic Computation*, 19:353–391, 1995.

7. B. Beckert, R. Hähnle, and F. Manyà. Transformations between signed and classical clause logic. In *Proceedings, International Symposium on Multiple-Valued Logics, ISMVL'99, Freiburg, Germany*, pages 248–255. IEEE Press, Los Alamitos, 1999.

8. B. Beckert, R. Hähnle, and F. Manyà. The SAT problem of signed CNF formulas. In D. Basin, M. D'Agostino, D. Gabbay, S. Matthews, and L. Viganò, editors, *Labelled Deduction*, volume 17 of *Applied Logic Series*, pages 61–82. Kluwer, Dordrecht, 2000.

9. R. Béjar, A. Cabiscol, C. Fernández, F. Manyà, and C. P. Gomes. Capturing structure with satisfiability. In *7th International Conference on Principles and Practice of Constraint Programming, CP-2001,Paphos, Cyprus*, pages 137–152. Springer LNCS 2239, 2001.

10. R. Béjar, R. Hähnle, and F. Manyà. A modular reduction of regular logic to classical logic. In *Proceedings, 31st International Symposium on Multiple-Valued Logics (ISMVL), Warsaw, Poland*, pages 221–226. IEEE CS Press, Los Alamitos, 2001.

11. R. Béjar and F. Manyà. A comparison of systematic and local search algorithms for regular CNF formulas. In *Proceedings of the 5th European Conference on Symbolic and Quantitative Approaches to Reasoning with Uncertainty, ECSQARU'99, London, England*, pages 22–31. Springer LNAI 1638, 1999.

12. J. Culberson. Graph coloring page: The flat graph generator. See http://web.cs.ualberta.ca/~joe/Coloring/Generators/flat.html, 1995.

13. G. Escalada-Imaz and F. Manyà. The satisfiability problem for multiple-valued Horn formulæ. In *Proceedings, International Symposium on Multiple-Valued Logics, ISMVL'94, Boston/MA, USA*, pages 250–256. IEEE Press, Los Alamitos, 1994.

14. A. M. Frisch and T. J. Peugniez. Solving non-boolean satisfiability problems with stochastic local search. In *Proceedings of the International Joint Conference on Artificial Intelligence, IJCAI-2001*, pages 282–288, 2001.

15. R. Hähnle. Towards an efficient tableau proof procedure for multiple-valued logics. In *Selected Papers from Computer Science Logic (CSL'90), Heidelberg, Germany*, LNCS 533, pages 248–260. Springer, 1991.

16. R. Hähnle. *Automated Deduction in Multiple-Valued Logics*, volume 10 of *International Series of Monographs in Computer Science*. Oxford University Press, 1994.

17. R. Hähnle. Advanced many-valued logic. In D. Gabbay and F. Guenthner, editors, *Handbook of Philosophical Logic*, volume 2. Kluwer, second edition, 2001.

18. F. Manyà. *Proof Procedures for Multiple-Valued Propositional Logics*. PhD thesis, Universitat Autònoma de Barcelona, 1996.

19. F. Manyà. The 2-SAT problem in signed CNF formulas. *Multiple-Valued Logic. An International Journal*, 5(4):307–325, 2000.

20. F. Manyà, R. Béjar, and G. Escalada-Imaz. The satisfiability problem in regular CNF-formulas. *Soft Computing: A Fusion of Foundations, Methodologies and Applications*, 2(3):116–123, 1998.

21. M. Moskewicz, C. Madigan, Y. Zhao, L. Zhang, and S. Malik. Chaff: Engineering an efficient sat solver. In *39th Design Automation Conference*, 2001.

22. N. V. Murray and E. Rosenthal. Resolution and path-dissolution in multiple-valued logics. In *Proceedings of the International Symposium on Methodologies for Intelligent Systems, ISMIS'91, Charlotte, NC*, pages 570–579. Springer LNAI 542, 1991.
23. S. D. Prestwich. Local search on SAT-encoded colouring problems. In *Proceedings of the 6th International Conference on the Theory and Applications of Satisfiability Testing*, pages 105–109. Springer LNCS 2919, 2003.
24. B. Smith and M. Dyer. Locating the phase transition in binary constraint satisfaction problems. *Artificial Intelligence*, 81:155–181, 1996.
25. T. Walsh. SAT v CSP. In *Proceedings of the 6th International Conference on Principles of Constraint Programming, CP-2000, Singapore*, pages 441–456. Springer LNCS 1894, 2000.

A SAT-Based Decision Procedure for the Boolean Combination of Difference Constraints

Alessandro Armando[1], Claudio Castellini[1],
Enrico Giunchiglia[2], and Marco Maratea[2]

[1] AILab, DIST - University of Genova viale Francesco Causa,
13 — 16145 Genova (Italy)
http://www.ai.dist.unige.it
{armando, drwho}@dist.unige.it

[2] STARLab, DIST - University of Genova viale Francesco Causa,
13 — 16145 Genova (Italy)
http://www.star.dist.unige.it
{enrico, marco}@dist.unige.it

Abstract. The problem of deciding satisfiability of Boolean combinations of difference constraints is at the core of many important techniques such as planning, scheduling and bounded model checking of real-time systems. Efficient decision procedures for this class of formulas are, therefore, strongly needed. In this paper we present TSAT++, a system implementing a SAT-based decision procedure for this problem, and the techniques implemented in it; in particular, TSAT++ takes full advantage of recent SAT techniques. Comparative experimental results indicate that TSAT++ outperforms its competitors both on randomly generated, hand-made and real world problems.

1 Introduction

In temporal reasoning, one of the best known and studied formalisms is the so-called Simple Temporal Problem (STP) [DMP91], consisting of a conjunction of *difference constraints*, i.e., constraints of the form $x - y \leq c$ where x and y are variables ranging over a fixed numeric domain (typically the integers or the reals) and c is a numeric constant.

STPs are tractable, but their expressiveness is rather limited; therefore, recently, several extensions to them have been introduced, allowing propositional atoms, disjunctions and negations of binary constraints. Such extensions are more expressive than the STP but retain most of its conciseness and clarity; still, they are in general able to express complex problems such as planning, scheduling and verification of real-time hardware.

It is then clear that efficient decision procedures for these extensions are strongly needed; in fact, in the last five years, at least six systems have been proposed that are able to deal with disjunctions of difference constraints, four of which in the AI literature and two in the formal verification literature, meaning that the topic is hot and interdisciplinary. Not surprisingly, five out of these six systems are SAT-based or CSP-based; this means that the satisfiability of a problem ϕ is determined by

H.H. Hoos and D.G. Mitchell (Eds.): SAT 2004, LNCS 3542, pp. 16–29, 2005.

1. (enumeration) generating a set of propositional atoms and difference constraints "propositionally satisfying" ϕ, using SAT or CSP techniques, and
2. (satisfiability checking) testing the satisfiability of each generated set using standard techniques (such as, e.g., the Bellman-Ford procedure).

As we will see, the enumeration, and thus the specific SAT/CSP technique being used in the enumeration phase, is crucial to the efficiency of the procedure. Despite this, none of the aforementioned five systems take advantage of the recent developments in the SAT field.

In this paper we first survey the techniques and optimizations that have been realized in the aforementioned solvers, trying to highlight their pros and cons, meanwhile introducing our system TSAT++. TSAT++ is able to deal with arbitrary conjunctions, disjunctions, and negations of difference constraints and propositional atoms. We then briefly describe how we have integrated in TSAT++ most of the above mentioned techniques, plus some new ones. Lastly we show the outcome of an extensive comparative analysis among the solvers, including TSAT++.

TSAT++ integrates the latest techniques proposed in the SAT field (and, in particular, those proposed in [MMZ+01]) and uses new ideas designed to take maximum advantage from the techniques used in the enumeration phase. Thanks to these methods and their fruitful integration, we show that TSAT++ has a clear edge over its competitors; in fact, the analysis shows that TSAT++ is

- at least 2 orders of magnitude faster in the hard region on real-valued randomly generated problems;
- at least 6 times faster in the hard region on integer-valued randomly generated problems;
- on average at least a factor of 4 faster on instances coming from real world problems; and
- up to 3 orders of magnitude faster on hand-made problems

than its fastest competitor in each category. These results are even more interesting if one considers that, unlike most of the competitors, TSAT++ is not tuned nor customized on any particular class of problems.

The paper is structured as follows: after some preliminaries (Section 2), we introduce TSAT++ and describe the techniques implemented in it (Section 3); then we show our experimental results (Section 4) and finally draw some conclusions.

2   Preliminaries

2.1   Temporal Reasoning Problems

Let D be either the set of the integer or real numbers. Let also $\mathcal{V}$ and $\mathcal{P}$ be two disjoint sets of symbols, called *variables* and *propositional atoms* respectively. Then a *difference constraint* is an expression of the form $x - y \leq c$ where $x, y \in \mathcal{V}$ and c is a numeric constant.

An *atom* is either a difference constraint or a propositional atom; a *literal* is either an atom or its negation (if a is an atom, then $\bar{a}$ abbreviates $\neg a$ and $\overline{\neg a}$ stands for a); and lastly, a *Temporal Reasoning Problem* (TRP) is a Boolean combination of literals.

A *TRP-assignment* (on D), or *assignment* when it is not ambiguous, is a function mapping each variable to an element of D, and each propositional atom to the truth values $\{\bot, \top\}$. An assignment σ is extended to map a TRP to $\{\bot, \top\}$ by defining

- $\sigma(x - y \leq c) = \top$ if and only if $\sigma(x) - \sigma(y) \leq c$, and
- $\sigma(\phi) = \top$ (with ϕ being a TRP) according to the truth tables of propositional logic.

Let ϕ be a TRP. We say that an assignment σ *satisfies* ϕ if and only if $\sigma(\phi) = \top$; such an assignment will be called a *TRP-model*, or *model* when it is not ambiguous, of ϕ. A TRP is *satisfiable* (in D) if and only if there exists an assignment (on D) which satisfies it. A finite set of literals is satisfiable (in D) if and only if their conjunction, as a TRP, is.

Here we deal with the problem of determining whether a TRP is satisfiable or not, having fixed the domain of interpretation D. Clearly the problem is NP-complete, no matter which of the above mentioned sets D is.

In the following, we will use the term *valuation* to mean a mapping from atoms to $\{\bot, \top\}$, extended to arbitrary TRPs according to the truth tables of propositional logic; a valuation *satisfies* a TRP if and only if it makes the TRP true. On the other side, we will represent a valuation as the set of literals in it assigned to true; then an *(un)satisfiable* valuation is such that the set of difference constraints represented by the associated literals is (un)satisfiable.

Also, we restrict our attention to TRPs in CNF. This is not a limitation since any TRP can be efficiently reduced to an equi-satisfiable formula in CNF. With this assumption, we represent a TRP as a set of clauses, each clause being a set of literals. As is customary, clauses with one literal only will be called *unit clauses*.

TRPs actually represent an interesting extension to the STP, introducing disjunction and negation of difference constraints and propositional atoms; this added value makes them as expressive as Separation Logic [SSB02], which employs the predicates $<$, $\geq$, $>$, $=$, and $\neq$. Another well-known related framework, the Disjunctive Temporal Problem (DTP), is limited to two disjuncts per clause and admits no propositional atoms, but it is as expressive as a TRP (introducing new variables). In general, valuations of TRPs, restricted to difference constraints, can be seen as Simple Temporal Problems — they are equivalent to conjunctive sets of difference constraints.

2.2   SAT and CSP-Based Procedures

Beside TSAT++, the systems we will be considering here are: the system presented in [SK98] (that we will call SK), Tsat [ACG00], CSPi [OC00], Epilitis[TP03], SEP [SSB02] and MathSAT [ABC+02]. As far as we know, these systems represent a fair snapshot of the current state-of-the-art; and, with the exception of SEP (see [SSB02] for more details), all approaches and systems built so far for TRPs and similar problems are quite alike from an algorithmic point of view. In fact, given a TRP ϕ, they all work by

1. (enumeration) generating all the valuations μ which satisfy ϕ,
2. (satisfiability checking) for each μ, testing whether it is satisfiable.

Enumeration can be done as search in a Constraint Satisfaction Problem (CSP) associated to the basic temporal reasoning problem (systems which do this are SK, CSPi, Epilitis) or by solving the corresponding SAT problem (Tsat, MathSAT, TSAT++). In the first approach, search is performed in a meta-search space in which a new variable is associated with each clause, its domain being the set of disjuncts in the clause. In the SAT-based approach, the given TRP is abstracted to a propositional formula obtained by substituting each distinct binary constraint with a newly introduced propositional atom.

As one can see, SAT- and CSP-based approaches are tightly connected, and it is therefore not surprising that in their basic versions and starting from Tsat, all the systems perform the following steps:

1. assign to true the literals in unit clauses;[1]
2. if there are no more literals to assign according to the previous step, they branch on a literal l (i.e., assign true to l), and, upon failure of the subsequent search, add the negation of l to the current state and continue the search, till either a satisfying assignment is found, or backtrack occurs.

The similarity to the search performed by SAT solvers is apparent; despite this, none of the above systems incorporates the last advancements done in the SAT field.

As far as expressiveness is concerned, TSAT++ is able to deal with any TRP; SK, Tsat, CSPi and Epilitis are restricted to DTPs; SEP is as expressive as TSAT++, and neither is comparable to MathSAT, since MathSAT allows for arbitrary linear constraints as atoms and does not allow the integers to be considered as domain of interpretation.

3   TSAT++

In this section we describe the main ideas behind TSAT++. Most of the terminology we will be using from now on is customary in the AI and Formal Methods literature. Rather than as a monolithic system, TSAT++ is conceived as an open platform for the implementation of such ideas and techniques. For a more detailed description of the system, refer to [ACG⁺04].

In particular, we will describe (i) the computation done before the search starts (*preprocessing*), (ii) the way the search space is pruned after each branching node (*look-ahead*), (iii) the way recovery from failures happens (*look-back*); (iv) the heuristics used for picking the literal on which to branch (*branching rule*), and (v) the procedure used for checking the satisfiability of a set of literals (*satisfiability checking*).

TSAT++ employs an API-like modified version of SIMO [GMT03] for the enumeration phase.

---

[1] In CSPi and Epilitis, priority of unit clauses is embedded in the heuristics used for selecting the literal to branch on. Furthermore, these systems employ *forward checking*, which removes the binary constraints whose negation is entailed by the current valuation.

3.1   Pre-processing

One drawback of the generate-and-test approach is that (exponentially) many trivially unsatisfiable valuations can be generated and then discarded, essentially because the machinery in charge of the enumeration phase knows nothing about difference constraints. This may happen both in SAT and CSP-based approaches. For example, in the enumeration phase there is no constraint relating the truth values of, e.g., $x - y \leq 3$ and $x - y \leq 5$. Thus, many trivially unsatisfiable valuations (e.g., with $x - y \leq 3$ assigned to true and $x - y \leq 5$ to false) can be generated.

In order to reduce this effect, in TSAT++ for each pair c_1, c_2 of difference constraints in the same variables and occurring in the input formula, the satisfiability of all possible pairs of literals built out of them, i.e., $\{c_1, c_2\}$, $\{\neg c_1, c_2\}$, $\{c_1, \neg c_2\}$, and $\{\neg c_1, \neg c_2\}$, is checked.

Assuming, e.g., $\{c_1, c_2\}$ is unsatisfiable, the clause $\{\neg c_1, \neg c_2\}$ is added to the input formula *before the search starts*. In our example, we would add the clause $\{\neg x - y \leq 3, x - y \leq 5\}$.

This can dramatically speed up the search, especially on randomly generated problems. In fact, e.g., as soon as $x - y \leq 3$ is assigned to true, $x - y \leq 5$ gets also assigned to true by unit propagation. This technique is an extension of a technique called IS(2), introduced in [ACG00].

3.2   Look-Ahead

Consider a TRP ϕ and let S be the set of literals assigned to true so far. The idea behind look-ahead techniques is to try to detect new literals l that are *entailed* by ϕ and S, i.e., such that l is satisfied by each assignment satisfying ϕ and S. If l is one of such literal, we can (i) add l to S and (ii) simplify ϕ on the basis that l is true. This has the beneficial effect of postponing the branching phase and in doing so it may lead to huge savings.

The basic look-ahead technique common to all solvers is unit-propagation. A simple profiling of the code of TSAT++ on real world problems reveals that most of the CPU time is spent in the enumeration phase (often more than 80%, sometime close to 100%), within which most of the time is spent by unit-propagation ($> 90\%$ in most cases). Therefore, the choice of a good data-structure for unit-propagation is capital.

Two-literal watching is an efficient data-structure for unit-propagation (see, e.g., [MMZ+01]). With it, each clause maintains two fields meant to store two "watched" open (i.e. not assigned) literals. Assigning an atom and detecting new units, causes the visit of a sub-linear (in the number of occurrences of the atom) number of clauses. Further, following the same paper, when backtracking occurs, nothing needs to be undone, and thus backtracking takes constant time. On the other hand, by using standard *counters structures* as in, e.g., Tsat and MathSAT, assigning an atom and detecting new units has a cost which is at least linear in the number of occurrences of the atom. Furthermore, when backtracking occurs and an atom is unassigned, each operation needs be undone and this, again, has a cost which is linear in the number of occurrences of the atom. This alternative approach has a higher computational cost than two-literal watching, especially when managing large clauses, such as those emerging from learning. Still, even in SAT, it is not always true that two-literal watching is better than counters structures. Indeed, the former are a de-facto standard for SAT solvers designed for real world applications, while the latter are at the basis of the most efficient solvers on randomly generated problems [BS03].

Having in mind real world applications, we implemented two-literal watching in TSAT++: for these applications, look-back mechanisms based on learning are fundamental, and these imply handling large clauses (with hundreds of literals). It turns out, however, that TSAT++ outperforms the other solvers even on random problems.

Unit-propagation prunes valuations on the basis of propositional reasoning. However, it may be the case that the set of literals μ assigned at a certain point of the search tree, is already unsatisfiable. In this case, there is no point in continuing the search expanding μ. Because of this, before each branching we may check whether μ is satisfiable or not. If not, we can immediately force backtracking. This technique is called *early-pruning* and it is implemented by all the solvers we considered.

3.3   Look-Back

If recovery from a failure is performed by simple chronological backtracking, it is not infrequent to keep exploring a possibly large subtree whose leaves are all dead-ends, especially if the failure is due to some choices performed way up in the search tree. The solution to this problem is to jump back over the choices that do not belong to the reason for the failure. Intuitively, if S is a set of literals such that $S \cup \phi$ (where ϕ is the input CNF formula) is unsatisfiable, then a *reason* R for S is a subset of S such that $\phi \cup R$ is unsatisfiable. Reasons are initialized as soon as an unsatisfiability is detected, and updated while backtracking. The corresponding technique is known as *(Conflict-Directed) Back-jumping* (CBJ) [Pro93].

With *learning* (see [BM96, SS99]), each reason R computed while back-jumping is turned into the clause $\{\bar{l} \mid l \in R\}$ that may be added to the input formula. Learned clauses will prune the subsequent search space, thus avoiding the repetition of the same mistakes. On the other hand, exponentially many reasons can be learned, and each learned clause causes an overhead when assigning literals. In practice it is necessary to introduce criteria (i) for limiting the clauses that have to be learned and (ii) for removing some of them.

TSAT++ features *1-UIP learning* [MMZ+01]. This technique ensures that at each decision level of each branch at most one clause is added to the input formula. Still, an exponential blow-up may happen. To prevent this in TSAT++, each added clauses is analyzed with a given periodicity and (possibly) deleted. Standard alternatives to 1-UIP learning are [BM96]

1. *relevance-bounded learning* of order n (used in MathSAT with $n = 3, 4$) and
2. *size-bounded learning* of order n (used in Epilitis with $n = 10$).

Compared to the 1-UIP learning implemented in TSAT++, both MathSAT and Epilitis may store more than one clause per level.

3.4   Branching Rule

TSAT++ uses a conflict-based heuristic, whose basic idea is to select the literal mostly occurring in the most recently learned clauses. The rationale behind it is that learned clauses represent conflicts among the literals that have emerged during the search. By satisfying these clauses we avoid doing the same "mistake" over and over again. However, not all the learned clauses are equally important: Indeed, some of them, e.g., those discovered at the beginning of the search, may become obsolete for guiding the search in the current branch. Thus, the score associated with each literal is periodically divided by 2, giving more relevance to the atoms that will occur in the newly discovered conflicts.

Of course, such conflict-based heuristics make sense only for solvers with learning. Epilitis uses a similar heuristics. The main difference is that, in Epilitis, all conflicts are equally important, i.e., it does not focus on the atoms in the most recently learned clauses. MathSAT employs a wide variety of heuristics, some of which specifically designed for solving a specific class of problems. However, even though MathSAT uses learning and thus could employ a conflict-based heuristic, all its heuristics are MOMS-based (Maximum Occurrences in clauses of Minimal Size): They give higher scores to literals in shorter clauses. These heuristics have been mutuated from the SAT literature, and are used also by Tsat. In the CSP-based systems, MOMS-based heuristics correspond to the Minimum Remaining Value (MRV) heuristics, used in SK and CSPi.

It is not easy to compare these heuristics for TRPs. One could make some considerations on the basis of what is known in the SAT literature. For instance, in SAT, MOMS-based heuristics are known to be better than conflict-based heuristics on randomly generated problems. However, it is not clear whether such considerations still hold in this setting. For instance, the results in [TP03] point out that its conflict-based heuristics is better than its MOMS-based heuristics even on randomly generated problems. What is true is that MOMS-based heuristics are not compatible with a two literal watching data structure, since they require to know which clauses are active and their length.[2]

3.5   Satisfiability Checking

Consider a set S of literals. For all the procedures here considered, an effective method for checking the satisfiability of S is needed. Moreover, when S is unsatisfiable, it is important to be able to extract a reason of its unsatisfiability, i.e., an unsatisfiable subset S' of S. Of course, a naïve selection of such a set S' is the set S itself; however, applying this selection is seldom a good idea since S' is to be used by the look-back mechanisms, e.g., to backjump over irrelevant nodes. It is thus of fundamental importance to keep S' as "small" as possible in order to try and maximize the benefits of the look-back.

We now describe how we compute such a small set S'. For the time being, let us assume that S is just a set of difference constraints, i.e., that we are facing a STP. We will see later how to generalize the discussion to arbitrary literals. The standard method to check the satisfiability of a STP S is the Bellman-Ford procedure (BF — see, e.g., [CLR98]). The basic idea is to associate with S a *constraint graph*, whose nodes are the variables in S, and which has an edge from y to x with weight c, for each constraint $x - y \leq c$ in S. Then, an extra node s (the "source") connected to all the other nodes with weight 0 is added, and BF is used to compute the "single source shortest-paths"

---

[2] One could argue that two literal watching structures are compatible with unit-based heuristics [LA97]. However, as a preliminary step, unit-based heuristics use a MOMS-based criteria in order to select the variables to score with a unit-based heuristics.

problem. If S is satisfiable, there are no negative cycles in the graph, and BF returns true; otherwise, a minimal (under a suitable set ordering, the simplest of which being cardinality), unsatisfiable subset S' of S can be easily detected by inspection of the constraint graph.

Notice that the constraint graph of S may have several different negative cycles, each one corresponding to a minimal unsatisfiable subset of S. The standard approach amounts to stopping the search as soon as one such negative cycle is detected. TSAT++ instead continues the search in order to determine a negative cycle involving the smallest number of nodes (corresponding to an unsatisfiable set with minimal cardinality). This modification does not alter the overall complexity of BF, which remains $O(n \times m)$, where n and m are the numbers of variables and constraints in S respectively. On the other hand, the advantage of returning an unsatisfiable set with fewer constraints is that it possibly leads to pruning a larger portion of the search space.

Furthermore, when S is a valuation satisfying the input TRP ϕ, some of the literals in S may be not necessary to satisfy ϕ. In other words, there may be a literal l in S such that, for each clause $C \in \phi$ with $l \in C$, there is another literal l' in $S \cap C$. If this is the case, also $(S \setminus \{l\}) \cup \{\bar{l}\}$ satisfies ϕ, and we can safely check the satisfiability of $S \setminus \{l\}$ instead of S. TSAT++ may recursively eliminate such literals l from S. If $S' \subseteq S$ is the resulting set, it will then check the satisfiability of S'. We call the above procedure *reduction*, and it may be useful because

- if S is satisfiable, so is S', and we are done;
- if S is unsatisfiable, it may nevertheless be the case that S' is satisfiable, and we interrupt the search and exit with a satisfying assignment;
- if S and S' are both unsatisfiable, checking the satisfiability of S instead of S' can cause exponentially many more satisfiability checks. In fact, any valuation extending S' also satisfies ϕ, and each could be generated and then rejected by TSAT++.

The last two cases are of particular relevance in TSAT++. In fact, because of the two-literal watching data structure, the generated valuations satisfying ϕ are always total. Thus, it is very often the case that huge portions of the difference constraints in S are irrelevant for satisfying ϕ and, by removing them, we end-up in a set S' with many less difference constraints. Notice that the reduction procedure is not to be applied when early pruning is enabled. With early pruning, the hope is that S is unsatisfiable in order to stop the search. If S' turns out to be satisfiable, we cannot conclude about S, and we have to go on expanding S.

So far, we have been using the assumption that S is a set of difference constraints. The problem is how to deal with the negation of difference constraints. Assume we have $\neg x - y \leq c$ in S. Then, such a literal is equivalent to $y - x < -c$, and we can replace every such constraint in S with a constraint $y - x \leq d$, where d is

- the maximum integer strictly smaller than $-c$, if variables range over the integers; and
- $-c - \frac{1}{10^p(n^2+1)}$, otherwise. In the expression, n is the number of variables in S, and p is the maximum number of digits appearing to the right of the decimal point (assuming that there are no useless "0"), in any of the constants of the input TRP. If all the constants are integers, $p = 0$.

The resulting set does not contain any negation of difference constraint, and it is satisfiable if and only if the initial set is (this follows from Theorem 3 in [GC97]).

4   Experimental Analysis

4.1   Experimental Setting

In order to thoroughly compare TSAT++ with the state-of-the-art we have considered a wide variety of publicly available random, real world and hand-made TRPs (the classification has been done following what is standard practice in the SAT competition [BS03]). As far as the solvers are concerned, we have initially considered all the publicly available systems, that is, the above mentioned SK, CSPi, Epilitis, MathSAT, SEP and Tsat plus, of course, TSAT++. After a first run, we have discarded SK, because it is clearly not competitive with respect to the others.

Each solver has been run on all the benchmarks it can deal with, not only on the benchmarks the solver was analyzed on by the authors. In particular, Epilitis can only handle DTPs with binary clauses and integer valued variables; CSPi and Tsat can only handle DTPs with real valued variables; MathSAT can handle arbitrary TRPs with real valued variables; SEP and TSAT++ can handle arbitrary TRPs.

Each solver has been run using the settings or the version of the solver suggested by the authors for the *specific* problem instances; when not publicly available, we directly asked the authors for the best setting. TSAT++ has many possibilities, also beyond those described in this paper. Of the features described in this paper, only preprocessing, early pruning and reduction of satisfying assignments can be enabled and disabled at the command line.

All experiments were run on a Pentium IV 2.4GHz processor with 1GB of RAM. CPU time is given in seconds; timeout was set to 1000 seconds.

4.2   Comparative Evaluation on Random DTPs

We start our analysis considering randomly generated DTPs as introduced in [SK98] and since then used as a benchmark in [ACG00, OC00, ABC$^+$02, TP03]. DTPs are randomly generated by fixing the number k of disjuncts per clause, the number n of arithmetic variables, a positive integer L such that all the constants are taken in $[-L, L]$.[3] Then, (i) the number of clauses m is increased in order to range from satisfiable to unsatisfiable instances, (ii) for each tuple of values of the parameters, 100 instances are generated and then given to the solvers, and (iii) the median of the CPU time is plotted against the m/n ratio. The results for $k = 2$, $L = 100$ and $n = 35$ are given in Figure 1: Plots (a) and (b) shows the performance when the variables are real and integer valued respectively.

When $m/n \geq 6$, TSAT++ clearly outperforms all other systems: In the peak region, the solver that is closer to TSAT++ in this domain, namely Epilitis, is a factor of 6 slower on 35 variables (cf. plot (b)). This is a very positive result, taking into account

---

[3] A random DTP generator is available at the URI of the SMT-LIB initiative, http://goedel.cs.uiowa.edu/smtlib.

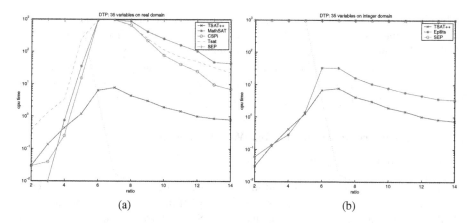

Fig. 1. Comparative analysis on (a) randomly generated DTPs with 35 real valued variables (b) randomly generated DTPs with 35 integer valued variables. Systems are stopped after 1000 seconds. Back: satisfiability percentage

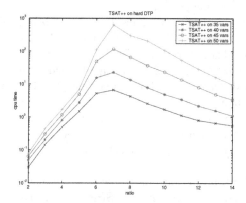

Fig. 2. Scalability of TSAT++. Curves represent TSAT++'s performance on DTPs with 35, 40, 45 and 50 real valued variables

that Epilitis only works on DTP with $k = 2$, and it has been thoroughly tested and optimized on this type of problems (see [TP03]).

All other systems are about 2 orders of magnitude slower than TSAT++ in the peak region. TSAT++ has been run with early pruning and pre-processing enabled, and these are fundamental for its performance on this test set: Without early pruning (resp. pre-processing) TSAT++ on the peak is slower by 2 (resp. 1) orders of magnitude. The fact that these two techniques are important comes at no surprise, and confirm previous results in [ACG00]. The new look-ahead, heuristics and look-back mechanisms used by TSAT++ explain the 2 orders gap with respect to Tsat.

Lastly, in order to evaluate the scalability of TSAT++, we tested it against DTPs with a larger number of variables, namely 40, 45, and 50; as far as we know, the larger problems tackled so far in literature have 35 variables (see [TP03]). The results (in-cluding also the results for 35 variables) are plotted in Figure 2. The plots show that TSAT++ does not timeout even with 50 variables and that it pays around a factor 4 on the hardest point when adding five variables. All other systems we considered were not able to deal with problems with 50 variables; also, the performance gap between TSAT++ and the other systems increased with the number of variables.

4.3   Comparative Evaluation on Real World Problems

In this Subsection we consider

1. the 40 post-office benchmarks introduced in [ABC+02], coming in 4 series (con-sisting of 7, 9, 11, and 13 instances respectively) of increasing difficulty, and
2. the 16 hardware verification problems from [SSB02], 9 (resp. 7) of which are with real (resp. integer) valued variables.

The post-office benchmarks represent bounded model checking for timed automata; the hardware verification suite include scheduling, cache coherence protocol, load-store unit and out-of-order execution problems.

Consider Figure 3. By looking at the results of MathSAT, SEP and TSAT++ on the post-office problems, our first observation is that SEP is not competitive on these problems: On 13 of the hardest instances, SEP had a segmentation fault in 11 cases, and on the other 2 hardest instances SEP is outperformed by orders of magnitude by TSAT++ and MathSAT. Our second observation is that TSAT++ (with pre-processing and assignment reduction) performs better than MathSAT up to a factor of 6, on *each single instance*: This is particularly remarkable given that the authors have customized a version of MathSAT explicitly for this kind of problems. Without pre-processing TSAT++'s performance is worse of about a factor of 3 and of about 10% worse without reduction.

Considering the hardware verification problems, all of them are easy to solve (less than 3s) for all the three solvers, except for SEP that timeouts on one instance. Of

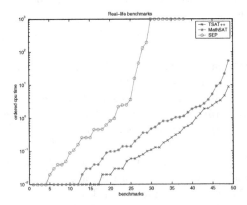

Fig. 3. Comparative analysis on real world problems. Systems are stopped after 1000 seconds

the 9 (resp. 16) runs of MathSAT (resp. SEP and TSAT++), only 3 take more than 0.1s. These observations are confirmed by Figure 3, which gives the overall picture of the results for MathSAT, SEP and TSAT++ on the 49 instances with real valued variables: The x-axis is the number of instances solved by each solver within the CPU time specified on the y-axis.

4.4 Comparative Evaluation on Hand-Made Problems

Finally, we consider the "hand-made" diamonds problems shown in [SSB02]. Given a parameter D, these problems are characterized by an exponentially large (2^D) number of satisfying valuations, some of which correspond to TRP-models; hard instances with a unique TRP-model can be generated. A second parameter, S, is used to make TRP-models larger, further increasing the difficulty. Variables range over the reals.

Table 1. Diamonds problems

D	S	unique	TSAT++	TSAT++^{p}	SEP	SEP (no c.m.)	MathSAT
50	4	N	0	0.02	0.03	0.12	0.05
50	4	Y	0.01	0.14	0.84	0.07	TIME
100	5	N	0.01	0.11	0.13	1.18	0.61
100	5	Y	0.04	7.57	10.20	0.17	TIME
250	5	N	0.08	0.76	0.95	52.20	5.4
250	5	Y	0.21	194.99	288.30	0.77	TIME
500	5	N	0.29	4.46	5.92	742.99	21.22
500	5	Y	1.05	TIME	TIME	4.85	TIME

Table 1 shows comparative results on the diamonds problems. The third column denotes whether the problem has a unique TRP-model; the remaining columns show CPU times for TSAT++ with reduction of assignment enabled, TSAT++ plain version (denoted as TSAT++^{p}), SEP with and without conjunction matrix and MathSAT.[4]

TSAT++ with reduction of assignment enabled clearly performs best, often by orders of magnitude; instances with a unique solution are more difficult than non-unique ones, as expected, except for SEP without conjunction matrix.

For this test set, the good interplay between look-back and satisfiability checking engines is fundamental. In particular, the reduction of assignment step is crucial, and this is clear from the comparison between TSAT++ and TSAT++^{p} columns in Table 1: Without reduction, TSAT++ performs significantly worse, up to the point that problems that are solved in 1 second, are not solved without reduction within the time limit.

Related Work. Two further systems which tackle the same problem are LPSAT and CPLEX. LPSAT ([WW99]) has been excluded from the comparative analysis since

---

[4] The configurations employed were suggested by the authors of SEP and MathSAT.

it is algorithmically quite similar to MathSAT and it employs the SAT solver RelSat ([BS97]), similar to SIM, which is used in MathSAT. Therefore we do not expect LP-SAT to be competitive.

CPLEX ([Cpl93]) is a commercial linear programming system not freely available; it is the object of our future research to compare with it.

5   Conclusions

In this paper we have presented TSAT++, an effective system for temporal reasoning that improves the state-of-the-art both on randomly generated, real world and hand-made problems. TSAT++ enforces a number of reasoning techniques and optimizations, both borrowed from the AI and Formal Methods literature and new, and takes full advantage of recent SAT improvements. Thanks to this, TSAT++ outperforms all its competitors, *in each different problem class*. This is particularly remarkable, given that most competitors are optimized or even customized for solving specific classes of problems.

Acknowledgments. This research is partially supported by MIUR (Italian Ministry of Education, University and Research) under the project RoboCare – A Multi-Agent System with Intelligent Fixed and Mobile Robotic Components. Also, we wish to thank Gilles Audemard, Angelo Oddi, Ofer Strichman and Ioannis Tsamardinos for providing assistance with their solvers.

References

[ABC⁺02]   Gilles Audemard, Piergiorgio Bertoli, Alessandro Cimatti, Artur Kornilowicz, and Roberto Sebastiani. A SAT based approach for solving formulas over boolean and linear mathematical propositions. In Andrei Voronkov, editor, *Automated Deduction – CADE-18*, volume 2392 of *Lecture Notes in Computer Science*, pages 195–210. Springer-Verlag, July 27-30 2002.

[ACG00]   Alessandro Armando, Claudio Castellini, and Enrico Giunchiglia. SAT-based procedures for temporal reasoning. In Susanne Biundo and Maria Fox, editors, *Proceedings of the 5th European Conference on Planning (Durham, UK)*, volume 1809 of *Lecture Notes in Computer Science*, pages 97–108. Springer, 2000.

[ACG⁺04]   Alessandro Armando, Claudio Castellini, Enrico Giunchiglia, Massimo Idini, and Marco Maratea. TSAT++: an open platform for satisfiability modulo theories. Elsevier Science Publishers, 2004. Proceedings of PDPAR 2004 - Pragmatics of Decision Procedures in Automated Reasoning, Cork, Ireland. To appear.

[BM96]   R. J. Bayardo, Jr. and D. P. Miranker. A complexity analysis of space-bounded learning algorithms for the constraint satisfaction problem. In *Proc. AAAI*, pages 298–304, 1996.

[BS97]   Roberto J. Bayardo, Jr. and Robert C. Schrag. Using CSP look-back techniques to solve real-world SAT instances. In *Proceedings of the 14th National Conference on Artificial Intelligence and 9th Innovative Applications of Artificial Intelligence Conference (AAAI-97/IAAI-97)*, pages 203–208, Menlo Park, July 27–31 1997. AAAI Press.

[BS03]   Le Berre and Simon. The essentials of the SAT 2003 competition. In *International Conference on Theory and Applications of Satisfiability Testing (SAT), LNCS*, volume 6, 2003.

[CLR98]   Thomas H. Cormen, Charles E. Leiserson, and Ronald R. Rivest. *Introduction to Algorithms*. MIT Press, 1998.

[Cpl93]   CPLEX user's guide. Manual, CPLEX Optimization, Inc., Incline Village, NV, USA, 1993.

[DMP91]   R. Dechter, I. Meiri, and J. Pearl. Temporal constraint networks. *Artificial Intelligence*, 49(1-3):61–95, January 1991.

[GC97]   Alfonso Gerevini and Matteo Cristani. On finding a solution in temporal constraint satisfaction problems. In *Proceedings of the 15th International Joint Conference on Artificial Intelligence (IJCAI-97)*, pages 1460–1465, San Francisco, August 23–29 1997. Morgan Kaufmann Publishers.

[GMT03]   E. Giunchiglia, M. Maratea, and A. Tacchella. Look-ahead vs. look-back techniques in a modern SAT solver. In *Proceedings of the 6th International Conference on Theory and Applications of Satisfiability Testing (SAT).*, Portofino, Italy, May 5–8 2003.

[LA97]   Chu Min Li and Anbulagan. Heuristics based on unit propagation for satisfiability problems. In *Proceedings of the 15th International Joint Conference on Artificial Intelligence (IJCAI-97)*, pages 366–371, San Francisco, August 23–29 1997. Morgan Kaufmann Publishers.

[MMZ$^+$01] Matthew W. Moskewicz, Conor F. Madigan, Ying Zhao, Lintao Zhang, and Sharad Malik. Chaff: Engineering an Efficient SAT Solver. In *Proceedings of the 38th Design Automation Conference (DAC'01)*, June 2001.

[OC00]   A. Oddi and A. Cesta. Incremental forward checking for the disjunctive temporal problem. In *Proceedings of the 14th European Conference on Artificial Intelligence (ECAI-2000)*, pages 108–112, Berlin, 2000.

[Pro93]   Patrick Prosser. Hybrid algorithms for the constraint satisfaction problem. *Computational Intelligence*, 9(3):268–299, 1993.

[SK98]   Kostas Stergiou and Manolis Koubarakis. Backtracking algorithms for disjunctions of temporal constraints. In *Proc. AAAI*, 1998.

[SS99]   Joao P. Marques Silva and Karem A. Sakallah. GRASP: A search algorithm for propositional satisfiability. *IEEETC: IEEE Transactions on Computers*, 48, 1999.

[SSB02]   Ofer Strichman, Sanjit A. Seshia, and Randal E. Bryant. Deciding separation formulas with SAT. *Lecture Notes in Computer Science*, 2404:209–222, 2002.

[TP03]   Ioannis Tsamardinos and Martha Pollack. Efficient solution techniques for disjunctive temporal reasoning problems. *Artificial Intelligence*, 2003. To appear.

[WW99]   Steven Wolfman and Daniel Weld. The LPSAT-engine & its application to resource planning. In *Proc. IJCAI-99*, 1999.

An Algebraic Approach to the Complexity of Generalized Conjunctive Queries*

Michael Bauland[1], Philippe Chapdelaine[2],
Nadia Creignou[3], Miki Hermann[4], and Heribert Vollmer[1]

[1] Theoretische Informatik, Universität Hannover, Germany
{bauland, vollmer}@thi.uni-hannover.de
[2] GREYC (UMR 6072), Université de Caen, France
pchapdel@info.unicaen.fr
[3] LIF (UMR 6166), Univ. de la Méditerranée, France
creignou@lif.univ-mrs.fr
[4] LIX (UMR 7161), École Polytechnique, France
hermann@lix.polytechnique.fr

Abstract. Conjunctive-query containment is considered as a fundamental problem in database query evaluation and optimization. Kolaitis and Vardi pointed out that constraint satisfaction and conjunctive query containment are essentially the same problem. We study the Boolean conjunctive queries under a more detailed scope, where we investigate their counting problem by means of the algebraic approach through Galois theory, taking advantage of Post's lattice. We prove a trichotomy theorem for the generalized conjunctive query counting problem, showing this way that, contrary to the corresponding decision problems, constraint satisfaction and conjunctive-query containment differ for other computational goals. We also study the audit problem for conjunctive queries asking whether there exists a frozen variable in a given query. This problem is important in databases supporting statistical queries. We derive a dichotomy theorem for this audit problem that sheds more light on audit applicability within database systems.

1 Introduction

Constraint satisfaction is recognized as a fundamental problem in artificial intelligence, in automated deduction, in computer-aided verification, in operations research, etc. At the same time conjunctive-query containment is considered as a fundamental problem in database query evaluation and optimization [1]. Recent research points out that query containment is a central problem in several database and knowledge base applications, including data warehousing [26], data integration [15], query optimization, and (materialized) view maintenance [28]. Kolaitis and Vardi pointed out in [13] that constraint satisfaction and conjunctive-query containment are essentially the same problem.

---

* Supported by ÉGIDE 05835SH, DAAD D/0205776 and DFG VO 630/5-1.

H.H. Hoos and D.G. Mitchell (Eds.): SAT 2004, LNCS 3542, pp. 30–45, 2005.
© Springer-Verlag Berlin Heidelberg 2005

Constraints are usually specified by means of relations. The standard constraint satisfaction problem can therefore be parameterized by restricting the set of allowed relations. In particular, given a finite set S of Boolean relations, we consider conjunctive propositional formulas consisting of clauses built over relations from S, also called S-formulas. Deciding the satisfiability of such an S-formula is known as the *generalized satisfiability problem*, denoted by SAT(S), and was first investigated by Schaefer [20]. It turns out that the complexity of SAT(S) can be characterized by closure properties of S. This correspondence is obtained through a generalization of Galois theory. In order to get complexity results via this algebraic approach, conjunctive queries COQ(S) over a set of relations S turn out to be useful. Roughly speaking, a conjunctive query from COQ(S) is an S-formula with distinguished variables, where all non-distinguished variables are existentially quantified. These queries play an important role in database theory, since they represent a broad class of queries and their expressive power is equivalent to select-join-project queries in relational algebra. Thus they are also of interest in their own right and we study the complexity of some related computational problems. The algebraic approach is particularly well adapted to this study, yielding short and elegant proofs.

We focus here on the counting and the audit problems for conjunctive queries. In the former the problem is to count the number of entries in the database that match the query, i.e., the number of satisfying assignments. In the latter the problem is to audit a database to ensure protection of sensitive data, where the goal is to decide whether the conjunctive query evaluates to false or whether there is some distinguished variable that is frozen, i.e., that takes the same value in all satisfying assignments. This frozen variable would then be considered as not protected. This is a generalization of the audit problem for Boolean attributes defined in [11] (see also [14]), which is particularly interesting in databases supporting statistical queries. For both considered problems we obtain a complete complexity classification that indicates a difference with respect to satisfiability problems of Boolean constraints. Peter Jonsson and Andrei Krokhin ([10] manuscript, submitted for publication) independently examined a variant of our audit problem. Our results can be shown to follow from theirs.

Measures such as conditional probability (confidence) and correlation have been used to infer rules of the form "buying diapers causes you to buy beer". However, such rules indicate only a statistical relationship between items, but they do not specify the nature and causality of the relationship. In applications, knowing such causal relationship is extremely useful for enhancing understanding and effecting change. While distinguishing causality from correlation is a truly difficult problem, recent work in statistics and Bayesian learning provide some promising directions of attack. In this context, the ideas of Bayesian learning, where techniques are being developed to infer causal relationships from observational data, to mining large databases [21] trigger the necessity to study counting problems in connection with existing database applications. Yet another recent application of Bayesian learning based on counting is the task of spam elimination. Therefore we think that our results will have an impact on concrete database implementations and applications, since the considered formulas in our computational problems correspond better to the model of queries formulated within existing database systems than the so far mainly studied S-formulas.

2   Preliminaries

Throughout the paper we use the standard correspondence between predicates and relations. We use the same symbol for a predicate and its corresponding relation, since the meaning will always be clear from the context, and we say that the predicate *represents* the relation.

An n-ary *logical relation* R is a Boolean relation of arity n. Each element of a logical relation R is an n-ary Boolean vector $m = (m_1, \ldots, m_n) \in \{0, 1\}^n$. Let V be a set of variables. A *constraint* is an application of R to an n-tuple of variables from V, i.e., $R(x_1, \ldots, x_n)$. An assignment $I \colon V \to \{0, 1\}$ satisfies the constraint $R(x_1, \ldots, x_n)$ if $(I(x_1), \ldots, I(x_n)) \in R$ holds.

Example 1. Equivalence is the binary relation defined by $Eq = \{(0, 0), (1, 1)\}$. Given the ternary relations

$$R_{\mathrm{nae}} = \{0, 1\}^3 \setminus \{(0, 0, 0), (1, 1, 1)\} \quad \text{and}$$
$$R_{1/3} = \{(1, 0, 0), (0, 1, 0), (0, 0, 1)\},$$

the constraint $R_{\mathrm{nae}}(x, y, z)$ is satisfied if not all variables are assigned the same value and the constraint $R_{1/3}(x, y, z)$ is satisfied if exactly one of the variables x, y, and z is assigned to 1.

Throughout the text we refer to different types of Boolean constraint relations following Schaefer's terminology [20]. We say that a Boolean relation R is

- *1-valid* if $(1, \ldots, 1) \in R$ and it is *0-valid* if $(0, \ldots, 0) \in R$,
- *Horn* (*dual Horn*) if R can be represented by a conjunctive normal form (CNF) formula having at most one unnegated (negated) variable in each clause,
- *bijunctive* if it can be represented by a CNF formula having at most two variables in each clause,
- *affine* if it can be represented by a conjunction of linear functions, i.e., a CNF formula with $\oplus$-clauses (XOR-CNF),
- *complementive* if for each $(\alpha_1, \ldots, \alpha_n) \in R$, also $(\neg\alpha_1, \ldots, \neg\alpha_n) \in R$.

A set S of Boolean relations is called 0-valid (1-valid, Horn, dual Horn, affine, bijunctive, complementive) if every relation in S is 0-valid (1-valid, Horn, dual Horn, affine, bijunctive, complementive).

Let S be a non-empty finite set of Boolean relations. An *S-formula* is a finite conjunction of *S-clauses*, $\varphi = c_1 \wedge \cdots \wedge c_k$, where each S-clause c_i is a constraint application of some logical relation $R \in S$. An assignment I satisfies the formula φ if it satisfies all clauses c_i. We denote by $\mathrm{sol}(\varphi)$ the set of satisfying assignments of a formula φ.

Schaefer in his seminal paper [20] developed a complexity classification of the satisfiability problem of S-formulas, denoted by SAT(S). *Conjunctive queries* turn out to be useful in order to obtain this result. Given a set S of Boolean relations, we denote by COQ(S) the set of all formulas of the form

$$F(x_1, \ldots, x_k) = \exists y_1 \exists y_2 \cdots \exists y_l \; \varphi(x_1, \ldots, x_k, y_1, \ldots, y_l),$$

where φ is an S-formula. These existentially quantified formulas are called *conjunctive queries over S* [13], with $x = \{x_1, \ldots, x_k\}$ being the *distinguished variables*. We denote by SAT-COQ(S) the satisfiability problem of conjunctive queries over S.

3   Closure Properties of Constraints

There exist easy criteria to determine if a given relation is Horn, dual Horn, bijunctive, or affine. We recall these properties here briefly for completeness. An interested reader can find a more detailed description with proofs in the paper [20] or in the monograph [6]. The operations of conjunction, disjunction, majority, and addition applied coordinate-wise on n-ary Boolean vectors $m, m', m'' \in \{0, 1\}^n$ are defined as follows:

$$m \wedge m' = (m[1] \wedge m'[1], \ldots, m[n] \wedge m'[n])$$
$$m \vee m' = (m[1] \vee m'[1], \ldots, m[n] \vee m'[n])$$
$$\mathrm{maj}(m, m', m'') = (m \vee m') \wedge (m' \vee m'') \wedge (m'' \vee m)$$
$$m \oplus m' = (m[1] \oplus m'[1], \ldots, m[n] \oplus m'[n])$$

where $m[i]$ is the i-th coordinate of the vector m and $\oplus$ is the exclusive-or operator. Given a logical relation R, the following *closure properties* fully determine the structure of R.

- R is Horn if and only if $m, m' \in R$ implies $m \wedge m' \in R$.
- R is dual Horn if and only if $m, m' \in R$ implies $m \vee m' \in R$.
- R is bijunctive if and only if $m, m', m'' \in R$ implies $\mathrm{maj}(m, m', m'') \in R$.
- R is affine if and only if $m, m', m'' \in R$ implies $m \oplus m' \oplus m'' \in R$.

The notion of closure property of a Boolean relation has been defined more generally, see for instance [9, 16]. Let $f \colon \{0, 1\}^k \to \{0, 1\}$ be a Boolean function

$\mathrm{Pol}(R) \supseteq \mathrm{E}_2$	$\Leftrightarrow$ R is Horn	$\mathrm{Pol}(R) \supseteq \mathrm{V}_2$	$\Leftrightarrow$ R is dual Horn
$\mathrm{Pol}(R) \supseteq \mathrm{D}_2$	$\Leftrightarrow$ R is bijunctive	$\mathrm{Pol}(R) \supseteq \mathrm{L}_2$	$\Leftrightarrow$ R is affine
$\mathrm{Pol}(R) \supseteq \mathrm{N}_2$	$\Leftrightarrow$ R is complementive	$\mathrm{Pol}(R) \supseteq \mathrm{N}$	$\Leftrightarrow$ R is compl., 0- and 1-valid
$\mathrm{Pol}(R) \supseteq \mathrm{I}_0$	$\Leftrightarrow$ R is 0-valid	$\mathrm{Pol}(R) \supseteq \mathrm{I}_1$	$\Leftrightarrow$ R is 1-valid
$\mathrm{Pol}(R) \supseteq \mathrm{I}$	$\Leftrightarrow$ R is 0- and 1-valid	$\mathrm{Pol}(R) \supseteq \mathrm{I}_2$	$\Leftrightarrow$ R is Boolean

Fig. 1. Polymorphism correspondences

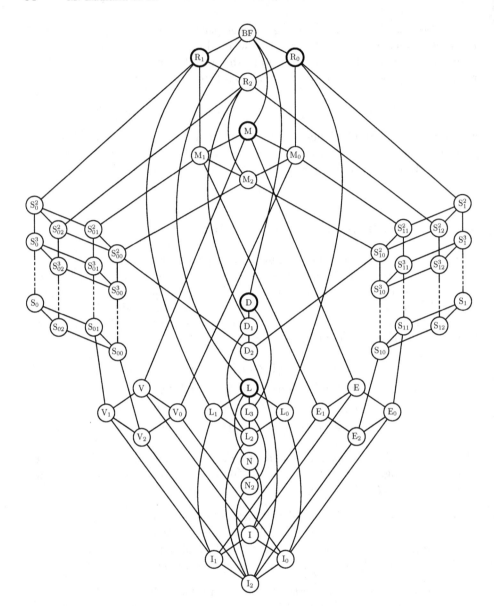

Fig. 2. Graph of all closed classes of Boolean functions

of arity k. We say that R is *closed under f*, or that f is a *polymorphism* of R, if for any choice of k vectors $m_1, \ldots, m_k \in R$, not necessarily distinct, we have that

$$\Big(f\big(m_1[1], \ldots, m_k[1]\big), \ f\big(m_1[2], \ldots, m_k[2]\big), \ \ldots, \ f\big(m_1[n], \ldots, m_k[n]\big) \Big) \in R,$$

i.e., that the new vector constructed coordinate-wise from $m_1, \ldots, m_k$ by means of f belongs to R.

We denote by $\mathrm{Pol}(R)$ the set of all polymorphisms of R and by $\mathrm{Pol}(S)$ the set of Boolean functions that are polymorphisms of every relation in S. It turns out that $\mathrm{Pol}(S)$ is a *closed set of Boolean functions* for every set of relations S. All closed classes of Boolean functions were identified by Post [19]. Post also detected the inclusion structure of these classes, which is now referred to as *Post's lattice*, presented in Fig. 2 with the notation from [2]. We did not use the previously accepted notation for the clones, as in [16, 18], since we think that the new one used in [2, 3] is better suited mnemotechnically and also scientifically than the old one. The correspondence of the most studied classes with respect to the polymorphisms of a relation R is presented in Fig. 1. The class I_2 is the closed class of Boolean functions generated by the identity function, thus for every Boolean relation R we have $\mathrm{Pol}(R) \supseteq I_2$.

An interesting Galois correspondence has been exhibited between the sets of Boolean functions $\mathrm{Pol}(S)$ and the sets of Boolean relations S. A basic introduction to this correspondence can be found in [16, 17] and a comprehensive study in [18]. This theory helps us to get elegant and short proofs for results concerning the complexity of conjunctive queries. Indeed, it shows that the smaller the set of polymorphisms is, the more expressive the corresponding conjunctive queries are, which is the cornerstone for applying the algebraic method to complexity (see [3] for a survey). The following proposition can be found, e.g., in [16, 18].

Proposition 2. *Let S_1 and S_2 be two finite sets of Boolean relations. If the relation $\mathrm{Pol}(S_2) \subseteq \mathrm{Pol}(S_1)$ holds, then $\mathrm{COQ}(S_1 \cup \{\mathrm{Eq}\}) \subseteq \mathrm{COQ}(S_2 \cup \{\mathrm{Eq}\})$.*

4   Complexity Results

The only difference between conjunctive queries and S-formulas is that the former contain some existentially quantified variables, thus distinguishing the remaining ones. While this certainly does not lead to a different complexity of the satisfiability problem, this is not any more the case for other computational goals, such as counting the number of satisfying assignments. The algebraic correspondence described above is useful to determine the complexity of the satisfiability problem, since it proves that the complexity of $\mathrm{SAT}\text{-}\mathrm{COQ}(S)$ strongly depends on the set $\mathrm{Pol}(S)$, as shown in Proposition 2. It provides a polynomial-time reduction from the problem $\mathrm{SAT}\text{-}\mathrm{COQ}(S_1)$ to $\mathrm{SAT}\text{-}\mathrm{COQ}(S_2 \cup \{\mathrm{Eq}\})$ by locally replacing each S_1-clause by its equivalent constraint in $\mathrm{COQ}(S_2 \cup \{\mathrm{Eq}\})$. Moreover, the equivalence relation is actually superfluous. Indeed, from a set of equivalent variables we choose one variable, say z. Then we can delete the corresponding equivalence constraints and substitute the equivalent variables by z in the rest of the formula. Note that we must choose z to be a distinguished variable if an existentially quantified variable occurs in the equivalence set. This proves that $\mathrm{SAT}\text{-}\mathrm{COQ}(S_1)$ is polynomial-time reducible to $\mathrm{SAT}\text{-}\mathrm{COQ}(S_2)$. We will show in the sequel that the algebraic approach is helpful to study the complexity of the counting and the audit problems for conjunctive queries.

4.1   Introduction to Counting Problems and Their Reducibilities

A *counting problem* is typically presented using a suitable *witness* function which for every input x, returns a set of *witnesses* for x. Formally, a *witness* function is a function $w \colon \Sigma^* \longrightarrow \mathcal{P}^{<\omega}(\Gamma^*)$, where Σ and Γ are two alphabets, and $\mathcal{P}^{<\omega}(\Gamma^*)$ is the collection of all finite subsets of Γ^*. Every such witness function gives rise to the following *counting problem*: given a string $x \in \Sigma^*$, find the cardinality $|w(x)|$ of the *witness* set $w(x)$.

Let Σ, Γ be two alphabets and let $R \subseteq \Sigma^* \times \Gamma^*$ be a binary relation between strings such that, for each $x \in \Sigma^*$, the set $R(x) = \{y \in \Gamma^* \mid R(x, y)\}$ is finite. We write $\#R$ to denote the following counting problem: given a string $x \in \Sigma^*$, find the cardinality $|R(x)|$ of the witness set $R(x)$ associated with x. It is easy to see that every counting problem is of the form $\#R$ for some R.

Valiant [24, 25] was the first to investigate the computational complexity of counting problems. To this effect, he introduced the class $\#P$ of counting functions that count the number of accepting paths of nondeterministic polynomial-time Turing machines. The prototypical problem in $\#P$ is $\#SAT$, which is the counting version of Boolean satisfiability. Valiant [24] showed that $\#SAT$ is $\#P$-*complete* via *parsimonious* reductions, that is, every counting problem in $\#P$ can be reduced to $\#SAT$ via a polynomial-time reduction that preserves the cardinalities of the witness sets. Creignou and Hermann [5] proved that the complexity of the counting problem $\#SAT(S)$ of S-formulas is dichotomic: $\#SAT(S)$ is in FP if S is a set of affine relations, otherwise the problem is $\#P$-complete under Turing reductions.

Hemaspaandra and Vollmer [8] have introduced higher complexity counting classes using a predicate-based framework that focuses on the complexity of membership in the witness sets. Specifically, if $\mathcal{C}$ is a complexity class of decision problems, then $\#\cdot\mathcal{C}$ is the class of all counting problems whose witness function w satisfies the following conditions:

1. There is a polynomial $p(n)$ such that for every x and every $y \in w(x)$, we have that $|y| \leq p(|x|)$, where $|x|$ is the length of x and $|y|$ is the length of y.
2. The witness recognition problem "given x and y, is $y \in w(x)$?" is in $\mathcal{C}$.

In particular, $\#\cdot NP$ is the class of counting problems associated with decision problems, for which the witness size is polynomially bounded and the witness recognition problem is in NP. Following Toda [22], the inclusions $\#\cdot\Sigma_k P \subseteq \#\cdot\Pi_k P$ and $\#\cdot\Pi_k P \subseteq \#\cdot\Sigma_{k+1} P$ among counting classes hold for each k. In particular, we have the inclusion $\#P \subseteq \#\cdot NP$.

Following Valiant [24], we say that a reduction is *parsimonious* if it is a polynomial-time many-one reduction preserving the number of solutions. However, this reduction does not allow to prove completeness of many known $\#P$-complete problems. Valiant [25] used counting reductions in his $\#P$-completeness proofs, but the aforementioned counting classes are not closed under this reduction, following Toda and Watanabe [23]. Their result implies that every problem hard for $\#P$ under Turing reduction is also hard for $\#\cdot NP$ under the same reduction. However, since the closure of $\#P$ under Turing reductions is the whole counting counterpart of the polynomial hierarchy, this does not say anything about the actual complexity of the problem in terms of counting classes. Therefore we have to aim at a result involving a reducibility that preserves (or almost preserves) the relevant classes. More useful for counting problems are *subtractive reductions* [7]. They allow us to obtain many completeness results and at the same time they leave the $\#\cdot\Pi_k\mathrm{P}$ classes closed. Nevertheless, these reductions do not seem to be well-suited for our purposes. Indeed, we need to express the operation of halving the witness set, which is quite delicate if we require the closure of the counting classes under these reductions. For this purpose, we define the *complementive reductions* which satisfy the aforementioned requirements, provided that every witness set of the target counting problem is complementive.

A finite alphabet Γ is called *even* if $|\Gamma| = 2k$ for some $k \in \mathbb{N}$. A permutation π on an even alphabet Γ is called *bipartite* if there exists a partition of Γ into two disjoint sets Γ_0 and Γ_1 such that the following conditions hold:

- $\Gamma = \Gamma_0 \cup \Gamma_1$, $\quad \Gamma_0 \cap \Gamma_1 = \emptyset$, $\quad$ and $\quad |\Gamma_0| = |\Gamma_1|$
- for all $x \in \Gamma_i$ we have $\pi(x) \in \Gamma_{1-i}$ for each $i = 0, 1$.

We homomorphically enlarge every permutation π on Γ to the strings in Γ^* by means of the identity $\pi(x_1 \cdots x_k) = \pi(x_1) \cdots \pi(x_k)$ for each string $x_1 \cdots x_k \in \Gamma^*$.

A set of strings $E \subseteq \Gamma^*$ over an even alphabet Γ is called *complementive* if there exists a bipartite permutation π_E on Γ such that $x \in E$ holds if and only if $\pi_E(x) \in E$. If we know that a set of strings E is complementive, we always assume that we are effectively given the permutation π_E. Given Σ, Γ two alphabets with Γ being even, a binary relation B between strings from Σ and Γ is said to be *complementive* if the sets $B(y)$ for each string $y \in \Sigma^*$ are complementive with respect to the same bipartite permutation π_B.

Definition 3. *Let Σ, Γ be two alphabets, Γ being even, and let $\#A$ and $\#B$ be two counting problems determined by the binary relations A and B between the strings from Σ and Γ, where B is complementive.*

- *We say that the counting problem $\#A$ reduces to the counting problem $\#B$ via a **strong complementive reduction**, if there exists two polynomial-time computable functions f and g such that for every string $x \in \Sigma^*$:*
  - $B(g(x)) \subseteq B(f(x))$
  - $2 \cdot |A(x)| = |B(f(x))| - |B(g(x))|$
- *A **complementive reduction** $\#A \leq_{cr} \#B$ from a counting problem $\#A$ to $\#B$ is a transitive closure of strong complementive and parsimonious reductions.*

It is clear that complementive reductions present a special case of counting reductions, the most frequently used reductions among counting problems.

Theorem 4. *$\#\mathrm{P}$ and all higher complexity classes $\#\cdot\Pi_k\mathrm{P}$, $k \geq 1$, are closed under complementive reductions.*

Proof. Let k be a fixed nonnegative integer. We prove that the class $\#\cdot\Pi_k P$ is closed under strong complementive reductions. The result will follow by induction on the number of strong complementive and parsimonious reductions used to compose the final complementive reduction. Recall that Toda [22] showed that $\#\cdot\Pi_k P = \#\cdot P^{\Sigma_k P}$.

Let $\#A$ and $\#B$ be two counting problems such that $\#B \in \#\cdot\Pi_k P$, B is complementive, and $\#A$ reduces to $\#B$ via a strong complementive reduction. We will show that $\#A$ belongs to $\#\cdot\Pi_k P$ by constructing a predicate A' in $P^{\Sigma_k P}$ such that for each string x we have $2\cdot|A'(x)| = |B(f(x))|-|B(g(x))| = 2\cdot|A(x)|$, where f and g are the required polynomial-time computable functions.

Let $*$ be a delimiter symbol not in the alphabets Σ and Γ. Let Γ_0 and Γ_1 be the partition sets defined by the bipartite permutation π_B on Γ. The predicate A' consists of all pairs (x, y') of strings x and y', such that y' is of the form

$$f(x) * g(x) * y \text{ with } (f(x), y) \in B, \ (g(x), y) \notin B, \text{ and } \operatorname{last}(y) \in \Gamma_0,$$

where $\operatorname{last}(y)$ denotes the last symbol of the string y. Thus, a pair (x, y') belongs to A' if and only if (x, y') is accepted by the following algorithm:

1. extract $f(x)$, $g(x)$, and y from y';
2. check that $\operatorname{last}(y)$ belongs to Γ_0;
3. check that $(f(x), y)$ belongs to B;
4. check that $(g(x), y)$ does not belong to B.

Steps 1 and 2 take polynomial time. The test in Step 3 is in $\Pi_k P$, therefore also in $P^{\Sigma_k P}$. The test in Step 4 is in $\Sigma_k P$, hence it can be done in $P^{\Sigma_k P}$. Therefore the predicate A' is in $P^{\Sigma_k P}$. It is clear from the construction that the identity $2\cdot|A'(x)| = |B(f(x))| - |B(g(x))|$ holds, since B is complementive, and this implies $|A'(x)| = |A(x)|$. It follows that the counting problem $\#A$ is in the counting class $\#\cdot P^{\Sigma_k P} = \#\cdot\Pi_k P$. If we take $k = 0$ in the proof, we get also the closure for the class $\#P$. □

In view of the preceding Theorem 4, it is quite natural to ask whether the classes $\#\cdot\Sigma_k P$ are also closed under complementive reductions. The following proposition provides the evidence that *no* class $\#\cdot\Sigma_k P$ is closed under complementive reductions.

Proposition 5. *For every $k \in \mathbb{N}$, the counting class $\#\cdot\Sigma_k P$ is not closed under complementive reductions, unless $\#\cdot\Sigma_k P = \#\cdot\Pi_k P$.*

Proof. Following Wrathal [27], we must perform a case analysis, whether k is even or odd, To obtain completeness for levels of the polynomial hierarchy we have to use CNF or DNF, according to whether we are in an odd or even level. In the even case, take a $\Pi_{2i} P$-formula $\varphi(x_1, \ldots, x_n)$ and construct the formulas

$$\tau(x_0, x_1, \ldots, x_n) = x_0 \vee x_1 \vee \cdots \vee x_n \vee \neg x_0 \vee \neg x_1 \vee \cdots \vee \neg x_n$$
$$\psi(x_0, x_1, \ldots, x_n) = \big(x_0 \wedge \neg\varphi(x_1, \ldots, x_n)\big) \vee \big(\neg x_0 \wedge \neg\varphi(\neg x_1, \ldots, \neg x_n)\big)$$

where $\neg\varphi$ is formed from φ by de Morgan's laws. For the odd case, take a $\Pi_{2i+1}P$-formula φ, maintain the same formula τ, and construct the formula

$$\psi(x_0, x_1, \ldots, x_n) = \left(x_0 \vee \neg\varphi(x_1, \ldots, x_n)\right) \wedge \left(\neg x_0 \vee \neg\varphi(\neg x_1, \ldots, \neg x_n)\right)$$

Both τ and ψ are complementive formulas, hence $\mathrm{sol}(\tau)$ and $\mathrm{sol}(\psi)$ are complementive sets of strings with $\Gamma_0 = \{0\}$ and $\Gamma_1 = \{1\}$.

The non-quantified part of the $\Pi_{2i}P$ formula φ is in CNF, therefore the formulas $\neg\varphi(x_1, \ldots, x_n)$ and $\neg\varphi(\neg x_1, \ldots, \neg x_n)$ are in DNF. Using the distributive law, both formulas $x_0 \wedge \neg\varphi(x_1, \ldots, x_n)$ and $\neg x_0 \wedge \neg\varphi(\neg x_1, \ldots, \neg x_n)$ can be transformed into DNF in polynomial time and linear space. Hence, the formula ψ is equivalent to a DNF-formula, which can be obtained in polynomial time and linear space. Similarly for the odd case, the non-quantified part of the $\Pi_{2i+1}P$ formula φ is in DNF, therefore the formulas $\neg\varphi(x_1, \ldots, x_n)$ and $\neg\varphi(\neg x_1, \ldots, \neg x_n)$ are in CNF. Using the distributive law, we can show that the final formula ψ can be transformed in polynomial time and linear space into an equivalent CNF formula.

In both cases, it is clear that $\mathrm{sol}(\psi) \subseteq \mathrm{sol}(\tau)$, $|\mathrm{sol}(\tau)| = 2 \cdot 2^n$, and $|\mathrm{sol}(\psi)| = 2 \cdot |\mathrm{sol}(\neg\varphi)| = 2 \cdot \left(2^n - |\mathrm{sol}(\varphi)|\right)$. Thus we conclude that $2 \cdot |\mathrm{sol}(\varphi)| = |\mathrm{sol}(\tau)| - |\mathrm{sol}(\psi)|$. Hence, we have a complementive reduction from a $\#\cdot\Pi_k P$-complete problem to a counting problem in $\#\cdot\Sigma_k P$. $\qquad\square$

4.2   The Counting Problem of Conjunctive Queries

The counting problem associated with the satisfiability of generalized conjunctive queries is defined as follows.

Problem: $\#\mathrm{SAT\text{-}COQ}(S)$
Input: A conjunctive query $F(\boldsymbol{x}) = \exists \boldsymbol{y}\, \varphi(\boldsymbol{x}, \boldsymbol{y})$ from $\mathrm{COQ}(S)$.
Output: Number of different satisfying assignments to the distinguished variables $\boldsymbol{x}$.

We used the notation $\#\mathrm{SAT\text{-}COQ}$ to point out the importance of conjunctive queries, contrary to the cryptic notation $\#\Sigma_1\mathrm{SAT}$ used on a more theoretical level in [7]. Our ultimate goal is to determine the complexity of $\#\mathrm{SAT\text{-}COQ}(S)$ for all possible sets S. Observe first that $\#\mathrm{SAT\text{-}COQ}(S)$ is in $\#\cdot\mathrm{NP}$ for every set of Boolean relations S. A central result for our development is the following easy consequence of Proposition 2.

Proposition 6. *Let S_1 and S_2 be two finite sets of Boolean relations. If the inclusion $\mathrm{Pol}(S_2) \subseteq \mathrm{Pol}(S_1)$ holds, then there exists a parsimonious reduction from $\#\mathrm{SAT\text{-}COQ}(S_1)$ to $\#\mathrm{SAT\text{-}COQ}(S_2)$.*

This result, together with Post's lattice, allows us to prove the following trichotomy complexity classification. We need two propositions whose predecessors can already be found in a slightly different form in [4] and which provide two basic $\#\cdot\mathrm{NP}$-complete problems.

Proposition 7. $\#\text{SAT}(R_{1/3})$ *is* $\#\text{P-complete and} \#\text{SAT-COQ}(R_{1/3})$ *is* $\#\cdot\text{NP-complete, both via parsimonious reductions.}$

Proof. From Valiant's original results [24] follows that $\#\text{SAT}$ is the generic $\#\text{P-}$complete problem via parsimonious reductions. From the same reference and also from [7] it follows that $\#\text{SAT-COQ}$ is the generic $\#\cdot\text{NP-complete counting prob-}$lem under parsimonious reductions (see also [12]). It is clear that $\#\text{SAT}(R_{1/3})$ is in $\#\text{P}$ and $\#\text{SAT-COQ}(R_{1/3})$ is in $\#\cdot\text{NP}$.

The standard reduction from SAT to 3SAT is also a parsimonious reduction from $\#\text{SAT}$ to $\#3\text{SAT}$, and it gives rise to a parsimonious reduction from $\#\text{SAT-COQ}$ to $\#3\text{SAT-COQ}$. Each clause $c = l_1 \vee l_2 \vee l_3$ of a 3SAT formula defines one of the following four relations.

$$\begin{aligned}
\text{OR}_0(x_1, x_2, x_3) &= \text{sol}(x_1 \vee x_2 \vee x_3) &&= \{0,1\}^3 \setminus \{(0,0,0)\} \\
\text{OR}_1(x_1, x_2, x_3) &= \text{sol}(\neg x_1 \vee x_2 \vee x_3) &&= \{0,1\}^3 \setminus \{(1,0,0)\} \\
\text{OR}_2(x_1, x_2, x_3) &= \text{sol}(\neg x_1 \vee \neg x_2 \vee x_3) &&= \{0,1\}^3 \setminus \{(1,1,0)\} \\
\text{OR}_3(x_1, x_2, x_3) &= \text{sol}(\neg x_1 \vee \neg x_2 \vee \neg x_3) &&= \{0,1\}^3 \setminus \{(1,1,1)\}
\end{aligned}$$

We will show that every relation OR_i can be represented as a conjunction of relations $R_{1/3}$. Note first that the relation $Z(v_1, v_2) = R_{1/3}(v_1, v_1, v_2)$ forces the variables v_1 to be assigned the value 0. Therefore the relation $N(x, y, v_1, v_2) = R_{1/3}(x, y, v_1) \wedge Z(v_1, v_2)$ forces y to be the negation of x. For each $c = \text{OR}_i$ we construct now the corresponding formula $r(\text{OR}_i)$ by means of $R_{1/3}$. We obtain the following constructions.

$$\begin{aligned}
r(\text{OR}_0)(x_1, x_2, x_3) &= R_{1/3}(x_1, z_1, z_2) \wedge R_{1/3}(y_2, z_1, z_3) \wedge R_{1/3}(y_3, z_2, z_4) \wedge \\
&\quad R_{1/3}(z_2, z_3, z_5) \wedge N(x_2, y_2, v_1, v_2) \wedge N(x_3, y_3, v_1, v_2) \\
r(\text{OR}_1)(x_1, x_2, x_3) &= r(\text{OR}_0)(u_1, x_2, x_3) \wedge N(x_1, u_1, v_1, v_2) \\
r(\text{OR}_2)(x_1, x_2, x_3) &= r(\text{OR}_1)(x_1, u_2, x_3) \wedge N(x_2, u_2, v_1, v_2) \\
r(\text{OR}_3)(x_1, x_2, x_3) &= r(\text{OR}_2)(x_1, x_2, u_3) \wedge N(x_3, u_3, v_1, v_2)
\end{aligned}$$

where $u_1, \ldots, u_3, v_1, v_2, y_2, y_3, z_1, \ldots, z_5$ are new variables. In the case of conjunctive queries, these new variables will be existentially quantified. The resulting formula is the conjunction of these partial formulas $r(c)$ for all clauses c. This proves the required parsimonious reductions from $\#\text{SAT}$ to $\#\text{SAT}(R_{1/3})$ and from $\#\text{SAT-COQ}$ to $\#\text{SAT-COQ}(R_{1/3})$ □

Remark 8. There exists an alternative and shorter proof of Proposition 7 making use of algebraic arguments. We mention this proof here, since one of our goals is to promote the algebraic approach. The drawback of the proof is that it does not provide an explicit parsimonious reduction and that it is valid only for $\#\text{SAT-COQ}$.

Proof. Since $\text{Pol}(R_{1/3}) = I_2$ and $I_2 \subseteq S$ for every clone S, we conclude by Proposition 6 that $\#\text{SAT-COQ}(S)$ reduces to $\#\text{SAT-COQ}(R_{1/3})$ via parsimonious reductions. □

Proposition 9. $\#\text{SAT}(R_{\text{nae}})$ *is* $\#\text{P-complete and} \ \#\text{SAT-COQ}(R_{\text{nae}})$ *is* $\#\cdot\text{NP-complete, both via complementive reductions.}$

Proof. It is clear that $\#\text{SAT}(R_{\text{nae}})$ is in $\#\text{P}$ and $\#\text{SAT-COQ}(R_{\text{nae}})$ is in $\#\cdot\text{NP}$, respectively. To prove completeness, we will reduce $\#\text{SAT}(R_{1/3})$ to $\#\text{SAT}(R_{\text{nae}})$. Observe that the algebraic approach is of no use here. Indeed, since R_{nae} is complementive, whereas $R_{1/3}$ is not, we have $\text{Pol}(R_{1/3}) \subset \text{Pol}(R_{\text{nae}})$, which does not provide the desired reduction. Therefore we have to construct an explicit reduction. For each clause $c = R_{1/3}(x_1, x_2, x_3)$ of a $\{R_{1/3}\}$-formula φ, we construct the formula

$$q(c) \ = \ R_{\text{nae}}(x_1, x_2, z) \wedge R_{\text{nae}}(x_2, x_3, z) \wedge R_{\text{nae}}(x_3, x_1, z) \wedge R_{\text{nae}}(x_1, x_2, x_3)$$

where z is a new variable. The resulting formula $q(\varphi)$ is the conjunction of these partial formulas $q(c)$ for all clauses c. Observe that if an assignment I satisfies φ, then the dual assignment $\bar{I}$ does not. Observe also that the set of satisfying assignments for the formula $q(c)$ is complementive, therefore the resulting formula $q(\varphi)$ will have twice as many satisfying assignments as the original formula φ. This proves the required complementive reduction from $\#\text{SAT}(R_{1/3})$ to $\#\text{SAT}(R_{\text{nae}})$.

In case of conjunctive queries, z will be an existentially quantified variable. In order to be allowed to apply the same argument as above, we have to make sure that if an assignment I on the distinguished variables $\boldsymbol{x}$ satisfies the conjunctive query $F(\boldsymbol{x}) = \exists \boldsymbol{y} \varphi(\boldsymbol{x}, \boldsymbol{y})$, then the dual assignment $\bar{I}$ does not. Since it is not necessarily the case, we have to introduce two new variables u and v, and to consider first a new conjunctive query $F'(\boldsymbol{x}, u, v) = \exists \boldsymbol{y} \varphi(\boldsymbol{x}, \boldsymbol{y}) \wedge R_{1/3}(u, u, v)$. The number of satisfying assignments for F' is equal to the number of satisfying assignments for F. Moreover, F' has the desired property mentioned above. Therefore the previous construction, namely $q(F')$, provides a complementive reduction from $\#\text{SAT-COQ}(R_{1/3})$ to $\#\text{SAT-COQ}(R_{\text{nae}})$. Using Proposition 7, this proves the result. $\qquad\square$

Theorem 10. *Let S be a non-empty finite set of Boolean relations.*

- *If S is affine, then $\#\text{SAT-COQ}(S)$ is in* FP.
- *Else if S is bijunctive, or Horn, or dual Horn, then $\#\text{SAT-COQ}(S)$ is $\#$P-complete under counting reductions.*
- *Otherwise $\#\text{SAT-COQ}(S)$ is $\#\cdot\text{NP}$-complete under complementive reductions.*

Proof. If S is affine, then the Gaussian elimination algorithm used in [5] for $\#\text{SAT}(S)$ can also be used to construct a corresponding polynomial-time algorithm for $\#\text{SAT-COQ}(S)$.

If S is Horn, dual Horn, or bijunctive, then $\text{SAT}(S)$ is in P following [20] and therefore $\#\text{SAT-COQ}(S)$ is in $\#\text{P}$. Moreover, we know from [5] that in this case $\#\text{SAT}(S)$ is $\#\text{P-hard}$. Hence, the trivial (parsimonious) reduction from $\#\text{SAT}(S)$ to $\#\text{SAT-COQ}(S)$ finally shows that $\#\text{SAT-COQ}(S)$ is $\#\text{P-complete}$.

It remains to treat the case where $\text{Pol}(S) = \text{N}$. In fact, observe that all the other nonconsidered classes N_2, I, I_0, I_1 or I_2 are subsets of N. Therefore according to Proposition 6 and Post's lattice, it suffices to exhibit a set S of Boolean relations, such that $\text{N} \subseteq \text{Pol}(S)$ but $\#\text{SAT-COQ}(S)$ is $\#\cdot\text{NP}$-complete.

According to Proposition 9 we know that $\#\text{SAT-COQ}(R_{\text{nae}})$ is $\#\cdot\text{NP}$-complete via complementive reductions. Construct now the relations

$$
\begin{aligned}
R''(u, v, x, y, z) &= (\neg u \wedge \neg v \wedge \neg x \wedge \neg y \wedge \neg z) \vee (u \wedge v \wedge x \wedge y \wedge z) \quad \text{and} \\
R'(u, v, x, y, z) &= R''(u, v, x, y, z) \vee \\
&\quad (u \wedge \neg v \wedge R_{\text{nae}}(x, y, z)) \vee (\neg u \wedge v \wedge R_{\text{nae}}(x, y, z)).
\end{aligned}
$$

Consider now the formula $F(\boldsymbol{x}) = \exists \boldsymbol{y} \bigwedge_{i=1}^{m} R_{\text{nae}}(x_1^i, x_2^i, x_3^i)$ being an instance of $\#\text{SAT-COQ}(R_{\text{nae}})$, where x_1^i, x_2^i, x_3^i are variables from the vector $\boldsymbol{x}$. Build the formulas

$$
F'(\boldsymbol{x}, u, v) = \exists \boldsymbol{y} \bigwedge_{i=1}^{m} R'(u, v, x_1^i, x_2^i, x_3^i) \quad \text{and}
$$

$$
F''(\boldsymbol{x}, u, v) = \exists \boldsymbol{y} \bigwedge_{i=1}^{m} R''(u, v, x_1^i, x_2^i, x_3^i)
$$

from the relations R' and R''. The satisfying assignments of the query F' include those of F''. If q is the number of satisfying assignments of F then those of F' is $2q + 2$ and those of F'' is 2. Hence, we have the equality $2\,|\text{sol}(F)| = |\text{sol}(F')| - |\text{sol}(F'')|$, implying a complementive reduction from the counting problem $\#\text{SAT-COQ}(R_{\text{nae}})$ to $\#\text{SAT-COQ}(\{R', R''\})$, proving that $\#\text{SAT-COQ}(\{R', R''\})$ is $\#\cdot\text{NP}$-complete. Moreover, both R' and R'' are 0-valid, 1-valid, and complementive, since R_{nae} is complementive. Hence $\text{Pol}(\{R', R''\})$ contains N. □

4.3   The Audit Problem

Another problem of interest, defined by Kleinberg *et al.* [11] and studied from a complexity standpoint by Jonsson and Krokhin [10, 14], is the *audit problem*. This problem is related to databases that support statistical queries. It can be generalized to conjunctive queries in the following way.

Problem: AUDIT-COQ(S)
Input: A conjunctive query $F(\boldsymbol{x}) = \exists \boldsymbol{y}\, \varphi(\boldsymbol{x}, \boldsymbol{y})$ from COQ(S).
Question: Is F unsatisfiable or is there some variable among $\boldsymbol{x}$ that is frozen, i.e., that takes the same value in all satisfying assignments?

Note that our AUDIT-COQ(S) problem is different from the 1-AUDIT problem studied in [10], since we do not include the variable candidate to be frozen as part of the input. Nevertheless, our result can be shown to follow from those in [10]. We want to insist here on the clarity and simplicity of our proof.

It is easy to see that this problem belongs to the class coNP. We prove that the algebraic approach applies to study the complexity of this problem. The following result follows again immediately from Proposition 2 (see also Proposition 6).

Proposition 11. *Let S_1 and S_2 be two finite sets of Boolean relations. If the inclusion* $\mathrm{Pol}(S_2) \subseteq \mathrm{Pol}(S_1)$ *holds, then* AUDIT-COQ(S_1) *is polynomial-time many-one reducible to* AUDIT-COQ(S_2).

Once more, this result together with Post's lattice allows us to get a complete complexity classification.

Theorem 12. *Let S be a non-empty finite set of Boolean relations.*

 - *If S is both 0- and 1-valid, or affine, or Horn, or dual Horn or bijunctive, then* AUDIT-COQ(S) *is in* P.
 - *Otherwise* AUDIT-COQ(S) *is* coNP-*complete.*

Proof. If S is both 0- and 1-valid, i.e., $I \subseteq \mathrm{Pol}(S)$, then the problem is trivial.

If S is affine, Horn, dual Horn, or bijunctive, then observe that given an S-formula and a variable x, we can check in polynomial time whether both $F \wedge x$ and $F \wedge \neg x$ are satisfiable. Therefore, in this case AUDIT-COQ(S) is in P.

If S is complementive, but neither 0-valid, nor included in the four previous cases, i.e., $\mathrm{Pol}(S) = N_2$, then no variable can be frozen. Therefore in this case the problem AUDIT-COQ(S) is equivalent to the coNP-complete problem UNSAT(S), asking whether an S-formula is unsatisfiable.

The remaining cases are those for which $\mathrm{Pol}(S) = I_0$, I_1 or I_2. According to Proposition 6 and Post's lattice, in order to conclude the proof it suffices to exhibit a Boolean relation R_0 (resp. R_1) such that $I_0 \subseteq \mathrm{Pol}(R_0)$ (resp. $I_1 \subseteq \mathrm{Pol}(R_1)$) and AUDIT-COQ$(R_0)$ (resp. AUDIT-COQ(R_1)) is coNP-complete. Recall first that SAT$(R_{1/3})$ is NP-complete, so UNSAT$(R_{1/3})$ is coNP-complete. Consider an instance of UNSAT$(R_{1/3})$ defined by the formula $F(\boldsymbol{x}) = \bigwedge_{i=1}^{m} R_{1/3}(x_1^i, x_2^i, x_3^i)$. Construct the 0-valid relation

$$R_0(v, x, y, z) \quad = \quad (\neg v \wedge x \wedge y \wedge z) \vee (\neg v \wedge \neg x \wedge \neg y \wedge \neg z) \vee (v \wedge R_{1/3}(x, y, z))$$

and build the formula $F'(\boldsymbol{x}, v) = \bigwedge_{i=1}^{m} R_0(v, x_1^i, x_2^i, x_3^i)$. Clearly, the inclusion $I_0 \subseteq \mathrm{Pol}(\{R_0\})$ holds since the relation R_0 is 0-valid.

Observe that F' is always satisfiable, that no variable among the $\boldsymbol{x}$ is frozen, and that F is unsatisfiable if and only if the variable v is frozen to 0 in F'. So, we have a reduction from UNSAT$(R_{1/3})$ to AUDIT-COQ(R_0), therefore the problem AUDIT-COQ(R_0) is coNP-complete. The proof is similar for $\mathrm{Pol}(S) = I_1$, with a 1-valid relation R_1 similar to R_0, just flip the polarity of the variable v. □

5   Conclusion

While the complexity of conjunctive-query evaluation and constraint satisfaction is the same, we determined that this is not any more the case for other computational goals. We have shown that the counting problem for conjunctive queries has a different structure than that for conjunctive formulas. The latter displays a dichotomy behavior between the affine formulas in FP and the

#P-complete other cases, as it was shown in [5], whereas the former presents a trichotomy structure between the affine cases in FP, the Horn, dual Horn, and bijunctive #P-complete cases, and finally the general #·NP-complete case. This shows that, under the more fine grained analysis presented by counting, the conjunctive queries present three different levels of (in)tractability. As a byproduct, we developed a new kind of reductions among counting problems, called the complementive reductions, that allow to use halving functions within the counting classes under certain circumstances , i.e., when every instance of the target set is complementive. Since there are many counting problems presenting this structure, we think that the complementive reductions will have a broader impact.

We have also shown that the corresponding audit problem for conjunctive queries displays a dichotomic behavior, where the cases of Horn, dual Horn, bijunctive, or both 0 and 1-valid constraints are in P, whereas the other cases are coNP-complete.

Acknowledgment. We thank Elmar Böhler, Matthias Galota, and Steffen Reith for helpful discussions.

References

1. S. Abiteboul, R. Hull, and V. Vianu. *Foundation of databases.* Addison-Wesley, 1995.
2. E. Böhler, N. Creignou, S. Reith, and H. Vollmer. Playing with Boolean blocks, part I: Post's lattice with applications to complexity theory. *SIGACT News, Complexity Theory Column 42*, 34(4):38–52, 2003.
3. E. Böhler, N. Creignou, S. Reith, and H. Vollmer. Playing with Boolean blocks, part II: Constraint satisfaction problems. *SIGACT News, Complexity Theory Column 43*, 35(1):22–35, 2004.
4. N. Creignou and M. Hermann. On #P-completeness of some counting problems. Research report 2144, Institut de Recherche en Informatique et en Automatique, December 1993. URL = http://www.lix.polytechnique.fr/~hermann/publications/satcount.ps.gz.
5. N. Creignou and M. Hermann. Complexity of generalized satisfiability counting problems. *Information and Computation*, 125(1):1–12, 1996.
6. N. Creignou, S. Khanna, and M. Sudan. *Complexity Classifications of Boolean Constraint Satisfaction Problems*, volume 7 of *SIAM Monographs on Discrete Mathematics and Applications*. SIAM, Philadelphia (PA), 2001.
7. A. Durand, M. Hermann, and P. G. Kolaitis. Subtractive reductions and complete problems for counting complexity classes. In M. Nielsen and B. Rovan, editors, *Proceedings 25th International Symposium on Mathematical Foundations of Computer Science (MFCS 2000), Bratislava (Slovakia)*, volume 1893 of *Lecture Notes in Computer Science*, pages 323–332. Springer-Verlag, August 2000. To appear in *Theoretical Computer Science*.
8. L. A. Hemaspaandra and H. Vollmer. The satanic notations: Counting classes beyond #P and other definitional adventures. *SIGACT News, Complexity Theory Column 8*, 26(1):2–13, March 1995.

9. P. Jeavons, D. Cohen, and M. Gyssens. Closure properties of constraints. *Journal of the Association for Computing Machinery*, 44(4):527–548, 1997.

10. P. Jonsson and A. Krokhin. Computational complexity of auditing finite attributes in statistical databases. In *Proceedings Structural Theory of Automata, Semigroups and Universal Algebra, Montreal (Canada)*, July 2003.

11. J. Kleinberg, C. Papadimitriou, and P. Raghavan. Auditing Boolean attributes. *Journal of Computer and System Science*, 66(1):244–253, 2003.

12. J. Köbler, U. Schöning, and J. Torán. On counting and approximation. *Acta Informatica*, 26(4):363–379, 1989.

13. P. G. Kolaitis and M. Y. Vardi. Conjunctive-query containment and constraint satisfaction. *Journal of Computer and System Science*, 61(2):302–332, 2000.

14. A. Krokhin and P. Jonsson. Recognizing frozen variables in constraint satisfaction problems. Technical Report TR03-062, Electronic Colloquium on Computational Complexity, 2003.

15. M. Lenzerini. Data integration: a theoretical perspective. In *Proceeding 21st Symposium on Principles of Database Systems (PODS 2002), Madison (Wisconsin, USA)*, pages 233–246. SIGACT-SIGMOD-SIGART, ACM Press, June 2002.

16. N. Pippenger. *Theories of Computability*. Cambridge University Press, Cambridge, 1997.

17. R. Pöschel. Galois connection for operations and relations. Technical Report MATH-AL-8-2001, Technische Universität Dresden, 2001.

18. R. Pöschel and L. A. Kalužnin. *Funktionen- und Relationenalgebren*. Deutscher Verlag der Wissenschaften, Berlin, 1979.

19. E. L. Post. The two-valued iterative systems of mathematical logic. *Annals of Mathematical Studies*, 5:1–122, 1941.

20. T. J. Schaefer. The complexity of satisfiability problems. In *Proceedings 10th Symposium on Theory of Computing (STOC'78), San Diego (California, USA)*, pages 216–226, 1978.

21. C. Silberstein, S. Brin, R. Motwani, and J. D. Ullman. Scalable techniques for mining causal structures. *Data Mining and Knowledge Discovery*, 4(2-3):163–192, 2000.

22. S. Toda. *Computational complexity of counting complexity classes*. PhD thesis, Tokyo Institute of Technology, Department of Computer Science, Tokyo, Japan, 1991.

23. S. Toda and O. Watanabe. Polynomial-time 1-Turing reductions from #PH to #P. *Theoretical Computer Science*, 100(1):205–221, 1992.

24. L. G. Valiant. The complexity of computing the permanent. *Theoretical Computer Science*, 8(2):189–201, 1979.

25. L. G. Valiant. The complexity of enumeration and reliability problems. *SIAM Journal on Computing*, 8(3):410–421, 1979.

26. J. Widom. Research problems in data warehousing. In *Proceedings 4th International Conference on Information and Knowledge Management (CIKM'95), Baltimore (Maryland, USA)*, pages 25–30. Association for Computing Machinery, 1995.

27. C. Wrathall. Complete sets and the polynomial-time hierarchy. *Theoretical Computer Science*, 3(1):23–33, 1976.

28. Y. Zhuge, H. Garcia-Molina, J. Hammer, and J. Widom. View maintenance in a warehousing environment. In M. J. Carey and D. A. Schneider, editors, *Proceedings SIGMOD International Conference on Management of Data, San Jose (California, USA)*, pages 316–327. ACM Press, May 1995.

Incremental Compilation-to-SAT Procedures

Marco Benedetti[§] and Sara Bernardini

Istituto per la Ricerca Scientifica e Tecnologica (IRST),
Via Sommarive 18, 38055 Povo, Trento, Italy
{benedetti, bernardini}@itc.it

Abstract. We focus on *incremental compilation-to-SAT procedures* (iCTS), a promising way to push standard SAT-based approaches beyond their limits. We propose the first comprehensive framework that encompasses all the aspects of an *incremental decision procedure*, from the encoding to the incremental solver. We apply our guidelines to a real-world CTS approach (*Bounded Model Checking*) and show how to modify both the generation mechanism of a real BMC tool (*NuSMV*) and the solving engine of a public-domain SAT solver (*SIM*). Related approaches and experimental results are discussed as well.

1 Introduction

Many decision and search problems may be successfully tackled by generating and solving a chain of increasingly complex SAT instances. A *compilation-to-SAT* (CTS) algorithm specifies the mapping between the original problem and the sequence of satisfiability instances.

Well known examples of CTS approaches exist: computer-aided design of integrated circuits [18, 16], planning [15], model checking for dynamic systems [6], operations research, scheduling [8], and cryptography [19], just to name a few. These techniques share an underlying working schema. They first establish an ordering among classes of potential solutions. Small and short solutions come first. More and more complex candidates follow. Each class is then mapped onto a SAT instance solved by a general purpose solver [20, 13].

One remarkable strength of this family of techniques is *modularity*: state-of-the-art SAT solvers can be picked off-the-shelf and applied to the solution step. Thus, every advance from the SAT community is possibly transferred to the above procedures with a minimum effort. Also, advancements proceed the other way around: a great part of the renewed interest in propositional decision procedures (and of the boost of performances of SAT solvers during the last ten years) is due to the relevance and generality of the above family of techniques.

As usual, modularity shows an unpleasant side: solvers have to be treated as almost completely black boxes. This choice limits the amount of information exchanged between the generating and the solving side during *one single round* of the procedure, besides preventing information exchange among *subsequent rounds*. The solver thus

---

[§] This work is funded by PAT (*Provincia Autonoma di Trento*, Italy) under grant n. 3248/2003.

H.H. Hoos and D.G. Mitchell (Eds.): SAT 2004, LNCS 3542, pp. 46–58, 2005.

misses the key point that it is presented with a *chain* of strictly related instances. A further underestimated duty one pays for easily plug standard solvers in, is the flattening of highly structured instances down to a conjunctive normal form (the standard input format for general SAT solvers).

Some approaches have recently emerged to exploit the crucial observation that neither a given instance in a chain is unrelated to the previous ones, nor the solver is approaching a completely different search problem every time it is invoked [16, 24, 23, 5]. These approaches aim both to increase the efficiency of the overall decision procedure and to allow kinds of reasoning that don't fit well within the usual CTS framework.

In this paper, we present the first comprehensive framework that encompasses all the aspects of an *incremental decision procedure* based on propositional satisfiability. After a few notation and preliminaries (Section 2 and 3), we characterize a large family of CTS approaches that are eligible for *incrementalisation*, and stress the often overlooked issue of turning a standard encoding machinery into an incremental one (Section 4). Also, issues arising on the solving side are addressed, and two detailed examples are developed during the presentation. In Section 5 we apply our guidelines to the incrementalisation of a specific CTS approach (*Bounded Model Checking*, or BMC for short [6]) and show how to modify both the generation mechanism employed by a real BMC tool (*NuSMV* [7]) and a public-domain SAT solver (*SIM* [13]).

We carefully review the related literature in Section 6, and then present our conclusions and future work in Section 7. A more thorough presentation of our technique and the proofs of all the results are given in [2].

2   Notation

Given a *conjunctive normal form* (CNF) formula f and a set of literals Δ on the variables $var(f)$ of f, we denote by $f * \Delta$ the propositional formula obtained from f after the assignment Δ is made, i.e. the clause set obtained after unit subsumption ad unit resolution have been performed against each literal in Δ considered as a unit clause. Given two propositional formulas f_1 and f_2 and a set of propositional variables $V \subseteq var(f_1) \cap var(f_2)$, we write $f_1 \equiv_V f_2$ when the set of models of f_1 projected onto V is equal to the set of models of f_2 projected onto V. We graphically represent formulas by means of direct acyclic graphs that avoid sub-formula replications (known as RBC and extensions thereof, see [1]). Yet, propositional solvers often require a conjunctive normal form to work. We denote by $cnf(f)$ the set of clauses obtained from f according to the guidelines described in [10, 21]. For this set, it holds that $cnf(f) \equiv_{var(f)} f$. The cnf function is omitted whenever the context suffices to understand that a CNF formula is required.

3   CTS Approaches

Most CTS frameworks tacitly exploit the deduction theorem over a language $\mathcal{L}$ (more expressive than propositional logic), by stating that $\mathcal{T} \models P \Leftrightarrow \not\models \mathcal{T} \wedge \neg P$, where $\mathcal{T}$ is a consistent theory that models a relevant phenomenon or system or protocol, while P

expresses an (un)desired property over that phenomenon/system/protocol. The problem is to decide the consistency of $W = T \wedge \neg P$.

A mechanism purposely designed to get rid of the excess of expressive power of $\mathcal{L}$ w.r.t. pure propositional logic ($Prop$) stays at the very heart of every CTS framework. This mechanism allows resorting to $Prop$ by considering chains of *bounded* versions of the original problem obtained through a function $[\![.]\!] : \mathcal{L} \times \mathbb{N} \rightarrow Prop$ - called *encoding function* - that maps a formula $W \in \mathcal{L}$ and a bound k onto a propositional formula $[\![W]\!]_k$ on variables $V_k = var([\![W]\!]_k)$.

From the point of view of a state-space search, things work as follows: (1) The space of possible solutions to the problem is partitioned according to a bound k identifying finite classes C_k of possible models for W (the larger the bound k the more complex the solutions in C_k); (2) an encoding $[\![W]\!]_k$ - satisfiable iff a solution for W happens to lay in C_k - is computed together with a decoding function mapping propositional models of satisfiable encodings onto solutions to W in C_k; (3) $[\![W]\!]_k$ is solved; should it come out to be satisfiable, the decoding function would play its role in reconstructing a solution to W. Step 2 is selected for another round with a higher bound when no solution exists in C_k. The loop is exited when either some resource limit is exhausted or it is possible to prove that none of the remaining C_i, $i > k$ contains solutions. So, the problem of finding out whether or not W has models is answered by deciding a sequence of SAT problems on $\{[\![W]\!]_i, i = 0, 1, 2, ...\}$.

The peculiar structure of W - due to application of the deduction theorem within $\mathcal{L}$ - is maintained after the propositional translation, provided the encoding function is commutative w.r.t. negation and distributive w.r.t. conjunction. So, we manage a *structured* sequence $[\![W]\!]_k = [\![T \wedge \neg P]\!]_k = [\![T]\!]_k \wedge \neg[\![P]\!]_k$, where $[\![P]\!]_k$ is usually by far smaller than $[\![T]\!]_k$ and nonetheless responsible for potential inconsistencies in $[\![W]\!]_k$.

As an example of a CTS approach, let us consider SAT-based classical planning [15]. It works by encoding into $[\![W]\!]_k$ two components: (1) an instance $[\![T]\!]_k$ of the theory describing the planning domain in terms of the interconnected preconditions and effects *of at most k layers of actions* (together with other constraints such as mutual exclusion conditions between pairs of actions in the same layer), and (2) the condition or goal $[\![P]\!]_k$ to be reached after the last layer of actions has been executed.

Concepts out of reach for raw propositional logic here are the universal quantifiers in front of the action schemata, and the existence of fluent predicates along the infinite timeline of the modeled world. Both of them are dealt with by propositionally *instantiating* state variables and action schemata as many times as needed.

In the basic encoding, each operator is instantiated with all the possible combinations of arguments to obtain several parameterless (boolean) actions. As the number of objects in classical planning domains doesn't change over time, this *groundization* can be done once for all and doesn't require incrementality. Conversely, the unrolling of plans over the time line (in terms of the number of action layers) is potentially infinite. Indeed, the bound for this CTS approach represents the maximal number of action layers in the solution plan we are currently looking for, and C_i is the set of feasible plans with exactly i layers of actions.

As classical planning domains have a finite state space in spite of the infinite number of feasible plans, it is also possible to check the set of reached states for saturation, thus ensuring that no solution exists for unfeasible goals. In case a satisfiable instance is encountered, the resulting plan immediately grows out of the given model as filtered by the decoding function, that remembers (1) which layer of action and status are associated with each propositional state variables, and (2) which layer and parameters are associated with propositional action instantiations.

4   Incremental Compilation-to-SAT (iCTS)

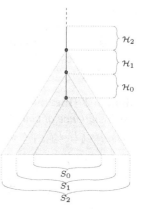

Every iCTS approach is made up of an incremental solver, an incremental generation mechanism and an architecture that explains how these components interact. As opposite to classical SAT solvers, an incremental solver is a persistent object partly aware of its surroundings that addresses the problem of deciding a chain of related satisfiability instances as a whole, thus re-using information gathered from past search.

Let us consider a SAT solver as a search engine in the space of truth assignments over $V = var(f)$ attempting to make f evaluate to true. Then, an iSAT solver is a search engine that explores a search space S defined only *once per chain*, not once per instance.

Each instance f_i in a chain $\{f_i, i = 0, 1, ...\}$ specifies which portion S_i of the whole search space has to be searched for a solution. When a subspace is proved empty, a larger subspace (monotonically containing the previous ones) is considered. As depicted in the picture aside, subspaces are connected to one another by means of some special sets of propositional hypotheses $\mathcal{H}_i$ that mark the boundary between S_i and S_{i+1}, in so as S_i is just the subspace of S_{i+1} rooted at the branch $\mathcal{H}_i$.

Definition 1   (iSAT problem). *An iSAT instance is a sequence of couples $\{\langle f_i, \mathcal{H}_i \rangle, i = 0, 1, ...\}$ where f_i is a CNF formula, $\mathcal{H}_i \in var(f_i)$ is a set of propositional hypotheses, and $\forall i. f_i \subseteq f_{i+1}$. The iSAT problem consists of deciding whether $\exists i. SAT(f_i * \mathcal{H}_i)$.*

An iSAT instance is passed to an incremental solver step by step by repeatedly invoking the primitive "enlargeSearchSpace($\Delta f_i, \mathcal{H}_i$)" to notify the dimension $|var(f_i) \setminus var(f_{i-1})|$ and the "shape" f_i of the new subspace to be explored, together with the position $\mathcal{H}_i$ where it is attached as a subspace of S_{i+1} (with $\Delta f_i = f_i \setminus f_{i-1}$).

   When the time for implementation comes, it is by far convenient to modify an existing DPLL solver (thus retaining state-of-the-art technology) at the expense of performing some modifications.

   Each instance f_i is considered under the hypotheses $\mathcal{H}_i$, placed at the very bottom of the search stack. Standard solvers are allowed to withdraw every stacked hypothesis as soon as it comes out to be responsible for inconsistencies. An incremental solver behaves in the same way in all the cases but when the hypothesis to be removed is within $\mathcal{H}_i$. By removing such hypothesis it would indeed escape from S_i. Rather, it stops working and waits for the next enlargement of the search space. Search is then restarted across the newly added subspace by removing the selected source of inconsistency. The hypotheses $\mathcal{H}_i$ loose their inviolability, which is inherited by $\mathcal{H}_{i+1}$.

We modified the SIM solver [13] to obtain i-SIM [2] by (1) slightly revising the standard LIFO policy employed by the stack of hypotheses to allow the insertion of $\mathcal{H}_i$, (2) substituting *"the stack only contains (a subset of) $\mathcal{H}_i$, wait!"* for *"the stack is empty, quit!"* as a stop condition and (3) making eligible for dynamic enlargement all the internal data structures whose size depends on the number of variables and/or clauses in the formula (taking care to keep consistency between all mutual references).

We now briefly show which kind of connections among adjacent instances can be leveraged during the solving process. For a more thorough description of our technique we refer the reader to [2].

Let us consider a structured CTS problem on $W = \langle \mathcal{T}, P \rangle$ that generates a sequence of SAT instances $[\![W]\!]_i = [\![\mathcal{T}]\!]_i \wedge \neg[\![P]\!]_i$ with a *monotone encoding* for the background theory $(\forall i.[\![\mathcal{T}]\!]_i \subseteq [\![\mathcal{T}]\!]_{i+1})$.

Definition 2 (Incremental Encoding). *An* incremental encoding *for* $\{[\![W]\!]_i, i = 0, 1,$ $...\}$ *is a sequence of couples* $\{\langle [\![\mathcal{T}, P]\!]_i^+, \mathcal{H}_i \rangle, i = 0, 1, ...\}$*, with* $[\![\mathcal{T}, P]\!]_0^+ = I_0 \wedge P_0$*,* $[\![\mathcal{T}, P]\!]_i^+ = [\![\mathcal{T}, P]\!]_{i-1}^+ \wedge \Delta[\![\mathcal{T}, P]\!]_{i-1}^i$*,* $\Delta[\![\mathcal{T}, P]\!]_i^{i+1} \doteq \Delta[\![\mathcal{T}]\!]_i^{i+1} \wedge \Delta[\![P]\!]_i^{i+1}$ *and*

$$\begin{cases} [\![\mathcal{T}]\!]_0^+ \doteq I_0 \\ [\![\mathcal{T}]\!]_i^+ \doteq [\![\mathcal{T}]\!]_{i-1}^+ \wedge \Delta[\![\mathcal{T}]\!]_{i-1}^i \ i > 0 \end{cases} \qquad \begin{cases} [\![P]\!]_0^+ \doteq P_0 \\ [\![P]\!]_i^+ \doteq [\![P]\!]_{i-1}^+ \wedge \Delta[\![P]\!]_{i-1}^i \ i > 0 \end{cases}$$

where $[\![\mathcal{T}]\!]_i^+$ *($\Delta[\![\mathcal{T}]\!]_{i-1}^i$) is the* incremental (differential) *encoding for a monotone background theory* $\mathcal{T}$ *and is such that* $\forall i.[\![\mathcal{T}]\!]_i^+ \equiv [\![\mathcal{T}]\!]_i$*, while* $[\![P]\!]_i^+$ *($\Delta[\![P]\!]_{i-1}^i$) is the* incremental (differential) *encoding for the property* $P \in \mathcal{L}$ *and is such that* $\forall i. [\![P]\!]_i^+ *$ $\mathcal{H}_i \equiv_{V_i} [\![P]\!]_i$ *for a sequence* $\{\mathcal{H}_i, i = 0, 1, ...\}$ *of sets of literals called* closing set of hypotheses *over* $[\![P]\!]_i^+$*. The incremental property encoding is built after an* open encoding *over* $\mathcal{L}$*, which is a function mapping a formula* $Q \in \mathcal{L}$ *and a couple of indexes* k, k' *($k \leq k'$) onto a propositional formula* $]\!Q]\!_k^j$ *in such a way that* $[\![P]\!]_0^+ = P_0 =]\!P]\!_0^0$*,* $\mathcal{H}_k = \{\neg\phi_k^Q | \phi_k^Q \in [\![P]\!]_k^+\}$*, and* $\Delta[\![P]\!]_k^{k'} = \bigwedge_{\phi_k^Q \in [\![P]\!]_k^+} \phi_k^Q \rightarrow]\!Q]\!_{k'}^{k+1}$*.*

The key property of an incremental encoding is that *it defines an iSAT instance satisfiability equivalent to the corresponding sequence of standard encodings*, provided a valid open encoding over $\mathcal{L}$ is defined. Open encodings mimic the usual encodings but introduce a number of additional literals ϕ_k^Q (associated to a bound k and to a formula $Q \in \mathcal{L}$) used by the subsequent open encodings as coupling points to have the overall formula grow up: after $]\!Q]\!_k^k$ inserts a closing literals $\phi_k^{Q'}$, $\Delta[\![P]\!]_k^{k+1}$ expands its meaning $]\!Q']\!_{k+1}^{k+1}$ and this in turn creates coupling points for the next round.

At the clause level, a valid open encoding for adjacent bounds may be obtained by posing $]\!P]\!_k^k := (\Delta_k^+ \setminus \Delta_{k+1}^-) \cup \{\phi_k^P \vee \Gamma | \Gamma \in \Delta_{k+1}^-\}$, given the sequences of sets of clauses such that $cnf([\![P]\!]_{i+1}) = cnf([\![P]\!]_i) \setminus \Delta_i^- \cup \Delta_i^+$.

Alternatively, we may *incrementally CNF-ize* any encoding exhibiting the two properties $]\!P]\!_{k'}^0 * \mathcal{H}_{k'} \equiv [\![P]\!]_{k'}$ and $]\!P]\!_{k'}^k \cdot \sigma_{k'} =]\!P]\!_{k'+1}^k$ for the substitution $\sigma_{k'} = \{]\!f]\!_{k'+1}^{k'+1} / \phi_k^f | \phi_k^f \in pure(]\!P]\!_{k'}^k)\}$. These conditions are met by open encodings built over the family of *disjunctive* property chains. For these chains it is $\forall i \geq 0.[\![P]\!]_i \rightsquigarrow$

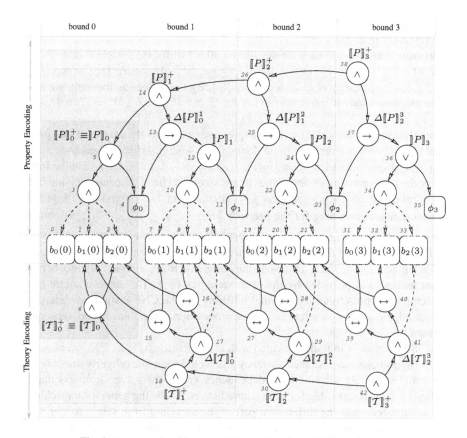

Fig. 1. An example of incremental encoding for a BMC problem

$[P]_{i+1}$, where $g \rightsquigarrow f$ (f is *disjunctively expanded* from g) iff a formula h with pure literals on $\{v_1, ..., v_n\}$ and two substitutions $\sigma_g = \{g_1/v_1, ..., g_n/v_n\}$ and $\sigma_f = \{(g_1 \vee f_1)/v_1, ..., (g_n \vee f_n)/v_n\}$ exist such that $g = h \cdot \sigma_g$ and $f = h \cdot \sigma_f$. Thereafter, incremental CNF-ization amounts to ensure consistency across the clause versions of all the formulas undergoing CNF-ization by maintaining the same meaning for variables shared among differential encodings.

The intuition behind the incremental encoding is that to connect subsequent meshes of the chain we just need to focus on the property encoding, whose open encoding indeed exhibits the forethought of spreading place-holders across the formula as coupling points between adjacent instances. When the solver stops searching and a resolution tree rooted at the empty clause is found, either it is independent from stacked hypotheses (in this case, not only the current SAT instance but the whole iSAT instance is inconsistent), or some closing hypothesis lays among its leaves. In the latter case, the empty clause fails to survive the backtrack step over the closing hypothesis, so the search can restart over the enlarged problem defined by gathering new constraints.

A simple example that captures many relevant aspects of the iCTS framework is the following. We incrementally test a *shift register* with n bits and the entry bit always equal to 0 against the (false) property "*the register never becomes empty if the two most significant bits are initially set*".

Formally, if $b_j(t)$ is the value of the j-th bit after t shifts, the system is described by $\mathcal{T} = Init \wedge \forall t > 0.\neg b_0(t) \wedge \forall j > 0.b_j(t) \leftrightarrow b_{j-1}(t-1)$, where $Init = b_{n-1}(0) \wedge b_{n-2}(0)$, and the property is $\overline{P} = \neg \exists t.\forall j.\neg b_j(t)$. By the deduction theorem, we assert that the property holds iff no model exists for $\mathcal{T} \wedge \neg\overline{P}$, i.e.: for $W = \mathcal{T} \wedge P$ where $P = \neg\overline{P}$.

Even though the above theory is too expressive to be directly translated into propositional logic (t ranges over an infinite domain) conjunctions/disjunctions over a finite set of propositional variables $V_k^i = \{b_j(t), j = 0, ..., n-1, t = 0, ..., k\}$ can be substituted for the quantifiers given any finite time horizon k. The incremental version of the resulting propositional theory (up to step 3) is depicted in the bottom half of Figure 1.

An incremental encoding for the property $[\![P]\!]_k \equiv \bigvee_{t=0}^{k} \bigwedge_{j=0}^{n-1} \neg b_j(t)$ is obtained by choosing $]\![P]\!]_t^t = \wedge_{j=0}^{n-1}\neg b_j(t) \vee \phi_i$ (upper half in the figure), and it is easy to check that $[\![P]\!]_t^+ * \neg\phi_t \equiv_{V_t} [\![P]\!]_t$. As the instance is built incrementally, argument-arrows never end in a region lighter than the one they originate from, and the only link between adjacent instances is provided by the open variables $\{\phi_i\}$. The small numeric labels near each node represent an incremental labelling for the CNF-ization procedure that generates adjacent and consistent ranges of propositional variables across subsequent encodings.

The iSAT solver is initially provided with the problem $cnf([\![\mathcal{T}, P]\!]_0^+)$ to be solved under the hypothesis $\neg\phi_0$. An inconsistency is detected soon (the property does not hold in the initial state). By traversing the dependency graph, the solver discovers that the hypothesis $\neg\phi_0$ is responsible for such contradiction. It tells the generation machinery, which in turn generates the differential part of the encoding from step 0 to step 1 and produces the clauses arising from the incremental CNF-ization of the subgraph rooted at $[\![\mathcal{T}, P]\!]_1^+$. The solver is notified of the new size of the problem, and is then given both the clause-set $cnf(\Delta[\![\mathcal{T}, P]\!]_0^1)$ and the new working hypothesis $\neg\phi_1$. Then, it dismisses $\neg\phi_0$ (it also notices that $\neg\phi_0$ was at the very bottom of the stack, so the unit clause ϕ_0 is learned, and this amounts to learn that $[\![\mathcal{T}, P]\!]_0$ has no model).

This incremental generate-and-solve loop goes on encountering other contradictions until step 3, when $[\![\mathcal{T}, P]\!]_3^+$ is considered under the hypothesis $\neg\phi_3$. A model for $[\![\mathcal{T}, P]\!]_3^+ * \neg\phi_3$ is found and used to reconstruct a 3-step witness falsifying the property. The solver maintains its internal state through the whole process and also retains all the consequences of the already performed search. In this simple example, the necessary truth value of many variables and some unit clauses are inherited from previous runs.

5   Bounded Model Checking as a Testbed

BMC [6] is a SAT-based automatic technique to verify a reactive system modelled as a finite state automaton M against a property f expressed in *linear temporal logic* (LTL). The semantic entailment $M \models_k \mathbf{E}\neg f$ to be checked is dealt with by solving $[\![M, f]\!]_k := [\![M]\!]_k \wedge [\![\neg f]\!]_k$, where $[\![M]\!]_k$ is a k-step long boolean encoding of the transition relation associated with M, while $[\![\neg f]\!]_k$ unrolls the semantics of $\neg f$ over a path of length k by representing all the possible behaviours violating f on such path.

We refer the reader to [6, 2] for a detailed description of this technique and to [3] for the PLTL (an extension of LTL) standard encoding we start from.

To incrementalise this technique we first generate a differential encoding for the monotone background theory: $I_0 \doteq I(s_0)$ and $\Delta[\![M]\!]_k \doteq T(s_{k-1}, s_k)$ (see [6, 2, 5]). According to [3], two adjacent property encodings only differ as each future time operator unrolls its semantics to the newly added time step.

From a CNF point of view, the clauses to be added/removed are obtained by recursively considering such operators, and are obtained by traversing the syntactic tree of the PLTL formula and conjuncting the set of clauses to be added/removed due to each node labeled by a future time operator.

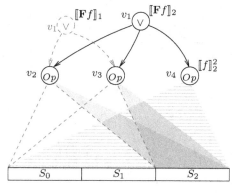

The picture aside shows that shifting from $[\![\mathbf{F}f]\!]_1$ to $[\![\mathbf{F}f]\!]_2$ the CNF translation of the node v_1 changes from $cnf_a = cnf(v_1 \leftrightarrow (v_2 \vee v_3)) = \{\{\neg v_1, v_2, v_3\}, \{\neg v_2, v_1\}, \{\neg v_3, v_1\}\}$ to $cnf_b = cnf(v_1 \leftrightarrow (v_2 \vee v_3 \vee v_4))$ $= \{\{\neg v_1, v_2, v_3, v_4\}, \{\neg v_2, v_1\}, \{\neg v_3, v_1\}, \{\neg v_4, v_1\}\}$ and that the clauses $cnf_c = cnf([\![f]\!]_2^2)$ appear, so $\Delta_2^+ = \{\{\neg v_1, v_2, v_3, v_4\}, \{\neg v_4, v_1\}\} \cup cnf_c \cup \Gamma_2^+$ and $\Delta_2^- = \{\{\neg v_1, v_2, v_3\}\} \cup \Gamma_2^-$, where Γ_2^+ and Γ_2^- are computed by recursively looking for nested future time operators within the subtrees rooted at v_2 and v_3. These expressions can be easily generalized to shift from k to $k+1$.

Rather than working at the clause level, we may construct a higher-level incremental procedure that exploits the semantics of time operators. This procedure acts *before* the CNF converter is presented with the formula, thus yielding a more intuitive encoding. It follows the guidelines given in the previous section and consists of defining a valid open encoding (see [3] for details) for PLTL formulas.

Definition 3 (Open PLTL Encoding). *The open translation of a PLTL formula from bound i to bound k ($i \leq k$) is a propositional formula inductively defined as follows.*

$$[\![q]\!]_k^i \doteq q^i \quad [\![\neg q]\!]_k^i \doteq \neg q^i \quad [\![f \wedge g]\!]_k^i \doteq [\![f]\!]_k^i \wedge [\![g]\!]_k^i \quad [\![f \vee g]\!]_k^i \doteq [\![f]\!]_k^i \vee [\![g]\!]_k^i$$

$$[\![\mathbf{X}f]\!]_k^i \doteq \begin{cases} \phi_k^f & i = k \\ [\![f]\!]_k^{i+1} & i < k \end{cases} \qquad [\![\mathbf{F}f]\!]_k^i \doteq \bigvee_{j \in [i,k]} [\![f]\!]_k^j \vee \phi_k^{\mathbf{F}f} \qquad [\![\mathbf{G}f]\!]_k^i \doteq \bot$$

$$[\![f\mathbf{U}g]\!]_k^i \doteq \bigvee_{j \in [i,k]} \left([\![g]\!]_k^j \wedge \bigwedge_{h \in [i,j)} [\![f]\!]_k^h \right) \vee \left(\phi_k^{f\mathbf{U}g} \wedge \bigwedge_{h \in [i,k]} [\![f]\!]_k^h \right)$$

$$[\![f\mathbf{R}g]\!]_k^i \doteq \bigwedge_{j \in [i,k]} \left([\![g]\!]_k^j \vee \bigvee_{h \in [i,j)} [\![f]\!]_k^h \right) \wedge \left(\neg\phi_k^{f\mathbf{R}g} \vee \bigvee_{h \in [i,j)} [\![f]\!]_k^h \right)$$

$$[\![\mathbf{Y}f]\!]_k^i \doteq \begin{cases} \bot & i = 0 \\ [\![f]\!]_k^{i-1} & i > 0 \end{cases} \qquad [\![\mathbf{Z}f]\!]_k^i \doteq \begin{cases} \top & i = 0 \\ [\![f]\!]_k^{i-1} & i > 0 \end{cases}$$

$$[\![\mathbf{O}f]\!]_k^i \doteq \bigvee_{j \in [0,i]} [\![f]\!]_k^j \qquad [\![f\mathbf{S}g]\!]_k^i \doteq \bigvee_{j \in [0,i]} \left([\![g]\!]_k^j \wedge \bigwedge_{h \in (j,i]} [\![f]\!]_k^h \right)$$

$$[\![\mathbf{H}f]\!]_k^i \doteq \bigwedge_{j \in [0,i]} [\![f]\!]_k^j \qquad [\![f\mathbf{T}g]\!]_k^i \doteq \bigwedge_{j \in [0,i]} \left([\![g]\!]_k^j \vee \bigvee_{h \in (j,i]} [\![f]\!]_k^h \right)$$

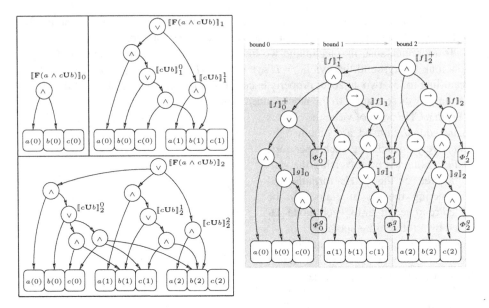

Fig. 2. Standard and incremental encoding for $f = \mathbf{F}(a \wedge g) = \mathbf{F}(a \wedge cUb)$ with $g = cUb$

Figure 2 compares this incremental encoding with the standard one for a sample PLTL formula. The former, instead of growing form right to left in a definitely non-incremental manner as the latter does, is such that $[\![f]\!]_0^+ \subseteq [\![f]\!]_1^+$ and $[\![f]\!]_1^+ \subseteq [\![f]\!]_2^+$. The semantics of the original encoding is nonetheless preserved according to $[\![f]\!]_k \equiv_{V_k} [\![f]\!]_k^+ * \mathcal{H}_k$.

Let us show with an example that the semantics of the original encoding is preserved by checking that $[\![f]\!]_1 \equiv_{V_1} [\![f]\!]_1^+ * \mathcal{H}_1$, with $V_1 = \{a(0), b(0), c(0), a(1), b(1), c(1)\}$ and $\mathcal{H}_1 = \{\neg\phi_1^g, \neg\phi_1^f\}$. The set of models of $[\![f]\!]_1$ on V_1 is obtained up by merging three pieces: (1) all the assignments where $\{a(0), b(0)\}$ holds (when a and b are true in the initial state, the subformula g is immediately satisfied, and $f = \mathbf{F}(a \wedge g)$ is satisfied as well), (2) all the assignments containing $\{a(0), c(0), b(1)\}$ (a is true in the initial state, while g is true because c holds until b becomes true), and (3) all the assignments where $\{a(1), b(0)\}$ holds (the argument of $\mathbf{F}$ is true at time 1 rather than at time 0).

The set of models of $[\![f]\!]_1^+ * \mathcal{H}_1$ on V_1 can be obtained by abstracting over the truth values of the two variables in $[\![f]\!]_1^+ * \mathcal{H}_1$ not in V_1, namely ϕ_0^g and ϕ_0^f. By playing with ϕ_0^g and ϕ_0^f (i.e.: by existentially quantifying over their truth values), we obtain that (1) $\{a(0), b(0)\}$ satisfies the formula once $\{\neg\phi_0^g, \neg\phi_0^f\}$ is assigned, (2) $\{a(0), c(0), b(1)\}$ makes the formula evaluate to true under $\{\phi_0^g, \neg\phi_0^f\}$, and (3) $\{a(1), b(0)\}$ is a model when $\{\phi_0^f\}$ is given. Hence, $[\![f]\!]_1^+ * \mathcal{H}_1$ has the same models as $[\![f]\!]_1$ (over V_1).

Let us also notice that if $[\![W]\!]_0^+$ is proven inconsistent under $\mathcal{H}_0$, the clause $\{\phi_f^0\}$ is learned (in general, a set Γ of theory-related clauses without closing variables is also learned) from $[\![T]\!]_0^+ : [\![T]\!]_0^+ \vdash \phi_f^0 \wedge \Gamma$, so that when $[\![W]\!]_1^+ * \mathcal{H}_1$ is approached, the system is actually facing $[\![T]\!]_1 \wedge \Gamma \wedge \neg[\![P]\!]_0 \wedge [\![P]\!]_1$.

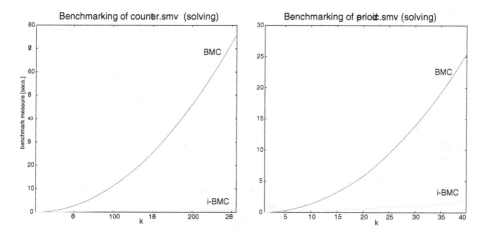

Fig. 3. Solving time compared on 2 BMC chains (290 SAT instances)

Our technique has been implemented within NuSMV [7], a state-of-the-art symbolic model checker used both to verify industrial designs of hardware and software systems and to test new formal verification techniques. NuSMV that integrates BDD-based and SAT-based model checking techniques. We modified the encoder/decoder modules according to Definition 3 and the CNF converter. Then, we experimented with several problems from the standard distribution of NuMSV using iSIM, experiencing a remarkable improvement in carrying out both the encoding and the solving task.

Figure 3 and 4 present some experimental results related to the cumulative solving time for chains of instances. The "counter" instance (false property with a deep counterexample at step 256 over a rather simple model) is particularly interesting as it isolates the contribute of the incremental machinery from the complexity of the underlying theory.
The "periodic" and "dme" instances test true specifications against more complex models (an asynchronous pipeline and a sequential logic network).

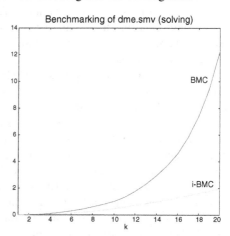

Fig. 4. Comparison on a "dme" model

6   Related Works

Several slightly different notions of incrementality have been proposed for the SAT problem during the last ten years. The first one was introduced by Hooker [14] in 1993. He addressed the problem of deciding whether $g \cup \{\Gamma\}$ is still satisfiable, given a satisfiable clause set g and an additional clause Γ. This basic mechanism is then exploited to decide a formula f, by adding one clause at a time and solving $|f|$ incremental sub-problems. The proposed algorithm is an adaptation of the basic DPLL procedure [9] that retains the position in the search tree (the path form the root to the last node examined) when a model for f is encountered. Then, if the assignment also satisfies $f \cup \{\Gamma\}$, nothing has to be done. Otherwise, the algorithm adds Γ to the clause set (and the possibly nonempty set $var(\Gamma) \setminus var(f)$ to the variable set), backtracks until Γ stops generating inconsistencies, and finally restarts to visit the search tree. This method was later extended to deal with the addition of multiple clauses at one time [4].

The Hooker's approach was conceived to solve problems arising from logic circuit verification, and, more generally, from problems related to the Electronic Design Automation (EDA). Significant improvements over the original proposal have been recently reported for applications on the same domain [16, 17]. These contributions describe a method to simultaneously solve a series of closely related SAT instances which is similar to the Hooker's one but also allows for the removal of sets of clauses. Formally, the proposed technique tackles the following problem [17]. Given a tree $G = (V, E)$ where each node $v \in V$ denotes a set of clauses $C(v)$ and each path from the root to a particular node v_i is associated with the SAT instance $\varphi(v_i) = \bigcup_{0 \leq j \leq i} C(v_j)$, decide the satisfiability of all the instances on the leaves. A former version of this formalization exists [16] where only trees of depth one are considered, in so as each formula $f_i := f_C \cup \Delta f_i$ on a leaf just adds some specific set of clauses Δf_i to a shared root subformula f_C. A DPLL-like algorithm is first applied to check the satisfiability of the root. If it is unsatisfiable then all the SAT problems at the leaves are unsatisfiable. Otherwise the algorithm recursively traverses the problem tree and checks the satisfiability of each node. The model of each satisfiable node is used as a starting point for all the child problems. Whenever the algorithm bumps into an unsatisfiable node it concludes that all the instances in the sub-tree rooted at that node are inconsistent and backtracks.

The same authors later proposed SATIRE [24], once more in the framework of EDA verification and optimization problems. SATIRE is a DPLL-like SAT solver that uses an incremental reasoning engine to decide n related SAT instances $\varphi_1, \ldots, \varphi_n$ where $\varphi_{i+1} = (\varphi_i \setminus \rho_i) \cup \alpha_{i+1}$ (ρ_i is the clauses to be removed and α_{i+1} the clauses to be added in order to transform φ_i in φ_{i+1}). As a major contribution this work enlightens the importance of learned clauses in the incremental solving process. SATIRE indeed tries to take advantage of the conflict clauses learned during the solution of the instances $\varphi_1, \ldots, \varphi_i$ while tackling φ_{i+1}. The main issue in reusing learned clauses is that the removal of clauses may clash with the validity of recorded conflict clauses. Whenever a clause belonging to the clause set generating a given conflict clause is removed the conflict clause does't hold any more and has to be removed. The problem is overcome in SATIRE by means of a detailed determination of the relationships between learned clauses and existing constraints performed during the conflict analysis. This mechanism requires an extra computation that can be very time consuming.

SATIRE's authors first enlightened that the reuse of learned clauses could be very effective in the context of the BMC procedure. They indeed experimented with some SAT formulas coming from BMC encodings and showed significant improvements as to solving time. The big potential of sharing learned clauses between similar BMC instances was independently investigated by Shtrichman [23], who already has worked on tuning generic SAT solvers for BMC instances by means of pre-computation of the variable ordering and some form of internal constraints replication to reduce the dimension of the search space [22]. He observed that the sets of learned clauses obtained for consecutive bounds are quite similar, though the additional problem of deciding which conflict clauses maintain validity still arises. The author proposes a DPLL algorithm augmented with a procedure to isolate the reusable conflict clauses. This procedure is based on a careful exploration of the implication graph used to perform the conflict analysis similar to the one introduced in [24] and it suffers from the same disadvantages coming from the additional book-keeping required. Experimental results show that constraints sharing generally has a positive effect on performances, but sometime its overhead overcomes the benefits.

Recently, one SAT solver (SATZOO [11]) has been implemented to incorporate the concept of incremental resolution for highly related SAT instances. It is based on a traditional DPLL-style procedure augmented with an interface which, given two subsequent related SAT instances, allows the second to be specified incrementally from the first by means of adding and removing constraints. The interface lets only unit clauses be removed from the clause database. This way, all the clauses learned during one run may be reused by the search procedure during the subsequent runs because the unit clauses can be considered as assumptions and learned clauses are independent of the assumptions under which they are deduced.

The potential of an incremental resolution of many related instances is so evident that also the state of the art SAT solver Chaff [20] has been integrated in his latest release with a module that shows an incremental solving capability. A more detailed description of the used technique can be found in [12].

7   Conclusions and Future Work

We proposed an integrated approach to incremental satisfiability that allows to *incrementalise* existing CTS procedures. We presented an incremental machinery that mainly retains the simplicity and strength of the original non-incremental one. In particular, we showed how to connect subsequent instances in a chain by relying on a tighter integration with the solver. We unveiled the importance of the deduction theorem and of the incremental property encoding to completely solve the learned-clause problem. We discussed the modifications needed to obtain an incremental solver and pointed out some details of the incremental generation step. We also presented an example of a complete iCTS implementation on top of real-world tools and gave a summary of the research on incremental decision procedures reported so far in the literature.

Our future work goes towards 1) the application of our framework to other domains, 2) a further enlargement of the iSAT solver interface, 3) the integration of *inductive learning* methods within our approach, and 4) some form of *validity checking* obtained by inductively reasoning on the structure of refutations rather than by explicit induction.

References

1. P. A. Abdulla, P. Bjesse, and N. Eén. Symbolic Reachability Analysis Based on SAT-Solvers. In *Proc. of TACAS 2000*, volume 1785, pages 411–425, 2000.
2. M. Benedetti and S. Bernardini. Incremental Compilation-to-SAT Procedures. Technical Report T03-12-13, sra.itc.it/people/benedetti/TR031213.pdf, Istituto per la Ricerca Scientifica e Tecnologica, 2003.
3. M. Benedetti and A. Cimatti. Bounded Model Checking for Past LTL. In *Proc. of TACAS 2003*, number 2619 in LNCS, pages 18–33, 2003.
4. H. Bennaceur, I. Gouachi, and G. Plateau. An Incremental Branch-and-Bound Method for Satisfiability Problem. *INFORMS Journal on Computing*, 10:301–308, 1998.
5. S. Bernardini. Structure and Satisfiability in Propositional Formulae. *AI*IA Notizie*, 4, 2003.
6. A. Biere, A. Cimatti, E. M. Clarke, M. Fujita, and Y. Zhu. Symbolic Model Checking without BDDs. In *Proc. of Design Automation Conference*, volume 1579, pages 193–207, 1999.
7. A. Cimatti, E. M. Clarke, E. Giunchiglia, F. Giunchiglia, M. Pistore, M. Roveri, R. Sebastiani, and A. Tacchella. NuSMV 2: An OpenSource Tool for Symbolic Model Checking. In *Proc. of CAV 2002*, volume 2404 of *LNCS*, 2002.
8. J. M. Crawford and A. D.Baker. Experimental results on the application of satisfiability algorithms to scheduling problems. In *Proc. of 12th AAAI '94*, pages 1092–1097, 1994.
9. M. Davis, G. Logemann, and D. Loveland. A machine program for theorem proving. *Journal of the ACM*, 5:394–397, 1962.
10. T. Boy de la Tour. Minimizing the Number of Clauses by Renaming. In *Proc. of the 10th Conference on Automated Deduction*, pages 558–572, 1990.
11. N. Eén and N. Sörensson. Temporal Induction by Incremental SAT Solving. In *Proc. of the First International Workshop on Bounded Model Checking*, 2003.
12. Z. Fu. zChaff. http://ee.princeton.edu/~chaff/zchaff.php, 2003.
13. E. Giunchiglia, M. Maratea, A. Tacchella, and D. Zambonin. Evaluating Search Heuristics and Optimization Techniques in Propositional Satisfiability. In *Proc. of IJCAR 2001*, 2001.
14. J.N. Hooker. Solving the Incremental Satisfiability Problem. *Journal of Logic Programming*, 15:177–186, 1993.
15. H. Kautz and B. Selman. Planning as satisfiability. In *Proc. of ECAI 1992*, pages 359–363.
16. J. Kim, J. Whittemore, J. P. M. Silva, and K. A. Sakallah. On Applying Incremental Satisfiability to Delay Fault Problem. In *Proc. of DATE 2000*, pages 380–384, 2000.
17. J. Kim, J. Whittemore, J. P. M. Silva, and K. A. Sakallah. On Solving Stack-Based Incremental Satisfiability Problems. In *Proc. of the ICCD 2000*, pages 379–382, 2000.
18. T. Larrabee. Test pattern generation using boolean satisfiability. In *IEEE Transaction on Computer-aided Design*, pages 4–15, 1992.
19. F. Massacci and L. Marraro. Logical Cryptanalysis as a SAT Problem. *Journal of Automated Reasoning*, 24, 2000.
20. M. Moskewicz, C. Madigan, Y. Zhao, L. Zhang, and S. Malik. Chaff: Engineering an Efficient SAT Solver. In *Proc. of the 38th DAC*, pages 530–535, 2001.
21. D. A. Plaisted and S. Greenbaum. A Structure-preserving Clause Form Translation. *Journal of Symbolic Computation*, 2:293–304, 1986.
22. O. Shtrichman. Tuning SAT checkers for Bounded Model Checking. In *Proc. of the 12th International Conference on Computer Aided Verification*, Lecture Notes in Computer Science. Springer Verlag, 2000.
23. O. Shtrichman. Pruning Techniques for the SAT-based Bounded Model Checking Problem. In *Proc. of CHARME'01*, pages 58–70, 2001.
24. J. Whittemore, J. Kim, and K. A. Sakallah. SATIRE: A New Incremental Satisfiability Engine. In *Proc. of the 38th Conference on Design Automation*, pages 542–545, 2001.

Resolve and Expand

Armin Biere

Johannes Kepler University, Institute for Formal Models and Verification,
Altenbergerstrasse 69, A-4040 Linz, Austria
biere@jku.at

Abstract. We present a novel expansion based decision procedure for quantified boolean formulas (QBF) in conjunctive normal form (CNF). The basic idea is to resolve existentially quantified variables and eliminate universal variables by expansion. This process is continued until the formula becomes propositional and can be solved by any SAT solver. On structured problems our implementation **quantor** is competitive with state-of-the-art QBF solvers based on DPLL. It is orders of magnitude faster on certain hard to solve instances.

1 Introduction

Recent years witnessed huge improvements in techniques for checking satisfiability of propositional logic (SAT). The advancements are driven by better algorithms on one side and by new applications on the other side. The logic of quantified boolean formulas (QBF) is obtained from propositional logic by adding quantifiers over boolean variables. QBF allows to represent a much larger class of problems succinctly.

The added expressibility unfortunately renders the decision problem PSPACE complete [20]. Nevertheless, various attempts have been made to lift SAT technology to QBF, in order to repeat the success of SAT. The goal is to make QBF solvers a versatile tool for solving important practical problems such as symbolic model checking [13] or other PSPACE complete problems.

For QBF the nesting order of variables has to be respected. Accordingly two approaches to solve QBF exist. Either variables are eliminated in the direction from the outermost quantifier to the innermost quantifier or vice versa. We call the first approach *top-down*, and the second one *bottom-up*.

Current state-of-the-art QBF solvers [4, 17, 12, 9, 22] are all top-down and implement a variant of the search-based Davis & Putnam procedure DPLL [7]. Additionally, QBF requires decision variables to be chosen in accordance with the quantifier prefix. Learning has to be adapted to cache satisfiable existential sub goals. DPLL also forms the basis of most state-of-the-art SAT solvers, and therefore it was most natural to use it for QBF as well.

Even for SAT, there are alternatives to DPLL, based on variable elimination, such as the resolution based Davis & Putnam procedure DP [8]. It has never been used much in practice, with the exception of [5], since usually too many clauses are generated.

Eliminating variables by resolution as in DP can be lifted from SAT to QBF as well. The result is a bottom-up approach for QBF called Q-resolution [10]. The only differ-

H.H. Hoos and D.G. Mitchell (Eds.): SAT 2004, LNCS 3542, pp. 59–70, 2005.

ence between Q-resolution and ordinary resolution is, that in certain cases universally quantified variables can be dropped from the resolvent.

In theory, Q-resolution is complete but impractical for the same reasons as resolution based DP [8]. It has not been combined with compact data structures either. In our approach, we apply Q-resolution to eliminate innermost existentially quantified variables. To make this practical, we carefully monitor resource usage, always pick the cheapest variable to eliminate, and invoke Q-resolution only if the size of the resulting formula does not increase much. We use *expansion* of universally quantified variables otherwise.

Expansion of quantifiers has been applied to QBF in [1] and used for model checking in [21, 2]. All three approaches work on formulae or circuit structure instead of (quantified) CNF. We argue that CNF helps to speed up certain computationally intensive tasks, such as the dynamic computation of the elimination schedule. First it is not clear how Q-resolution can be combined with this kind of structural expansion. In addition our goal is to eventually combine bottom-up and top-down approaches. CNF currently is the most efficient data structure for representing formulas in top-down approaches for SAT.

Another general bottom-up approach [16, 14, 15, 6] is also based on quantifier elimination. A SAT solver is used to eliminate multiple innermost variables in parallel. In practice these approaches have only been applied to SAT or model checking. In principle it would be possible to apply them directly to QBF. In our approach single variables are eliminated one after the other. We can also alternate between either eliminating existential variables of the innermost scope and eliminating universal variables of the enclosing universal scope.

2   Preliminaries

Given a set of variables V, a literal l over V is either a variable v or its negation $\neg v$. A *clause* is a disjunction of literals, also represented by the set of its literals. A conjunctive normal form (CNF) is a conjunction of clauses. Assume that the set of variables is partitioned into m non empty scopes $S_1, \ldots S_m, \subseteq V$, with $V = S_1 \cup \ldots \cup S_m$ and $S_i \cap S_j = \emptyset$ for $i \neq j$. Each variable $v \in V$ belongs to exactly one scope $\sigma(v)$. Scopes are ordered linearly $S_1 < S_2 \ldots < S_m$, with S_1 the outermost and S_m the innermost scope. For each clause C the maximal scope $\sigma(v)$ over all variables v in C is unique and defined as the scope $\sigma(C)$ of C. The scope order induces a pre-order on the variables which we extend to an arbitrary linear variable order.

Each scope is labelled as *universal* or *existential* by the labelling $\Omega(S_i) \in \{\exists, \forall\}$. Variables are labelled with the label of their scope as $\Omega(v) \equiv \Omega(\sigma(v))$. We further require that the ordered partition of V into scopes is maximal with respect to the labelling, or more precisely $\Omega(S_i) \neq \Omega(S_{i+1})$ for $1 \leq i < m$.

Now a quantified boolean formula (QBF) in CNF is defined as a CNF formula f together with an ordered partition of the variables into scopes. This definition matches the QDIMACS formats [11] very closely, with the additional restriction of maximality.

A variable v is defined to occur in positive (negative) phase, or just positively (negatively), in a clause C, if C contains the literal v ($\neg v$). A clause in which a variable occurs in both phases is *trivial* and can be removed from the CNF. Two clauses C, D, where v occurs positively in C and negatively in D, can be resolved to a resolvent clause. The resolvent consists of all literals from C except v and all literals from D except $\neg v$.

For a non-trivial clause C we define the process of *forall reduction* as follows. The set of *forall reducible variables* in C is defined as the set of universal variables in C for which there is no larger existential variable in C, with respect to the variable order. The clause D obtained from C by forall reduction contains all variables of C except forall reducible variables. For instance the two clauses in the following QBF

$$\exists x . \forall y . (x \vee y) \wedge (\neg x \vee \neg y)$$

are *not* forall reduced. Forall reduction results in removing the literal y in the first clauses and the literal $\neg y$ in the second, which results in two contradicting units. Also note, that forall reduction can result in an empty clause if the original clause contains universal variables only. Plain resolution followed by forall reduction is the same as Q-resolution [10].

Forall reduction is an equivalence preserving transformation. Thus without loss of generality we can assume that the CNF is in forall reduced form: by forall reduction no clause can be reduced further. This assumption establishes the important invariant, that $\Omega(\sigma(C)) = \exists$ for all clauses C. In other words, all clauses have an existential scope. There are no clauses with a universal scope. Particularly, the innermost scope is always existential ($\Omega(S_m) = \exists$). In our implementation, for each existential scope, we maintain a list of its clauses, and for each clause a reference to its scope.

3   Elimination

We eliminate variables until the formula is propositional and contains only existential quantifiers. Then it can be handed to a SAT solver. After establishing the invariant discussed above, a non-propositional QBF formula has the following structure

$$\Omega(S_1)\, S_1 \,.\, \Omega(S_2)\, S_2 \,.\, \ldots \forall S_{m-1} \,.\, \exists\, S_m \,.\, f \wedge g \qquad m \geq 2 \qquad (1)$$

where the formula f is *exactly* the conjunction of clauses with scope S_m. We either eliminate a variable in the innermost existential scope $S_\exists \equiv S_m$ by Q-resolution or a variable in the innermost universal scope $S_\forall \equiv S_{m-1}$ by expansion.

3.1   Resolve

An existential variable v of $S_\exists$ is eliminated as in [8, 10] by performing all resolutions on v, adding the forall reduced resolvents to the CNF, and removing all clauses containing v in either phase. As example consider the clauses in Fig. 1.

We assume that these 7 clauses are all clauses of a CNF in which the innermost existential variable v occurs. To eliminate v, we simply perform all 3×2 resolutions between a clause on the left side, in which v occurs positively, with all clauses on the right side, in which v occurs negatively. In this case 3 resolvents are trivial. The other three resolvents

$$(s \vee r), \qquad (x \vee y \vee r), \qquad \text{and} \quad (s \vee \neg x \vee \neg y \vee r)$$

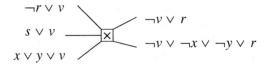

Fig. 1. Number of resolution pairs is quadratic

are added to the CNF and the original 5 clauses containing v in either phase are removed. As always, before adding one of the clauses, forall reduction is applied.

3.2   Expand

Expansion of a universal variable v in $S_\forall$ requires to generate a copy $S_\exists'$ of $S_\exists$, with a one-to-one mapping of variables $u \in S_\exists$ mapped to $u' \in S_\exists'$. With f' we denote the conjunction of clauses obtained from f by replacing all occurrences of $u \in S_\exists$ by u'. The result of expanding $v \in S_\forall$ in Eqn. (1) is as follows

$$\Omega(S_1)\, S_1 \cdot \Omega(S_2)\, S_2 \cdot \ldots \forall(S_\forall - \{v\}) \cdot \exists(S_\exists \cup S_\exists') \cdot f\{v/0\} \wedge f'\{v/1\} \wedge g$$

By $f\{v/0\}$ we denote the result of substituting v by the constant 0 in f. This is equivalent to removing all clauses in which v occurs in negative phase and removing the occurrences of v in those clauses in which v occurs positively, followed by forall reduction. The substitution by 1 is defined accordingly.

4   Optimizations

Before invoking one of the two costly elimination procedures described in Sec. 3, we first apply unit propagation, a simple form of equivalence reasoning, and the standard QBF version of the pure literal rule. These simplifications are repeated until saturation.

4.1   Equivalence Reasoning

To detect equivalences we search for pairs of dual binary clauses. A clause is called *dual* to another clause if it consists of the negation of the literals of its dual. If such a pair is found, we take one of the clauses and substitute the larger literal by the negation of the smaller one throughout the whole CNF.

The search for dual clauses can be implemented efficiently by hashing binary clauses. In more detail, whenever a binary clause is added, we also save a reference to it in a hash table and check, whether the hash table already contains a reference to its dual. If this is the case an equivalence is found. After an equivalence is found, it is used to eliminate one of the variables of the equivalence. Consider the following QBF formula:

$$\exists x \cdot \forall y \cdot \exists z \cdot \underline{(x \vee z)} \wedge (x \vee y \vee \neg z) \wedge (\neg x \vee \neg z) \wedge \underline{(\neg x \vee \neg z)}$$

The two underlined dual binary clauses involving x and z form an equivalence. After the last clause is added, the equivalence $x = \neg z$ is detected and z is replaced by $\neg x$, which results in the following QBF formula:

$$\exists x . \forall y . (x \vee \neg x) \wedge \underline{(x \vee y \vee x)} \wedge (\neg x \vee x) \wedge (\neg x \vee x)$$

After removal of 3 trivial clauses and forall reduction of the underlined clause, the only clause left is the unit clause x. In general, before searching for dual clauses, forall reduction has to be applied first. This way all substitutions triggered by equivalences will always replace existential variables by smaller literals. Replacing universal variables would be incorrect as the standard example $\exists x . \forall y . (x \vee \neg y) \wedge (\neg x \vee y)$ shows.

4.2   Subsumption

Expansion often needs to copy almost all clauses of the CNF. Moreover, the elimination procedures of Sec. 3 produce a lot of redundant subsumed clauses. Therefore, subsumed clauses should be removed. If a new clause is added, all old clauses are checked for being subsumed by this new clause. This check is called backward subsumption [19] and can be implemented efficiently on-the-fly, by using a signature-based algorithm. However, the dual check of forward subsumption [19] is very expensive and is only invoked periodically, for instance at each expansion step.

The subsumption algorithm is based on signatures, where a signature is a subset of a finite signature domain D. In our implementation $D = \{0, \dots, 31\}$ and a signature is represented by an unsigned 32-bit word. Each literal l is hashed to $h(l) \in D$. The signature $\sigma(C)$ of a clause C is the union of the hash values of its literals. Finally, the signature $\sigma(l)$ of a literal l is defined as the union of the signatures of the clauses in which it occurs, and is updated whenever a clause is added to the CNF.

Let C be a new clause, which is supposed to be added to the CNF. Further assume that the current CNF already contains a clause D which is subsumed by C, or more formally $C \subseteq D$. Then the signature of C is a subset of the signature of D, which in turn is a subset of the signatures of all the literals in D. Since all the literals of C are also literals of D, we obtain the necessary condition, $\sigma(C) \subseteq \sigma(l)$ for all literals $l \in C$. The signature $\sigma(l)$ is still calculated with respect to the current CNF, to which C has not been added yet.

If this necessary condition fails, then no clause in the current CNF can be backward subsumed by the new clause. In this case our caching scheme using signatures is successful and we call it a cache hit. Otherwise, in the case of a cache miss, we need to traverse all clauses D of an arbitrary literal in the new clause, and explicitly check for $C \subseteq D$. To minimize the number of visited clauses, we take the literal with the smallest number of occurrences. During the traversal, inclusion of signatures is a necessary condition again. This can easily be checked, since the signature of a clause is constant, and can be saved.

In practice, the overhead of maintaining signatures and checking for backward subsumption in the way just described turns out to be low. For forward subsumption no such efficient solution exists, and thus, forward subsumption, in our implementation, is only invoked before expensive operations, like expansion. Then we remove all clauses, flush signatures and add back the clauses in reverse chronological order.

Finally, if a clause C is added to the CNF, the signatures of all its literals $l \in C$ have to be updated. However, if a clause is removed, hash collision does not allow to subtract its signature from all the signatures of its literals. Therefore we just keep the old signatures as an over approximation instead. After a certain number of clauses are removed a recalculation of accurate clause signatures is triggered.

4.3   Tree-Like Prefix

We also realized that there are situations in which a linear quantifier prefix is not optimal and the basic expansion step as described above copies too many clauses. Consider the QBF

$$\exists x \, . \, \forall y, u \, . \, \exists z, v \, . \, f_1(x,y,z) \wedge f_2(x,u,v)$$

It is a linearization of the following formula with a tree-like prefix:

$$\exists x$$

$$\wedge$$

$$\begin{array}{cc} \forall y & \forall u \\ \exists z & \exists v \\ f_1(x,y,z) & f_2(x,u,v) \end{array}$$

The result of expanding y as described above would contain redundant copies of clauses from f_2 and vice versa redundant copies of f_1 when expanding u. In general, this problem can be coped with in the copying phase of expansion. The idea is to copy only those clauses that contain a variable connected to the expanded variable. In this context we call a variable *locally connected* to another variable if both occur in the same clause. The relation *connected* is defined as the transitive closure of *locally connected*, ignoring variables smaller than the expanded variable and all other universal variables in the same scope.

This technique is cheap to implement and avoids to pay the price for one single expansion. But we have not found an efficient way to use the information about tree like scopes to generate better elimination schedules on-the-fly.

5   Scheduling

The remaining problem, and one of our key contributions, is an efficient algorithm for on-the-fly generation of elimination schedules. Our scheduler has to answer the question, which of the variables in $S_\exists \cup S_\forall$ to eliminate next. As a cost function for choosing the next variable we try to minimize the size of the CNF after elimination. The size is measured in number of literals, which is equal to the sum of sizes of all clauses. We separately calculate for each variable a pessimistic but tight upper bound on the number of literals added, if the variable is eliminated. The variable with the smallest bound, which can be negative, is chosen.

For each literal l we maintain two counters reflecting the number of occurrences $o(l)$ and the sum $s(l)$ of the sizes of the clauses in which l occurs. These counters need to be updated only when a clause is added or removed. The update is linear in the clause size. This also shows that the obvious alternative cost function, which minimizes the number of added clauses instead of literals, is less precise, without improving complexity. For each existential scope S we maintain a counter reflecting the sum $s(S)$ of the sizes of its clauses.

5.1   Expansion Cost

For the expansion of $v \in S_\forall$ in Eqn. (1) according to Sec. 3.2 a tight upper bound on the number of added literals is calculated as follows. First f would be copied, which adds $s(S_\exists)$ literals. In f clauses are removed in which v occurs negatively, in the copy f' clauses are removed in which v occurs positively. This means subtracting both $s(v)$ and $s(\neg v)$ from $s(S_\exists)$. We also have to take care of the single literals removed, and the cost for eliminating v by expansion becomes

$$s(S_\exists) \quad - \quad \big(s(v) + s(\neg v) + o(v) + o(\neg v)\big)$$

For all $v \in S_\forall$ the term $s(S_\exists)$ is the same. Thus we only need to order these variables with respect to $-\big(s(v) + s(\neg v) + o(v) + o(\neg v)\big)$, which does not depend on other literals. This is essential for efficiency. In our implementation we use a separate heap based priority queue for each scope.

5.2   Resolving Cost

For the elimination of an existential variable $v \in S_\exists$ in Eqn. (1) according to Sec. 3.1 the calculation of a tight upper bound is similar but more involved. Consider Fig. 1. The literals on the left side, except v are copied $o(\neg v)$ times, which results in $o(\neg v) \cdot (s(v) - o(v))$ added literals. The number of copies of literals from the right side is calculated in the same way. Finally we have to remove all original literals, which all together results in the following cost, which again only depends on one variable:

$$o(\neg v) \cdot \big(s(v) - o(v)\big) \quad + \quad o(v) \cdot \big(s(\neg v) - o(\neg v)\big) \quad - \quad \big(s(v) + s(\neg v)\big)$$

As the example of Fig. 1 shows, this expression is only an upper bound on the cost of eliminating an existential variable by resolution. The bound is tight as the following example shows. Take the set of variables on each side of Fig. 1. If the intersection of these two sets only contain v, and all variables are existential, then the number of added literals exactly matches the bound.

Note that for bad choices of v calculating the multiplication may easily exceed the capacity of 32 bit integer arithmetic. Since variables with large costs can not be eliminated anyhow, we used saturating arithmetic with an explicit representation of infinity instead of arbitrary precision arithmetic.

5.3   Further Scheduling Heuristics

There are two exceptions to the scheduling heuristics just presented. First, as long as the minimal cost to eliminate an existential variable in $S_\exists$ is smaller than a given bound E, we eliminate the cheapest existential variable by resolution. This technique is also applied to pure propositional formulas. In this way **quantor** can be used as a preprocessor for SAT.

In our experiments, it turned out that in many cases, forcing the formula not to increase in size by setting $E = 0$, already reduces the final formula considerably. However, allowing small increases in size works even better. For scheduling purposes we use $E = 50$. This bound should probably be smaller if **quantor** is only used for preprocessing propositional formulas.

Another additional scheduling heuristics monitors the literals per clause ratio of the clauses with scope $S_\exists$. If it reaches a certain threshold, 4.0 in our implementation, an expansion is forced. After each forced expansion the threshold is increased by 10%. The reasoning behind forced expansion is as follows. A small literals per clause ratio increases the likelihood that the optimizations of Sec. 4 are applicable. In this sense, the scheduler should slightly bias decisions towards expansion instead of resolving, in particular, if the literals per clause ratio is high.

6   Experiments

We focus on structured instances, also called non-random, because we believe them to be more important for practical applications. As SAT solver we used **funex**, our own state-of-the-art SAT solver. It has not particularly been tuned towards our application. We have also seen rare cases where **funex** performs considerably worse than other SAT solvers, on SAT formulas generated by **quantor**.

In the first experiment we targeted the non random benchmarks of the SAT'03 evaluation of QBF [11] and compared **quantor** against **semprop** [12], the most efficient solver on these benchmarks in the evaluation [11]. We added **decide** [17] and **qube** with learning [9] as reference. In order to measure the effect of optimizations and using Q-resolution we also configured **quantor** in *expand* only mode. In this mode the scheduler always chooses expansion and all the optimizations are switched off. Exceptions are the pure literal rule, simplification by unit resolution, and forall reduction. This configuration, marked *expand* in Tab. 1, almost matches the original algorithm of the first version of **quantor**, which took part in SAT'03 evaluation of QBF [11].

As platform for this experiment we used an Intel Pentium IV 2.6 GHz with 1.5 GB main memory running Debian Linux. The results in Tab. 1 are clustered in families of benchmarks. For each family we count the number of instances solved in the given time limit of 32 seconds and memory limit of 1 GB. The numbers of families solved are printed in bold for best solvers. For a single best solver the numbers are underlined.

The comparison of the last two columns shows that expansion alone is very weak, and our new optimizations are essential to obtain an efficient state-of-the-art expansion based QBF solver. The number of cases in which **quantor** is among the best solvers for a family is the same as for **semprop**. There are four more families, for which **quantor** is the single best solver, three more than for **semprop**. Also note, that the families qbf* and R3CNF*, on which **quantor** performs poorly compared to the other solvers, can actually be considered to be randomized.

A detailed analysis revealed that **quantor** was able to solve 10 instances classified as *hard* in [11]. These hard formulas could not be solved by any solver in 900 seconds during the SAT'03 evaluation of QBF [11]. In a second experiment we restricted the benchmark set to these hard instances, a far smaller set.

The new time limit was set to 800 seconds to accommodate for the slightly faster processor (2.6 GHz instead of 2.4 GHz in [11]). As predicted by the evaluation results in [11] all solvers except **quantor** timed out on these instances. The results for **quantor** are presented in Tab. 2. Only solved instances are listed and are not clustered into families, e.g. C49*1.*_0_0* is the single instance with file name matching this pattern.

Table 1. Number solved instances for benchmarks families of the QBF evaluation 2003

	benchmark family	#inst	decide	qube	semprop	expand	quantor
1	adder*	16	2	2	2	1	**3**
2	Adder2*	14	2	2	2	2	**3**
3	BLOCKS*	3	**3**	**3**	**3**	**3**	**3**
4	C[0-9]*	27	2	3	2	3	**4**
5	CHAIN*	11	10	7	**11**	4	**11**
6	comp*	5	4	4	**5**	**5**	**5**
7	flip*	7	6	**7**	**7**	**7**	**7**
8	impl*	16	12	**16**	**16**	**16**	**16**
9	k*	171	77	91	97	60	**108**
10	logn*	2	**2**	**2**	**2**	**2**	**2**
11	mutex*	2	1	**2**	**2**	**2**	**2**
12	qbf*	695	518	565	**694**	130	210
13	R3CNF*	27	**27**	**27**	**27**	25	21
14	robots*	48	0	**36**	**36**	15	24
15	term1*	4	2	**3**	**3**	1	**3**
16	toilet*	260	187	**260**	**260**	259	259
17	TOILET*	8	**8**	6	**8**	**8**	**8**
18	tree*	12	10	**12**	**12**	8	**12**
19	vonN*	2	**2**	**2**	**2**	**2**	**2**
20	z4ml*	13	**13**	**13**	**13**	**13**	**13**
#(among best in family)			6	12	16	9	16
#(single best in family)			**0**	**0**	**1**	**0**	**4**

In all but two of the cases where the full version of **quantor** succeeded the *expand* only version quickly reached the memory limit of 1 GB. We note the time until the memory limit was reached in parentheses. It is also remarkable that the memory requirements for **quantor** have a large variance. The columns $\forall$ and $\exists$ contain the number of universal quantifications by expansion and existential quantifications by resolution respectively.

We added columns containing the numbers of unit simplifications, applications of the pure literal rule, subsumed clauses, applied substitutions, and number of removed literals due to forall reduction ($\forall$red). With the exception of subsumption, all optimizations are rather cheap with respect to run-time overhead, and as the data suggests, should be implemented. In particular the high number of pure literals in solving some instances is striking. Substitution does not seem to be important. More important, though also more costly, is subsumption.

For the two hard C[0-9]* instances covered in Tab. 2 more than 99% of the time was spent in the SAT solver. For the other solved hard instances no call to a SAT solver was needed. In an earlier experiment we used a slightly slower computer, an Alpha ES40 Server running at 666 MHz. The time limit was set to one hour, and the memory limit to 8 GB. In this setting, we were able to solve two more of the hard C[0-9]* benchmarks (with names matching C43*out*) in roughly 2500 seconds each. Again most time was

Table 2. Solved hard instances of SAT'03 evaluation of QBF

	expand			quantor								
hard instance	time	space	∀	time	space	∀	∃	units	pure	subsu.	subst.	∀red.
1 Adder2-6-s	(12.2)	m.o.	–	29.6	19.7	90	13732	126	13282	174081	0	37268
2 adder-4-sat	(12.1)	m.o.	–	0.2	2.8	42	1618	0	884	6487	0	960
3 adder-6-sat	(13.0)	m.o.	–	36.6	22.7	90	13926	0	7290	197091	0	54174
4 C49*1.*_0_0*	98.3	40.8	1	27.9	13.3	1	579	0	0	48	84	0
5 C5*1.*_0_0*	357.0	45.6	2	56.2	16.0	2	2288	10	0	4552	2494	0
6 k_path_n-15	(16.5)	m.o.	–	0.1	0.8	32	977	66	82	2369	2	547
7 k_path_n-16	(16.6)	m.o.	–	0.1	0.8	34	1042	69	85	2567	2	597
8 k_path_n-17	(16.2)	m.o.	–	0.1	0.9	36	1087	72	100	3020	2	639
9 k_path_n-18	(16.8)	m.o.	–	0.1	0.9	36	1146	76	106	3242	2	725
10 k_path_n-20	(21.4)	m.o.	–	0.1	0.9	38	1240	84	149	3967	2	855
11 k_path_n-21	(21.0)	m.o.	–	0.1	1.0	40	1318	84	130	4470	2	909
12 k_t4p_n-7	(16.8)	m.o.	–	15.5	105.8	43	88145	138	58674	760844	8	215
13 k_t4p_p-8	(21.4)	m.o.	–	5.8	178.6	29	12798	206	5012	85911	4	138
14 k_t4p_p-9	(21.2)	m.o.	–	0.3	4.5	32	4179	137	1389	23344	10	142
15 k_t4p_p-10	(17.3)	m.o.	–	27.9	152.9	35	130136	193	63876	938973	4	137
16 k_t4p_p-11	(17.3)	m.o.	–	86.0	471.5	38	196785	204	79547	1499430	4	140
17 k_t4p_p-15	(21.3)	m.o.	–	84.6	354.7	50	240892	169	181676	1336774	9	226
18 k_t4p_p-20	(20.9)	m.o.	–	3.6	16.1	65	27388	182	21306	197273	11	325

time in seconds, space in MB, m.o. = memory out (> 1 GB)

spent in the SAT solver. Except for those reported in Tab. 2, no further hard instance of [11] could be solved within these limits.

We also like to report on experiments involving benchmarks from QBFLIB, which turned out to be very simple for **quantor**. These include two families of benchmarks consisting of the 10 impl* instances and the 14 tree* instances. These 24 instances can be solved altogether in less than 0.1 seconds.

One of the most appealing aspects of QBF is, that an efficient QBF solver may also be used for *unbounded* model checking via the translation of [18, 20], also described in [17]. This translation needs only one copy of the transition relation but requires $2 \cdot l$ alternations of quantifiers, where $l = \lceil log_2 r \rceil$ and r is the initialized diameter (radius) of the model. In a boolean encoding l can be bounded by the number of state bits n. To check the hypothesis that QBF can be used for model checking in this way, we generated models of simple n-bit hardware counters, with reset and enable signal.

We check the invalid simple safety property, that the all-one state is not reachable from the initial state where all state bits are zero. This is the worst-case scenario for bounded model checking [3] since $2^n - 1$ steps are necessary to reach the state violating the safety property. Symbolic model checking [13] without iterative squaring needs 2^n fix point iterations. However, the size of the result of the translation of this problem to QBF is quadratic in n, the width of the counters.

With a time out of 60 seconds **decide** could only handle 3-bit-counters, **qube** and **semprop** up to 4 bits, while **quantor** solved 7 bits, matching the result by plain BMC with the same SAT solver. Since this example is very easy for BDD-based model checking, it is clear that QBF based model checking still needs a long way to go.

7 Conclusion

The basic idea of our QBF decision procedure is to resolve existential and expand universal variables. The key contribution is the resource-driven, pessimistic scheduler for dynamically choosing the elimination order. In combination with an efficient implementation of subsumption we obtain an efficient QBF solver for quantified CNF.

As future work we want to explore additional procedures for simplifying CNF and combine bottom-up elimination with top-down search. It may be also interesting to look into other representations, such as BDDs or ZBDDs.

Finally, we would like to thank Uwe Egly and Helmuth Veith for insisting on the argument that there is a benefit in not only focusing on a linear prefix normal form. Acknowledgements also go to Rainer Hähnle, whose comments triggered the optimization of our subsumption algorithm.

References

1. A. Ayari and D. Basin. QUBOS: deciding quantified boolean logic using propositional satisfiability solvers. In *Proc. 4th Intl. Conf. on Formal Methods in Computer-Aided Design (FMCAD'02)*, volume 2517 of *LNCS*. Springer, 2002.
2. P. Aziz Abdulla, P. Bjesse, and N. Eén. Symbolic reachability analysis based on SAT-solvers. In *Proc. 6th Intl. Conf. on Tools and Algorithms for the Construction and Analysis of Systems (TACAS'00)*, volume 1785 of *LNCS*. Springer, 2000.
3. A. Biere, A. Cimatti, E. M. Clarke, and Y. Zhu. Symbolic Model Checking without BDDs. In *Proc. 5th Intl. Conf. on Tools and Algorithms for the Construction and Analysis of Systems (TACAS'99)*, volume 1579 of *LNCS*. Springer, 1999.
4. M. Cadoli, A. Giovanardi, and M. Schaerf. An algorithm to evaluate quantified boolean formulae. In *Proc. 16th National Conference on Artificial Intelligence (AAAI-98)*, 1998.
5. P. Chatalic and L. Simon. ZRes: The old Davis-Putnam procedure meets ZBDDs. In *17th Intl. Conf. on Automated Deduction (CADE'17)*, volume 1831 of *LNAI*, 2000.
6. P. Chauhan, E. M. Clarke, and D. Kröning. Using SAT based image computation for reachability analysis. Technical Report CMU-CS-03-151, Carnegie Mellon University, 2003.
7. M. Davis, G. Logemann, and D. Loveland. A machine program for theorem-proving. *Communications of the ACM*, 5, 1962.
8. M. Davis and H. Putnam. A computing procedure for quantification theory. *Journal of the ACM*, 7, 1960.
9. E. Giunchiglia, M. Narizzano, and A. Tacchella. Learning for quantified boolean logic satisfiability. In *Proc. 18th National Conference on Artificial Intelligence (AAAI'02)*, 2002.
10. H. Kleine Büning, M. Karpinski, and A. Flögel. Resolution for quantified boolean formulas. *Information and Computation*, 117, 1995.
11. D. Le Berre, L. Simon, and A. Tacchella. Challenges in the QBF arena: the SAT'03 evaluation of QBF solvers. In *Proc. 6th Intl. Conf. on Theory and Applications of Satisfiability Testing (SAT'03)*, volume 2919 of *LNCS*. Springer, 2003.
12. R. Letz. Lemma and model caching in decision procedures for quantified boolean formulas. In *Proc. Intl. Conf. on Automated Reasoning with Analytic Tableaux and Related Methods (TABLEAUX'02)*, volume 2381 of *LNCS*. Springer, 2002.
13. K. L. McMillan. *Symbolic Model Checking: An approach to the State Explosion Problem*. Kluwer Academic Publishers, 1993.

14. K. L. McMillan. Applying SAT methods in unbounded symbolic model checking. In *Proc. 14th Intl. Conf. on Computer-Aided Verification (CAV'02)*, volume 2404 of *LNCS*. Springer, July 2002.
15. M. Mneimneh and K. Sakallah. Computing vertex eccentricity in exponentially large graphs: QBF formulation and solution. In *Proc. 6th Intl. Conf. on Theory and Applications of Satisfiability Testing (SAT'03)*, volume 2919 of *LNCS*. Springer, 2003.
16. D. Plaisted, A. Biere, and Y. Zhu. A satisfiability procedure for quantified boolean formulae. *Discrete Applied Mathematics*, 130(2), 2003.
17. J. Rintanen. Partial implicit unfolding in the Davis-Putnam procedure for quantified boolean formulae. In *International Conference on Logic for Programming, Artificial Intelligence and Reasoning (LPAR'01)*, 2001.
18. W. J. Savitch. Relation between nondeterministic and deterministic tape complexity. *Journal of Computer and System Sciences*, 4, 1970.
19. R. Sekar, I. V. Ramakrishnan, and A. Voronkov. Term indexing. In *Handbook of Automated Reasoning*, volume II. North-Holland, 2001.
20. L. J. Stockmeyer and A. R. Meyer. Word problems requiring exponential time. In *5th Annual ACM Symposium on the Theory of Computing*, 1973.
21. P. F. Williams, A. Biere, E. M. Clarke, and A. Gupta. Combining decision diagrams and SAT procedures for efficient symbolic model checking. In *Proc. 12th Intl. Conf. on Computer Aided Conf. Verification (CAV'00)*, volume 1855 of *LNCS*. Springer, 2000.
22. L. Zhang and S. Malik. Conflict driven learning in a quantified boolean satisfiability solver. In *Proc. Intl. Conf. on Computer-Aided Design (ICCAD'02)*, 2002.

Looking Algebraically at Tractable Quantified Boolean Formulas

Hubie Chen and Víctor Dalmau

Departament de Tecnologia, Universitat Pompeu Fabra,
Barcelona, Spain
{hubie.chen, victor.dalmau}@upf.edu

Abstract. We make use of the algebraic theory that has been used to study the complexity of constraint satisfaction problems, to investigate tractable quantified boolean formulas. We present a pair of results: the first is a new and simple algebraic proof of the tractability of quantified 2-satisfiability; the second is a purely algebraic characterization of models for quantified Horn formulas that were given by Kleine Büning, Subramani, and Zhao, and described proof-theoretically.

1 Introduction

An instance of the *generalized satisfiability* problem is a set of constraints, where a constraint is a relation over the two-element domain $\{0, 1\}$ paired with a variable tuple having the same arity as the relation; the question is to decide whether or not there is a $0 - 1$ assignment to all of the variables satisfying all of the constraints. A constraint is satisfied under an assignment if the variable tuple mapped under the assignment falls into the corresponding relation. Schaefer was the first to consider the generalized satisfiability problem [22]. He proved a now famous complexity classification theorem, showing that for any constraint language–a set of relations that can be used to express constraints–the generalized satisfiability problem over that constraint language is either in P or is NP-complete. This result has spawned a number of extensions and generalizations (see for example [12]). Analogous classification theorems have been proven for variants of the satisfiability problem such as quantified satisfiability [12]; also, there has been much recent work on establishing a complexity classification theorem for the general *constraint satisfaction problem (CSP)*, in which relations over domains of size greater than two are permitted.

In the nineties, an *algebraic viewpoint* on constraints was established that made it possible to approach the task of performing CSP complexity classification using tools from universal algebra [18, 16]; this viewpoint has produced a rich line of results, including [17, 13, 7, 4, 5, 6, 8, 3]. One fruit of this viewpoint has been a perspective on the tractable cases of satisfiability established by Schaefer's theorem, which include *2-satisfiability* and *Horn satisfiability*. For instance, an exact algebraic characterization of CSPs where it is possible to "go from local

H.H. Hoos and D.G. Mitchell (Eds.): SAT 2004, LNCS 3542, pp. 71–79, 2005.

to global consistency" has been given [17]; the 2-satisfiability problem yields the simplest non-trivial example from this class of CSPs.

In this paper, we demonstrate that the algebraic viewpoint that has been used to study the complexity of generalized satisfiability and the CSP can be used to derive new and interesting results concerning the quantified satisfiability problem–despite the fact that the complexity classification program for which the algebraic viewpoint was originally developed, has been completed in the two-element case (for both standard satisfiability [22] and quantified satisfiability [12]). It is our hope that this paper will stimulate further work on satisfiability that utilizes this algebraic viewpoint.

We present two results, one on quantified 2-satisfiability and the other on quantified Horn satisfiability; these two particular cases of the quantified satisfiability problem are known to be tractable [1, 19, 20]. Our results are as follows.

First, we give a new algebraic proof that quantified 2-satisfiability is tractable in polynomial time. In particular, we analyze an algorithm of Gent and Rowley [15]. From an implementation standpoint, this algorithm is extremely simple (and in the spirit of an algorithm for 2-satisfiability given by Del Val [14]): other than simple manipulations such as setting and removing variables, the only conceptual primitive used is unit resolution. We establish the correctness of this algorithm via a relatively simple and succinct proof which, unlike the proof of correctness given in [15], does not rely on the theory developed in [1]. In fact, we establish a more widely applicable result: we give a generalization of the algorithm given in [15] and demonstrate that it yields a general, algebraic sufficient condition for the tractability of the quantified constraint satisfaction problem. Our presentation of this new tractability result is self-contained, though the result was inspired by ideas in [11].

Second, we give a purely algebraic characterization of models for quantified boolean Horn formulas that were identified by Kleine Büning, Subramani, and Zhao [21]. They demonstrated that any true quantified Horn formula has a model of a particularly simple form–where every existentially quantified variable is set to either a constant or a conjunction of universally quantified variables. For any true quantified Horn formula Φ, they identified a model of this form described using the clauses derivable from Φ in Q-unit-resolution, a proof system known to be sound and complete for quantified Horn formulas [20]. We give an equivalent description of the models that they identified by making use of the semilattice structure possessed by the models of a quantified Horn formula, along with some natural homomorphisms among quantified Horn formulas that we introduce.

2   Tractability of Quantified 2-Satisfiability

The following is the basic terminology of quantified constraint satisfaction that we will use in this section. A *constraint* is an expression $R(v_1, \ldots, v_k)$ where each v_i is a variable, and $R \subseteq B^k$ is an arity k relation over a finite domain B. The constraint $R(v_1, \ldots, v_k)$ is true under an interpretation $f : V \to B$ defined on the variables v_i if $(f(v_1), \ldots, f(v_k)) \in R$. A quantified formula (over domain

B) is an expression of the form $Q_1 v_1 \ldots Q_n v_n \phi$ where each Q_i is a quantifier from the set $\{\forall, \exists\}$ and where ϕ is a conjunction of constraints over the variables $\{v_1, \ldots, v_n\}$. A constraint language is defined to be a set of relations over the same domain. The quantified constraint satisfaction problem over a constraint language Γ, denoted by $\mathsf{QCSP}(\Gamma)$, is the problem of deciding, given as input a quantified formula Φ having relations from Γ, whether or not Φ is true.

We now review some of the key elements of the algebraic viewpoint that has been fruitful in the study of constraint satisfaction [18, 16]. The central notion of this viewpoint is the concept of *polymorphism*. Let $f : B^m \to B$ be an m-ary operation on B, let R be a relation over B, and let k denote the arity of R. We say that f is a *polymorphism* of R, or that R is *invariant* under f, if for all (not necessarily distinct) tuples $(b_1^1, \ldots, b_k^1), \ldots, (b_1^m, \ldots, b_k^m)$ in R, the tuple

$$(f(b_1^1, \ldots, b_1^m), \ldots, f(b_k^1, \ldots, b_k^m))$$

belongs also to R. For example, let T be the binary boolean relation

$$\{(0,0), (0,1), (1,1)\}$$

having the property that the constraint $T(u, v)$ is equivalent to the clause ($\neg u \vee v$), and let $\mathsf{maj} : \{0,1\}^3 \to \{0,1\}$ be the ternary majority function on $\{0,1\}$. It is not difficult to verify that maj is a polymorphism of T. In fact, it is known that a boolean relation R is invariant under maj if and only if any constraint over R is equivalent to a 2-satisfiability formula.

In general, there is an intimate relationship between polymorphisms and conjunctive formulas: it has been shown that the complexity of $\mathsf{QCSP}(\Gamma)$ is determined by the set of all polymorphisms common to all relations in Γ, denoted by $\mathsf{Pol}(\Gamma)$. More precisely, if Γ_1 and Γ_2 are finite constraint languages such that $\mathsf{Pol}(\Gamma_1)$ and $\mathsf{Pol}(\Gamma_2)$ contain the same functions, then $\mathsf{QCSP}(\Gamma_1)$ and $\mathsf{QCSP}(\Gamma_2)$ are reducible to each other via polynomial-time many-one reductions [2]. Furthermore, the literature contains many results that link the complexity of $\mathsf{CSP}(\Gamma)$ and $\mathsf{QCSP}(\Gamma)$ with the presence (or absence) of functions of certain types in $\mathsf{Pol}(\Gamma)$.[1] As an example, it is known that the presence in $\mathsf{Pol}(\Gamma)$ of the ternary *dual discriminator* function $t : B^3 \to B$ defined by

$$t(x, y, z) = \begin{cases} x \text{ if } x = y \\ z \text{ otherwise} \end{cases}$$

implies that $\mathsf{QCSP}(\Gamma)$ is solvable in polynomial time [2]. In turn, this implies the tractability of quantified 2-satisfiability, since the dual discriminator function over a two-element domain $\{0, 1\}$ is equivalent to the maj function; however, the proof given in [2] is fairly involved.

In this section, we will use an algebraic approach slightly different from the polymorphism-based approach. Our approach is based on a new notion called

---

[1] By $\mathsf{CSP}(\Gamma)$, we denote the standard CSP over Γ–that is, the restriction of $\mathsf{QCSP}(\Gamma)$ to formulas having only existential quantifiers.

the *extended set polymorphism*, which we have studied also in the context of CSP complexity [11]. Let R be a relation, say k-ary, over a domain B. An extended set function f is any function with domain $\mathcal{P}(B) \times B$ and range B. (Here, we use $\mathcal{P}(B)$ to denote the set containing all non-empty subsets of B, that is, the power set of B excluding the empty set.) We say that an extended set function f is an *extended set polymorphism* of R (or, R is *invariant* under f) if for every $m \geq 1$ and all tuples $(b_1^1, \ldots, b_k^1), \ldots, (b_1^m, \ldots, b_k^m), (c_1, \ldots, c_k)$ in R, the tuple

$$(f(\{b_1^1, \ldots, b_1^m\}, c_1), \ldots, f(\{b_k^1, \ldots, b_k^m\}, c_k))$$

belongs also to R. As an example, consider again the boolean relation T as defined above, and let $g : \mathcal{P}(\{0, 1\}) \times \{0, 1\} \to \{0, 1\}$ be defined as

$$g(S, b) = \begin{cases} s \text{ if } |S| = 1 \text{ and } S = \{s\} \\ b \text{ otherwise} \end{cases}$$

It is immediate to verify that g is an extended set polymorphism of T. Indeed, with a little bit of effort it can be proven that a relation R is invariant under g if and only if any constraint over R is equivalent to a 2-satisfiability formula.

As with regular polymorphisms, we will say that a constraint language Γ is invariant under an extended set function f if every relation in Γ is invariant under f. We have the following general tractability result.

Theorem 1. *Let Γ be a constraint language over domain B invariant under an extended set function $f : \mathcal{P}(B) \times B \to B$ such that $f(B, b) = b = f(\{b\}, c)$ for all $b, c \in B$. Then, $\mathsf{QCSP}(\Gamma)$ is solvable in polynomial time.*

We can derive the tractability of quantified 2-satisfiability from Theorem 1 and the fact that the set of satisfying assignments of a 2-clause is invariant under the extended set function g described above. In fact, we can derive from Theorem 1 the tractability of any constraint language Γ invariant under the dual discriminator function $t : B^3 \to B$ described above.

The algorithm used to establish Theorem 1 makes use of the notion of arc consistency. We say that a conjunction of constraints ϕ is *arc consistent* if when $R(w_1, \ldots, w_k)$, $R'(w_1', \ldots, w_{k'}')$ are two constraints in ϕ such that $w_i = w_j'$, then the projection of R onto the ith coordinate is equal to the projection of R' onto the jth coordinate. This common projection is called the *domain* of the variable $w_i = w_j'$. We say that *arc consistency can be established* on a conjunction of constraints ϕ if tuples can be removed from the relations of ϕ in such a way that the resulting conjunction of constraints is logically equivalent to ϕ (that is, has the same satisfying assignments as ϕ), is arc consistent, and has no variable with empty domain. It is well-known that testing to see if arc consistency can be established is performable in polynomial time.

The algorithm (for Theorem 1), which generalizes the algorithm for quantified 2-satisfiability given in [15], is as follows. In each step of the algorithm, the outermost quantified variable is eliminated. Let $\Phi = Q_1 v_1 \ldots Q_n v_n \phi$ be a quantified formula. First suppose that Q_1 is an existential quantifier. In this case, the algorithm attempts to find a value b in the domain B such that arc consistency can be established on $\phi[v_1 = b]$ where every universal variable has a *full* domain, that is, domain equal to B. If there is no such value, the formula Φ is false. Otherwise, the algorithm sets v_1 to such a value, and continues. Next, suppose that Q_1 is a universal quantifier. In this case, the algorithm attempts to ensure that for *every* value b in the domain B, arc consistency can be established on $\phi[v_1 = b]$ where every universal variable (other than v_1) has a full domain. If there is any such value where this is not the case, the formula is false. Otherwise, the algorithm sets v_1 to *any* value, and continues.

To prove the correctness of this algorithm (for $\mathsf{QCSP}(\Gamma)$ satisfying the hypothesis of Theorem 1), we use the following characterization of true quantified formulas: a quantified formula $\Phi = Q_1 v_1 \ldots Q_n v_n \phi$ is true if there is a set $\mathcal{S}$ of satisfying assignments for ϕ satisfying the two following properties:

(a) For every partial assignment α to the universal variables, there exists an extension β of α in $\mathcal{S}$.
(b) Suppose that $\alpha, \beta \in \mathcal{S}$ are two assignments such that for every universal variable y preceding an existential variable x, we have that $\alpha(y) = \beta(y)$. Then, it holds that $\alpha(x) = \beta(x)$.

Proof. (Theorem 1) We establish the correctness of the algorithm by proving that whenever a variable is eliminated, truth of the formula is preserved. To demonstrate this, it suffices to prove the following fact: for any $b, b' \in B$, if arc consistency can be established on $\phi[v_1 = b]$ where every universal variable (coming after v_1) has full domain, and $\Phi[v_1 = b']$ is true, then $\Phi[v_1 = b]$ is true. Let $\mathcal{S}'$ be a set of satisfying assignments for $\phi[v_1 = b']$ with the above two properties. Let V denote the set of variables $\{v_1, \ldots, v_n\}$. By definition of arc consistency, there exists a mapping $a : V \to \mathcal{P}(B)$ such that $a(v_1) = \{b\}$, $a(y) = B$ for all universal variables y, and for any constraint $R(w_1, \ldots, w_k)$ in ϕ, there are tuples $(b_1^1, \ldots, b_k^1), \ldots, (b_1^m, \ldots, b_k^m) \in R$ where $a(w_i) = \{b_i^1, \ldots, b_i^m\}$ for all $i = 1, \ldots, k$. Let $\mathcal{S}$ be the set of assignments $h : V \setminus \{v_1\} \to B$ of the form $h(v) = f(a(v), h'(v))$ where $h' \in \mathcal{S}'$. It is straightforward to verify that $\mathcal{S}$ evidences the truth of $\Phi[v_1 = b]$. $\qquad\square$

3   Quantified Horn Formulas and KSZ Models

In this section, we give a purely algebraic characterization of the models for true quantified Horn formulas provided in [21], which we call KSZ models.

Before proceeding, we introduce some basic terminology and notation for this section. We will often, for sake of notation, restrict attention to quantified formulas of the form $\forall y_1 \exists x_1 \ldots \forall y_n \exists x_n \phi$, that is, formulas where there is a strict alternation between universal and existential quantifiers. In our quantified formulas, ϕ will always denote a boolean formula in conjunctive normal form, that is, a conjunction of clauses. Recall that a conjunction of clauses ϕ is a Horn formula if all of its clauses contain at most one positive literal. We define an *existential unit clause* to be a clause C such that there is exactly one existential literal $l \in C$, the literal l is a postive literal, and all literals in $C \setminus \{l\}$ come before l in the quantification order of the formula in which C appears.

A *strategy* for a quantified formula $\Phi = \forall y_1 \exists x_1 \ldots \forall y_n \exists x_n \phi$ is a sequence of mappings $\{\sigma_i : \{0,1\}^i \rightarrow \{0,1\}\}_{i \in [n]}$, where $[n]$ denotes the set containing the first n positive integers, $\{1, \ldots, n\}$. A strategy $\{\sigma_i : \{0,1\}^i \rightarrow \{0,1\}\}_{i \in [n]}$ is a *model* of Φ if for all $a_1, \ldots, a_n \in \{0,1\}$, it holds that the assignment mapping y_i to a_i and x_i to $\sigma_i(a_1, \ldots, a_i)$ (for all $i \in [n]$) satisfies ϕ. We consider a quantified formula Φ to be true if it has a model, and use $\mathcal{M}_\Phi$ to denote the set of all models of Φ.

Before defining the KSZ model for a quantified Horn formula, we need to introduce the following proof system for quantified Horn formulas.

Definition 1. *[20] The* Q-unit-resolution *proof system is defined as follows: Let $\Phi = Q_1 v_1 \ldots Q_m v_m \phi$ be a quantified boolean formula.*

- *For any clause $C \in \phi$, $\Phi \vdash C$.*
- *If $\Phi \vdash C$, $\Phi \vdash C'$, C is an existential unit clause with existential variable x, and C' is a clause containing $\neg x$, then $\Phi \vdash (C \cup C') \setminus \{x, \neg x\}$.*
- *If $\Phi \vdash C$ and $l \in C$ is a literal over a universal variable coming after all other literals in C in the quantification order, then $\Phi \vdash C \setminus \{l\}$.*

It has been shown that this proof system is sound and complete for the class of all quantified Horn formulas [20]: for any quantified Horn formula Φ, the empty clause is derivable from Φ (that is, $\Phi \vdash \emptyset$) if and only if Φ is false. Having defined this proof system, we can now define the KSZ model for a quantified Horn formula Φ, which is described in terms of the clauses derivable from Φ.

Definition 2. *[21] The* KSZ model *of a true quantified Horn formula $\Phi = \forall y_1 \exists x_1 \ldots \forall y_n \exists x_n \phi$ is defined as follows. Let U denote the set of all existential variables x such that there exists an existential unit clause C where $x \in C$ and $\Phi \vdash C$. For an existential variable x, let $W(x)$ denote the set containing all sets of negative universal literals C such that $C \cup \{x\}$ is an existential unit clause derivable from Φ (using Q-unit-resolution); and, let $V(x)$ denote the set $\cap_{C \in W(x)} C$.*

The KSZ model of Φ is the model $\{\sigma_i\}_{i \in [n]}$ such that

$$
\sigma_i(y_1, \ldots, y_i) = \begin{cases} 0 & \text{if } x_i \notin U \\ \wedge_{l \in V(x_i)} \neg l & \text{if } x_i \in U, V(x_i) \neq \emptyset \\ 1 & \text{if } x_i \in U, V(x_i) = \emptyset \end{cases}
$$

In what follows, we give a purely algebraic characterization of the KSZ model of a true quantified Horn formula. We first observe that if for any two models $\Sigma = \{\sigma_i : \{0,1\}^i \rightarrow \{0,1\}\}_{i \in [n]}$, $\Sigma' = \{\sigma_i' : \{0,1\}^i \rightarrow \{0,1\}\}_{i \in [n]}$ of a quantified Horn formula Φ, the strategy $\Sigma \wedge \Sigma'$ defined as $\{\sigma_i \wedge \sigma_i'\}_{i \in [n]}$ can be verified to also be a model for Φ. This follows from the fact that the operation $\wedge$ is a polymorphism of any Horn clause: when ϕ is a Horn clause (or more generally, a conjunction of Horn clauses) over a variable set V, and $f : V \to \{0,1\}$ and $f' : V \to \{0,1\}$ are interpretations both satisfying ϕ, the interpretation $f \wedge f'$ also satisfies ϕ. Because the operation $\wedge$ (applied to strategies as above) is associative, commutative, and idempotent, we have the following observation.

Proposition 1. *For all quantified Horn formulas Φ, $(\mathcal{M}_\Phi, \wedge)$ is a semilattice.*

For any quantified Horn formula $\Phi = \forall y_1 \exists x_1 \ldots \forall y_n \exists x_n \phi$, let $\Phi[y_k]$ be the formula obtained from Φ by removing all universal variables y_i from the quantifier prefix, except for y_k, and instantiating all instances of variables y_i (with $i \neq k$) in ϕ with the value 1. We define a mapping $\mathbf{d}_{y_k}$ from the set of strategies for Φ to the set of strategies for $\Phi[y_k]$ as follows. When $\Sigma = \{\sigma_i\}_{i \in [n]}$ is a strategy for Φ, define $\mathbf{d}_{y_k}(\Sigma)$ to be the strategy $\Sigma' = \{\sigma'_i\}_{i \in [n]}$ for $\Phi[y_k]$ such that for $i < k$, it holds that $\sigma'_i = \sigma_i(1, \ldots, 1)$, and for $i \geq k$, it holds that $\sigma'_i(y_k) = \sigma_i(1, \ldots, 1, y_k, 1, \ldots, 1)$, where in the right hand side expression, the kth coordinate contains y_k, and all other coordinates contain 1. It is clear that if Σ is a model of Φ, then $\mathbf{d}_{y_k}(\Sigma)$ is a model of $\Phi[y_k]$. We define $\mathbf{d}$ to be the mapping taking a strategy Σ of Φ to the n-tuple $(\mathbf{d}_{y_1}(\Sigma), \ldots, \mathbf{d}_{y_n}(\Sigma))$, and call $\mathbf{d}$ the deconstruction mapping, as it deconstructs a model for Φ into models for the various $\Phi[y_k]$.

Interestingly, given models for the various ϕ_{y_k}, we can construct a model for ϕ. Let $\mathbf{r}$ be the mapping taking an n-tuple of strategies $(\Sigma_1, \ldots, \Sigma_n)$, where $\Sigma_k = \{\sigma_i^k\}_{i \in [n]}$ is a strategy for ϕ_{y_k}, to the strategy $\Sigma = \{\sigma_i : \{0,1\}^i \to \{0,1\}\}_{i \in [n]}$ for ϕ, where for $i \in [n]$, the mapping σ_i is defined by

$$\sigma_i(y_1, \ldots, y_i) = (\sigma_i^1(y_1) \wedge \ldots \wedge \sigma_i^i(y_i)) \wedge (\sigma_i^{i+1} \wedge \ldots \wedge \sigma_i^n).$$

We call $\mathbf{r}$ the *reconstruction* mapping. We have the following result concerning the deconstruction and reconstruction mappings.

Theorem 2. *For all quantified Horn formulas $\Phi = \forall y_1 \exists x_1 \ldots \forall y_n \exists x_n \phi$,*

- *the deconstruction mapping $\mathbf{d}$ is a homomorphism from the semilattice $(\mathcal{M}_\Phi, \wedge)$ to the semilattice $(\mathcal{M}_{\Phi[y_1]}, \wedge) \times \cdots \times (\mathcal{M}_{\Phi[y_n]}, \wedge)$; and,*
- *the reconstruction mapping $\mathbf{r}$ is a homomorphism from the semilattice $(\mathcal{M}_{\Phi[y_1]}, \wedge) \times \cdots \times (\mathcal{M}_{\Phi[y_n]}, \wedge)$ to the semilattice $(\mathcal{M}_\Phi, \wedge)$.*

From Theorem 2, we can deduce the following corollary: if for a quantified Horn formula $\Phi = \forall y_1 \exists x_1 \ldots \forall y_n \exists x_n \phi$ it holds that all of the formulas $\Phi[y_1], \ldots, \Phi[y_n]$ are true, then Φ itself is true. This fact was observed, for instance, in [20]; however, the proof of Theorem 2 gives a purely algebraic justification of it.

It is well-known that every semilattice $(S, \oplus)$ has a maximal element m such that $m \oplus s = s \oplus m = m$ for all $s \in S$. For every quantified Horn formula Φ, we let Υ_Φ denote the maximal element of $(\mathcal{M}_\Phi, \wedge)$. Notice that for the pointwise ordering $\leq$ on functions where $1 \leq 0$, when $\Sigma = \{\sigma_i\}_{i \in [n]}$ is a model of Φ and $\{v_i\}_{i \in [n]}$ denotes Υ_Φ, it holds that $\sigma_i \leq v_i$, for all $i \in [n]$. The model Υ_Φ is rather canonical, but is not (in general) equal to the KSZ model. However, there is an intimate relationship between these two models that we can give, in terms of the two homomorphisms defined: deconstructing the model Υ_Φ and then reconstructing the resulting models yields the KSZ model.

Theorem 3. *For all true quantified Horn formulas Φ, it holds that the KSZ model of Φ is equal to* $\mathbf{r}(\mathbf{d}(\Upsilon_\Phi))$.

Theorem 3 gives a purely algebraic characterization of the KSZ model which, as we have seen, was originally defined proof-theoretically.

References

1. Bengt Aspvall, Michael F. Plass, and Robert Endre Tarjan. A Linear-Time Algorithm for Testing the Truth of Certain Quantified Boolean Formulas. Inf. Process. Lett. 8(3): 121-123 (1979).
2. F. Börner, A. Bulatov, A. Krokhin, and P. Jeavons. Quantified Constraints: Algorithms and Complexity. Computer Science Logic 2003.
3. A. Bulatov. Combinatorial problems raised from 2-semilattices. Manuscript.
4. Andrei A. Bulatov. A Dichotomy Theorem for Constraints on a Three-Element Set. FOCS 2002.
5. A. Bulatov. Malt'sev constraints are tractable. Technical report PRG-RR-02-05, Oxford University, 2002.
6. Andrei A. Bulatov. Tractable conservative Constraint Satisfaction Problems. LICS 2003.
7. Andrei A. Bulatov, Andrei A. Krokhin, and Peter Jeavons. Constraint Satisfaction Problems and Finite Algebras. ICALP 2000.
8. A. Bulatov and P. Jeavons. An Algebraic Approach to Multi-sorted Constraints Proceedings of 9th International Conference on Principles and Practice of Constraint Programming, 2003.
9. A. Bulatov, and P. Jeavons. Algebraic structures in combinatorial problems. Technical report MATH-AL-4-2001, Technische Universitat Dresden, 2001.
10. A. Bulatov, and P. Jeavons. Tractable constraints closed under a binary operation. Technical report PRG-TR-12-00, Oxford University, 2000.
11. Hubie Chen and Víctor Dalmau. (Smart) Look-Ahead Arc Consistency and the Pursuit of CSP Tractability. CP 2004.
12. Nadia Creignou, Sanjeev Khanna, and Madhu Sudan. Complexity Classifications of Boolean Constraint Satisfaction Problems. SIAM Monographs on Discrete Mathematics and Applications 7, 2001.
13. Víctor Dalmau and Justin Pearson. Set Functions and Width 1. Constraint Programming '99.
14. Alvaro del Val. On 2SAT and Renamable Horn. In AAAI'00, Proceedings of the Seventeenth (U.S.) National Conference on Artificial Intelligence, 279-284. Austin, Texas, 2000.
15. Ian Gent and Andrew Rowley. Solving 2-CNF Quantified Boolean Formulae using Variable Assignment and Propagation. APES Research Group Report APES-46-2002. 2002.
16. Peter Jeavons. On the Algebraic Structure of Combinatorial Problems. Theor. Comput. Sci. 200(1-2): 185-204, 1998.

17. P.G.Jeavons, D.A.Cohen and M.Cooper. Constraints, Consistency and Closure. Artificial Intelligence, 1998, 101(1-2), pages 251-265.
18. Peter Jeavons, David A. Cohen, and Marc Gyssens. Closure properties of constraints. J. ACM 44(4): 527-548 (1997).
19. Marek Karpinski, Hans Kleine Büning, and Peter H. Schmitt. On the Computational Complexity of Quantified Horn Clauses. CSL 1987: 129-137.
20. Hans Kleine Büning, Marek Karpinski, and Andreas Flögel. Resolution for Quantified Boolean Formulas. Information and Computation 117(1): 12-18 (1995).
21. Hans Kleine Büning, K. Subramani, Xishun Zhao. On Boolean Models for Quantified Boolean Horn Formulas. SAT 2003.
22. T. Schaefer. The complexity of satisfiability problems. Proceedings of the 10th Annual Symposium on Theory of Computing, ACM, 1978.

Derandomization of Schuler's Algorithm for SAT

Evgeny Dantsin and Alexander Wolpert

Roosevelt University, 430 S. Michigan Av., Chicago, IL 60605, USA
{edantsin, awolpert}@roosevelt.edu

Abstract. Recently Schuler [17] presented a randomized algorithm that solves SAT in expected time at most $2^{n(1-1/\log_2(2m))}$ up to a polynomial factor, where n and m are, respectively, the number of variables and the number of clauses in the input formula. This bound is the best known upper bound for testing satisfiability of formulas in CNF with no restriction on clause length (for the case when m is not too large comparing to n). We derandomize this algorithm using deterministic k-SAT algorithms based on search in Hamming balls, and we prove that our deterministic algorithm has the same upper bound on the running time as Schuler's randomized algorithm.

1 Introduction

Known Upper Bounds

A natural way to evaluate a satisfiability-testing algorithm is to find an upper bound on its worst-case running time. Such bounds can be also used to compare algorithms with each other. Since the mid 80s there has been a "competition" for the "record" upper bounds for different versions of SAT. Typically, bounds for SAT have the form α^n up to a polynomial factor, where n is the number of variables in the input formula. The exponent's base α may be a constant ($\alpha < 2$) or may depend on parameters of input formulas (such as the number of variables or the number of clauses). A challenging problem is to lower α as much as we can.

The currently best known upper bounds are discussed below (we give only the exponential terms of the bounds, omitting polynomial factors). Figure 1 summarizes these bounds.

Randomized Algorithms for k-SAT. All "record" randomized algorithms for k-SAT use one (or both) of the following two approaches:

- Random-assignment generation combined with unit clause elimination and bounded resolution (Paturi, Pudlák, Saks, Zane [12, 11]);
- Multistart random walk (Schöning [15, 16]).

The best known bounds for 3-SAT and 4-SAT are obtained using an algorithm based on a combination of both methods, namely: 1.324^n for 3-SAT and 1.474^n for 4-SAT [9]. Other recent algorithms for 3-SAT, e.g. [1, 8, 14], follow up the

H.H. Hoos and D.G. Mitchell (Eds.): SAT 2004, LNCS 3542, pp. 80–88, 2005.
© Springer-Verlag Berlin Heidelberg 2005

	Randomized algorithms	Deterministic algorithms
3-SAT	1.324^n [9]	1.481^n [2]
4-SAT	1.474^n [9]	1.6^n [2]
k-SAT $(k > 4)$	$2^{n\left(1-\frac{\mu_k}{k-1}\right)+o(n)}$ where $\lim_{k\to\infty}\mu_k = \pi^2/6$ [11]	$\left(2 - \frac{2}{k+1}\right)^n$ [2]
SAT	$2^{n\left(1-\frac{1}{\log(2m)}\right)}$ [17]	$2^{n\left(1-\frac{1}{\log(2m)}\right)}$ [this paper]

Fig. 1. "Record" worst-case upper bounds for k-SAT and SAT

multistart random walk approach. The bounds obtained using [11] are close: 1.362^n and 1.476^n for 3-SAT and 4-SAT respectively.

The best known bounds for $k > 4$ are due to the Paturi-Pudlák-Saks-Zane algorithm in [11]:

$$2^{n\left(1-\frac{\mu_k}{k-1}\right)+o(n)}$$

where $\mu_k \to \pi^2/6$ as $k \to \infty$. In particular, for $k = 5$ and $k = 6$, this gives 1.569^n and 1.637^n respectively. The multistart random walk algorithm [15] gives close bounds:

$$\left(2 - \frac{2}{k}\right)^n.$$

Deterministic Algorithms for k-SAT. Until recently, "record" upper bounds for k-SAT were obtained usind DPLL-like algorithms [6, 5], for example the 1.505^n bound for 3-SAT [10]. Newer deterministic algorithms borrow ideas from randomized approaches to testing satisfiability. The algorithms in [3, 2] that have the best known upper bounds for k-SAT are based on the derandomization of multistart random walk. They cover the Boolean cube $\{0, 1\}^n$ by Hamming balls and apply a local search method to find a satisfying assignment inside these balls. The "record" bound is

$$\left(2 - \frac{2}{k+1}\right)^n.$$

For $k = 3$, the bound can be improved to 1.481^n.

Randomized Algorithms for SAT (No Restriction on Clause Length). The best known bound was proved by R. Schuler in [17]. His algorithm uses a combination of the Paturi-Pudlák-Saks-Zane algorithm [11] and "clause shortening" (see Sect. 2 for details). The bound is

$$2^{n\left(1-\frac{1}{\log(2m)}\right)}$$

where m is the number of clauses in the input formula and $\log x$ denotes $\log_2 x$. Also, there is another bound: $2^{n-c\sqrt{n}}$, where c is a constant. This bound is due to two different algorithms. One algorithm [13] uses the Paturi-Pudlák-Saks-Zane algorithm in combination with the DPLL approach. The second algorithm [4] is based on multistart search in Hamming balls: Generate a random assignment and use local search to find a solution within a certain Hamming distance around this assignment. Schuler's bound [17] is more interesting than the $2^{n-c\sqrt{n}}$ bound because it is better for the case when m is not too large comparing to n, namely when $m = o(2^{\sqrt{n}})$. Note that for longer formulas, both bounds are worse than the trivial bound 2^n.

Deterministic Algorithms for SAT (No Restriction on Clause Length). Up until now, the only non-trivial upper bound for deterministic SAT algorithms has been given in [4]:

$$2^{n\left(1-\frac{2}{\sqrt{n\log n}}\right)}.$$

The corresponding algorithm is a derandomized version of multistart search in Hamming balls. The derandomization is based on covering codes. In the case of deterministic algorithms for SAT, there are also other types of bounds that are "more" dependent on the number of clauses or other input parameters, e.g., 1.239^m [7].

In this paper we give a deterministic algorithm that has the same bound

$$2^{n\left(1-\frac{1}{\log(2m)}\right)}$$

as in the case of randomized algorithms for SAT.

Our Result

We prove that SAT can be solved by a deterministic algorithm with the same upper bound on the running time as Schuler's randomized algorithm, i.e., with the bound $2^{n(1-1/\log_2(2m))}$ up to a polynomial factor.

Like Schuler's algorithm, our deterministic algorithm can be described in terms of two algorithms $\mathcal{M}$ (stands for *Main*) and $\mathcal{S}$ (stands for *Subroutine*). The algorithm $\mathcal{S}$ is used to test satisfiability of formulas with "short" clauses (of length at most $\log(2m)$). The algorithm $\mathcal{M}$ is the main algorithm that transforms an input formula F into F' by "shortening" the clauses in F. Then $\mathcal{M}$ invokes $\mathcal{S}$ to check whether F' is satisfiable. If so, we are done. Otherwise, the algorithm $\mathcal{M}$ simplifies F and recursively invokes itself on the results of simplification.

Theorem 1 in Sect. 3 gives an upper bound on the running time of the algorithm $\mathcal{M}$ under an assumption on the running time of the subroutine $\mathcal{S}$. More exactly, the assumption is that $\mathcal{S}$ runs in time at most $2^{n(1-1/k)}$ up to a polynomial factor, where k is the maximum length of clauses in F. Then $\mathcal{M}$ runs in time at most $2^{n(1-1/\log(2m))}$ up to a polynomial factor. Does there exist any deterministic subroutine $\mathcal{S}$ that meets this assumption? The answer is positive (Theorem 2): the algorithms [2] have the required upper bound on the running time. Thus, taking any of them as the subroutine $\mathcal{S}$, we obtain a deterministic algorithm that solves SAT with the bound

$$2^{n\left(1-\frac{1}{\log(2m)}\right)}.$$

Notation

By a *formula* we mean Boolean formulas in conjunctive normal form (CNF) defined as follows. A *literal* is a Boolean variable x or its negation $\neg x$. A *clause* is a finite set C of literals such that C contains no opposite literals. The *length* of C (denoted by $|C|$) is the number of literals in C. A *formula* is a set of clauses. An *assignment* to variables $x_1, \ldots, x_n$ is a mapping from $\{x_1, \ldots, x_n\}$ to the truth values $\{\text{TRUE}, \text{FALSE}\}$. This mapping is extended to literals: each literal $\neg x_i$ is mapped to the truth value opposite to the value assigned to x_i. We say that a clause C is *satisfied* by an assignment A if A assigns TRUE to at least one literal in C. The formula F is *satisfied* by A if every clause in F is satisfied by A. In this case, A is called a *satisfying* assignment for F.

By *SAT* we mean the following computational problem: Given a formula F in CNF, decide whether F is satisfiable or not. The *k-SAT* problem is the restricted version of SAT that allows only clauses of length at most k.

Here is a summary of the notation used in the paper.

- F denotes a formula;
- n denotes the number of variables in F;
- m denotes the number of clauses in F;
- k denotes the maximum length of clauses in F;
- $|C|$ denotes the length of clause C;
- $\log x$ denotes $\log_2 x$.

2   Algorithms Based on Clause Shortening

Schuler's Algorithm

We first sketch Schuler's algorithm [17]. More exactly, we describe a polynomial-time randomized procedure $\mathcal{R}$ that finds a satisfying assignment (if any) with probability at least $2^{-n(1-1/\log(2m))}$. This probability can be increased to a constant by repetitions in the usual way. The procedure $\mathcal{R}$ tests satisfiability in two steps:

1. Convert the input formula to a formula in k-CNF where $k = \log(2m)$;
2. Use a k-SAT algorithm to test satsfiability of the resulting formula.

Let F be an input formula consisting of clauses $C_1, \ldots, C_m$. Assuming that F is satisfied by an (unknown) assignment A, we show how $\mathcal{R}$ finds A. The procedure $\mathcal{R}$ starts with shortening the clauses in F as follows:

1. For each clause C_i such that $|C_i| > \log(2m)$, choose any $\log(2m)$ literals in C_i and delete the other literals.
2. Leave the shorter clauses as is.

Let $F' = \{D_1, \ldots, D_m\}$ be the result of the shortening. Obviously, any satisfying assignment for F' also satisfies F. The formula F' is in k-CNF where $k = \log(2m)$. Therefore, a satisfying assignment for F' (if any) can be found using a k-SAT algorithm. The procedure $\mathcal{R}$ uses the Paturi-Pudlák-Zane method from [12] to find a satisfying assignment for F' in polynomial time with probability at least $2^{-n(1-1/k)}$. There are two possible cases:

Case 1. A satisfies F'. Then the Paturi-Pudlák-Zane method finds A in polynomial time with probability at least $2^{-n(1-1/\log(2m))}$.

Case 2. A does not satisfy F'. Then there is a clause D_i such that all of its literals are false under A (but C_i is true under A). Therefore, if we "guess" this clause correctly, we may simplify F by assigning FALSE to all literals occurring in D_i. We choose a clause in F' uniformly at random. The probability that we have "guessed" the clause correctly (i.e., we have chosen D_i) is at least $1/m$. Then we simplify F' as follows:

   1. For each literal l in the chosen clause, remove all clauses that contain $\neg l$;
   2. Delete l from the remaining clauses.

Finally, we recursively apply $\mathcal{R}$ to the result of the simplification.

The analysis of $\mathcal{R}$ in [17] shows that $\mathcal{R}$ finds A with the required probability. Note that the same bound holds if the Paturi-Pudlák-Zane method (used as a subroutine in $\mathcal{R}$) is replaced by another procedure that finds a satisfying assignment in polynomial time with the same or higher probability, for example by Schöning's random-walk method [15].

Algorithms $\mathcal{M}$ (for Main) and $\mathcal{S}$ (for Subroutine)

Schuler's algorithm invokes the Paturi-Pudlák-Saks-Zane procedure [11] for testing satisfiability of formulas with "short" clauses. Our derandomized version will also use a subroutine to check formulas with "short" clauses. However, we first describe our algorithm without specifying the invoked subroutine. That is, assuming that $\mathcal{S}$ is an arbitrary procedure that tests satisfiability of formulas in k-CNF in time $2^{n(1-\frac{1}{k})}$ up to a polynomial factor, we define our main algorithm $\mathcal{M}$ as an algorithm that invokes $\mathcal{S}$ as a subroutine.

Algorithm $\mathcal{S}$

Input: Formula F (with no restriction on clause length).
Output: Satisfying assignment or "no".
Any algorithm that tests satisfiability of a formula F in time $2^{n(1-\frac{1}{k})}$ up to a polynomial factor, where n is the number of variables in F and k is the maximum length of clauses in F.

Algorithm $\mathcal{M}$

Input: Formula F with clauses $C_1, \ldots, C_m$ over n variables.
Output: Satisfying assignment or "no".

   1. Change each clause C_i to a clause D_i as follows: If $|C_i| > \log(2m)$ then choose any $\log(2m)$ literals in C_i and delete the other literals; otherwise leave C_i as is, i.e., $D_i = C_i$. Let F' denote the resulting formula.

2. Test satisfiability of F' using the algorithm $\mathcal{S}$.
3. If F' is satisfiable, return the satisfying assignment found in the previous step. Otherwise, F could be still satisfiable: there may exist an assignment A that satisfies all clauses $C_1, \ldots, C_m$ but falsifies some D_i. Therefore, A can be found by successively falsifying the reduced clauses in F'. Namely, for each clause D_i different from C_i, do the following:
   (a) Convert F to F_i by assigning FALSE to all literals in D_i. Namely, for each literal l in D_i, remove all clauses containing $\neg l$ and delete l from the remaining clauses.
   (b) Recursively invoke $\mathcal{M}$ on F_i.
4. Return "no".

3   Bound for SAT

We prove an upper bound for $\mathcal{M}$ assuming that $\mathcal{S}$ exists. Then we choose $\mathcal{S}$ such that this subroutine runs in the required time. As a result, we obtain the claimed upper bound for the main algorithm $\mathcal{M}$.

Theorem 1. *The running time of the algorithm $\mathcal{M}$ is at most*

$$2^{n\left(1 - \frac{1}{\log(2m)}\right)}$$

up to a polynomial factor.

Proof. Let $t_\mathcal{S}(F)$ and $t_\mathcal{M}(F)$ be, respectively, the running times of the algorithms $\mathcal{S}$ and $\mathcal{M}$ on a formula F. It is not difficult to see that $t_\mathcal{M}(F)$ can be estimated (up to a polynomial factor) as follows:

$$t_\mathcal{M}(F) \leq t_\mathcal{S}(F') + m \cdot t_\mathcal{M}(F_i) \tag{1}$$

where F' and F_i are as described in the algorithm $\mathcal{M}$. Let $T_\mathcal{M}(n, m)$ denote the maximum of the running time of $\mathcal{M}$ on formulas with m clauses over n variables. For the subroutine $\mathcal{S}$, we define $T_\mathcal{S}(n, m)$ as the maximum running time on a different set of formulas, namely let $T_\mathcal{S}(n, m)$ be the maximum of the running time of $\mathcal{S}$ on the set of formulas F such that each F has m clauses over n variables and the maximum length of clauses is not greater than $\log(2m)$. Let L denote $\log(2m)$. Then for any n and m, the inequality (1) implies the following recurrence relation:

$$T_\mathcal{M}(n, m) \leq T_\mathcal{S}(n, m) + m \cdot T_\mathcal{M}(n - L, m)$$

Iterating this recurrence and using the bound on $t_\mathcal{S}(F)$ with $k \leq L$, we get (again up to a polynomial factor)

$$T_\mathcal{M}(n, m) \leq \sum_{i=0}^{n/L} m^i \cdot T_\mathcal{S}(n - iL, m)$$

$$\leq \sum_{i=0}^{n/L} m^i \cdot 2^{(n - iL)(1 - 1/L)} = 2^{n(1 - 1/L)} \sum_{i=0}^{n/L} \left(m \cdot 2^{1-L}\right)^i$$

Since $L = \log(2m)$, we have $m\, 2^{1-L} = 1$. Therefore,

$$t_{\mathcal{M}}(F) \leq T_{\mathcal{M}}(n, m) \leq 2^{n(1-1/L)}$$

up to a polynomial factor.   □

Theorem 2 (based on [2]). *There exists a deterministic algorithm that tests satisfiability of an input formula F in time at most*

$$2^{n\left(1-\frac{1}{k}\right)}$$

up to a polynomial factor, where n is the number of variables in F, and k is the maximum length of clauses in F.

Proof. Paper [2] defines two algorithms that can be applied to any formula. Their running times are estimated in terms of the maximum length of clauses in the input formula (thus, they can be viewed as algorithms for k-SAT). Both algorithms cover the Boolean cube $\{0,1\}^n$ by Hamming balls and search for a satisfying assignment inside these balls. The first algorithm runs in time at most $2^{n(1-\log(1+1/k))}$ up to a polynomial factor (Theorem 1 in [2]). Since

$$\log\left(1 + \frac{1}{k}\right) = \frac{\log e}{k} + o\left(\frac{1}{k}\right),$$

this algorithm meets the claim. The second algorithm has a parameter δ; its running time is at most

$$2^{n\left(1-\log\left(1+\frac{1}{k}\right)+\delta\right)}$$

up to a polynomial factor (Theorem 2 in [2]). Taking $\delta \leq \frac{\log(e/2)}{k}$, we have

$$2^{n\left(1-\log\left(1+\frac{1}{k}\right)+\delta\right)} \leq 2^{n\left(1-\frac{\log e}{k}+\frac{\log(e/2)}{k}\right)} \leq 2^{n\left(1-\frac{1}{k}\right)}.$$

Hence, the second algorithm also meets the claim.

The algorithms differ in the construction of the covering of $\{0,1\}^n$ by Hamming balls. The first algorithm uses a greedy method to construct the covering that is minimal up to a polynomial factor. The construction requires an exponential space (approximately $2^{n/6}$). The second algorithm constructs a "nearly minimal" covering, i.e., a covering that is minimal up to a factor of $2^{\delta n}$, where δ can be chosen arbitrary small.

To estimate the space used by the second algorithm, we have to consider details of how it constructs the covering of $\{0,1\}^n$. Each ball center is the concatenation of n/b blocks of length b (Lemma 7 in [2]). The algorithm constructs a covering code $\mathcal{C}$ of length b for blocks. Then, keeping this code in memory, the algorithm generates code words of length n (centers of balls) one by one. An upper bound on the space can be estimated as the cardinality of the covering code $\mathcal{C}$ for blocks. Using Lemma 4 in [2], we can estimate the cadrinality $|\mathcal{C}|$ as follows:

$$|\mathcal{C}| \leq b\sqrt{b}\, 2^{b\left(1-H\left(\frac{1}{k+1}\right)\right)} \tag{2}$$

where $H(x) = -x \log x - (1-x) \log(1-x)$ is the binary entropy function. To obtain the desired upper bound on the running time, we should choose b so that

$$|\mathcal{C}|^{n/b} \leq 2^{n\left(1-H\left(\frac{1}{k+1}\right)+\delta\right)}. \tag{3}$$

Using the bound (2) on $|\mathcal{C}|$ and the inequality (3), we get the following constraint on b:

$$\left(b\sqrt{b}\, 2^{b\left(1-H\left(\frac{1}{k+1}\right)\right)}\right)^{n/b} \leq 2^{n\left(1-H\left(\frac{1}{k+1}\right)+\delta\right)} \tag{4}$$

which is equivalent to $(b\sqrt{b})^{1/b} \leq 2^{\delta}$. Now we substitute

$$\delta = \frac{\log(e/2)}{k}$$

and take $b = 4k \log k$. Then (4) holds for all sufficiently large k. Therefore, we can use blocks of length $4k \log k$. In fact, the algorithm will be applied to formulas with $k = \log(2m)$, which gives the upper bound $(2m)^{4 \log \log(2m)}$ on the space.

$\square$

Theorem 3. *Suppose that the algorithm $\mathcal{M}$ uses the algorithm from Theorem 2 as the subroutine $\mathcal{S}$. Then $\mathcal{M}$ tests satisfiability of an input formula F with m clauses over n variables in time at most*

$$2^{n\left(1-\frac{1}{\log(2m)}\right)}$$

up to a polynomial factor.

Proof. Immediately follows from Theorems 1 and 2. $\square$

Acknowledgments

We thank Edward A. Hirsch for useful discussions. We also thank anonymous referees for their comments that helped to improve the paper.

References

1. S. Baumer and R. Schuler. Improving a probabilistic 3-SAT algorithm by dynamic search and independent clause pairs. Electronic Colloquium on Computational Complexity, Report No. 10, February 2003.
2. E. Dantsin, A. Goerdt, E. A. Hirsch, R. Kannan, J. Kleinberg, C. Papadimitriou, P. Raghavan, and U. Schöning. A deterministic $(2 - 2/(k + 1))^n$ algorithm for k-SAT based on local search. *Theoretical Computer Science*, 289(1):69–83, October 2002.
3. E. Dantsin, A. Goerdt, E. A. Hirsch, and U. Schöning. Deterministic algorithms for k-SAT based on covering codes and local search. In *Proceedings of the 27th International Colloquium on Automata, Languages and Programming, ICALP 2000*, volume 1853 of *Lecture Notes in Computer Science*, pages 236–247. Springer, July 2000.

4. E. Dantsin, E. A. Hirsch, and A. Wolpert. Algorithms for SAT based on search in Hamming balls. In *Proceedings of the 21st Annual Symposium on Theoretical Aspects of Computer Science, STACS 2004*, volume 2996 of *Lecture Notes in Computer Science*, pages 141–151. Springer, March 2004.

5. M. Davis, G. Logemann, and D. Loveland. A machine program for theorem-proving. *Communications of the ACM*, 5:394–397, 1962.

6. M. Davis and H. Putnam. A computing procedure for quantification theory. *Journal of the ACM*, 7:201–215, 1960.

7. E. A. Hirsch. New worst-case upper bounds for SAT. *Journal of Automated Reasoning*, 24(4):397–420, 2000.

8. T. Hofmeister, U. Schöning, R. Schuler, and O. Watanabe. A probabilistic 3-SAT algorithm further improved. In *Proceedings of the 19th Annual Symposium on Theoretical Aspects of Computer Scienceg, STACS 2002*, volume 2285 of *Lecture Notes in Computer Science*, pages 192–202. Springer, March 2002.

9. K. Iwama and S. Tamaki. Improved upper bounds for 3-SAT. Electronic Colloquium on Computational Complexity, Report No. 53, July 2003.

10. O. Kullmann. New methods for 3-SAT decision and worst-case analysis. *Theoretical Computer Science*, 223(1-2):1–72, 1999.

11. R. Paturi, P. Pudlák, M. E. Saks, and F. Zane. An improved exponential-time algorithm for k-SAT. In *Proceedings of the 39th Annual IEEE Symposium on Foundations of Computer Science, FOCS'98*, pages 628–637, 1998.

12. R. Paturi, P. Pudlák, and F. Zane. Satisfiability coding lemma. In *Proceedings of the 38th Annual IEEE Symposium on Foundations of Computer Science, FOCS'97*, pages 566–574, 1997.

13. P. Pudlák. Satisfiability — algorithms and logic. In *Proceedings of the 23rd International Symposium on Mathematical Foundations of Computer Science, MFCS'98*, volume 1450 of *Lecture Notes in Computer Science*, pages 129–141. Springer, 1998.

14. D. Rolf. 3-SAT in $RTIME(O(1.32793^n))$ — improving randomized local search by initializing strings of 3-clauses. Electronic Colloquium on Computational Complexity, Report No. 54, July 2003.

15. U. Schöning. A probabilistic algorithm for k-SAT and constraint satisfaction problems. In *Proceedings of the 40th Annual IEEE Symposium on Foundations of Computer Science, FOCS'99*, pages 410–414, 1999.

16. U. Schöning. A probabilistic algorithm for k-SAT based on limited local search and restart. *Algorithmica*, 32(4):615–623, 2002.

17. R. Schuler. An algorithm for the satisfiability problem of formulas in conjunctive normal form. To appear in *Journal of Algorithms*, 2003.

Polynomial Time SAT Decision, Hypergraph Transversals and the Hermitian Rank

Nicola Galesi[1,*] and Oliver Kullmann[2,**]

[1] Departament de Llenguatges i Sistemes Informàtics,
Universitat Politécnica de Catalunya, Barcelona - Spain
galesi@lsi.upc.es
http://www.lsi.upc.es/~galesi
[2] Computer Science Department, University of Wales Swansea,
Swansea, SA2 8PP, UK
O.Kullmann@Swansea.ac.uk
http://cs-svr1.swan.ac.uk/~csoliver/

Abstract. Combining graph theory and linear algebra, we study SAT problems of low "linear algebra complexity", considering formulas with bounded hermitian rank. We show polynomial time SAT decision of the class of formulas with hermitian rank at most one, applying methods from hypergraph transversal theory. Applications to heuristics for SAT algorithms and to the structure of minimally unsatisfiable clause-sets are discussed.

1 Introduction

Connections between graphs, clause-sets and matrices based on conflicting literals have been investigated in [13], introducing the notions "hermitian rank" and "hermitian defect" to the field of SAT-related problems. Here we continue these investigations, and we present new SAT decision algorithms for some classes of conjunctive normal forms, combining graph theory and linear algebra.

The *conflict multigraph* cmg(F) of a clause-set F ([14, 13]) has the clauses of F as vertices, and as many (parallel) edges joining two vertices as the clauses have conflicts, i.e., clashing literals. The *hermitian rank* $h(F)$, as adopted in [14, 13] from [9], is the hermitian rank of the adjacency matrix of cmg(F), where the hermitian rank $h(A)$ of a symmetric real matrix A is the maximum of the number of positive and negative eigenvalues of A, and can also be naturally computed from the sign changes in the characteristic polynomial of A ([15]). The *hermitian defect* $m - h(A)$, where m is the dimension of A, equals the Witt index of the quadratic form associated with A ([15]).

In this paper we explore the use of the conflict multigraph cmg(F) of a clause-set F for SAT algorithms. On the one hand, we exploit structural properties of

* Supported by grant CICYT TIC2001-1577-C03-02.
** Supported by grant EPSRC GR/S58393/01.

H.H. Hoos and D.G. Mitchell (Eds.): SAT 2004, LNCS 3542, pp. 89–104, 2005.

$\mathrm{cmg}(F)$, namely bipartiteness and variations, and on the other hand we use the hermitian rank $h(F)$ as a complexity measure for SAT decision.

1.1   Using the Hermitian Rank as Complexity Measure

We investigate, whether $h(F)$ can be used as a measure of problem complexity for SAT decision, yielding a SAT decision algorithm whose running time depends mainly and monotonically on $h(F)$. The hermitian rank is at most the number of variables, i.e., $h(F) \leq n(F)$, as shown in [14, 13], reformulating the original Graham-Pollak theorem [8]. It would be interesting to prove an upper bound $2^{h(F)}$ on time complexity of SAT decision (ignoring polynomial factors). Given that $h(F)$ can be computed in polynomial time using the symmetric form of Gaussian elimination ([2]), the upper bound $2^{h(F)}$ would follow using the framework proposed in [11, 17, 18]: Use $h(F)$ as the heuristic of a DLL-like SAT algorithm, i.e., chose a branching variable where in both branches $h(F)$ strictly decreases (as much as possible). In Section 5 we further discuss this approach.

In this article we concentrate on the case of SAT decision for clause-sets F with $h(F) = 1$, which, besides the trivial case $h(F) = 0$ (that is, F only has pure literals), constitutes the base case of the above approach. Let $\mathcal{CLS}_h(1)$ denote the set of clause-sets F with $h(F) \leq 1$. Since the hermitian rank is computable in polynomial time, membership in $\mathcal{CLS}_h(1)$ is decidable in polynomial time. In [14, 13] it was shown that application of partial assignments does not increase the hermitian rank (using Cauchy's interlacing inequalities), i.e., $h(\varphi * F) \leq h(F)$ holds for any clause-set F and partial assignment φ. In the special case of $h(F) = 1$ we thus either have $h(\varphi * F) = 0$ (i.e., $\varphi * F$ has no clashing literals) or $h(\varphi * F) = 1$, and it follows, that to decide satisfiability for $F \in \mathcal{CLS}_h(1)$ we can not use splitting on a variable as considered above (since except of trivial cases in both branches the hermitian rank does not decrease). So a different approach is needed, exploiting the special structure of $F \in \mathcal{CLS}_h(1)$.

The hermitian rank $h(A)$ of a symmetric real matrix is the minimal rank of a matrix B with $A = B + B^t$ ([14, 13]). This property is basic to infer in Lemma 2 a characterisation of symmetric matrices with zero diagonal and hermitian rank one. As an application, in **Theorem 3** we see that after elimination of blocked clauses ([12, 11]) every $F \in \mathcal{CLS}_h(1)$ has a complete bipartite *graph* (without parallel edges) as conflict multigraph. This special structure is exploited for SAT decision of F as explained in Subsection 1.2.

Following the program of characterising the (hereditary) classes of graphs with at most k positive respectively negative eigenvalues via forbidden induced subgraphs, in [19, 20] it is shown that a connected *graph* G (i.e., its adjacency matrix A) has at most one negative eigenvalue (i.e., $h(A) \leq 1$) if and only if it is a complete bipartite graph (see Theorem 1.34 in [4]). Our characterisation in Theorem 3 differs by its use of linear algebra, and by considering the general case of *multi*graphs, where no characterisation via a finite number of forbidden induced subgraphs is possible and thus elimination of blocked clauses is essential.

1.2   Using the Transversal Hypergraph Problem for SAT Decision

In **Theorem 3** we showed that clause-sets with hermitian rank 1 after elimination of blocked clauses have a conflict multigraph which is a complete bipartite graph. We will use this characterisation for fast SAT decision.

In Section 4 we define *bipartite clause-sets* as clause-sets with a bipartite conflict multigraph, and we characterise them using basic results from algebraic graph theory. The main case of bipartite clause-sets is given by *positive-negative clause-sets* (PN-clause-sets) where every clause either is positive or negative. The SAT problem for the class of PN-clause-sets is easily seen to be NP-complete. Therefore we need to refine the class of bipartite clause-sets to obtain a class with feasible SAT decision capturing clause-sets of hermitian rank one.

A *bi-hitting clause-set* has a conflict multigraph which is complete bipartite (that is, every pair of vertices from different parts is connected by at least one edge). Bi-hitting PN-clause-sets constitute again the core of the class of bi-hitting clause-sets. From a bi-hitting PN-clause-set F we extract two hypergraphs $\mathcal{H}_P$ resp. $\mathcal{H}_N$ by considering the set of positive clauses resp. the set of negative clauses. If F does not contain subsumed clauses, then F is unsatisfiable if and only if $(\mathcal{H}_P, \mathcal{H}_N)$ is a *transversal hypergraph pair*, that is, $\mathcal{H}_N$ is the set of minimal transversals of $\mathcal{H}_P$ (and vice versa). Whether the transversal hypergraph problem is solvable in polynomial time is an important open problem ([6]). Recently it was shown to be solvable in quasi-polynomial time ([7]), and thus the SAT problem for bi-hitting clause-sets is solvable in quasi-polynomial time.

The final step exploits, that Theorem 3 not only yields complete bipartite multigraphs, but complete bipartite *graphs* (no parallel edges). Clause-sets where the conflict multigraph is a complete bipartite graph are called 1-*uniform bi-hitting clause-sets*, and in Lemma 11 we show that the SAT problem for this class is essentially the same as the *exact transversal hypergraph problem*. This problem was investigated in [5] and shown to be solvable in polynomial time; hence the SAT problem for clause-sets with hermitian rank one is decidable in polynomial time (**Theorem 13**).

The paper is organised as follows: We start with the preliminaries in Section 2. In Section 3 we characterise the class of clause-sets with hermitian rank one. In Section 4 the relations between (uniform) bi-hitting clause-sets and the (exact) hypergraph transversal problem is examined. Finally open problems and directions for future research are discussed in Section 5.

2   Preliminaries

We assume a universe $\mathcal{VA}$ of *variables*, from which *literals* (negated and unnegated variables) are constructed, using $\overline{x}$ for the negation of a literal x. A *clause-set* is a finite sets of *clauses*, where a *clause* is a finite and complement-free set of literals, denoting the set of all clause-sets by $\mathcal{CLS}$. For $F \in \mathcal{CLS}$ the number of variables is $n(F)$ and the number of clauses is $c(F)$. A *partial assignment* is a map $\varphi : V \rightarrow \{0, 1\}$ for some $V \subseteq \mathcal{VA}$, and the application of φ to F is denoted by $\varphi * F \in \mathcal{CLS}$. Given some class $\mathcal{C} \subseteq \mathcal{CLS}$ of clause-sets and some "measure" $f : \mathcal{C} \to \mathbb{R}$, by $\mathcal{C}_f(b) := \{F \in \mathcal{C} : f(F) \le b\}$ we denote the set of clause-sets in $\mathcal{C}$ with measure at most $b \in \mathbb{R}$.

A hypergraph is a pair $(V, \mathbb{H})$, where V is the set of vertices and $\mathbb{H}$ is a set of subsets of V (the "hyperedges"). A transversal T of a hypergraph $(V, \mathbb{H})$ is a subset $T \subseteq V$ with $T \cap H \ne \emptyset$ for all $H \in \mathbb{H}$, while a minimal transversal is a transversal such that no strict subset is also a transversal. The set of all minimal transversals of $(V, \mathbb{H})$ is $\mathrm{Tr}(V, \mathbb{H})$, and fulfils $\mathrm{Tr}(\mathrm{Tr}(\mathbb{H})) = \mathbb{H}$ if $\mathbb{H}$ is simple, that is, does not contain subsumed hyperedges. An independent set of $(V, \mathbb{H})$ is a subset $I \subseteq V$ such that there is no hyperedge $H \in \mathbb{H}$ with $H \subseteq I$. For more information on hypergraphs see for example [1]. A permutation matrix is a square matrix over $\{0, 1\}$ such that every row and every column contains exactly one entry equal to 1. Transposition of matrices A is denoted by A^t.

The *conflict multigraph* of $F \in \mathcal{CLS}$ is denoted by $\mathrm{cmg}(F)$; the vertices of $\mathrm{cmg}(F)$ are the clauses of F, and clauses $C, D \in F$ are joined by exactly $|C \cap \overline{D}|$ parallel edges, using $\overline{D} = \{\overline{x} : x \in D\}$. The *symmetric conflict matrix* $\mathrm{CM}(F)$ is the adjacency matrix of $\mathrm{cmg}(F)$.

The *hermitian rank* $h(F)$ for a clause-set F is defined as $h(F) := h(\mathrm{CM}(F))$, where the hermitian rank $h(A)$ of a symmetric real matrix can be computed in the following ways (see [14, 13, 15] for proofs and references; let m denote the dimension of A):

1. $h(A) = \max(i_-(A), i_+(A))$, where $i_\pm(A)$ can be computed as follows:
   (a) $i_\pm(A)$ is the number of positive resp. negative eigenvalues of A;
   (b) $i_\pm(A)$ is the maximum dimension of a positive resp. negative definite subspace of the quadratic (or bilinear) space associated with A;
   (c) $i_\pm(A)$ is the number of positive resp. negative diagonal entries in any matrix A' congruent to A (that is, there exists an invertible matrix T with $A' = T^t A T$).
   (d) $i_+(A)$ is the number of sign changes for the coefficients of the characteristic polynomial $\chi_A(x)$ of A, while $i_-(A)$ is the number of sign changes in the coefficients of $\chi_A(-x)$; alternatively one can use $i_-(A) = m - i_+(A) - k$, where k is the minimal exponent of $\chi_A(x)$.
2. The *hermitian defect* $\delta_h(A) := m - h(A)$ equals the Witt index of the quadratic (or bilinear form) associated with A (the maximum dimension of a null subspace of the associated quadratic (bilinear) space).
3. $h(A)$ is equal to the following four quantities:
   (a) the minimal number $k \in \mathbb{N}_0$ such that there are real matrices X, Y with $A = X^t X - Y^t Y$, where X, Y both have k rows;
   (b) the minimal number $k \in \mathbb{N}_0$ such that there are real matrices X, Y with $A = Y^t X + X^t Y$, where X, Y both have k rows;
   (c) the minimal $\mathrm{rank}(B)$ for real matrices B with $A = B + B^t$;
   (d) the minimal $k \in \mathbb{N}_0$ such that there are real symmetric matrices $B_1, \ldots, B_k$ of rank 2 with $A = B_1 + \cdots + B_k$.

In [14, 13] *multi*-clause-sets have been considered instead of clause-sets, since for example for applications of matching theory it is important to avoid contraction of multiple clauses, and furthermore for example the symmetric conflict number $\mathrm{bcp}(A)$ of a matrix A is defined as the minimal $n(F)$ for *multi*-clause-sets with $\mathrm{CM}(F) = A$, since possible repetition of clauses is needed here to apply algebraic methods. However, in our context there is no need for using multi-clause-sets (and no advantage), and thus in this paper only clause-sets are considered. One last comment on a potential difference between clause-sets and multi-clause-sets: Consider a clause-set $F \in \mathcal{CLS}$ and a partial assignment φ, and let us denote by $F' \in \mathcal{MCLS}$ the multi-clause-set corresponding to F. Now $\varphi * F \in \mathcal{CLS}$ is obtained from $\varphi * F' \in \mathcal{MCLS}$ by contracting multiple clauses (in the computation of $\varphi * F'$ no contraction takes place, and thus $c(\varphi * F')$ equals $c(F') = c(F)$ minus the number of clauses in F satisfied by φ, while $c(\varphi * F) \leq c(\varphi * F')$), which could make a difference for certain measures. However it can easily be seen that contraction of multiple clauses does not affect $i_{\pm}$, and thus can be applied freely in our context.

3   Characterisation of Clause-Sets with Hermitian Rank One

Since $h(F)$ can be computed in polynomial time, polynomial time decision of the class $\mathcal{CLS}_h(1)$ follows. The following basic lemma on "combinatorial linear algebra" is used for our characterisation of matrices with hermitian rank 1.

Lemma 1. *Consider a real square matrix A with zeros on the diagonal and with $\mathrm{rank}(A) = 1$. Then there is a permutation matrix P and a matrix B with $\mathrm{rank}(B) = 1$ having no zero entry such that $P^t \cdot A \cdot P = \left(\begin{smallmatrix} 0 & B \\ 0 & 0 \end{smallmatrix} \right)$.*

Proof. The elementary matrix operation we use is row resp. column exchange followed by the corresponding column resp. row exchange, i.e., we exchange rows i and j, immediately followed by exchange of columns i and j, or vice versa — in both ways we get the same result. We speak of a *combined row/column exchange* resp. a *combined column/row exchange*. Eventually, when we transformed A into $\left(\begin{smallmatrix} 0 & B \\ 0 & 0 \end{smallmatrix} \right)$, multiplying all matrices together representing the column exchanges we obtain P, and multiplying all matrices together representing the row exchanges we obtain P^t. Note that matrices obtained by applying combined row/column exchanges from A have zero diagonal.

Let the order of A be $m \geq 1$. Our aim is to move all non-zero entries of A to the upper right corner. There exists a non-zero row in A, and by a combined row/column exchange we get A' having a non-zero first row. Now we apply combined column/row exchanges for columns i, j with $A'_{1,i} \neq 0$, $A'_{1,j} = 0$ and $i < j$ until the first row is ordered in such a way that first come all zero entries and then all non-zero entries (note that due to $A_{1,1} = 0$ we always have $i, j \geq 2$ in this process, and thus the first row as a whole stays untouched). We obtain a matrix A'' having a column index $2 \leq k^* \leq m$ with the property that $A''_{1,i} = 0$ for all $1 \leq i \leq k^*$ and $A''_{1,i} \neq 0$ for all $k^* \leq i \leq m$. For the purpose of this proof, we call a matrix A^* with this property a k^*-matrix. Any k^*-matrix A of rank one has the following properties:

Every row of A is a multiple of the first row, and thus we have
$$A_{i,j} = 0 \text{ for all indices } i, j \text{ with } j < k^*.$$
If some row i contains a non-zero entry $A_{i,j} \neq 0$ (thus $j \geq k^*$),
then actually for all $k^* \leq j' \leq m$ we have $A_{i,j'} \neq 0$,
which can be seen as follows:
There is $\lambda \in \mathbb{R}$ with $\lambda \cdot A_{1,*} = A_{i,*}$. If there would be some $k^* \leq j' \leq m$ with
$A_{i,j'} = 0$, then due to $\lambda \cdot A_{1,j'} = A_{i,j'}$ and $A_{1,j'} \neq 0$ we would have $\lambda = 0$
contradicting $\lambda \cdot A_{1,j} = A_{i,j} \neq 0$.

Now consider a zero row $A''_{i,*}$ and some non-zero row $A''_{i',*}$ with $i < i'$. Since
$A''_{i',i'} = 0$ and A'' has the k^*-property, $i' < k^*$ must hold. Performing the combined row/column exchange on A'' for rows i and i' maintains the k^*-property. Repeating this process until every zero row is below any non-zero row we obtain a matrix of the form $\left(\begin{smallmatrix} 0 & B \\ 0 & 0 \end{smallmatrix}\right)$ with all entries of B non-zero. $\square$

As already mentioned, the hermitian rank $h(A)$ of a symmetric real matrix is the minimal rank of some matrix B with $A = B + B^t$.

Lemma 2. *Consider a symmetric real matrix A with zero diagonal. Then $h(A) \leq$ 1 iff there is a real matrix B with* $\mathrm{rank}(B) \leq 1$ *and a permutation matrix P with* $A = P^t \cdot \left(\begin{smallmatrix} 0 & B \\ B^t & 0 \end{smallmatrix}\right) \cdot P$.

Proof. First assume $h(A) \leq 1$. If $h(A) = 0$, then $A = 0$, and thus we can choose $B := 0$ and $P := I$. So assume $h(A) = 1$. Thus there exists a matrix B_0 with $\mathrm{rank}(B_0) = 1$ and $B_0 + B_0^t = A$. It has B_0 a zero diagonal, and thus by Lemma 1 there exists B with $\mathrm{rank}(B) = 1$ and only non-zero entries, and a permutation matrix P such that $P^t \cdot B_0 \cdot P = \left(\begin{smallmatrix} 0 & B \\ 0 & 0 \end{smallmatrix}\right)$. Since also $P^t \cdot B_0 \cdot P$ has a zero diagonal, B is located in $P^t \cdot B_0 \cdot P$ above the diagonal, and thus B^t in $(P^t \cdot B_0 \cdot P)^t = \left(\begin{smallmatrix} 0 & 0 \\ B^t & 0 \end{smallmatrix}\right)$ is located below the diagonal:

$$\begin{pmatrix} 0 & B \\ 0 & 0 \end{pmatrix} + \begin{pmatrix} 0 & B \\ 0 & 0 \end{pmatrix}^t = \begin{pmatrix} 0 & 0 & B \\ 0 & 0 & 0 \\ B^t & 0 & 0 \end{pmatrix}.$$

Using $P^{-1} = P^t$ we have

$$P \cdot \begin{pmatrix} 0 & 0 & B \\ 0 & 0 & 0 \\ B^t & 0 & 0 \end{pmatrix} \cdot P^t = P \cdot \left(\begin{pmatrix} 0 & B \\ 0 & 0 \end{pmatrix} + \begin{pmatrix} 0 & B \\ 0 & 0 \end{pmatrix}^t \right) \cdot P^t =$$

$$P \cdot \begin{pmatrix} 0 & B \\ 0 & 0 \end{pmatrix} \cdot P^t + P \cdot \begin{pmatrix} 0 & B \\ 0 & 0 \end{pmatrix}^t \cdot P^t = B_0 + B_0^t = A.$$

On the other hand, if $A = P^t \cdot \left(\begin{smallmatrix} 0 & B \\ B^t & 0 \end{smallmatrix}\right) \cdot P$ for some B with $\mathrm{rank}(B) \leq 1$ and some permutation matrix P, then with $B_0 := P^t \cdot \left(\begin{smallmatrix} 0 & B \\ 0 & 0 \end{smallmatrix}\right) \cdot P$ obviously $\mathrm{rank}(B_0) = \mathrm{rank}(B) \leq 1$ and $B_0 + B_0^t = A$, thus $h(A) \leq \mathrm{rank}(B_0) \leq 1$. $\square$

We remind at the notion of a *blocked clause* w.r.t. a clause-set F ([11, 12]), which is a clause C containing a literal $x \in C$ such that for all clauses $D \in F$ with $\bar{x} \in D$ there exists a literal $y \in C \setminus \{x\}$ with $\bar{y} \in D$. Blocked clauses can be eliminated satisfiability-equivalently, and the result of eliminating all blocked clauses does not depend on the order of eliminations. If a clause-set F not containing the empty clause does not contain a blocked clause, then obviously every row and every column of $CM(F)$ contains at least one entry equal to 1. In the proof of the following first main result we use the fact, that a multigraph G is a complete bipartite graph iff there exists a permutation matrix P and a matrix B with all entries equal to 1 such that the adjacency matrix of G is $P^t \cdot \left(\begin{smallmatrix} 0 & B \\ B^t & 0 \end{smallmatrix} \right) \cdot P$.

Theorem 3. *Consider $F \in \mathcal{CLS}_h(1)$ with $\perp \notin F$, and obtain $F' \in \mathcal{CLS}_h(1)$ from F by iterated elimination of (all) blocked clauses. Then the conflict multigraph $\mathrm{cmg}(F')$ is a complete bipartite graph (without parallel edges).*

Proof. We have $F' \in \mathcal{CLS}_h(1)$, since elimination of clauses can not increase the hermitian rank ([14, 13]). If $F' = \top$, then the assertion is trivial. So we assume that F' contains at least one clause. Since F' does not contain a blocked clause, every row and every column of $CM(F')$ contains at least one entry equal to 1. Applying Lemma 2 to $CM(F')$ we obtain $CM(F') = P^t \cdot \left(\begin{smallmatrix} 0 & B \\ B^t & 0 \end{smallmatrix} \right) \cdot P$ for some permutation matrix P and a matrix B with $\mathrm{rank}(B) = 1$. There are no zero rows or columns in B, and since the rank of B is one, every row (resp. column) of B is a non-zero multiple of every other row (resp. column). We want to show that every entry of B is 1. So assume that there are indices i, j with $B_{i,j} \neq 1$. We have $B_{i,j} \neq 0$, since otherwise every entry in row i would be zero, using that every column j' of B is a multiple of column j. Thus $B_{i,j} \geq 2$. We know that there is a column index j' with $B_{i,j'} = 1$ and a row index i' with $B_{i',j} = 1$. Thus column j multiplied with $1/B_{i,j}$ yields column j', and thus $B_{i',j}/B_{i,j} = B_{i',j'}$, but $0 < B_{i',j}/B_{i,j} = 1/B_{i,j} < 1$ contradicting the integrality of $B_{i',j'}$. $\qquad\square$

4   SAT Decision, Conflict Multigraphs and the Transversal Hypergraph Problem

In this section we investigate how to exploit for efficient SAT decision the fact, that the conflict multigraph of a clause-set is a complete bipartite graph. We proceed in three stages: First we consider clause-sets with bipartite conflict multigraphs in general, then the case of *complete* bipartite conflict multigraphs is investigated, and finally we turn to the case where the conflict multigraph is a complete bipartite *graph*.

4.1   Bipartite Clause-Sets and PN-Pairs

$F \in \mathcal{CLS}$ is called **bipartite** if the conflict multigraph of F is bipartite. This is easily seen to be equivalent to the existence of a permutation matrix P and a matrix B such that $CM(F) = P^t \cdot \left(\begin{smallmatrix} 0 & B \\ B^t & 0 \end{smallmatrix} \right) \cdot P$ holds. Immediately from a well-known characterisation of bipartite (multi-)graphs in algebraic graph theory (see for example Theorem 3.11 in [4] together with the footnote) we get

Lemma 4. *A clause-set $F \in \mathcal{CLS}$ is bipartite if and only if the following two conditions hold:*

1. *$i_-(F) = i_+(F) = h(F)$;*
2. *for $1 \le i \le h(F)$ we have $\theta_i(F) = \theta_{c(F)-i+1}(F)$.*

Since the sum of the eigenvalues of a symmetric real matrix is equal to the trace of the matrix (the sum of the diagonal elements), and (symmetric) conflict matrices have a zero diagonal, it is easy to see that a clause-set F fulfils $h(F) = 1$ iff $i_+(F) = i_-(F) = 1$ holds and the absolute values of the positive and the negative eigenvalue coincide. Thus Lemma 4 yields an alternative proof that clause-sets in $\mathcal{CLS}_h(1)$ are bipartite as proven directly in Theorem 3. Actually Theorem 3 proved much more, and we will now see that the property of a clause-set being bipartite alone does not help much for satisfiability decision.

A clause is called *positive* resp. *negative* if it only contains positive resp. negative literals. A clause-set $F \in \mathcal{CLS}$ is called *positive-negative* ("PN-clause-set" for short) if for every $C \in F$ we have, that C is positive or negative. Obviously, every positive-negative clause-set is bipartite. The Pigeonhole formulas are examples of positive-negative clause-sets. By introducing new variables, every clause-set can be transformed in linear time into a satisfiability-equivalent positive-negative clause-set, and thus satisfiability decision for bipartite clause-sets is NP-complete.

Intuitively, the class of PN-clause-sets is the core of the class of bipartite clause-sets. We make this more precise to clarify the relationship to the hypergraph transversal problem. A *PN-pair* is a pair $(\mathbb{H}_1, \mathbb{H}_2)$, where each $\mathbb{H}_i \subseteq \mathbb{P}(\mathcal{VA})$ is a set of hyperedges considered as a hypergraph with vertex set the set of variables appearing in it. For a PN-pair $(\mathbb{H}_1, \mathbb{H}_2)$ we define the clause-set $F(\mathbb{H}_1, \mathbb{H}_2) := \mathbb{H}_1 \cup \{\overline{H} : H \in \mathbb{H}_2\}$, that is, the positive clauses of the PN-clause-set $F(\mathbb{H}_1, \mathbb{H}_2)$ are given by the elements of $\mathbb{H}_1$, while the negative clauses are given by the elements of $\mathbb{H}_2$ with elementwise complemented literals. For example $F(\{\{a, b\}, \{a, c\}\}, \{\{b, c\}\}) = \{\{a, b\}, \{a, c\}, \{\overline{b}, \overline{c}\}\}$. As noticed in [6]:

Lemma 5. *Consider a PN-pair $(\mathbb{H}_1, \mathbb{H}_2)$. Then $F(\mathbb{H}_1, \mathbb{H}_2)$ is unsatisfiable if and only if for all $H_1 \in \mathrm{Tr}(\mathbb{H}_1)$ there exists $H_2 \in \mathbb{H}_2$ with $H_2 \subseteq H_1$, or equivalently, $F(\mathbb{H}_1, \mathbb{H}_2)$ is satisfiable iff there exists $H_1 \in \mathrm{Tr}(\mathbb{H}_1)$ such that H_1 is an independent set of $\mathbb{H}_2$.*

Proof. If $F := F(\mathbb{H}_1, \mathbb{H}_2)$ is satisfiable, then there is a partial assignment φ with $\varphi * F = \top$. Let $V_1 := \{v \in \mathrm{var}(\varphi) : \varphi(v) = 1\}$. It is V_1 a transversal of $\mathbb{H}_1$, and there is no $H_2 \in \mathbb{H}_2$ with $H_2 \subseteq V_1$, i.e., V_1 is an independent set of $\mathbb{H}_2$. If on the other hand there is a transversal T of $\mathbb{H}_1$ which is an independent set of $\mathbb{H}_2$, then the assignment $\varphi := \langle v \to 1 : v \in T \rangle \cup \langle v \to 0 : v \in \mathrm{var}(F) \setminus T \rangle$ is a satisfying assignment for F. □

Now a pair $((\mathbb{H}_1, \mathbb{H}_2), \zeta)$, where $(\mathbb{H}_1, \mathbb{H}_2)$ is a PN-pair and $\zeta : \mathcal{VA} \to \{-1, +1\}$ is a sign-flip, *represents* a clause-set F if $F' = \zeta * F(\mathbb{H}_1, \mathbb{H}_2)$, where "$*$" denotes the application of the sign flip, and F' is obtained from F by removal of pure literals (setting them to false, not to true(!)). If a clause-set can be represented in this way, then it is bipartite. In the reverse direction we have

Lemma 6. *For bipartite clause-sets a representation can be computed in quadratic time.*

Since we want to apply the algorithm of Lemma 6 to some subclasses of the class of bipartite clause-sets, the following notion will be useful. Given a class $\mathcal{C}$ of bipartite clause-sets and a class $\mathcal{H}$ of PN-pairs, we say that $\mathcal{H}$ *represents* $\mathcal{C}$ if for all $F \in \mathcal{CLS}$ the following statements are equivalent:

1. $F \in \mathcal{C}$;
2. for all representations $((\mathbb{H}_1, \mathbb{H}_2), \zeta)$ of F we have $(\mathbb{H}_1, \mathbb{H}_2) \in \mathcal{H}$;
3. there exists a representation $((\mathbb{H}_1, \mathbb{H}_2), \zeta)$ of F with $(\mathbb{H}_1, \mathbb{H}_2) \in \mathcal{H}$.

4.2   Bi-hitting Clause-Sets and Bi-hitting PN-Pairs

A clause-set $F \in \mathcal{CLS}$ is called a **bi-hitting clause-set** if the conflict multigraph of F is complete bipartite, or in other words, if there exists a permutation matrix P and a matrix B with no zero-entry such that $CM(F) = P^t \cdot \left(\begin{smallmatrix} 0 & B \\ B^t & 0 \end{smallmatrix} \right) \cdot P$. As before, where we used PN-pairs as the "essential" representations of bipartite clause-sets, we now consider "bi-hitting PN-pairs" as the essential representations of bi-hitting clause-sets. A **bi-hitting PN-pair** is a PN-pair $(\mathbb{H}_1, \mathbb{H}_2)$ such that every $H \in \mathbb{H}_2$ is a transversal of $\mathbb{H}_1$. If $(\mathbb{H}_1, \mathbb{H}_2)$ is a bi-hitting PN-pair, then so is $(\mathbb{H}_2, \mathbb{H}_1)$.

Lemma 7. *The class of bi-hitting PN-pairs represents the class of bi-hitting clause-sets.*

A **transversal PN-pair** is a PN-pair $(\mathbb{H}_1, \mathbb{H}_2)$ with $\mathbb{H}_2 = \mathrm{Tr}(\mathbb{H}_1)$ and $\mathbb{H}_1 = \mathrm{Tr}(\mathbb{H}_2)$. Every transversal PN-pair is a bi-hitting PN-pair, and if $(\mathbb{H}_1, \mathbb{H}_2)$ is a transversal PN-pair, then so is $(\mathbb{H}_2, \mathbb{H}_1)$. We want to recognise transversal PN-pair as the representations of unsatisfiable bi-hitting clause-sets, and to do so, we need to remove subsumed clauses. A **simple PN-pair** is a PN-pair $(\mathbb{H}_1, \mathbb{H}_2)$ such that $\mathbb{H}_1$ and $\mathbb{H}_2$ are simple hypergraphs, that is, they do not contain subsumed hyperedges. Every transversal PN-pair is simple. If $(\mathbb{H}_1, \mathbb{H}_2)$ is a simple PN-pair, then $(\mathbb{H}_1, \mathbb{H}_2)$ is a transversal PN-pair iff $\mathbb{H}_2 = \mathrm{Tr}(\mathbb{H}_1)$.

Lemma 8. *Consider a simple bi-hitting PN-pair $(\mathbb{H}_1, \mathbb{H}_2)$. Then $F(\mathbb{H}_1, \mathbb{H}_2)$ is unsatisfiable if and only if $(\mathbb{H}_1, \mathbb{H}_2)$ is a transversal PN-pair.*

Proof. First assume $F(\mathbb{H}_1, \mathbb{H}_2)$ is unsatisfiable. We have to show that $\mathrm{Tr}(\mathbb{H}_1) = \mathbb{H}_2$. By the definition of bi-hitting every $T \in \mathbb{H}_2$ is a transversal of $\mathbb{H}_1$. If there would exist a transversal T' of $\mathbb{H}_1$ with $T' \subset T$, then by Lemma 5 there would exist some $T'' \in \mathbb{H}_2$ with $T'' \subseteq T' \subset T$ contradicting the simplicity of $\mathbb{H}_2$. Thus $\mathbb{H}_2 \subseteq \mathrm{Tr}(\mathbb{H}_1)$, and by Lemma 5 in fact equality holds. So we have shown that $(\mathbb{H}_1, \mathbb{H}_2)$ is a transversal PN-pair. If on the other hand $(\mathbb{H}_1, \mathbb{H}_2)$ is a transversal PN-pair, then immediately by Lemma 5 unsatisfiability of $F(\mathbb{H}_1, \mathbb{H}_2)$ follows.   □

The decision problem whether a PN-pair is a transversal PN-pair is known in the literature under the name of the *hypergraph transversal problem*. In [7] it has been shown that the hypergraph transversal problem is decidable in quasi-polynomial time (more precisely in time $O(s^{o(\log s)})$, where s is the sum of the sizes of the two hypergraphs). Thus by lemmata 6, 7 and 8 we get

Theorem 9. *The satisfiability problem for the class of bi-hitting clause-sets is solvable in quasi-polynomial time (more precisely in time $O(\ell(F)^{o(\log \ell(F))})$ for $F \in \mathcal{CLS}$, where $\ell(F) := \sum_{C \in F} |C|$ is the number of literal occurrences in F).*

It is an important open problem whether the hypergraph transversal problem is decidable in polynomial time, which is equivalent to the problem, whether the satisfiability problem for the class of bi-hitting clause-sets is decidable in polynomial time.

4.3   Uniform Bi-hitting Clause-Sets and Exact Bi-hitting PN-Pairs

We have seen in Theorem 9 that satisfiability for clause-sets with complete bi-partite conflict multigraph and thus also for clause-sets in $\mathcal{CLS}_h(1)$ (see Theorem 3) is decidable in quasi-polynomial time. Now we will look at the case where the conflict multigraph of a clause-set $F \in \mathcal{CLS}$ is a complete bipartite *graph*, which is equivalent to the existence of a permutation matrix P and a matrix J with all entries equal to 1 such that $\mathrm{CM}(F) = P^t \cdot \left(\begin{smallmatrix} 0 & J \\ J^t & 0 \end{smallmatrix} \right) \cdot P$. We will show polynomial time satisfiability decision for this class.

$F \in \mathcal{CLS}$ is called a k-**uniform bi-hitting clause-set** for $k \geq 0$ if $\mathrm{CM}(F) = P^t \cdot \left(\begin{smallmatrix} 0 & k \cdot J \\ k \cdot J^t & 0 \end{smallmatrix} \right) \cdot P$, while **uniform bi-hitting** means k-uniform bi-hitting for some k. A k-uniform bi-hitting clause-set F for $k \geq 2$ is unsatisfiable iff $\bot \in F$, and thus in the remainder we consider only the case $k = 1$. A transversal T of a hypergraph $\mathbb{H}$ is called *exact* ([5]) if for all $H \in \mathbb{H}$ we have $|T \cap H| = 1$. And $\mathbb{H}$ is called *exact* if every minimal transversal of $\mathbb{H}$ is exact. An **exact bi-hitting PN-pair** is a PN-pair $(\mathbb{H}_1, \mathbb{H}_2)$ such that every $H \in \mathbb{H}_2$ is an exact transversal of $\mathbb{H}_1$. Every exact bi-hitting PN-pair is a bi-hitting PN-pair, and if $(\mathbb{H}_1, \mathbb{H}_2)$ is an exact bi-hitting PN-pair, then so is $(\mathbb{H}_2, \mathbb{H}_1)$.

Lemma 10. *Exact bi-hitting PN-pairs represent 1-uniform bi-hitting clause-sets.*

An **exact transversal PN-pair** is a transversal PN-pair $(\mathbb{H}_1, \mathbb{H}_2)$ where $\mathbb{H}_1, \mathbb{H}_2$ are exact hypergraphs. If $(\mathbb{H}_1, \mathbb{H}_2)$ is an exact transversal PN-pair, then so is $(\mathbb{H}_2, \mathbb{H}_1)$. If $(\mathbb{H}_1, \mathbb{H}_2)$ is a simple PN-pair, then $(\mathbb{H}_1, \mathbb{H}_2)$ is an exact transversal PN-pair iff $\mathbb{H}_2 = \mathrm{Tr}(\mathbb{H}_1)$ and $\mathbb{H}_1$ is exact. Since a transversal PN-pair $(\mathbb{H}_1, \mathbb{H}_2)$ is an exact transversal PN-pair iff $(\mathbb{H}_1, \mathbb{H}_2)$ is an exact bi-hitting PN-pair, immediately from Lemma 8 we get

Lemma 11. *Consider a simple exact bi-hitting PN-pair $(\mathbb{H}_1, \mathbb{H}_2)$. Then the clause-set $F(\mathbb{H}_1, \mathbb{H}_2)$ is unsatisfiable if and only if $(\mathbb{H}_1, \mathbb{H}_2)$ is an exact transversal PN-pair.*

The *exact transversal hypergraph problem* is the problem to decide, whether a given PN-pair is an exact transversal PN-pair. In [5] it is shown that the exact transversal hypergraph problem can be decided in polynomial time, whence

Theorem 12. *The SAT problem for uniform bi-hitting clause-sets is decidable in polynomial time.*

By Theorem 3, modulo blocked clauses clause-sets with hermitian rank 1 are 1-uniform bi-hitting clause-sets, and thus

Theorem 13. *SAT decision for $\mathcal{CLS}_h(1)$ can be done in polynomial time.*

For the sake of completeness we present the polynomial time algorithm from [5] (Theorem 3.3) to decide, whether a simple hypergraph $\mathbb{H}$ is an "exact transversal hypergraph", that is, whether $(\mathbb{H}, \mathrm{Tr}(\mathbb{H}))$ is an exact transversal PN-pair. This together with Theorem 4.3 of [5], that for exact transversal hypergraphs the set of minimal transversals can be enumerated with polynomial delay (i.e., the time between two consecutive outputs is polynomially bounded in the size of the input), implies Theorem 12.

Let $\mathbb{H}(v) := \{H \in \mathbb{H} : v \in H\}$ be the star of vertex v, and let $V(\mathbb{H}) := \bigcup \mathbb{H} = \bigcup_{H \in \mathbb{H}} H$ denote the vertex set of $\mathbb{H}$. First we observe that a transversal T of a hypergraph $\mathbb{H} \neq \emptyset$ is minimal iff there exists $H \in \mathbb{H}$ with $|T \cap H| = 1$. Now it follows easily that $\mathbb{H}$ is exact transversal if and only if for all vertices $v \in V(\mathbb{H})$, for all hyperedges $H \in \mathbb{H}(v)$ and for all transversals $T \in \mathrm{Tr}(\mathbb{H})$ in case of $v \in T$ and $T \cap H = \{v\}$ we have $T \cap V(\mathbb{H}(v)) = \{v\}$. This condition is equivalent to

$$\forall v \in V(\mathbb{H}) \; \forall H \in \mathbb{H}(v) :$$
$$V(\mathbb{H}(v)) \cap V(\{T \in \mathrm{Tr}(\mathbb{H}) : T \cap H = \{v\}\}) = \{v\}. \quad (1)$$

For any hypergraph $\mathbb{H}$ and any hyperedge H of $\mathbb{H}$ by definition we have

$$\{T \in \mathrm{Tr}(\mathbb{H}) : T \cap H = \{v\}\} = \{T \cup \{v\} : T \in \mathrm{Tr}(\{H' \setminus H : H' \in \mathbb{H} \wedge v \notin H'\})\},$$

and thus (1) is equivalent to

$$\forall v \in V(\mathbb{H}) \; \forall H \in \mathbb{H}(v) :$$
$$V(\mathbb{H}(v)) \cap V(\mathrm{Tr}(\{H' \setminus H : H' \in \mathbb{H} \wedge v \notin H'\})) = \emptyset. \quad (2)$$

Using $\min(\mathbb{H})$ for the set of inclusion-minimal elements of a hypergraph $\mathbb{H}$, the trick is now to exploit the observation $V(\mathrm{Tr}(\mathbb{H})) = V(\min(\mathbb{H}))$ for any hypergraph $\mathbb{H}$, which follows from $\mathrm{Tr}(\mathrm{Tr}(\mathbb{H})) = \min(\mathbb{H})$, and which yields, that (2) is equivalent to

$$\forall v \in V(\mathbb{H}) \; \forall H \in \mathbb{H}(v) :$$
$$V(\mathbb{H}(v)) \cap V(\min(\{H' \setminus H : H' \in \mathbb{H} \wedge v \notin H'\})) = \emptyset,$$

where this final criterion obviously is decidable in polynomial time. The idea of this nice proof can be motivated as follows: By definition $\mathbb{H}$ is exact transversal iff for all $v \in V(\mathbb{H})$ we have

$$V(\mathbb{H}(v)) \cap V(\{T \in \mathrm{Tr}(\mathbb{H}) : v \in T\}) = \{v\}.$$

The problem is to compute $V(\{T \in \mathrm{Tr}(\mathbb{H}) : v \in T\})$, where we would like to recognise $\{T \in \mathrm{Tr}(\mathbb{H}) : v \in T\}$ as the transversal hypergraph of some $\mathbb{H}'$, which would yield $V(\{T \in \mathrm{Tr}(\mathbb{H}) : v \in T\}) = V(\min(\mathbb{H}'))$, avoiding the computation of the transversal hypergraph. Selecting exactly the minimal transversals T of $\mathbb{H}$ containing v seems not possible, but if we fix a hyperedge $H \in \mathbb{H}$ with $v \in H$, then the minimal transversals of $\mathbb{H}$ containing v but not any other vertex from H are exactly the $T \cup \{v\}$, where T is a minimal transversals of the hyperedges of $\mathbb{H}$ not containing v with all other vertices from H removed, and we arrive at the above proof.

5   Open Problems

5.1   Polynomial Time SAT Decision for Bounded Hermitian Rank

As mentioned in the introduction, it would be interesting to prove an upper bound $2^{h(F)}$ on time complexity of satisfiability decision (ignoring polynomial factors). We would achieve this aim, if we can find a polynomial time reduction $r : \mathcal{CLS} \to \mathcal{CLS}$ and some class $\mathbb{E} \subseteq \mathcal{CLS}$ which is decidable and satisfiability decidable in polynomial time, such that for all clause-sets $F \in \mathcal{CLS}$ and $F' := r(F)$ we have $h(F') \leq h(F)$, and we have $F' \in \mathbb{E}$ or there exists a variable $v \in \mathrm{var}(F')$ such that for both truth values $\varepsilon \in \{0,1\}$ we have $h(\langle v \to \varepsilon \rangle * F') < h(F')$. For $r = \mathrm{id}_{\mathcal{CLS}}$ and $\mathbb{E} = \mathcal{CLS}_h(1)$ this property does not hold.

Since application of partial assignments does not increase the hermitian rank, when allowing a logarithmic factor in the exponent it actually suffices to find a variable $v \in \mathrm{var}(F)$ such that for just *one truth value* $\varepsilon \in \{0,1\}$ we have $h(\langle v \to \varepsilon \rangle * F) < h(F)$ (following [10, 16]). Using computer experiments, we did not find a counterexample of small dimension, and so we conjecture

Conjecture 14. For all $F \in \mathcal{CLS}$ with $h(F) \geq 2$ there exists $v \in \mathrm{var}(F)$ and $\varepsilon \in \{0,1\}$ with $h(\langle v \to \varepsilon \rangle * F) < h(F)$.

If Conjecture 14 is true, then by Lemma 3.7 in [10] or Theorem 4.3 in [16] we get satisfiability decision for $\mathcal{CLS}$ in time $n(F)^{2h(F)}$ (ignoring polynomial factors). And furthermore the hardness $h_{\mathcal{CLS}_h(1)}(F)$ of clause-sets F as studied in [10, 16], using the polynomial time satisfiability decision for $\mathcal{CLS}_h(1)$ as oracle, would be bounded by $h_{\mathcal{CLS}_h(1)}(F) \leq h(F) - 1$. Since thus Conjecture 14 implies polynomial time satisfiability decision for formulas with bounded hermitian rank, from Conjecture 14 it follows:

Conjecture 15. For fixed $k \geq 0$ satisfiability decision of the class $\mathcal{CLS}_h(k)$ can be done in polynomial time.

5.2   Characterising (Minimally) Unsatisfiable Uniform Bi-hitting Clause-Sets

We now have a look at the bearings of the investigations of this paper on minimally unsatisfiable clause-sets (all proofs can be found in [15]). Every unsatisfiable clause-set $F \in \mathcal{USAT}$ contains some minimally unsatisfiable sub-clause-set $F' \in \mathcal{MUSAT}$. Let us say that F has a *unique core*, if there is exactly one

$F' \subseteq F$ with $F' \in \mathcal{MUSAT}$, in which case we call F' the *core* of F. Related to the formula classes considered in this article are three classes of clause-sets with unique core:

1. Every unsatisfiable bi-hitting clause-set F has a unique core given by the set of subsumption-minimal clauses of F, obtained from F by *elimination of subsumed clauses*.
2. For unsatisfiable uniform bi-hitting clause-sets the unique core can also be obtained by *elimination of pure literals* (i.e., the unique core is the lean kernel w.r.t. pure autarkies).
3. Every unsatisfiable clause-set F with $h(F) \leq 1$ and $\bot \notin F$ has a unique core, obtained from F by *elimination of blocked clauses*.

The class of cores of unsatisfiable bi-hitting clause-sets, i.e., the set of minimally unsatisfiable bi-hitting clause-sets, is the set of clause-sets which can be represented by transversal PN-pairs. Let $\mathcal{TPN}$ be the set of transversal PN-pairs. The class of all minimally unsatisfiable bi-hitting clause-sets is represented by $\mathcal{TPN}$, and thus characterising minimally unsatisfiable bi-hitting clause-sets amounts to characterise $\mathcal{TPN}$, which seems to be an elusive task, so we have to consider simpler cases. But before we do this, let us make some remarks on "splittings" of minimally unsatisfiable clause-sets.

For any set $\mathcal{P}$ of PN-pairs let $\mathcal{CLS}(\mathcal{P})$ denote the set of clause-sets $F(\mathbb{H}_1, \mathbb{H}_2)$ for some $(\mathbb{H}_1, \mathbb{H}_2) \in \mathcal{P}$. Thus $\mathcal{CLS}(\mathcal{TPN})$ is contained in the set of minimally unsatisfiable bi-hitting clause-sets, and for every minimally unsatisfiable bi-hitting clause-set F there exists a sign-flip ζ with $\zeta * F \in \mathcal{CLS}(\mathcal{TPN})$. Following [3], a *splitting* of $F \in \mathcal{MUSAT}$ on a variable $v \in \mathrm{var}(F)$ is a pair $(F_0, F_1) \in \mathcal{MUSAT}^2$ with $F_\varepsilon \subseteq \langle v \rightarrow \varepsilon \rangle * F$ for $\varepsilon \in \{0, 1\}$. A class $\mathcal{C} \subseteq \mathcal{MUSAT}$ is called *closed under splitting* if for every $F \in \mathcal{C}$ and every splitting (F_0, F_1) of F we have $F_0, F_1 \in \mathcal{C}$. It is $\mathcal{CLS}(\mathcal{TPN})$ closed under splitting, and every element of $\mathcal{CLS}(\mathcal{TPN})$ has a unique splitting (the same holds for the larger class of minimally unsatisfiable bi-hitting clause-sets).

The class of cores of unsatisfiable uniform bi-hitting clause-sets is identical to the class of cores of unsatisfiable clause-sets with hermitian rank at most one, and can be described as the class of clause-sets representable by some $((\mathbb{H}_1, \mathbb{H}_2), \zeta)$, where $(\mathbb{H}_1, \mathbb{H}_2)$ is an exact transversal PN-pair. Let $\mathcal{ETPN}$ be the set of exact transversal PN-pairs. The class of all minimally unsatisfiable uniform bi-hitting clause-sets is represented by $\mathcal{TPN}$. The class $\mathcal{CLS}(\mathcal{ETPN})$ is again closed under splitting (as is the larger class of minimally unsatisfiable uniform bi-hitting clause-sets). Characterising the class of minimally unsatisfiable uniform bi-hitting clause-sets amounts to characterising $\mathcal{ETPN}$. We can construct elements of $\mathcal{ETPN}$ as follows:

1. For variables $v_1, \ldots, v_n$, $n \in \mathbb{N}_0$ we have

$$(\{\{v_1, \ldots, v_n\}\}, \{\{v_i\} : i \in \{1, \ldots, n\}\}) \in \mathcal{ETPN}.$$

   (Actually the cases $n \geq 2$ can be simulated by rule 3.)
2. If $(\mathbb{H}_1, \mathbb{H}_2) \in \mathcal{ETPN}$, then also $(\mathbb{H}_2, \mathbb{H}_1) \in \mathcal{ETPN}$.

3. If $(\mathbb{H}_1, \mathbb{H}_2), (\mathbb{H}'_1, \mathbb{H}'_2) \in \mathcal{ETPN}$ with $\emptyset \notin \mathbb{H}_1 \cup \mathbb{H}'_1$ and $V(\mathbb{H}_1) \cap V(\mathbb{H}'_1) = \emptyset$, then
$$(\mathbb{H}_1 \cup \mathbb{H}'_1, \{T \cup T' : T \in \mathbb{H}_2 \wedge T' \in \mathbb{H}'_2\}) \in \mathcal{ETPN}.$$

Let $\mathcal{ETPN}_0$ be the set of exact transversal PN-pairs created by the above three rules. For example, from the first rule we obtain the pairs $(\{\emptyset\}, \emptyset)$, $(\{\{v\}\}, \{\{v\}\})$ and $(\{\{v, w\}\}, \{\{v\}, \{w\}\})$ in $\mathcal{ETPN}_0$ (for $n = 0, 1, 2$). The second rule creates $(\emptyset, \{\emptyset\})$ and $(\{\{v\}, \{w\}\}, \{\{v, w\}\})$ in $\mathcal{ETPN}_0$, and with the third rule then we can obtain for example the two pairs $(\{\{a, b\}, \{c, d\}\}, \{\{a, c\}, \{a, d\}, \{b, c\}, \{b, d\}\})$ and $(\{\{a, b\}, \{c\}, \{d\}\}, \{\{a, c, d\}, \{b, c, d\}\})$ in $\mathcal{ETPN}_0$. It is easy to see, that clause-sets represented by PN-pairs from $\mathcal{ETPN}_0$ have unbounded clause-lengths and unbounded deficiency (the difference between the number of clauses and the number of variables), and they are not renamable to Horn formulas except of trivial cases (since minimally unsatisfiable Horn formulas have deficiency 1). Given a PN-pair $(\mathbb{H}_1, \mathbb{H}_2)$ we can efficiently decide membership in $\mathcal{ETPN}_0$ as follows:

1. If $(\mathbb{H}_1, \mathbb{H}_2)$ is not a simple PN-pair, then $(\mathbb{H}_1, \mathbb{H}_2) \notin \mathcal{ETPN}_0$.
2. Otherwise check whether $(\mathbb{H}_1, \mathbb{H}_2)$ can be created by rule 1 (in which case we are done).
3. In the remaining case now $(\mathbb{H}_1, \mathbb{H}_2) \in \mathcal{ETPN}_0$ can only be the case if there is $i \in \{1, 2\}$ such that $\mathbb{H}_i$ is disconnected. Splitting $\mathbb{H}_i$ into its connected components and reverting the construction from rule 3, we can reduce the problem to smaller problems and proceed recursively.

We did not find an example for $F \in \mathcal{ETPN} \setminus \mathcal{ETPN}_0$, and thus

Conjecture 16. $\mathcal{ETPN}_0 = \mathcal{ETPN}$.

If Conjecture 16 is true, then we can draw the following conclusions:

1. The above procedure for deciding membership in $\mathcal{ETPN}_0$ yields a more efficient satisfiability decision procedure for uniform bi-hitting clause-sets (and for $\mathcal{CLS}_h(1)$) than the one outlined at the end of Subsection 4.3.
2. The elements of $\mathcal{CLS}(\mathcal{ETPN}_0)$ have a read-once resolution refutation (the simplest possible resolution refutations: a tree-resolution refutation where every node is labelled with a unique clause), and thus in fact every unsatisfiable uniform bi-hitting clause-set (as well as every element of $\mathcal{CLS}_h(1) \cap \mathcal{USAT}$) would have a read-once resolution refutation.

As a partial result towards Conjecture 16 we can completely characterise exact transversal PN-pairs $(\mathbb{H}_1, \mathbb{H}_2)$, where the rank of $\mathbb{H}_1$ or $\mathbb{H}_2$ is at most 2 (i.e., there is $i \in \{1, 2\}$ such that for all $H \in \mathbb{H}_i$ we have $|H| \leq 2$). Consider a simple hypergraph $\mathbb{H}$. Call $\mathbb{H}$ *exact transversal* if $(\mathbb{H}, \mathrm{Tr}(\mathbb{H}))$ is an exact transversal PN-pair. If $\mathbb{H} = \emptyset$, then $\mathbb{H}$ is exact transversal. If $\emptyset \in \mathbb{H}$, then $\mathbb{H} = \{\emptyset\}$ and $\mathbb{H}$ is exact transversal. If $\{x\} \in \mathbb{H}$, then $\mathbb{H}$ is exact transversal iff $\mathbb{H} \setminus \{x\}$ is exact transversal. So w.l.o.g. we assume that $\mathbb{H}$ is not empty and the size of a smallest hyperedge is at least 2.

Assume furthermore that the rank of $\mathbb{H}$ is 2. Now $\mathbb{H}$ constitutes a graph. And if $\mathbb{H}$ as a graph is connected, then $\mathbb{H}$ is exact transversal if and only if $\mathbb{H}$ as a graph is complete bipartite. Complete bipartite graphs with at least two vertices are exactly the transversal hypergraphs of hypergraphs $\{A, B\}$ for non-empty disjoint A, B. It follows that for an exact transversal PN-pairs $(\mathbb{H}_1, \mathbb{H}_2) \in \mathcal{ETPN}$ with rank of $\mathbb{H}_1$ or $\mathbb{H}_2$ at most 2 we have $(\mathbb{H}_1, \mathbb{H}_2) \in \mathcal{ETPN}_0$.

Acknowledgements

Part of this work was done while we were visiting the ICTP in Trieste participating in the Thematic Institute of the Complex Systems Network of Excellence (EXYSTENCE): "Algorithms and challenges in hard combinatorial problems and in optimization under 'uncertainty'". We want to thank Riccardo Zecchina for inviting us and for providing a highly stimulating environment.

References

1. Claude Berge. *Hypergraphs: Combinatorics of Finite Sets*, volume 45 of *North-Holland Mathematical Library*. North Holland, Amsterdam, 1989. ISBN 0 444 87489 5; QA166.23.B4813 1989.
2. James R. Bunch and Linda Kaufman. Some stable methods for calculating inertia and solving symmetric linear systems. *Mathematics of Computation*, 31(137):163–179, January 1977.
3. Hans Kleine Büning and Xishun Zhao. On the structure of some classes of minimal unsatisfiable formulas. *Discrete Applied Mathematics*, 130:185–207, 2003.
4. Dragoš M. Cvetković, Michael Doob, Ivan Gutman, and Aleksandar Torgašev. *Recent Results in the Theory of Graph Spectra*, volume 36 of *Annals of Discrete Mathematics*. North-Holland, Amsterdam, 1988. ISBN 0-444-70361-6; QA166.R43 1988.
5. Thomas Eiter. Exact transversal hypergraphs and application to boolean μ-functions. *Journal of Symbolic Computation*, 17:215–225, 1994.
6. Thomas Eiter and Georg Gottlob. Identifying the minimal transversals of a hypergraph and related problems. *SIAM Journal on Computing*, 24(6):1278–1304, 1995.
7. Michael L. Fredman and Leonid Khachiyan. On the complexity of dualization of monotone disjunctive normal forms. *Journal of Algorithms*, 21(3):618–628, 1996.
8. Ronald L. Graham and H.O. Pollak. On the addressing problem for loop switching. *Bell System Technical Journal*, 50(8):2495–2519, 1971.
9. David A. Gregory, Valerie L. Watts, and Bryan L. Shader. Biclique decompositions and hermitian rank. *Linear Algebra and its Applications*, 292:267–280, 1999.
10. Oliver Kullmann. Investigating a general hierarchy of polynomially decidable classes of CNF's based on short tree-like resolution proofs. Technical Report TR99-041, Electronic Colloquium on Computational Complexity (ECCC), October 1999.
11. Oliver Kullmann. New methods for 3-SAT decision and worst-case analysis. *Theoretical Computer Science*, 223(1-2):1–72, July 1999.
12. Oliver Kullmann. On a generalization of extended resolution. *Discrete Applied Mathematics*, 96-97(1-3):149–176, 1999.
13. Oliver Kullmann. On the conflict matrix of clause-sets. Technical Report CSR 7-2003, University of Wales Swansea, Computer Science Report Series, 2003.

14. Oliver Kullmann. The combinatorics of conflicts between clauses. In Enrico Giunchiglia and Armando Tacchella, editors, *Theory and Applications of Satisfiability Testing 2003*, volume 2919 of *Lecture Notes in Computer Science*, pages 426–440, Berlin, 2004. Springer. ISBN 3-540-20851-8.

15. Oliver Kullmann. The conflict matrix of (multi-)clause-sets — a link between combinatorics and (generalised) satisfiability problems. In preparation; continuation of [13], 2004.

16. Oliver Kullmann. Upper and lower bounds on the complexity of generalised resolution and generalised constraint satisfaction problems. *Annals of Mathematics and Artificial Intelligence*, 40(3-4):303–352, March 2004.

17. Oliver Kullmann and Horst Luckhardt. Deciding propositional tautologies: Algorithms and their complexity. Preprint, 82 pages, January 1997.

18. Oliver Kullmann and Horst Luckhardt. Algorithms for SAT/TAUT decision based on various measures. Preprint, 71 pages, December 1998.

19. M.M. Petrović. The spectrum of an infinite labelled graph. Master's thesis, University Beograd, Faculty of Science, 1981.

20. Aleksandar Torgašev. Graphs with the reduced spectrum in the unit interval. *Publ. Inst. Math. (Beograd)*, 36(50):15–26, 1984.

QBF Reasoning on Real-World Instances

Enrico Giunchiglia, Massimo Narizzano, and Armando Tacchella*

DIST, Università di Genova, Viale Causa, 13 – 16145 Genova, Italy
{enrico, mox, tac}@dist.unige.it

Abstract. During the recent years, the development of tools for deciding Quantified Boolean Formulas (QBFs) satisfiability has been accompanied by a steady supply of real-world instances, i.e., QBFs originated by translations from application domains such as formal verification and planning. QBFs from these domains showed to be challenging for current state-of-the-art QBF solvers, and, in order to tackle them, several techniques and even specialized solvers have been proposed. Among these techniques, there are (i) efficient detection and propagation of unit and monotone literals, (ii) branching heuristics that leverages the information extracted during the learning phase, and (iii) look-back techniques based on learning.

In this paper we discuss their implementation in our state-of-the-art solver QUBE, pointing out the non trivial issues that arised in the process. We show that all the techniques positively contribute to QUBE performances *on average*. In particular, we show that monotone literal fixing is the most important technique in order to improve capacity, followed by learning and the heuristics. The situation is reversed if we consider productivity. These and other observations are detailed in the body of the paper. For our analysis, we consider the formal verification and planning benchmarks from the 2003 QBF evaluation.

1 Introduction

During the recent years, the development of tools for deciding Quantified Boolean Formulas (QBFs) has been accompanied by a steady supply of real-word instances, i.e., QBFs originated by translations from application domains such as formal verification [1, 2] and planning [3, 4]. QBFs from these domains showed to be challenging for current state-of-the-art QBF solvers, and, in order to tackle them, several techniques and even specialized QBF solvers [5] have been proposed. Among the techniques that have been proposed for improving performances, there are (i) efficient detection and propagation of unit and monotone literals [6, 7], (ii) branching heuristics that leverage the information extracted during the learning phase [8], and (iii) look-back techniques based on conflict and solution learning [9, 10, 11].

In this paper we discuss the implementation of the above mentioned techniques in our state-of-the-art QBF solver QUBE [8], pointing out the non trivial issues that arised in the process. We show that all the techniques positively contribute to QUBE

---

* The authors wish to thank MIUR, ASI and the Intel Corporation for their financial support, and the reviewers who helped to improve the original manuscript.

H.H. Hoos and D.G. Mitchell (Eds.): SAT 2004, LNCS 3542, pp. 105–121, 2005.

performances. In particular, we show that monotone literal fixing is the most important technique in order to improve capacity (i.e., the ability to solve problems [12]), followed by learning and the heuristics. The situation is reversed if we consider productivity (i.e., the ability to quickly solve problems [12]). All these considerations obviously hold *on average*. Indeed, for each technique, there are instances in which it does not produce any benefit or, even worse, in which it causes a degradation in the performances. These and other observations are detailed in the body of the paper. For our analysis, we consider the formal verification and planning benchmarks from the 2003 QBF evaluation [13].[1] Our analysis is, to the best of our knowledge, the first which analyzes the contributions of (i) monotone literal fixing, (ii) the branching heuristic, and (iii) conflict and solution learning, for solving real-world problems. Though, strictly speaking, our results holds only for QUBE, we expect that the same will carry over to the majority of the other systems, which, like QUBE are based on [14]. We expect this to be true especially for solvers implementing learning as look-back mechanism, i.e., for solvers targeted to solve real-world benchmarks.

The paper is structured as follows. We first give some formal preliminaries, and the background knowledge representing the starting point of our analysis. We devote three sections to the improvements on lookahead techniques, the branching strategy and learning, respectively. In each section, we first briefly describe its implementation in QUBE, and then we present and discuss experimental results showing the positive contribution of the technique. We end the paper with some remarks.

2   Preliminaries

In this section, we first give some formal preliminaries. Then, we present the basic algorithm of QUBE, and finally the setting that we have used for our experimental analysis.

2.1   Formal Preliminaries

Consider a set P of propositional letters. An *atom* is an element of P. A *literal* is an atom or the negation thereof. Given a literal l, $|l|$ denotes the atom of l, and $\bar{l}$ denotes the *complement* of l, i.e., if $l = a$ then $\bar{l} = \neg a$, and if $l = \neg a$ then $\bar{l} = a$, while $|l| = a$ in both cases. A *propositional formula* is a combination of atoms using the k-ary ($k \geq 0$) connectives $\wedge$, $\vee$ and the unary connective $\neg$. In the following, we use $\top$ and $\bot$ as abbreviations for the empty conjunction and the empty disjunction respectively. A *QBF* is an expression of the form

$$\varphi = Q_1 x_1 Q_2 x_2 \ldots Q_n x_n \Phi \qquad (n \geq 0) \qquad (1)$$

where every Q_i ($1 \leq i \leq n$) is a quantifier, either existential $\exists$, or universal $\forall$; $x_1 \ldots x_n$ are distinct atoms in P, and Φ is a propositional formula. $Q_1 x_1 Q_2 x_2 \ldots Q_n x_n$ is the

---

[1] At the time of this writing, the 2004 QBF comparative evaluation has just finished. Unfortunately, a now corrected bug in QUBE look-ahead caused QUBE to incorrectly decide a few randomly generated benchmarks, and thus it did not enter in the second stage of the evaluation.

bool SOLVE($Q, \Sigma, \Delta, \Pi, S$)
1 $\langle Q, \Sigma, \Delta, \Pi, S \rangle \leftarrow$ LOOKAHEAD($Q, \Sigma, \Delta, \Pi, S$)
2 **if** ($\Delta = \emptyset$ **or** $\emptyset_\forall \in \Pi$) **then return** TRUE
3 **if** ($\emptyset_\exists \in \Sigma \cup \Delta$) **then return** FALSE
4 $l \leftarrow$ CHOOSE-LITERAL(Q, Σ, Δ, Π)
5 $\langle Q, \Sigma, \Delta, \Pi \rangle \leftarrow$ ASSIGN($l, Q, \Sigma, \Delta, \Pi$)
6 $V \leftarrow$ SOLVE($Q, \Sigma, \Delta, \Pi, S \cup \{l\}$)
7 **if** ((l is existential **and** V is TRUE) **then**
8   **return** TRUE
9 **else if** ((l is universal **and** V is FALSE) **then**
10   **return** FALSE
11 **else**
12     $\langle Q, \Sigma, \Delta, \Pi \rangle \leftarrow$ ASSIGN($\bar{l}, Q, \Sigma, \Delta, \Pi$)
13   **return** SOLVE($Q, \Sigma, \Delta, \Pi, S \cup \{\bar{l}\}$)

set LOOKAHEAD($Q, \Sigma, \Delta, \Pi, S$)
14 **do**
15     $\langle Q', \Sigma', \Delta', \Pi', S' \rangle \leftarrow \langle Q, \Sigma, \Delta, \Pi, S \rangle$
16   **for each** l s.t. $\{l\}_\exists \in \Sigma \cup \Delta$ **or** $\{\bar{l}\}_\forall \in \Pi$ **do**
17         $S \leftarrow S \cup \{l\}$
18         $\langle Q, \Sigma, \Delta, \Pi \rangle \leftarrow$ ASSIGN($l, Q, \Sigma, \Delta, \Pi$)
19   **for each** l s.t. $\{k \in \Sigma \cup \Delta \cup \Pi \mid \bar{l} \in k\} = \emptyset$ **do**
20   **if** (l is existential) **then**
21             $\langle Q, \Sigma, \Delta, \Pi \rangle \leftarrow$ ASSIGN($l, Q, \Sigma, \Delta, \Pi$)
22   **else**
23             $\langle Q, \Sigma, \Delta, \Pi \rangle \leftarrow$ ASSIGN($\bar{l}, Q, \Sigma, \Delta, \Pi$)
24 **while** $\langle Q, \Sigma, \Delta, \Pi, S \rangle \neq \langle Q', \Sigma', \Delta', \Pi', S' \rangle$
25 **return** $\langle Q, \Sigma, \Delta, \Pi, S \rangle$

set ASSIGN($l, Q, \Sigma, \Delta, \Pi$)
26 $Q \leftarrow$ REMOVE($Q, |l|$)
27 **for each** $c \in \Sigma$ s.t. $l \in c$ **do**
28     $\Sigma \leftarrow \Sigma \setminus \{c\}$
29 **for each** $c \in \Delta$ s.t. $l \in c$ **do**
30     $\Delta \leftarrow \Delta \setminus \{c\}$
31 **for each** $t \in \Pi$ s.t. $\bar{l} \in t$ **do**
32     $\Pi \leftarrow \Pi \setminus \{t\}$
33 **for each** $c \in \Sigma$ s.t. $\bar{l} \in c$ **do**
34     $\Sigma \leftarrow (\Sigma \setminus \{c\}) \cup \{c \setminus \{\bar{l}\}\}$
35 **for each** $c \in \Delta$ s.t. $\bar{l} \in c$ **do**
36     $\Delta \leftarrow (\Delta \setminus \{c\}) \cup \{c \setminus \{\bar{l}\}\}$
37 **for each** $t \in \Pi$ s.t. $l \in t$ **do**
38     $\Pi \leftarrow (\Pi \setminus \{t\}) \cup \{t \setminus \{l\}\}$
39 **return** $\langle Q, \Sigma, \Delta, \Pi \rangle$

Fig. 1. Basic search algorithm of QUBE

prefix and Φ is the *matrix* of (1). A literal l is *existential*, if $\exists |l|$ is in the prefix, and *universal* otherwise. We say that (1) is in *Conjunctive Normal Form* (CNF) when Φ is a conjunction of *clauses*, where each clause is a disjunction of literals in $x_1 \ldots x_n$; we say that (1) is in *Disjunctive Normal Form* (DNF) when Φ is a disjunction of *terms* (or *cubes*), where each term is a conjunction of literals in $x_1 \ldots x_n$. We use the term *constraints* when we refer to clauses and terms indistinctly. The semantics of a QBF φ can be defined recursively as follows. If the prefix is empty, then φ's satisfiability is defined according to the truth tables of propositional logic. If φ is $\exists x \psi$ (resp. $\forall x \psi$), φ is satisfiable if and only if $\{\varphi\}_x$ or (resp. and) $\{\varphi\}_{\neg x}$ are satisfiable. If $\varphi = Qx\psi$ is a QBF and l is a literal with $|l| = x$, $\{\varphi\}_l$ is the QBF obtained from ψ by substituting l with $\top$ and $\bar{l}$ with $\bot$.

2.2 QUBE Basic Algorithm

In Figure 1 we present the pseudo-code of SOLVE, the basic search algorithm of QUBE. SOLVE generalizes the standard backtracking algorithm for QBFs as introduced in [14], by taking into account that clauses and terms can be dynamically learned during the search. Given a QBF (1) in CNF, SOLVE takes five parameters:

1. Q is the prefix, i.e., the list $Q_1x_1, \ldots, Q_nx_n$.
2. Σ is a set of clauses. Initially Σ is empty, but clauses are added to Σ as the search proceeds as result of the learning process, see [9].
3. Δ is the set of clauses corresponding to the matrix of the input formula.
4. Π is a set of terms. As for Σ, initially Π is empty, but terms are added to Π as the search proceeds as result of the learning process, see [9].
5. S is a consistent set of literals called *assignment*. Initially $S = \emptyset$.

In the following, as customary in search algorithms, we deal with constraints as if they were sets of literals, and assume that forall constraint c, it is not the case that x and $\neg x$ belong to c, for some atom x. Further, given $\langle Q, \Sigma, \Delta, \Pi \rangle$ corresponding to a QBF (1), we assume that the clauses and terms added to Σ and Π respectively, do not alter the correctness of the procedure. This amounts to say that, for any sequence of literals $l_1; \ldots; l_m$ with $m \leq n$ and $|l_i| = x_i$, the following three formulas are equi-satisfiable:

$$\{\ldots\{\{Q_1x_1Q_2x_2\ldots Q_nx_n(\wedge_{c\in\Delta} \vee_{l\in c} l)\}_{l_1}\}_{l_2}\ldots\}_{l_m},$$
$$\{\ldots\{\{Q_1x_1Q_2x_2\ldots Q_nx_n((\wedge_{c\in\Delta} \vee_{l\in c} l) \wedge (\wedge_{c\in\Sigma} \vee_{l\in c} l))\}_{l_1}\}_{l_2}\ldots\}_{l_m},$$
$$\{\ldots\{\{Q_1x_1Q_2x_2\ldots Q_nx_n((\wedge_{c\in\Delta} \vee_{l\in c} l) \vee (\vee_{t\in\Pi} \wedge_{l\in t} l))\}_{l_1}\}_{l_2}\ldots\}_{l_m}.$$

Consider a QBF φ corresponding to $\langle Q, \Sigma, \Delta, \Pi \rangle$.

At a high level of abstraction, SOLVE can be seen as a procedure generating the semantic tree corresponding to φ. The basic operation is thus that of assigning a literal l and simplifying φ accordingly, i.e., compute $\{\varphi\}_l$ and simplify it. In Figure 1, this task is performed by the function ASSIGN($l, Q, \Sigma, \Delta, \Pi$), which

- Removes $|l|$ and its bounding quantifier from the prefix (line 26),
- Removes all the clauses (resp. terms) to which l (resp. $\bar{l}$) pertains (lines 27-32). These clauses are said to be *eliminated* by l.
- Removes $\bar{l}$ (resp. l) from all the clauses (resp. terms) to which $\bar{l}$ (resp. l) pertains (lines 33-38). These clauses are said to be *simplified* by l.

In the following we say that a literal l is:

- *open* if $|l|$ is in Q, and *assigned* otherwise;
- *unit* if there exist a clause $c \in \Sigma \cup \Delta$ (resp. a term $t \in \Pi$) such that l is the only existential in c (resp. universal in t) and there are no literals in c (resp. in t) with an higher level. The *level of an atom* is 1 + the number of expressions $Q_jx_jQ_{j+1}x_{j+1}$ in Q with $j \geq i$ and $Q_j \neq Q_{j+1}$. The *level of a literal* l is the level of $|l|$.
- *monotone* if for all constraints $k \in (\Sigma \cup \Delta \cup \Pi), \bar{l} \notin k$.

Now consider the routine LOOKAHEAD in Figure 1: $\{l\}_\exists$ (resp. $\{l\}_\forall$) denotes a constraint which is unit in l. The function LOOKAHEAD has the task of simplifying its input QBF by finding and assigning all unit (lines 16-18) and monotone literals (lines 19-23). Since assigning unit or monotone literals may cause the generation of new unit or monotone literals (this is different from the SAT case in which assigning monotone literals cannot generate new unit literals) LOOKAHEAD loops till no further simplification is possible (lines 14-15, 24).

The function SOLVE works in four steps:

1. Simplify the input instance with LOOKAHEAD (line 1).
2. Check if the termination condition is met (lines 2-3): if the test in line 2 is true, then S is a *solution*, while if the test in line 3 is true, then a S is a *conflict*; $\emptyset_\exists$ (resp. $\emptyset_\forall$) stands for the *empty clause* (resp. *empty term*), i.e., a constraint without existential (resp. universal) literals.
3. Choose heuristically a literal l (line 4) such that $|l|$ is at the highest level in the prefix. The literal returned by CHOOSE-LITERAL is called *branching literal*.
4. Assign the chosen literal (line 5), and recursively evaluate the resulting QBF (line 6):
   (a) if l is existential and the recursive evaluation yields TRUE then TRUE is returned (lines 7-8), otherwise
   (b) if l is universal and the recursive evaluation yields FALSE then FALSE is returned (lines 9-10), otherwise
   (c) $\bar{l}$ is assigned (line 12), and the result of the recursive evaluation of the resulting QBF is returned (line 13).
   It is easy to see that the execution of the code in lines 5-13 causes the generation of an AND-OR tree, whose *OR nodes* correspond to existential literals, while *AND nodes* correspond to universal literals.

SOLVE returns TRUE if the input QBF is satisfiable and FALSE otherwise.

For the sake of clarity we have presented SOLVE with recursive chronological backtracking. To avoid the expensive copying of data structures that would be needed to save Σ, Δ and Π at each node, QUBE (and most of the available systems as well) features a non-recursive implementation of the lookback procedure. The implementation is based on an explicit search stack and on data structures that can assign a literal during lookahead and then retract the assignment during lookback, i.e., restore Σ, Δ and Π to the configuration before the assignment was made. Further, as we already said in the introduction, QUBE differs from the above high-level description in that it features the techniques that are the subject of the next three sections.

2.3   Experimental Setting

In order to evaluate the contribution of the different components to QUBE performances, we considered the 450 formal verification and planning instances that constituted part of the 2003 QBF evaluation[2]: 25% of these instances are from formal verification problems [1, 2], and the remaining are from planning domains [3, 4]. As we said in the introduction, these instances showed to be challenging for the 2003 QBF solvers comparative evaluation. Further, a subset of this testset was also included in the 2004 evaluation, and, according to some preliminary results, some of instances showed to be still hard to solve.

All the experiments were run on a farm of identical PCs, each one equipped with a PIV 3.2GHz processor, 1GB of RAM, running Linux Debian 3.0. Finally, each system had a timeout value of 900s per instance.

---

[2] With respect to the non-random instances used in the 2003 QBF comparative evaluation, our test set does not include the QBF encodings of the modal K formulas submitted by Guoqiang Pan [15].

3   Lookahead

Goal of the LOOKAHEAD function is to simplify the formula by propagating unit and monotone literals till no further simplification is possible. Any efficient implementation of a QBF solver has to rely on an efficient implementation of LOOKAHEAD. Indeed, most of the literals are assigned inside the function LOOKAHEAD: on our test set, if we consider the ratio R between

- the number of calls to ASSIGN made within LOOKAHEAD (lines 18, 21, 23).
- the number of calls to ASSIGN made within SOLVE (lines 5, 12), and

We have that R, on the problems that QUBE takes more than 1s to solve, is 514 on average, i.e., assigning one branching literal, causes hundreds of literals to be assigned inside LOOKAHEAD. Further, by running a profiler on QUBE, we have seen that on all the instances that we have tried, lookahead always amounted to more than 70% of the total runtime: this result echoes analogous remarks made in the SAT literature (see, e.g., [16]). Finally, the need for a fast lookahead procedure is accentuated by the use of learning [9], where the solver adds (possibly very long) constraints to the initial set given in input.

3.1   Algorithm

The implementation of LOOKAHEAD in QUBE is based on an extension of the lazy data structures as presented in [6], to detect and assign unit and monotone literals.

In particular, for detecting existential unit literals, we use the three literal watching (3LW) schema as described in [6]. For detecting universal unit literals, 3LW can be easily adapted to the case. Similar to the SAT case, all the literal watching schemes in [6] do not require to perform elimination of constraints, but only some of the simplifications (those being "watched") are performed when assigning a literal. However, compared to the other watching literal schemes in [6], 3LW performs less operations if in each constraint the existential and universal literals are listed separately, as it is the case in QUBE.

For monotone literal fixing (MLF), we implemented clause watching (CW) as described in [6]. However, the implementation of CW in QUBE posed some problems due to the interaction between monotone literals and learning (see [7]). The first observation is that the detection of monotone literals requires, when assigning a literal l, to perform also all the associated eliminations: These operations at least in part obscure the advantages of 3LW. Then, in order to reduce the burden of eliminating constraints, it is important to reduce the set of constraints to be considered. Problems arise because, given a QBF $\varphi = \langle Q, \Sigma, \Delta, \Pi \rangle$, it may be the case that for a literal l, the condition

$$\{k \in \Delta \mid \bar{l} \in k\} = \emptyset \tag{2}$$

is satisfied, while

$$\{k \in \Sigma \cup \Delta \cup \Pi \mid \bar{l} \in k\} = \emptyset \tag{3}$$

is not. Thus, a literal l may be assigned as monotone because it does not occur in the matrix of the input formula, but assigning l to true may cause (i) the simplification of some learned constraint, (ii) the generation of an empty clause or an empty term, and (iii) the presence of $\bar{l}$ in the reason associated to the empty clause/term. This last fact would require the computation of a reason for monotone literals, to be used while backtracking. While from a theoretical point of view it is possible to compute such a reason, it is still an open issue how to do it efficiently, and [7] describes some problems that point out that it cannot be done in all cases.

Summing up, problems arise when

- l is monotone, existential and $\bar{l}$ belongs to a working reason originating from a conflict, or
- l is monotone, universal and $\bar{l}$ belongs to a working reason originating from a solution.

As discussed in [9], if condition (3) is satisfied then assigning l as monotone is not problematic meaning that it is always possible to compute the working reason wr in order to avoid $\bar{l} \in wr$. Still, checking condition (3) is not practical because, when assigning a literal l, would require to eliminate all the corresponding constraints in $\Sigma \cup \Delta \cup \Pi$ (or at least, from $\Sigma \cup \Delta$ if l is existential, and from $\Sigma \cup \Pi$ if l is universal).

The solution that we have adopted for MLF in QUBE is to assign a literal as monotone when condition (2) is satisfied, and then to temporarily delete from $\Sigma \cup \Pi$ all the clauses where $\bar{l}$ occurs. This solves all the above mentioned problems, and it has the following advantages:

1. It allows to assign as monotone more literals than those that would be assigned according to condition (3), and
2. From a computational point of view, it is far less expensive since, when assigning a literal l it only requires to eliminate constraints in Δ.

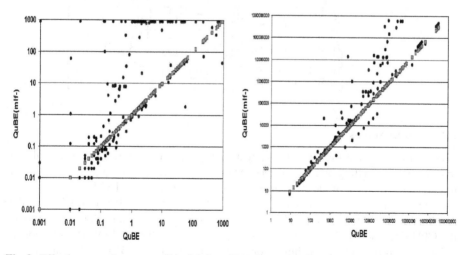

Fig. 2. Effectiveness of monotone literal fixing: CPU time (left) and number of assignment (right)

Table 1. Comparison among various versions of QUBE. In each table, the comparison considers a system taken as reference and written in the top left box in the table: QUBE in Table 1.a, and QUBE(RND)[3] in Table 1.b. In each table, if A is the system taken as reference in it, and $B \neq A$ is a solver in the first column, then the other columns report the number of problems that: "=", A and B solve in the same time; "<", A and B solve but A takes less time than B; ">", A and B solve but A takes more time than B; "≪", A solves while B does not; "≫", A does not solve while B does; "⋈", A and B do not solve; "×10<", both A and B solve but on which A is at least one order of magnitude faster; "×0.1<", both A and B solve but on which A is at least one order of magnitude slower; "TO", B does not solve. The number of timeouts for QUBE and QUBE(RND)[3] is 76 and 89 respectively

Table 1.a

QUBE	=	<	>	≪	≫	⋈	×10<	×0.1>	TO
QUBE(MLF⁻)	141	73	95	65	1	75	31	3	140
QUBE(RND)[1]	137	112	106	19	9	67	41	10	86
QUBE(RND)[2]	134	139	80	21	9	67	44	6	88
QUBE(RND)[3]	126	166	60	22	9	67	53	5	89
QUBE(RND)[4]	124	189	37	24	9	67	62	4	91
QUBE(RND)[5]	108	213	24	29	8	68	79	4	97
QUBE(CBJ,SBJ)	146	181	31	16	1	75	39	0	91

Table 1.b

QUBE(RND)[3]	=	<	>	≪	≫	⋈	×10<	×0.1>	TO
QUBE(RND)[1]	136	0	225	0	3	86	0	43	86
QUBE(RND)[2]	169	0	192	0	1	88	0	19	88
QUBE(RND)[4]	156	203	0	2	0	89	27	0	91
QUBE(RND)[5]	109	244	0	8	0	89	61	0	97
QUBE(RND,CBJ,SBJ)[1]	131	145	72	13	7	82	27	20	95
QUBE(RND,CBJ,SBJ)[2]	137	164	43	17	2	87	43	7	104
QUBE(RND,CBJ,SBJ)[3]	123	192	25	21	2	87	68	2	108
QUBE(RND,CBJ,SBJ)[4]	110	205	17	29	2	87	83	1	116
QUBE(RND,CBJ,SBJ)[5]	84	222	10	45	2	87	99	1	132
QUBE(RND,CBJ,SLN)[1]	130	96	128	7	5	84	20	26	91
QUBE(RND,CBJ,SLN)[2]	133	134	82	12	5	84	27	14	96
QUBE(RND,CBJ,SLN)[3]	129	169	48	15	3	86	40	5	101
QUBE(RND,CBJ,SLN)[4]	115	209	20	17	1	88	54	1	105
QUBE(RND,CBJ,SLN)[5]	86	245	6	24	1	88	87	0	112
QUBE(RND,CLN,SBJ)[1]	135	78	142	6	4	85	7	36	91
QUBE(RND,CLN,SBJ)[2]	151	110	90	10	4	85	15	15	95
QUBE(RND,CLN,SBJ)[3]	169	134	39	19	1	88	29	5	107
QUBE(RND,CLN,SBJ)[4]	141	183	11	26	0	89	51	0	115
QUBE(RND,CLN,SBJ)[5]	103	218	2	38	0	89	69	0	127

3.2   Effectiveness

It is well known that unit literal propagation is fundamental for efficiency, while, at least in the SAT setting, MLF is often considered to be inefficient, and indeed, it is not implemented by most of the state-of-the-art SAT solvers like ZCHAFF.

The performances of QUBE when run with and without MLF (we will call the resulting system QUBE(MLF⁻)) is shown in Figure 2. In the left plot, the x-axis is the CPU-time of QUBE and the y-axis is the CPU-time of QUBE(MLF⁻). A plotted point $\langle x, y \rangle$ represents a benchmark on which QUBE and QUBE(MLF⁻) take x and

y seconds respectively.[3] For convenience, we also plot the points $\langle x, x \rangle$, each representing the benchmarks solved by QUBE in x seconds. As it can be seen from the figure, QUBE(MLF$^-$) is faster than QUBE on many benchmarks (96), represented by the points below the diagonal. However, these points are mostly located at the beginning of the plot, and represent instances that are solved in less than a second. Indeed, assigning a literal when MLF is enabled is more expensive and, on easy instances, MLF does not pay off. Still, MLF can greatly cut the search tree. This is evident from the right plot in the figure, showing the number of assignments made by QUBE(MLF$^-$) wrt QUBE when considering the instances solved by both solvers: There are only 10 clearly visible points below the diagonal, meaning that the 96 problems on which QUBE(MLF$^-$) is faster than QUBE are due to the burden of MLF. Of these 96 problems,

- there is only one problem (represented by the point on the vertical axis at the extreme right) which is solved by QUBE(MLF$^-$) and on which QUBE timeouts, and
- among the instances that are solved by both solvers, QUBE(MLF$^-$) is faster than QUBE of at least one order of magnitude on only 3 instances, two of which solved by QUBE in less than 0.1s.

On the other hand, (i) QUBE is able to solve 65 instances not solved by QUBE(MLF$^-$), and (ii) QUBE is at least one order of magnitude faster than QUBE(MLF$^-$) on 31 other problems. These and other numbers are reported in Table 1.a.

4   Heuristic

In SAT, it is well known that the branching heuristic is important for efficiency. In the QBF setting, the situation is different. Indeed, we are only allowed to choose a literal l if $|l|$ is at the highest level in the prefix. Thus, on a QBF of the form

$$\exists x_1 \forall x_2 \exists x_3 ... \forall x_{n-1} \exists x_n \Phi \tag{4}$$

the heuristic is likely to be (almost) useless: unless atoms are removed from the prefix because unit or monotone, the atom to pick at each node is fixed. On the other hand, on a QBF of the form

$$\exists x_1 \exists x_2 ... \exists x_m \Phi \tag{5}$$

corresponding to a SAT instance, it is likely that the heuristic will play an important role. Indeed, we expect the role of the heuristic to have more and more importance as the number of alternations (i.e., expressions of the form $Q_j x_j Q_{j+1} x_{j+1}$ with $Q_j \neq Q_{j+1}$) in Q is small compared to the number of variables. In (4) the number of alternations is the number of variables -1, while in (5) is 0. In practice, the number of alternations is in between the two extreme cases represented by the above two equations. In many cases, it is 1 or 2, and thus we expect the heuristic to be important.

---

[3] In principle, one point $\langle x, y \rangle$ could correspond to many benchmarks solved by QUBE and QUBE(MLF$^-$) in x and y seconds respectively. However, in this and the other scatter diagrams that we consider, each point (except for the point $\langle 900, 900 \rangle$, representing the instances on which both solvers time-out) corresponds to a single instance in most cases.

In this section, we show that this is the case for QUBE heuristic, meaning that it performs consistently better, on average, than a simple random heuristic.

4.1   Algorithm

Even in SAT, where the number of alternations is 0, the design of an heuristic has to be a trade-off between accuracy and speed. VSIDS (Variable State Independent Decaying Sum) [16] is now at the basis of most recent SAT solvers for real world problems. The basic ideas of VSIDS are to (i) initially rank literals on the basis of the occurrences in the matrix, (ii) increment the weight of the literals in the learned constraints, and (iii) periodically divide by a constant the weight of each literal.

In our QBF setting, the above needs to be generalized taking into account the prefix and also the presence of both learned constraints and terms. In QUBE this is done by periodically sorting literals according to (i) the prefix level of the corresponding atom, (ii) their score, and (iii) their numeric ID. The score of each literal l is computed as follows:

- initially, it is set to the number of clauses in which l belongs,
- at the $i + 1$ step, the score is computed by summing the score at the previous step divided by two, and the number of constraints c such that
  - $l \in c$, if l is existential, and
  - $\bar{l} \in c$, if l is universal.

When the input QBF corresponds to a SAT formula, the above heuristic boils down to VSIDS.

4.2   Effectiveness

To evaluate the role of the heuristic in QUBE, we compared it with QUBE(RND), i.e., QUBE with a heuristic which randomly select a literal at the highest prefix level. Because of the randomness, we run QUBE(RND) 5 times on each instance. Given an instance φ, if we order the time spent to solve φ from the best (1) to the worst (5), then we can define QUBE(RND)[i] to be the system whose performance on φ is the i-th best among the 5 results. The results of QUBE, QUBE(RND)[1-5] are plotted in Figure 3 left. In the figure, the results of each solver are (i) sorted independently and in ascending order, (ii) filtered out by removing the first 148 and the last 76 values, and (iii) plotted against an ordinal in the range [1-226] on the x-axis. The filtering has been done in order to increase the readability of the figure. Indeed, each solver (i) is able to solve at least 148 problems in a time ≥ 0.02s, and (ii) timeouts on at least 76 values. Thus, if a point $\langle x, y \rangle$ belongs to the plot of a system S, this means that $x + 148$ instances are solved in less than y seconds by S.

Several observations are in order. The first one is that the heuristic plays a role. This is evident if we consider the five plots of QUBE(RND)[1-5], which show that there can be significant differences among different runs of QUBE with a random heuristics. The second observation is that QUBE is better than QUBE(RND)[1], i.e., the solver among QUBE(RND)[1-5] having the best result on each single instance:

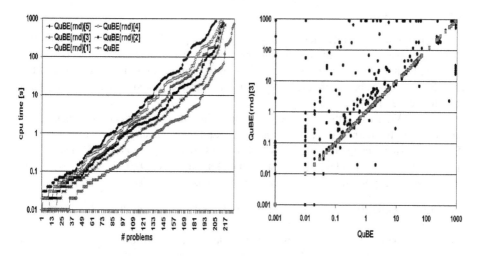

Fig. 3. Effectiveness of the heuristics: overall (left) and best (right)

- QuBE (resp. QuBE(RND)[1]) is able to solve 19 (resp. 9) instances that are not solved by QuBE(RND)[1] (resp. QuBE),
- among the instances solved by both solvers, QuBE (resp. QuBE(RND)[1]) is at least one order of magnitude faster than QuBE(RND)[1] (resp. QuBE) on 41 (resp. 10) instances.

The above data can be seen on the right plot, representing the performances of QuBE versus QuBE(RND)[1]. From the right plot, it also emerges that QuBE(RND)[1] is faster on many instances (115). Of course, this number goes down to 69 and 32 if we consider QuBE(RND)[3] or QuBE(RND)[5]. Still, the presence of 32 problems (roughly 8% of the problems that are solved by at least one solver) in which QuBE(RND)[5] is faster than QuBE points out that there are QBFs in which the heuristic does not seem to play any role. On these problems, QuBE pays the overhead of periodically computing the score of each literal, and sort them according to the above outlined criteria.

For more data, see Table 1.

5   Learning

Learning is a look-back strategy whose effectiveness for solving real-world problems is a consolidated result in the SAT literature (see, e.g., [17, 18, 16]). In the QBF setting, mixed results have been so far obtained, see, e.g., [9, 11, 10, 13, 19]. In particular, in [19] it is argued that while learning "conflicts" (computed while backtracking from an empty clause) leads to the expected positive results, learning "solutions" (computed while backtracking from an empty term or the empty matrix) does not always produce positive results, especially on real-world benchmarks.

Here we show that both conflict and solution learning are essential for QuBE performances.

5.1   Algorithm

Learning amounts to store clauses (resp. terms) computed while backtracking by performing clause (resp. term) resolution[4] between a "working reason" that is initially computed when an empty clause (resp. an empty term or the empty matrix) is found, and the "reason" corresponding to unit literals that are stored while descending the search tree, see [21].

As in SAT, two of the key issues are the criteria used for deciding when a constraint has to be learned (i.e., stored in Σ/Π), and then unlearned (i.e., removed from Σ/Π). Learning in QUBE works as follows. Assume that we are backtracking on a literal l having decision level n, i.e., such that there are n AND-OR nodes before l. The constraint corresponding to the current working reason wr is learned if and only if:

- l is existential (resp. universal) if we are backtracking from a conflict (resp. solution),
- all the assigned literals in wr but l, have a decision level strictly smaller than n, and
- there are no open universal (resp. existential) literals in wr that are before l in the prefix.

These conditions ensure that l is unit in the constraint corresponding to the reason. Once the constraint is learned, QUBE backjumps to the node corresponding to the literal $l' \neq l$ in wr with maximum decision level. Notice that on a SAT instance, QUBE learning mechanism behaves similarly to the "1-UIP-learning" scheme used in ZCHAFF.

For unlearning, QUBE uses a *relevance bounded learning (of order r)* schema as introduced in SAT by [18]: The set of learned constraints is periodically scanned and the constraints having more than r open literals are deleted.

In QBF, another key issue is the initialization of the working reason wr, especially when a solution is found (when a conflict is found, we can do as in SAT). In the case of a solution, in QUBE, our approach is to first set wr to a set of literals having the following properties:

- it is a subset of the assignment S that led to the current solution,
- it is a prime implicant of the matrix of the input QBF,
- is such that there does not exist another set of literals satisfying the first two properties and
  - with a smaller (under set inclusion) set of universal literals, or
  - causing a deeper backtrack.

Then, some extra computation is done in order to try to further reduce the universal literals in wr. See [21] for more details.

5.2   Effectiveness

Figure 4 left shows the performances of QUBE versus QUBE(CBJ,SBJ), i.e., QUBE without learning but both conflict and solution backjumping enabled [22]. The first consideration is that learning pays off:

---

[4] Clause resolution is called Q-resolution in [20].

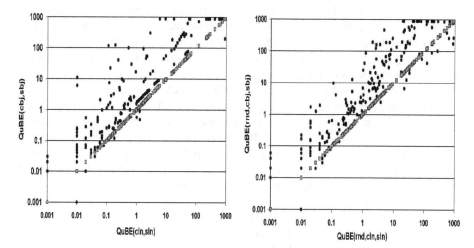

Fig. 4. Effectiveness of learning: with a VSIDS (left) and a random heuristic (right)

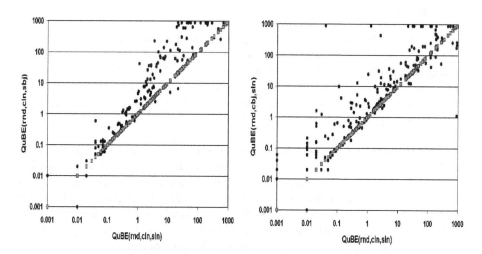

Fig. 5. Effectiveness of conflict (left) and solution (right) learning

- QUBE (resp. QUBE(CBJ,SBJ)) is able to solve 16 (resp. 1) instances that are not solved by QUBE(CBJ,SBJ) (resp. QUBE),
- among the instances solved by both solvers, QUBE (resp. QUBE(CBJ,SBJ)) is at least one order of magnitude faster than QUBE(CBJ,SBJ) (resp. QUBE) on 39 (esp. 0) instances.

Still, the above data are not entirely satisfactory for two reasons.

First, in QUBE learning and the heuristic are tightly coupled: Whenever QUBE learns a constraint, it also increments the score of the literals in it. In QUBE(CBJ,SBJ) no constraint is ever learned. As a consequence, in QUBE(CBJ,SBJ), (i) literals are initially sorted on the basis of their occurrences in the input QBF, and (ii) the score of each literal is periodically halved till it becomes 0. When all the literals have score 0, then literals at the same prefix level are chosen according to their lexicographic order.

Second, independently from the heuristic being used, a plot showing the performances of QuBE with and without learning, does not say which of the two learning schemes (conflict, solution) is effective [19].

To address the first problem, we consider QuBE with a random heuristic, with and without learning: We call the resulting systems QuBE(RND) and QuBE(RND,CBJ,SBJ) respectively. As in the previous section, we run each system 5 times on each benchmark, and we introduce the systems QuBE(RND)[i] and QuBE(RND,CBJ,SBJ)[i] ($1 \leq i \leq 5$)). The results for QuBE(RND)[3] and QuBE(RND,CBJ,SBJ)[3] are plotted in Figure 4 right. From the plot, it is easy to see that QuBE(RND)[3] is faster than QuBE(RND,CBJ,SBJ)[3] in most cases. To witness this fact

- QuBE(RND) (resp. QuBE(RND,CBJ,SBJ)) is able to solve 21 (resp. 2) instances that are not solved by QuBE(CBJ,SBJ) (resp. QuBE),
- among the instances solved by both solvers, QuBE (resp. QuBE(CBJ,SBJ)) is at least one order of magnitude faster than QuBE(CBJ,SBJ) (resp. QuBE) on 68 (esp. 2) instances.

Comparing with the results on the left plot, it seems that with a random heuristic learning becomes more important. This fact witnesses also in our setting the well-known tension between look-ahead and look-back techniques: Using a "smart" look-ahead makes the look-back less important, and viceversa. Still, because of the randomness in the systems, 5 runs are too few in order to draw precise quantitative conclusions. Still, at from a qualitative point of view, it is clear that learning can produce significant speed-ups.

To address the second problem, we considered the systems QuBE(RND,CBJ,SLN) and QuBE(RND,CLN,SBJ), i.e., the systems obtained from QuBE(RND) by disabling conflict learning and solution learning respectively. As usual we run each system 5 times on each instance, and we define QuBE(RND,CBJ,SLN)[i] and QuBE(RND,CLN,SBJ)[i] ($1 \leq i \leq 5$) as before. Figure 5 shows the performances of QuBE(RND)[3] versus QuBE(RND,CBJ,SLN)[3] (left plot) and QuBE(RND,CLN,SBJ)[3] (right plot). From the plots, we see that both conflict and solution learning pay off. In each plot, there are only a few points well below the diagonal. Further, by comparing the two plots, it seems that solution learning is more effective than conflict learning, but again we have to take this as qualitative indication. Some detailed quantitative information is reported in Table 1.b. From the table, we can see that the overall performances of QuBE(RND)[i] are better than the performances of QuBE(RND,CBJ,SLN)[i], QuBE(RND,CLN,SBJ)[i] and QuBE(RND,CBJ,SBJ)[i]. The above positive results for solution learning are confirmed if we compare the number of solutions found by QuBE(RND,CBJ,SLN)[i] and QuBE(RND,CBJ,SBJ)[i] as in [19]: For example, considering the instances solved by both solvers in more than 1s and for which at least one solution is found by both, the average (resp. maximum) of the ratio between the number of solutions found by QuBE(RND,CBJ,SBJ)[3] and QuBE(RND,CBJ,SLN)[3] is 5.4 (resp. 42.4). The negative results reported in [19] for solution-based look-back mechanisms are not comparable with ours, given the different mechanisms implemented by the respective solvers (e.g., for computing the initial solution and for monotone literal fixing), and the different experimental setting (e.g., the testset).

6   Concluding Remarks

Given the results in Table 1.a, we can say that all the techniques contribute to the effectiveness of QUBE. In terms of improving the *capacity* (i.e., the ability to solve problems [12]), the most effective one seems to be MLF, followed by learning and the heuristic: QUBE solves 64, 15 and 13 instances more than QUBE(MLF⁻), QUBE(CBJ,SBJ) and QUBE(RND) respectively. In terms of improving the *productivity* (i.e., the ability to quickly solve problems [12]) the picture seems to be the opposite. The most effective technique is the heuristic, then learning and finally MLF: the difference between

- the number of problems in which QUBE is at least 1 order of magnitude faster, and
- the number of problems in which QUBE is at least 1 order of magnitude slower

than QUBE(RND), QUBE(CBJ,SBJ) and QUBE(MLF⁻) is 48, 39 and 28 respectively. In the above statements, we used the phrase "seems to be" to stress once more that the numbers in the table have to be taken as indications of average behaviors because of the randomness of some of the solvers that we considered. Still, the fact that MLF increases capacity more than productivity, and that for the heuristic the situation is the opposite matches our intuition: Indeed, the bottleneck of QUBE and of all the solvers based on [14] is that the search tree is explored taking into account the prefix. Any look-ahead mechanism that, like MLF, overrides the "prefix-rule" may greatly improve the capacity of the solver, because it may allow to assign literals (i) that are fundamental for quickly deciding the formula (un)satisfiability, and (ii) that would have been assigned much later without it. Still, such look-ahead mechanisms have always an associated overhead, which may worsen the productivity. Other examples of look-ahead mechanisms having the above characteristics are the partial instantiation techniques described in [23]: Indeed, the results reported in [23] support our conclusions. The heuristic on the other hand, may still produce exponential speed-ups, but it does not address the main bottleneck of search algorithms based on [14]. Thus, the heuristic may improve performances and thus productivity, but we expect that in some domains many instances (that would be solvable with a proper look-ahead mechanism) will remain unsolvable no matter the heuristic being used.

References

1. C. Scholl and B. Becker. Checking equivalence for partial implementations. In *38th Design Automation Conference (DAC'01)*, 2001.
2. Abdelwaheb Ayari and David Basin. Bounded model construction for monadic second-order logics. In *12th International Conference on Computer-Aided Verification (CAV'00)*, number 1855 in LNCS, pages 99–113. Springer-Verlag, 2000.

3. Jussi Rintanen. Constructing conditional plans by a theorem prover. *Journal of Artificial Intelligence Research*, 10:323–352, 1999.

4. Claudio Castellini, Enrico Giunchiglia, and Armando Tacchella. Improvements to SAT-based conformant planning. In *Proc. ECP*, 2001.

5. Maher N. Mneimneh and Karem A. Sakallah. Computing vertex eccentricity in exponentially large graphs: QBF formulation and solution. In *Theory and Applications of Satisfiability Testing, 6th International Conference, (SAT)*, volume 2919 of *LNCS*, pages 411–425. Springer, 2004.

6. I. Gent, E. Giunchiglia, M. Narizzano, A. Rowley, and A. Tacchella. Watched data structures for QBF solvers. In Enrico Giunchiglia and Armando Tacchella, editors, *Theory and Applications of Satisfiability Testing, 6th International Conference, (SAT)*, volume 2919 of *LNCS*, pages 25–36. Springer, 2004.

7. E. Giunchiglia, M. Narizzano, and A. Tacchella. Monotone literals and learning in QBF reasoning. In *Tenth International Conference on Principles and Practice of Constraint Programming, CP 2004*, 2004.

8. E. Giunchiglia, M. Narizzano, and A. Tacchella. QUBE: an efficient QBF solver. In *5th International Conference on Formal Methods in Computer-Aided Design, FMCAD 2004*, 2004.

9. Enrico Giunchiglia, Massimo Narizzano, and Armando Tacchella. Learning for Quantified Boolean Logic Satisfiability. In *Proc. 18th National Conference on Artificial Intelligence (AAAI) (AAAI'2002)*, pages 649–654, 2002.

10. L. Zhang and S. Malik. Conflict driven learning in a quantified boolean satisfiability solver. In *Proceedings of International Conference on Computer Aided Design (ICCAD'02)*, 2002.

11. R. Letz. Lemma and model caching in decision procedures for quantified Boolean formulas. In *Proceedings of Tableaux 2002*, LNAI 2381, pages 160–175. Springer, 2002.

12. Fady Copty, Limor Fix, Enrico Giunchiglia, Gila Kamhi, Armando Tacchella, and Moshe Vardi. Benefits of bounded model checking at an industrial setting. In *Proc. 13th International Computer Aided Verification Conference (CAV)*, 2001.

13. D. Le Berre, L. Simon, and A. Tacchella. Challenges in the QBF arena: the SAT'03 evaluation of QBF solvers. In *Sixth International Conference on Theory and Applications of Satisfiability Testing (SAT 2003)*, volume 2919 of *LNCS*. Springer Verlag, 2003.

14. M. Cadoli, A. Giovanardi, and M. Schaerf. An algorithm to evaluate quantified boolean formulae. In *Proc. AAAI*, 1998.

15. Guoqiang Pan and Moshe Y. Vardi. Optimizing a BDD-based modal solver. In *Proceedings of the 19th International Conference on Automated Deduction*, 2003.

16. Matthew W. Moskewicz, Conor F. Madigan, Ying Zhao, Lintao Zhang, and Sharad Malik. Chaff: Engineering an Efficient SAT Solver. In *Proceedings of the 38th Design Automation Conference (DAC'01)*, June 2001.

17. J. P. Marques-Silva and K. A. Sakallah. GRASP - A New Search Algorithm for Satisfiability. In *Proceedings of IEEE/ACM International Conference on Computer-Aided Design*, pages 220–227, November 1996.

18. Roberto J. Bayardo, Jr. and Robert C. Schrag. Using CSP look-back techniques to solve real-world SAT instances. In *Proceedings of the 14th National Conference on Artificial Intelligence and 9th Innovative Applications of Artificial Intelligence Conference (AAAI-97/IAAI-97)*, pages 203–208, Menlo Park, July 27–31 1997. AAAI Press.

19. Ian P. Gent and Andrew G.D. Rowley. Solution learning and solution directed backjumping revisited. Technical Report APES-80-2004, APES Research Group, February 2004. Available from http://www.dcs.st-and.ac.uk/~apes/apesreports.html.

20. H. Kleine-Büning, M. Karpinski, and A. Flögel. Resolution for quantified Boolean formulas. *Information and computation*, 117(1):12–18, 1995.

21. E. Giunchiglia, M. Narizzano, and A. Tacchella. Clause-term resolution and learning in quantified Boolean logic satisfiability, 2004. Submitted.
22. Enrico Giunchiglia, Massimo Narizzano, and Armando Tacchella. Backjumping for Quantified Boolean Logic Satisfiability. *Artificial Intelligence*, 145:99–120, 2003.
23. Jussi Rintanen. Partial implicit unfolding in the Davis-Putnam procedure for Quantified Boolean Formulae. In *Proc. LPAR*, volume 2250 of *LNCS*, pages 362–376, 2001.

Automatic Extraction of Functional Dependencies

Éric Grégoire, Richard Ostrowski, Bertrand Mazure, and Lakhdar Saïs

CRIL CNRS & IRCICA – Université d'Artois,
rue Jean Souvraz SP-18, F-62307 Lens Cedex France
{gregoire, ostrowski, mazure, sais}@cril.univ-artois.fr

Abstract. In this paper, a new polynomial time technique for extracting functional dependencies in Boolean formulas is proposed. It makes an original use of the well-known Boolean constraint propagation technique (BCP) in a new preprocessing approach that extracts more hidden Boolean functions and dependent variables than previously published approaches on many classes of instances.

Keywords: SAT, Boolean function, propositional reasoning and search.

1 Introduction

Recent impressive progress in the practical resolution of hard and large SAT instances allows real-world problems that are encoded in propositional clausal normal form (CNF) to be addressed (see e.g. [13, 8, 20]). While there remains a strong competition about building more efficient provers dedicated to hard random k-SAT instances [7], there is also a real surge of interest in implementing powerful systems that solve difficult large real-world SAT problems. Many benchmarks have been proposed and regular competitions (e.g. [5, 2, 16, 17]) are organized around these specific SAT instances, which are expected to encode structural knowledge, at least to some extent.

Clearly, encoding knowledge under the form of a conjunction of propositional clauses can flatten some structural knowledge that would be more apparent in more expressive propositional logic representation formalisms, and that could prove useful in the resolution step [15, 10].

In this paper, a new pre-processing step is proposed in the resolution of SAT instances, that extracts and exploits some structural knowledge that is hidden in the CNF. The technique makes an original use of the well-known Boolean constraint propagation (BCP) process. Whereas BCP is traditionally used to produce implied and/or equivalent literals, in this paper it is shown how it can be extended so that it delivers an hybrid formula made of clauses together with a set of equations of the form $y = f(x_1, \ldots, x_n)$ where f is a standard connective operator among $\{\vee, \wedge\}$ and where y and x_i are Boolean variables of the initial SAT instance. These Boolean functions allow us to detect a subset of dependent variables, that can be exploited by SAT solvers.

H.H. Hoos and D.G. Mitchell (Eds.): SAT 2004, LNCS 3542, pp. 122–132, 2005.
© Springer-Verlag Berlin Heidelberg 2005

This paper extends in a significant way the preliminary results that were published in [14] in that it describes a technique that allows more dependent variables and hidden functional dependencies to be detected in several classes of instances. We shall see that the set of functional dependencies can underlie cycles. Unfortunately, highlighting actual dependent variables taking part in these cycles can be time-consuming since it coincides to the problem of finding a minimal cycle cutset of variables in a graph, which is a well-known NP-hard problem. Accordingly, efficient heuristics are explored to cut these cycles and deliver the so-called dependent variables.

The paper is organized as follows. After some preliminary definitions, Boolean gates and their properties are presented. It is then shown how more functional dependencies than [14] can be deduced from the CNF, using Boolean constraint propagation. Then, a technique allowing us to deliver a set of dependent variables is presented, allowing the search space to be reduced in an exponential way. Experimental results showing the interest of the proposed approach are provided. Finally, promising paths for future research are discussed in the conclusion.

2   Technical Preliminaries

Let $\mathcal{B}$ be a Boolean (i.e. propositional) language of formulas built in the standard way, using usual connectives ($\vee$, $\wedge$, $\neg$, $\Rightarrow$, $\Leftrightarrow$) and a set of propositional variables.

A *CNF formula* Σ is a set (interpreted as a conjunction) of *clauses*, where a clause is a set (interpreted as a disjunction) of *literals*. A literal is a positive or negated propositional variable. We note $\mathcal{V}(\Sigma)$ (resp. $\mathcal{L}(\Sigma)$) the set of variables (resp. literals) occurring in Σ. A *unit clause* is a clause formed with one unique literal. A *unit literal* is the unique literal of a unit clause.

In addition to these usual set-based notations, we define the negation of a set of literals ($\neg\{l_1, \ldots, l_n\}$) as the set of the corresponding opposite literals ($\{\neg l_1, \ldots, \neg l_n\}$).

An *interpretation* of a Boolean formula is an assignment of truth values $\{true, false\}$ to its variables. A *model* of a formula is an interpretation that satisfies the formula. Accordingly, SAT consists in finding a model of a CNF formula when such a model does exist or in proving that such a model does not exist.

Let c_1 be a clause containing a literal a and c_2 a clause containing the opposite literal $\neg a$, one *resolvent* of c_1 and c_2 is the disjunction of all literals of c_1 and c_2 less a and $\neg a$. A resolvent is called *tautological* when it contains opposite literals.

Let us recall here that any Boolean formula can be translated thanks to a linear time algorithm into CNF, equivalent with respect to SAT (but that can use additional propositional variables). Most satisfiability checking algorithms operate on clauses, where the structural knowledge of the initial formulas is thus flattened. In the following, CNF formulas will be represented as Boolean gates.

3   Boolean Gates

A *(Boolean) gate* is an expression of the form $y = f(x_1, \ldots, x_n)$, where f is a standard connective among $\{\vee, \wedge, \Leftrightarrow\}$ and where y and x_i are propositional literals, that is defined as follows :

- $y = \wedge(x_1, \ldots, x_n)$ represents the set of clauses $\{y \vee \neg x_1 \vee \ldots \vee \neg x_n, \neg y \vee x_1, \ldots, \neg y \vee x_n\}$, translating the requirement that the truth value of y is determined by the conjunction of the truth values of x_i s.t. $i \in [1..n]$;
- $y = \vee(x_1, \ldots, x_n)$ represents the set of clauses $\{\neg y \vee x_1 \vee \ldots \vee x_n, y \vee \neg x_1, \ldots, y \vee \neg x_n\}$;
- $y = \Leftrightarrow (x_1, \ldots, x_n)$ represents the following *equivalence chain* (also called *biconditional formula*) $y \Leftrightarrow x_1 \Leftrightarrow \ldots \Leftrightarrow x_n$, which is equivalent to 2^n clauses.

In the following, we consider gates of the form $y = f(x_1, \ldots, x_n)$ where y is a variable or the Boolean constant *true*, only.

Indeed, any clause can be represented as a gate of the form $true = \vee(x_1, \ldots, x_n)$. Moreover, a gate $\neg y = \wedge(x_1, \ldots, x_n)$ (resp. $\neg y = \vee(x_1, \ldots, x_n)$) is equivalent to $y = \vee(\neg x_1, \ldots, \neg x_n)$ (resp. $y = \wedge(\neg x_1, \ldots, \neg x_n)$). According to the well-known property of equivalence chain asserting that every equivalence chain with an odd (resp. even) number of negative literals is equivalent to the chain formed with the same literals, but all in positive (resp. except one) form, every gate of the form $y = \Leftrightarrow (x_1, \ldots, x_n)$ can always be rewritten into a gate where y is a positive literal. For example, $\neg y = \Leftrightarrow (\neg x_1, x_2, x_3)$ is equivalent to $y = \Leftrightarrow (x_1, x_2, x_3)$ and $\neg y = \Leftrightarrow (\neg x_1, x_2, \neg x_3)$ is equivalent to e.g. $y = \Leftrightarrow (x_1, x_2, \neg x_3)$.

A propositional variable y (resp. $x_1, \ldots, x_n$) is an *output variable* (resp. are *input variables*) of a gate of the form $y = f(x'_1, \ldots, x'_n)$, where $x'_i \in \{x_i, \neg x_i\}$.

A propositional variable z is an *output (dependent) variable of a set of gates* iff z is an output variable of at least one gate in the set. An *input (independent) variable of a set of gates* is an input variable of a gate which is not an output variable of the set of gates.

A gate is satisfied under a given Boolean interpretation iff the left and right hand sides of the gate are simultaneously *true* or *false* under this interpretation. An interpretation satisfies a set of gates iff each gate is satisfied under this interpretation. Such an interpretation is called a model of this set of gates.

4   From CNF to Gates

Practically, we want to find a representation of a CNF Σ using gates that highlights a *maximal* number of dependent variables, in order to decrease the actual computational complexity of checking the satisfiability of Σ. Actually, we shall describe a technique that extracts gates that can be deduced from Σ, and that thus *cover* a subset of clauses of Σ. Remaining clauses of Σ will be represented as or-gates of the form $true = \vee(x_1, \ldots, x_n)$, in order to get a uniform representation.

More formally, assume that a set G of gates whose corresponding clauses $Cl(G)$ are logical consequences of a CNF Σ, the set $\Sigma_{uncovered(G)}$ of uncovered clauses of Σ w.r.t. G is the set of clauses of $\Sigma \backslash Cl(G)$.

Accordingly, $\Sigma \equiv \Sigma_{uncovered(G)} \cup Cl(G)$.

Not trivially, we shall see that the additional clauses $Cl(G) \backslash \Sigma$ can play an important role in further steps of deduction or satisfiability checking.

Knowing output variables can play an important role in solving the consistency status of a CNF formula. Indeed, the truth-value of an y output variable of a gate depends on the truth value of the corresponding x_i input variables. The truth value of such output variables can be obtained by propagation, and they can be omitted by selection heuristics of DPLL-like algorithms [4]. In the general case, knowing n' output variables of a gate-oriented representation of a CNF formula using n variables allows the size of the set of interpretations to be investigated to decrease from 2^n to $2^{n-n'}$. Obviously, the reduction in the search space increases with the number of detected dependent variables.

Unfortunately, to obtain such a reduction in the search space, one might need to address the following difficulties:

– Extracting gates from a CNF formula can be a time-consuming process in the general case, unless some depth-limited search resources or heuristic criteria are provided. Indeed, showing that $y = f(x_1, \ldots, x_i)$ (where $y, x_1, \ldots, x_i$ belong to Σ) follows from a given CNF Σ, is coNP-complete.
– when the set of detected gates contains recursive definitions (like $y = f(x, t)$ and $x = g(y, z)$), assigning truth values to the set of independent variables is not sufficient to determine the truth values of all the dependent ones. Handling such recursive definitions coincides to the well-known NP-hard problem of finding a minimal cycle cutset in a graph.

In this paper, these two computationally-heavy problems are addressed. The first one by restricting deduction to Boolean constraint propagation, only. The second one by using graph-oriented heuristics.

Let us first recall some necessary definitions about Boolean constraint propagation.

5   Boolean Constraint Propagation (BCP)

Boolean constraint propagation or unit resolution, is one of the most used and useful lookahead algorithm for SAT.

Let Σ be a CNF formula, $BCP(\Sigma)$ is the CNF formula obtained by *propagating* all unit literals of Σ. Propagating a unit literal l of Σ consists in suppressing all clauses c of Σ such that $l \in c$ and replacing all clauses c' of Σ such that $\neg l \in c'$ by $c' \backslash \{\neg l\}$. The CNF obtained in such a way is equivalent to Σ with respect to satisfiability.

The *set of propagated unit literals* of Σ using BCP is noted $UP(\Sigma)$. Obviously, we have that $\Sigma \models UP(\Sigma)$. BCP is a restricted form of resolution, and can be performed in linear time. It is also complete for Horn formulas. In addition to its use in DPLL procedures, BCP is used in many SAT solvers as a processing step to deduce further interesting information such as implied [6] and equivalent literals [3][11]. Local processing based-BCP is also used to deliver promising branching variables (heuristic UP [12]).

In the sequel, it is shown that BCP can be further extended, allowing more general functional dependencies to be extracted.

6   BCP and Functional Dependencies

Actually, BCP can be used to detect hidden functional dependencies. The main result of the paper is the practical exploitation of the following original property: gates can be computed using BCP only, while checking whether a gate is a logical consequence of a CNF is coNP-complete in the general case.

Property 1. Let Σ be a CNF formula, $l \in \mathcal{L}(\Sigma)$, and $c \in \Sigma$ s.t. $l \in c$. If $c\backslash\{l\} \subset \neg UP(\Sigma \wedge l)$ then $\Sigma \vDash l = \wedge(\neg\{c\backslash\{l\}\})$.

Proof. Let $c = \{l, \neg l_1, \neg l_2, \ldots, \neg l_m\} \in \Sigma$ s.t. $c\backslash\{l\} = \{\neg l_1, \neg l_2, \ldots, \neg l_m\} \subset \neg UP(\Sigma \wedge l)$. The Boolean function $l = \wedge(\neg\{c\backslash\{l\}\})$ can be written as $l = \wedge(l_1, l_2, \ldots, l_m)$. To prove that $\Sigma \vDash l = \wedge(l_1, l_2, \ldots, l_m)$, we need to show that every model of Σ, is also a model of $l = \wedge(l_1, l_2, \ldots, l_m)$. Let I be a model of Σ, then

1. l is either *true* in I : I is also a model of $\Sigma \wedge l$. As $\{\neg l_1, \neg l_2, \ldots, \neg l_m\} \subset \neg UP(\Sigma \wedge l)$, we have $\{l_1, l_2, \ldots, l_m\} \subset UP(\Sigma \wedge l)$, then $\{l_1, l_2, \ldots, l_m\}$ are *true* in I. Consequently, I is also a model of $l = \wedge(l_1, l_2, \ldots, l_m\}\})$;
2. or l is *false* in I : as $c = \{l, \neg l_1, \neg l_2, \ldots, \neg l_m\} \in \Sigma$ then I satisfies $c = \{\neg l_1, \neg l_2, \ldots, \neg l_m\} \in \Sigma$. So, at least one of the literals $l_i, i \in \{1, \ldots, m\}$ is *true* in I. Consequently, I is also a model of $l = \wedge(l_1, l_2, \ldots, l_m\}\})$

Clearly, depending on the sign of the literal l, and-gates or or-gates can be detected. For example, the and-gate $\neg l = \wedge(l_1, l_2, \ldots, l_n)$ is equivalent to the or-gate $l = \vee(\neg l_1, \neg l_2, \ldots, \neg l_n)$. Let us also note that this property covers binary equivalence since $a = \wedge(b)$ is equivalent to $a \Leftrightarrow b$.

Actually, this property allows gates to be detected, which were not in the scope of the technique described in [14]. Let us illustrate this by means of an example.

Example 1. Let $\Sigma_1 \supseteq \{y \vee \neg x_1 \vee \neg x_2 \vee \neg x_3, \neg y \vee x_1, \neg y \vee x_2, \neg y \vee x_3\}$.

According to [14], Σ_1 can be represented by a graph where each vertex represents a clause and where each edge corresponds to the existence of tautological resolvent between the two corresponding clauses. Each connected component might be a gate. As we can see the first four clauses belong to a same connected component. This is a necessary condition for such a subset of clauses to represent a gate. Such a restricted subset of clauses (namely, those appearing in the same connected component) is then checked syntactically to determine if it represents an and/or gate. Such a property can be checked in polynomial time. In the above example, we thus have $y = \wedge(x_1, x_2, x_3)$.

Now, let us consider, the following example,

Example 2. $\Sigma_2 \supseteq \{y \vee \neg x_1 \vee \neg x_2 \vee \neg x_3, \neg y \vee x_1, \neg x_1 \vee x_4, \neg x_4 \vee x_2, \neg x_2 \vee x_5, \neg x_4 \vee \neg x_5 \vee x_3\}$.

Clearly, the graphical representation of this later example is different and the above technique does not help us in discovering the $y = \wedge(x_1, x_2, x_3)$ gate. Indeed, the above necessary but not sufficient condition is not satisfied.

Now, according to Property 1, both the and-gates behind Example 1 and Example 2 can be detected. Indeed, in example 1, $UP(\Sigma_1 \wedge y) = \{x_1, x_2, x_3\}$ and $\exists c \in \Sigma_1$, $c = (y \vee \neg x_1 \vee \neg x_2 \vee \neg x_3)$ such that $c \backslash \{y\} \subset \neg UP(\Sigma_1 \wedge y)$. Moreover, in example 2, $UP(\Sigma_2 \wedge y) = \{x_1, x_4, x_2, x_5, x_3\}$ and $\exists c' \in \Sigma_2$, $c' = (y \vee \neg x_1 \vee \neg x_2 \vee \neg x_3)$ such that $c' \backslash \{y\} \subset \neg UP(\Sigma_2 \wedge y)$.

Accordingly, a preprocessing technique to discover gates consists in checking the Property 1 for any literal occurring in Σ. A further step consists in finding dependent variables of the original formulas, as they can be recognised in the discovered gates. A gate clearly exhibits one dependent literal with respect to the inputs which are considered independent, as far as a single gate is considered. Now, when several gates share literals, such a characterisation of dependent variables does not apply anymore. Indeed, forms of cycle can occur as shown in the following example.

Example 3. $\Sigma_3 \supseteq \{x = \wedge(y, z), y = \vee(x, \neg t)\}$.

Clearly, Σ_3 contain a cycle. Indeed, x depends on the variables y and z, whereas y depends on the variables x and t. When a single gate is considered, assigning truth values to input variables determines the truth value of the output, dependent, variable. As in Example 3, assigning truth values to input variables that are not output variables for other gates is not enough to determine the truth value of all involved variables. In the example, assigning truth values to z and t is not sufficient to determine the truth value of x and y. However, in the example, when we assign a truth value to an additional variable (x, which is called a *cycle cutset variable*) in the cycle, the truth value of y is determined. Accordingly, we need to cut such a form of cycle in order to determinate a sufficient subset of variables that determines the values of all variables. Such a set is called a *strong backdoor* in [19]. In Example 3, the strong backdoor corresponds to the set of $\{x\} \cup \{z, t\}$. In this context, a strong backdoor is the union of the set of independent variables and of the variables of the cycle cutset. Finding the minimal set of variables that cuts all the cycles in the set of gates is an NP-hard problem. This issue is investigated in the next section.

7   Searching for Dependent Variables

In the following, a graph representation of the interaction of gates is considered. More formally,

A set of gates can be represented by a bipartite graph $G = (O \cup I, E)$ as follows:

 - for each gate we associate two vertices, the first one $o \in O$ represents the output of the gate, and the second one $i \in I$ represents the set of its input variables. So the number of vertex is less than $2 \times \#gates$, where $\#gates$ is the number of gates;

- For each gate, an edge (o, i) between the two vertices o and i representing the left and the right hand sides of a gate is created. Additional edges are created between $o \in O$ and $i \in I$ if one of the literals of the output variable associated to the vertex o belongs to the set of input literals associated to the vertex i.

Finding a *smallest* subset V' of O s.t. the subgraph $G' = (V' \cup I, E')$ is acyclic is a well-known NP-hard problem.

Actually, any subset V' that makes the graph acyclic is the representation of the set of variables, which together with all the independent ones, allows all variables to be determined. When V' is of size c, and the set of dependent variables is of size d, then the search space is reduced from 2^n to $2^{n-(d-c)}$, where n is the number of variable occurring in the original CNF formula.

We thus need to find a trade-off between the size of V', which influences the computational cost to find it, and the expected time gain in the subsequent SAT checking step.

In the following, two heuristics are investigated in order to find a cycle-cut set V'. The first-one is called *Maxdegree*. It consists in building V' incrementally by selecting vertices with the highest degree first, until the remaining subgraph becomes acyclic.

The second one is called *MaxdegreeCycle*. It consists in building V' incrementally by selecting first a vertex with the highest degree among the vertices that belong to a cycle. This heuristic guarantees that each time a vertex is selected, then at least one cycle is cut.

In the next section, extensive experimental results are presented and discussed, involving the preprocessing technique described above. It computes gates and cuts cycles when necessary in order to deliver a set of dependent variables. Two strategies are explored: in the first one, each time a gate is discovered, the covered clauses of Σ are suppressed; in the second one, covered clauses are eliminated at the end of the generation of gates, only. While the first one depends on the considered order of propagated literals, the second one is order-independent. These two strategies will be compared in terms of number of discovered gates, of the size of the cycle cutsets, of dependent variables and of the final uncovered clauses.

8   Experimental Results

Our preprocessing software is written in C under Linux Redhat 7.1 (available at : http://www.cril.univ-artois.fr/~ostrowski/Binaries/llsatpreproc). All experimentations have been conducted on Pentium IV, 2.4 Ghz. Description of the benchmarks can be found on SATLib (http://www.satlib.org).

We have applied both [14] and our proposed technique on all benchmarks from the last SAT competition [17, 18], covering e.g. model-checking, VLSI and planning instances. Complete results are available at : http://www.cril.univ-artois.fr/~ostrowski/result-llsatpreproc.ps.

Table 1. #G: Number of gates detected (average[standard deviation])

Family of Instances Name (#Inst.,#V[min-Max],#C[min-Max])	[14]'s technique #G	No cl. remov. #G	Cl. remov. #G	#C remov.
Blocks (3,484[283-758],27423[9690-47820])	10[3]	236[134]	18[5]	271[142]
Logistics (8,994[116-3016],12706[953-50457])	380[265]	437[417]	169[213]	630[585]
Pipe (6,1642[834-2577],18624[6695-33270])	1312[679]	1407[697]	1240[639]	13898[9083]
Facts (13,3178[2218-4315],48737[22539-90646])	713[147]	1601[541]	497[170]	1731[510]
Parity (30,1044[64-3176],3614[254-10325])	568[828]	510[594]	328[455]	663[870]
Qg (10,969[512-1331],33747[9685-64054])	310[91]	1828[652]	298[80]	1708[601]
Ca (7,637[26-2282],1835[70-6586])	419[547]	459[592]	414[542]	1233[1615]
Dp (11,1427[213-3193],3580[376-8308])	1117[856]	1468[1211]	915[812]	2534[2298]
Bmc2 (5,1952[316-4089],6908[1002-13531])	895[714]	1025[850]	744[623]	2082[1824]
Rand (6,2217[2000-2500],6568[5921-7401])	2133[236]	2444[381]	2103[252]	6212[692]
Ezfact (40,1441[193-3073],9169[1113-19785])	40[18]	268[127]	68[33]	68[33]
Med (3,761[341-1159],20154[5556-36291])	66[32]	316[162]	14[5]	319[164]
Avg-checker (4,917[648-1188],28661[17087-40441])	324[105]	1098[595]	304[101]	1092[373]
nw/nc/fw (13,3997[2756-5074],15829[10886-20123])	89[40]	468[136]	125[38]	125[38]
Am (4,2011[433-4264],6925[1458-14751])	989[835]	772[585]	393[276]	927[625]
Cnf (2,2424[2424-2424],14812[14812-14812])	2336[0]	3280[0]	2301[6]	13703[149]

Table 2. Size of backdoor with no remove option

Family of Instances (#V[min-Max])	Maxdregre #D	#CS	#B	MaxdegreeCycle #D	#CS	#B
Blocks (484[283-758])	38[13]	198[123]	353[215]	39[9]	197[124]	352[216]
Logistics (994[116-3016])	113[158]	245[218]	441[532]	143[164]	214[194]	410[522]
Pipe (1642[834-2577])	980[768]	265[219]	582[201]	764[449]	481[192]	798[348]
Facts (3178[2218-4315])	738[237]	813[256]	1964[604]	487[124]	1064[362]	2216[623]
Parity (1044[64-3176])	243[388]	84[46]	573[528]	287[410]	40[21]	528[505]
Qg (969[512-1331])	303[202]	228[236]	228[236]	11[6]	521[194]	521[194]
Ca (637[26-2282])	290[434]	130[142]	344[403]	265[341]	155[206]	369[481]
Dp (1427[213-3193])	513[463]	451[485]	725[625]	551[496]	412[343]	686[498]
Bmc2 (1952[316-4089])	662[716]	27[22]	886[874]	660[696]	30[10]	888[893]
Rand (2217[2000-2500])	1777[301]	357[339]	440[343]	1152[134]	981[111]	1064[115]
Ezfact (1441[193-3073])	28[35]	66[45]	1370[1073]	55[27]	39[18]	1343[1060]
Med (761[341-1159])	205[102]	110[72]	110[72]	14[4]	302[157]	302[157]
Avg-checker (917[648-1188])	209[357]	606[283]	606[283]	276[94]	539[187]	539[187]
nw/nc/fw (3997[2756-5074])	39[48]	151[47]	3899[854]	94[24]	96[23]	3844[855]
Am (2011[433-4264])	327[263]	97[68]	413[241]	298[206]	126[99]	441[287]
Cnf (2424[2424-2424])	472[564]	1801[564]	1953[564]	1170[2]	1103[2]	1255[2]

In the following, we illustrate some typical ones. On each class of instances, average and standard deviation results are provided with respect to the corresponding available instances.

In Table 1, for each considered class, the results of applying both [14]'s technique and the two new ones described above (in the first one, covered clauses are not suppressed as soon as they are discovered whereas they are suppressed in the second one) in terms of the mean number of discovered gates (#G). The results clearly show that our approach allows one to discover more gates. Not surprisingly, removing clauses causes the number of detected gates to decrease.

Table 3. Size of backdoor with remove option

Family of Instances	(#V[min-Max])	Maxdegree			MaxdegreeCycle		
		#D	#CS	#B	#D	#CS	#B
Blocks	(484[283-758])	18[4]	0[0]	373[219]	18[4]	0[0]	373[219]
Logistics	(994[116-3016]	135[147]	25[48]	419[539]	152[178]	7[13]	401[509]
Pipe	(1642[834-2577])	1020[735]	219[215]	543[223]	956[513]	282[124]	606[283]
Facts	(3178[2218-4315])	488[127]	0[0]	2214[621]	488[127]	0[0]	2214[621]
Parity	(1044[64-3176])	318[426]	0[0]	497[480]	318[426]	0[0]	497[480]
Qg	(969[512-1331])	122[99]	138[87]	410[189]	181[60]	80[25]	351[140]
Ca	(637[26-2282])	317[433]	94[113]	317[392]	302[388]	109[151]	332[434]
Dp	(1427[213-3193])	724[643]	149[151]	513[357]	728[641]	145[143]	509[353]
Bmc2	(1952[316-4089])	680[706]	1[1]	868[883]	680[705]	1[1]	868[884]
Rand	(2217[2000-2500])	1591[418]	495[396]	625[401]	1200[129]	886[102]	1016[111]
Ezfact	(1441[193-3073])	48[23]	10[5]	1350[1064]	49[23]	9[5]	1349[1064]
Med	(761[341-1159])	14[4]	0[0]	302[157]	14[4]	0[0]	302[157]
Avg-checker	(917[648-1188])	302[100]	0[0]	512[181]	302[100]	0[0]	512[181]
nw/nc/fw	(3997[2756-5074])	73[14]	40[22]	3864[857]	95[24]	18[10]	3842[856]
Am	(2011[433-4264])	367[254]	0[0]	373[239]	367[254]	0[0]	373[239]
Cnf	(2424[2424-2424])	1988[12]	285[12]	437[12]	2210[6]	63[6]	215[6]

In Table 2, we took the no-remove option. We explored the above two heuristics for cutting cycles (*Maxdregre* and *MaxdegreeCycle*). For each class of instances, we provide the average number of detected dependent variables ($\#D$), the size of the cycle cutsets ($\#CS$) and the size of the discovered backdoor ($\#B$), and the cumulated CPU time in seconds for discovering gates and computing these results. On some classes, the backdoor can be 10% of the number of variables, only.

In Table 3, the remove option was considered. The number of gates is often lower than with the no-remove option. On the other hand, the size of the cycle cutset is generally lower with the remove option.

Accordingly, no option is preferable to the other one in the general case. Indeed, finding a smaller backdoor depends both on the considered class of instances and the considered option.

However, in most cases, the remove option and the *MaxdegreeCycle* heuristic lead to smaller backdoors.

We are currently experimenting how such a promising preprocessing step can be grafted to the most efficient SAT solvers, allowing them to focus directly on the critical variables of the instances (i.e. the backdoor). Let us stress that our preprocessing step has been implemented in a non-optimized way. However, it takes less than 1 second in most cases. So time is omitted in different tables.

9   Future Works

Let us here simply motivate another interesting path for future research, related to the actual expressiveness of discovered clauses. Actually, our gate-oriented representation of a Boolean formula exhibits additional information that can prove powerful with respect to further steps of deduction or satisfiability checking. To illustrate this, let us consider Example 2 again. From the CNF Σ, the gate $y = \wedge(x_1, x_2, x_3)$ is extracted. The clausal representation of the gate is given by $\{y \vee \neg x_1 \vee \neg x_2 \vee \neg x_3, \neg y \vee x_1, \neg y \vee x_2, \neg y \vee x_3\}$.

Clearly, the additional clauses $\{\neg y \vee x_2, \neg y \vee x_3\}$ are resolvents from Σ, which can only be obtained using two and six basic steps of resolution, respectively. Accordingly, the gate representation of Σ involves non-trivial binary resolvents, which can ease further deduction or satisfiability checking steps. Taking this feature into account either in clausal-based or gate-based deduction of satisfiability solvers should be a promising path for future research. Also, some of the discovered gates represent equivalencies $(x \Leftrightarrow y)$, substituting equivalent literals might lead to further reductions with respect to the number of variables.

Another interesting path for future research concerns the analysis of the obtained graph and the use of e.g. decomposition techniques. To further reduce the size of the backdoor, we also plan to study how tractable parts of the formula (e.g. horn or horn-renommable) can be exploited.

10   Conclusions

Clearly, our experimental results are encouraging. Dependent variables can be detected in a preprocessing step at a very low cost. Cycles occur, and they can be cut. We are currently grafting such a preprocessing technique to efficient SAT solvers. Our preliminary experimentations show that this proves often beneficial. Moreover, we believe that the study of cycles and of dependent variables can be essential in the understanding of the difficulty of hard SAT instances.

Acknowledgements

This work has been supported in part by the CNRS, the FEDER, the *IUT de Lens* and the *Conseil Régional du Nord/Pas-de-Calais*. We thank the reviewers for valuable comments on a previous version of this paper.

References

1. F. Bacchus and J. Winter. Effective preprocessing with hyper-resolution and equality reduction. In *Sixth International Symposium on Theory and Applications of Satisfiability Testing (SAT'03)*, 2003.
2. First international competition and symposium on satisfiability testing, March 1996. Beijing (China).
3. L. Brisoux, L. Sais, and E. Grégoire. Recherche locale : vers une exploitation des propriétés structurelles. In *Actes des Sixièmes Journées Nationales sur la Résolution Pratique des Problèmes NP-Complets(JNPC'00)*, pages 243–244, Marseille, 2000.
4. Martin Davis, George Logemann, and Donald Loveland. A machine program for theorem proving. *Journal of the Association for Computing Machinery*, 5:394–397, 1962.

5. Second Challenge on Satisfiability Testing organized by the Center for Discrete Mathematics and Computer Science of Rutgers University, 1993. http://dimacs.rutgers.edu/Challenges/.

6. Olivier Dubois, Pascal André, Yacine Boufkhad, and Jacques Carlier. Sat versus unsat. In D.S. Johnson and M.A. Trick, editors, *Second DIMACS Challenge*, DIMACS Series in Discrete Mathematics and Theoretical Computer Science, American Mathematical Society, pages 415–436, 1996.

7. Olivier Dubois and Gilles Dequen. A backbone-search heuristic for efficient solving of hard 3–sat formulae. In *Proceedings of the Seventeenth International Joint Conference on Artificial Intelligence (IJCAI'01)*, volume 1, pages 248–253, Seattle, Washington (USA), August 4–10 2001.

8. E. Giunchiglia, M. Maratea, A. Tacchella, and D. Zambonin. Evaluating search heuristics and optimization techniques in propositional satisfiability. In *Proceedings of International Joint Conference on Automated Reasoning (IJCAR'01)*, Siena, June 2001.

9. Matti Järvisalo, Tommi Junttila, and Ilkka Niemelä. Unrestricted vs restricted cut in a tableau method for Boolean circuits. In *AI&M 2004, 8th International Symposium on Artificial Intelligence and Mathematics*, Fort Lauderdale, Florida, USA, January 4–6 2004.

10. Henry A. Kautz, David McAllester, and Bart Selman. Exploiting variable dependency in local search. In *Abstract appears in "Abstracts of the Poster Sessions of IJCAI-97"*, Nagoya (Japan), 1997.

11. Daniel Le Berre. Exploiting the real power of unit propagation lookahead. In *Proceedings of the Workshop on Theory and Applications of Satisfiability Testing (SAT2001)*, Boston University, Massachusetts, USA, June 14th-15th 2001.

12. Chu Min Li and Anbulagan. Heuristics based on unit propagation for satisfiability problems. In *Proceedings of the Fifteenth International Joint Conference on Artificial Intelligence (IJCAI'97)*, pages 366–371, Nagoya (Japan), August 1997.

13. Shtrichman Oler. Tuning sat checkers for bounded model checking. In *Proceedings of Computer Aided Verification (CAV'00)*, 2000.

14. Grégoire E. Mazure B. Ostrowski R. and Sais L. Recovering and exploiting structural knowledge from cnf formulas. In *Eighth International Conference on Principles and Practice of Constraint Programming (CP'2002)*, pages 185–199, Ithaca (N.Y.), 2002. LNCS 2470, Springer Verlag.

15. Antoine Rauzy, Lakhdar Saïs, and Laure Brisoux. Calcul propositionnel : vers une extension du formalisme. In *Actes des Cinquièmes Journées Nationales sur la Résolution Pratique de Problèmes NP-complets (JNPC'99)*, pages 189–198, Lyon, 1999.

16. Sat 2001: Workshop on theory and applications of satisfiability testing, 2001. http://www.cs.washington.edu/homes/kautz/sat2001/.

17. Sat 2002 : Fifth international symposium on theory and applications of satisfiability testing, May 2002. http://gauss.ececs.uc.edu/Conferences/SAT2002/.

18. Sat 2003 : Sixth international symposium on theory and applications of satisfiability testing, May 2003. http://www.mrg.dist.unige.it/events/sat03/.

19. Ryan Williams, Carla P. Gomez, and Bart Selman. Backdoors to typical case complexity. In *Proceedings of the Eighteenth International Joint Conference on Artificial Intelligence (IJCAI'03)*, pages 1173–1178, 2003.

20. L. Zhang, C. Madigan, M. Moskewicz, and S. Malik. Efficient conflict driven learning in a boolean satisfiability solver. In *Proceedings of ICCAD'2001*, pages 279–285, San Jose, CA (USA), November 2001.

Algorithms for Satisfiability Using Independent Sets of Variables

Ravi Gummadi[1], N.S. Narayanaswamy[1,*], and R. Venkatakrishnan[2]

[1] Department of Computer Science and Engineering,
Indian Institute of Technology Madras, Chennai-600036, India
gravi@peacock.iitm.ernet.in, swamy@shiva.iitm.ernet.in
[2] Department of Information Technology,
Crescent Engineering College, Vandalur, Chennai-600048, India
coolvenk@sancharnet.in

Abstract. An *independent set* of variables is one in which no two variables occur in the same clause in a given instance of k-SAT. Instances of k-SAT with an independent set of size i can be solved in time, within a polynomial factor of 2^{n-i}. In this paper, we present an algorithm for k-SAT based on a modification of the *Satisfiability Coding Lemma*. Our algorithm runs within a polynomial factor of $2^{(n-i)(1-\frac{1}{2k-2})}$, where i is the size of an independent set. We also present a variant of Schöning's randomized local-search algorithm for k-SAT that runs in time which is with in a polynomial factor of $(\frac{2k-3}{k-1})^{n-i}$.

1 Introduction

The Propositional Satisfiability Problem (SAT) is one of significant theoretical and practical interest. Historically, SAT was the first problem to be proven $\mathcal{NP}$-complete. No polynomial-time algorithm for a k-SAT problem ($k \geq 3$) is known, and no proof of its non-existence has been proposed, leaving open the question of whether $\mathcal{P} = \mathcal{NP}$?. The Satisfiability problem has important practical applications. For instance, in circuit design problems, a circuit that always produces an output of 0, can be eliminated from a larger circuit. This would reduce the number of gates needed to implement the circuit, thereby reducing cost. This problem naturally motivates the question of whether a given formula is satisfiable. Further, all the problems in the class $\mathcal{NP}$ can be reduced in polynomial-time to the Satisfiability problem. There are many practically important problems in this class. Therefore, a fast algorithm for SAT can also help to solve these problems efficiently. However, the existence of polynomial-time algorithms for $\mathcal{NP}$-complete problems is believed to be unlikely. Consequently, current research on SAT is focused on obtaining non-trivial exponential upper-bounds for SAT algorithms. For example, an algorithm running in $O(2^{n/r})$ for

---

* Partly Supported by DFG Grant No. Jo 291/2-1. Part of Work done when the author was at the Institut Für Informatik, Oettingenstrasse 67, 80538, Munich, Germany.

H.H. Hoos and D.G. Mitchell (Eds.): SAT 2004, LNCS 3542, pp. 133–144, 2005.
© Springer-Verlag Berlin Heidelberg 2005

instance, with large r could prove useful in solving many practical problems. Current research on SAT is focused on obtaining non-trivial exponential upper-bounds for SAT algorithms.

Algorithms for SAT. SAT algorithms are classified into *Splitting algorithms* and *Local Search algorithms* [DHIV01]. *Splitting algorithms* reduce the input formula into polynomially many formulae. The two families of Splitting algorithms are DPLL-like algorithms and PPSZ-like algorithms. *DPLL-like algorithms* [DP60, DLL62] replace the input formula F by two formulas $F[x]$ and $F[\neg x]$. This is done recursively. Using this technique, Monien and Speckenmeyer [MS85] gave the first non-trivial upper bounds for k-SAT. *PPSZ-like algorithms* [PPZ97, PPSZ98] use a different approach: variables are assigned values in a random order in the hope that the value of many variables can be obtained from the values of variables chosen prior to it. *Local Search algorithms* work by starting with an initial assignment and modifying it to come closer to a satisfying assignment. If, after a certain number of steps no satisfying assignment is found, the algorithm starts again with a new initial assignment. After repeating a certain number of times, if the algorithm does not find a satisfying assignment, it halts reporting "Unsatisfiable". *Greedy algorithms* [KP92] may be used to modify the current assignment in Local Search algorithms. These algorithms change the current assignment such that some function of the assignment increases as much as possible. *Random walk algorithms* [Pap91], on the other hand, modify the current assignment by flipping the value of a variable at random from an unsatisfied clause. 2-SAT can be solved in expected polynomial-time by a random walk algorithm [Pap91]. [Sch99] shows that k-SAT can be solved in time $(2 - 2/k)^n$ up to a polynomial factor. From the literature [Sch99, PPZ97, PPSZ98], it is clear that the current asymptotic performance of the local search algorithms is better than PPSZ-like algorithms.

Our Work and Results. Our main motivation is to explore further directions towards improving the performance of PPSZ-like algorithms. While the algorithm in [PPZ97] computes the values of variables in a random order, in the process shrinking the search space based on the formula, we observe that variable sets which have a special property with respect to the formula naturally shrink the search space. For example, if I is a set of variables in which no two of them occur in a clause, then the values to the variables of I can be computed very easily given an assignment to the variables outside I. Consequently, we could spend our effort on trying to find an assignment to variables outside I that can be extended to variables of I to obtain a satisfying assignment for the formula. This is precisely the approach of this paper. We first consider the brute force algorithm, and then modify the Satisfiability Coding Lemma, and Schöning's randomized algorithm to obtain an assignment to variables outside I. While we have not obtained an improved algorithm in general, we observe that our algorithms guarantee to be faster in the case when I is large enough in a given formula. On the other hand, it is also quite easy to construct formulae in which I is very small in which case the performance is the same as the algorithm in [PPZ97]. So the motivation for our work is our conjecture that random satisfiable formulae have large independent sets. This is reiterated by the benchmarks for satsolvers which have independent sets of size $\frac{n}{4}$.

Independent set like structures in the formula have been used to obtain better algorithms for 3-sat. In particular, the paper by [SSWH02] uses a set of disjoint clauses to identify the initial starting point of Schöning's randomized algorithm [Sch99]. Indeed the disjoint clauses form an independent set in the set of clauses. On the other hand, we use independent sets of variables in our attempt to improve the performance of algorithms for k-sat based on the Satisfiability Coding Lemma [PPZ97, PPSZ98].

Roadmap. Section 2 presents the preliminaries, the brute force algorithm in Section 2.1. The modified Satisfiability Coding Lemma is presented in Section 3, and the algorithm based on it is presented and analyzed in Section 4. The random walk algorithm is presented in Section 5. A discussion in Section 6, and a construction of formulae with small independent sets in Section 6.1 concludes the paper.

2   Preliminaries

We have used the usual notions associated with the k-SAT problem. The reader is referred to [DHIV01] for this. Let V denote the set of variables in an instance F of k-SAT. An *Independent Set* $I \subseteq V$, is a set of variables such that each clause contains at most one element of the set. In this paper, we consider independent sets that are maximal with respect to inclusion. I denotes a fixed maximal independent set of cardinality i in F. Given an assignment a' to the variables of $V - I$, we can check whether it can be extended to a satisfying assignment in polynomial time: when we substitute a' into the formula, then we get a conjunction of literals. Every variable in this conjunction is an element of I. Further, testing if a conjunction of literals is satisfiable is a trivial issue. A truth assignment to the variables of $V - I$ is said to be *extensible* if there is a truth assignment to the elements of I such that the resulting assignment to $\{x_1, \ldots, x_n\}$ is a satisfying assignment. An assignment that cannot be extended to a satisfying assignment is called a *non-extensible* assignment. An extensible assignment is said to be *isolated* along a direction $j, x_j \notin I$, if flipping the value of x_j results in a non-extensible assignment.

Isolated Extensible Assignments. For a truth assignment a, a_i is said to be *critical* for a clause C if the corresponding literal is the only true literal in C under the assignment a. Without loss of generality, let us consider the variables of $V - I$ to have indices from $[n - i] = \{1, \ldots n - i\}$. Further, for an assignment a' to the variables outside I, $F(a')$ is a conjunction of literals from I. Let $b = b_1 \ldots b_{n-i}$ be an extensible assignment that is isolated along directions indexed by the elements of $J \subseteq [n - i]$. Let b' be the assignment obtained by flipping $b_r, r \in J$. b' is non-extensible for one of the following two reasons:

1. The formula is falsified by b' as there is a clause with all its variables from $V - I$, and b_r is critical for this clause. An assignment is said to be *easy isolated* along x_r if this property is satisfied.
2. There exists an $x_l \in I$, two clauses C_1, C_2 such that $x_l \in C_1$, $x_r, \neg x_l \in C_2$, and only x_l occurs in $F(b)$, but both x_l and $\neg x_l$ occurs in $F(b')$. An assignment that is not easy isolated along x_r is said to be *hard isolated* along x_r if this condition is satisfied. We refer to the two clauses C_1 and C_2 as *falsifying clauses* for b along direction r. We refer to them as falsifying clauses, leaving out b and r, when there is no ambiguity. Clearly, if an extensible assignment is hard isolated along x_r, there exist two falsifying clauses.

2.1   The Brute Force Approach

The idea is to find the largest independent set I, and search through all possible assignments to $V - I$. If an assignment is extensible, we report that F is satisfiable, otherwise report unsatisfiable when all assignments to $V - I$ have been tried. This algorithm runs in $O(2^{n-i} poly(|F|))$. While finding a large enough independent set is a problem by itself, we propose to find the maximum independent set by using the algorithm due to Beigel [Bei99] that runs quite efficiently. The other approach is to permute the variables at random and consider the independent set obtained by considering variables all of whose neighbours occur to their left in the random permutation. Two variables are said to be neighbours, if they occur together in a clause. This approach yields an independent set whose expected size is $\frac{n}{\Delta+1}$, where each variable has at most Δ neighbours.

3   A Variant of Satisfiability Coding Lemma

In the Section 2.1 we have observed a simple brute force algorithm that finds extensible solutions given an independent set I. We now improve this brute force algorithm by modifying the satisfiability coding lemma suitably. The approach that we take is similar to the approach in [PPZ97]. We first consider the issue of encoding isolated extensible solutions and bound the expected length of an encoding. We then show that this encoding process is reversible and it does prune our search space yielding a randomized algorithm that performs better than the brute force approach in Section 2.1. However, this does not better the performance of [PPZ97] unless I is a sufficiently large set.

Encoding. We consider the set of j-isolated extensible solutions for a fixed independent set of variables I. Let $x_1, \ldots, x_{n-i}$ be the variables of $V - I$ in a k-SAT formula. Let σ be a permutation on the set $\{1, \ldots, n-i\}$. Let $A = a_1 \ldots a_{n-i}$ be a binary string visualized as an assignment of a_r to x_r, $1 \leq r \leq n-i$. Let A_σ denote the string $a_{\sigma(n-i)} a_{\sigma(n-i-1)} \cdots a_{\sigma(1)}$. In other words, A_σ is a permutation of A, according to σ. From A and σ, we construct an encoding $E(A, \sigma)$ as follows:

$E(A, \sigma)$ is the empty string.
```
for(r = n − i; r ≥ 1; r − −)
begin
        if  A is isolated along  σ(r)
                AND all other variables in a critical clause for  x_{σ(r)}
                        occur to the left of  x_{σ(r)} in  A_σ
                OR the variables ∉ I in two falsifying clauses occur to
                        the left of x_{σ(r)} in  A_σ
                then do not add  a_{σ(r)} to  E(A, σ).
        else
                add a_{σ(r)} to  E(A, σ).
end
```

The operation of adding a bit to $E(A, \sigma)$ is equivalent to concatenating to the right end. The bits of this string are assumed to be indexed from left to right starting with 1 for the leftmost bit, and using consecutive indices. The output of the loop is $E(A, \sigma)$. Another point of view on $E(A, \sigma)$ is that it is obtained from A_σ by deleting some bits which can be computed from *previous* information in A_σ. Obviously, its length cannot exceed $n - i$.

Reconstruction. Given a k-SAT formula F, an independent set I, a bit string E, and a permutation σ, we find a bit string A such that $E(A, \sigma) = E$, if such an A exists. To obtain A we find $A_\sigma = a_{\sigma(n-i)} a_{\sigma(n-i-1)} \cdots a_{\sigma(1)}$. The bit string E is considered from the leftmost bit. Each bit of E is assigned to at most one corresponding bit of A_σ. At each step the value of a bit of A_σ is inferred. It is inferred either by substituting the previously computed bits into the formula, or the current bit of E is assigned to A_σ.

Consider the case when A_σ has been computed up to the $r+1$-th bit, $n-i-1 \geq r \geq 1$. We substitute this partial assignment into F and consider the resulting formula. There are three cases:

$x_{\sigma(r)}$ *can be inferred from the previous values:* This can happen in two ways. The first, when $x_{\sigma(r)}$ occurs as a single literal. This means that there is a corresponding critical clause in which all other literals have been set to 0. $x_{\sigma(r)}$ is set appropriately to make the literal true. The second case is when a variable $x \in I$ occurs as $(x)(\neg x \vee y)$, where y is a literal of $x_{\sigma(r)}$. In this case, the value assigned to $x_{\sigma(r)}$ is inferred from the value that makes y true.

$x_{\sigma(r)}$ *takes its value from E:* This happens when $x_{\sigma(r)}$ does not satisfy either of the two conditions mentioned above. In this case, $x_{\sigma(r)}$ is to be assigned the current bit of E. If all the bits of E have been *used up* then halt reporting failure. At each step of the reconstruction, we keep track of whether a variable and its complement occur as single clauses. If this happens, we halt reporting failure. If A_σ is computed successfully, then it means that we have found an extensible assignment.

3.1   Quality of the Encoding

Here we discuss the expected length of $E(A, \sigma)$ when σ is chosen from a class of distributions on S_n, the set of permutations of $\{1, \ldots, n\}$. These distributions are characterized by γ and satisfy the following property

$$| Pr_{\pi \in \mathcal{F}}(\min\{\pi(X)\} = \pi(x)) - \frac{1}{|X|} | \leq \frac{\gamma}{|X|} \tag{1}$$

Here, $X \subseteq \{1, \ldots, n\}$ and $\pi(X)$ is the image of the set X under a permutation π. Clearly, the required probability is $\frac{1}{|X|}$ when π is chosen uniformly from the set of all permutations. The goal of identifying a smaller family of permutations that guarantee this property is well motivated and is studied by Charikar et. al in [MBFM00]. For each γ, D_γ is a probability distribution on S_n and $D_\gamma(\sigma)$ denotes probability of choosing σ in the distribution D_γ.

σ **Chosen from** D_γ. We now compute the average length of $E(A, \sigma)$ averaged over all $\sigma \in D_\gamma$. Clearly, the only directions that get eliminated are those along which A is either easy isolated or hard isolated. Let us assume that A is an extensible solution, easy isolated along j_e directions, and hard isolated along j_h directions. For a direction r along which A is easy isolated, we lower bound the probability that a_r is eliminated in the encoding of A with a randomly chosen permutation.

Since A is easy isolated along r, there is a corresponding critical clause all of whose variables are from $V - I$. a_r will be eliminated if all the $k - 1$ literals in the critical clause occur to the left in a randomly chosen permutation. This event happens with probability at least $\frac{1-\gamma}{k}$. It follows from the linearity of expectation that, for an A which is easy isolated along j_e directions, the expected number of variables eliminated is at least $\frac{j_e(1-\gamma)}{k}$. Similarly a direction r, along which A is hard isolated, will be eliminated if all the variables belonging to $V - I$ from corresponding falsifying clauses occur to the left of x_r in a randomly chosen permutation. The number of such variables from two falsifying clauses is at most $2k - 3$. Consequently, this event happens with probability at least $\frac{1-\gamma}{2k-2}$. By linearity of expectation, the expected number of hard isolated directions that get eliminated is at least $\frac{j_h(1-\gamma)}{2k-2}$. Therefore, the expected value of $E(A, \sigma)$ is at most $n - i - (1 - \gamma)(\frac{j_e}{k} + \frac{j_h}{2k-2})$.

Existence of a Good Permutation. We now use the above argument to show that there is a permutation $\sigma \in D_\gamma$ for which the average length $E(A, \sigma)$, over all extensible solutions A isolated along $j = j_e + j_h$ directions, is upper bounded by $n - i - (1 - \gamma)(\frac{j_e}{k} + \frac{j_h}{2k-2})$. For this we consider the following average,

$$\sum_\sigma D_\gamma(\sigma) \sum_{A \in J} \frac{1}{|J|} E(A, \sigma) = \sum_{A \in J} \frac{1}{|J|} \sum_\sigma D_\gamma(\sigma) E(A, \sigma) \tag{2}$$

This is upper bounded by $n - i - (1 - \gamma)(\frac{j_e}{k} + \frac{j_h}{2k-2})$ since we know from the above calculation that $\sum_\sigma D_\gamma(\sigma) E(A, \sigma) \leq n - i - (1 - \gamma)(\frac{j_e}{k} + \frac{j_h}{2k-2})$.

It now follows by the pigeon hole principle that for some σ, $\sum_{A \in J} \frac{1}{|J|} E(A, \sigma) \leq n - i - (1 - \gamma)(\frac{j_e}{k} + \frac{j_h}{2k-2})$. We state these bounds in the following theorem.

Theorem 1. *Let A be an extensible solution which is easy isolated along j_e directions, and hard isolated along j_h directions. The expected value of $E(A, \sigma)$ is at most $n - i - (1 - \gamma)(\frac{j_e}{k} + \frac{j_h}{2k-2})$. Consequently, for J, the set of extensible solutions, easy isolated along j directions,there is a permutation $\sigma \in D_\gamma$ such that $\sum_{A \in J} \frac{1}{|J|} E(A, \sigma) \leq n - i - (1 - \gamma)(\frac{j_e}{k} + \frac{j_h}{2k-2})$. Here i is the size of an independent set I.*

4   Algorithm to Find Satisfying Assignments

For a k-CNF $|F|$ with an independent set I, we use the result in Theorem 1 to design a randomized algorithm. Further, we set $\gamma = 0$, that is we use a family of permutations that guarantees exact min-wise independence. From now on, $\gamma = 0$. The effectiveness of this algorithm over the one presented in [PPZ97] depends on how large an independent set there is in the formula, and how much time is needed to find a reasonably large independent set. The algorithm that we present here, is quite similar to the randomized algorithm presented in [PPZ97]. In the description below, the word *forced* is a property of a variable whose value is determined by the values the previous variables. For example, a variable x_r is forced if it occurs as a single literal in $F(a_1, \ldots, a_{r-1})$. Here $a_1, \ldots, a_{r-1}$ are the assignments to the variables $x_1, \ldots, x_{r-1}$, respectively. x_r could also be forced if two falsifying clauses occur in $F(a_1, \ldots, a_{r-1})$.

```
Find an independent set I
Repeat n²2^(n-i)(1-1/(2k-2)) times
      While there is an unassigned variable in V - I
            select an unassigned variable y from V - I at random
            If y is forced, then set y as per the forcing
            Else set y to true or false at random
      end while
  If the assignment can be extended
    then output the satisfying assignment
End Repeat
```

We state the following lemma, a special case of the isoperimetric inequality, which is used to prove our main theorem. We present a complete proof here.

Lemma 1. *Let $S \subseteq \{0,1\}^n$, be a non-empty set. For $x \in S$, define $I_n(x)$ be the number of distance-1 neighbours of x that are not in S. Define $value(x) = 2^{(I_n(x)-n)}$. Then, $\sum_{x \in S} value(x) \geq 1$.*

Proof. The proof is by induction on n. The base case is when $n = 1$. If $I_1(x) = 0$, then we observe that $\sum_{x \in S} value(x) = 1$. If $I_1(x) = 1$, $\sum_{x \in S} value(x) = 1$.

For $n > 1$, and $i \in \{0, 1\}$, let S_i be a subset of $\{0, 1\}^{n-1}$ such that for each $x \in S_i, xi \in S$. Now we consider two cases:

Case I: If one of the two sets is empty, then we have a direction along which each element of S is isolated. Let us consider S' to be a subset of $\{0, 1\}^{n-1}$ obtained by projecting along the rightmost bit. By induction, $\Sigma_{x \in S'} value(x) \geq 1$. That is, $\Sigma_{x \in S'} 2^{I_{n-1}(x)-(n-1)} \geq 1$. Clearly, the number of directions along which an $x \in S$ is isolated in $\{0, 1\}^n$ is one more than the number of directions along which it's projection(along the rightmost bit) is isolated in $\{0, 1\}^{n-1}$. Consequently, $\Sigma_{x \in S'} 2^{I_{n-1}(x)+1-(n-1)}$ is exactly $\Sigma_{x \in S} 2^{I_n(x)-n}$. Hence the induction hypothesis is proved in this case.

Case II: If both S_i are non-empty. Then, by induction, $\Sigma_{x \in S_i} value(x) \geq 1$. Observe that, here $value(x)$ is defined with respect to S_i, for each i. Due of the induction hypothesis,

$$
\begin{aligned}
2 &\leq \Sigma_{x \in S_0} 2^{I_{n-1}(x)-(n-1)} + \Sigma_{x \in S_1} 2^{I_{n-1}(x)-(n-1)} \\
&\leq 2\Sigma_{x \in S_0} 2^{I_n(x0)-n} + 2\Sigma_{x \in S_1} 2^{I_n(x1)-n} \\
&= 2\Sigma_{x \in S} 2^{I_n(x)-n}
\end{aligned}
\tag{3}
$$

The equation 3 follows from the previous equation due to the fact that $I_{n-1}(x) \leq I_n(xi), i \in \{0, 1\}$. The induction hypothesis holds in this case too, and hence the lemma is proved.   □

The following theorem is proved using the Lemma 1 along the lines of a similar theorem in [PPZ97].

Theorem 2. *Let I be an independent set of variables in F, a satisfiable instance of k-SAT. The randomized algorithm in Section 4 finds a satisfying assignment with very high probability in time $O(n^2|F|2^{(n-i)(1-\frac{1}{2k-2})})$.*

Proof. Let S denote the set of extensible assignments. Let us assume that x is a j-isolated extensible solution of F. Among these let $j_e(x)$ and $j_h(x)$ be easy and hard isolated directions, respectively. The probability that x is output by the algorithm is the sum over all $d \geq 0$, probability that for a randomly chosen permutation, d directions are forced, and the remaining directions are chosen correctly. This is at least the probability that for a randomly chosen permutation, at least $\frac{j_e}{k} + \frac{j_h}{2k-2}$ directions are forced, and the remaining directions are guessed correctly. Recall that $\frac{j_e}{k} + \frac{j_h}{2k-2}$ is a lower bound expected number of directions that are eliminated by the process of encoding x. The probability of finding x is dependent on two events, one is to find a permutation that eliminates $\frac{j_e}{k} + \frac{j_h}{2k-2}$ directions, and the second is to make the correct choices on the remaining values. We now lower bound this probability by estimating the probability of finding a right permutation, and then conditioned on this event, estimate the probability of making the correct choices.

Probability of Finding a Right Permutation. Recall that a right permutation is one using which the process of encoding x eliminates at least $\frac{j_e}{k} + \frac{j_h}{2k-2}$ directions. We can now partition the permutation into the following sets: for $r < \frac{j_e}{k} + \frac{j_h}{2k-2}$, P_r consists of those permutation that eliminate r variables, and P_{av} consists of those permutations that eliminate at least $\frac{j_e}{k} + \frac{j_h}{2k-2}$ variables. The number of sets in this partition is at most $n - i$. Therefore, by the pigeon hole principle, one of these sets must have at least $\frac{1}{n-i}$ of the permutations. Following the argument in [PPZ97], P_{av} has at least $\frac{1}{n-i}$ of the permutations. Therefore, the probability of picking the right permutation is at least $\frac{1}{n-i} > \frac{1}{n}$.

Probability of Making the Right Choices on the Unforced Bits. Conditioned on the fact that a right permutation is chosen, we now estimate the probability that the right choices are made on the unforced bits so that we get x. The number of unforced bits is at most $n - i - \frac{j_e}{k} - \frac{j_h}{2k-2}$. The probability of making the correct choices is at least $2^{-(n-i-\frac{j_e}{k}-\frac{j_h}{2k-2})}$.

Therefore, the probability of picking x is at least $\frac{1}{n}2^{-(n-i-\frac{j_e}{k}-\frac{j_h}{2k-2})}$. The probability that the algorithm outputs some solution of F is given by the following:

$$\Sigma_{x \in S} \Pr(x \text{ is output}) \geq \Sigma_{x \in S} \frac{1}{n}2^{-(n-i-\frac{j_e(x)}{k}-\frac{j_h(x)}{2k-2})}$$

$$\geq \frac{1}{n}2^{-(n-i)(1-\frac{1}{2k-2})}\Sigma_{x \in S}2^{-(n-i)+I(x)}$$

$$\geq \frac{1}{n}2^{-(n-i)(1-\frac{1}{2k-2})} \tag{4}$$

The last inequality follows from Lemma 1. The repetition of the **while** loop $n^2 2^{(n-i)(1-\frac{1}{2k-2})}$ increases the probability of finding a satisfying assignment to a constant. Hence the theorem is proved. □

Comparison with the Randomized Algorithm in [PPZ97]. The randomized algorithm presented in [PPZ97] has a running time of $O(n^2|F|2^{n-n/k})$, and ours has a running time of $O(n^2|F|2^{(n-i)(1-\frac{1}{2k-2})})$. Our algorithm does better than the algorithm in [PPZ97] when $(n-i)(1-\frac{1}{2k-2}) < n(1-\frac{1}{k})$. This happens when $i > \frac{n(k-2)}{k(2k-3)}$. For $k = 3$, our algorithm does better when $i > \frac{n}{9}$.

5 Extensible Solutions via Local Search

In this section, we analyze a modification of Schöning's local search algorithm to find an extensible solution. As usual, let I denote an independent set of cardinality i. The algorithm is as follows:

```
    Let I be a maximal independent set of variables.
For  numtries   times
    Select a random partial assignment a ∈ {0,1}ⁿ⁻ⁱ
        Repeat 3n times
            Consider F(a) by substituting partial assignment a.
            if (C(a) = 0 for some C ∈ F)
                Randomly, flip the value of one of the literals from C
            else if (C₁(a) = x and C₂(a) = ¬x for C₁,C₂ ∈ F,x ∈ I)
                Randomly, flip one of the variables from C₁ ∪ C₂ - x
            else
                Extend a to a satisfying assignment s of F; return s;
        EndRepeat
EndFor
Return 'unsatisfiable'
```

Let a^* be an extensible assignment. We lower bound the probability that the above algorithm finds a^*, or some other extensible assignment. Let $a \in \{0,1\}^{n-i}$ be the initial random assignment, at a Hamming distance of j from a^*. To analyze the algorithm, we consider a walk on a Markov Chain, whose states are labelled $\{0, 1, 2, ..., n - i\}$. Each state represents the Hamming distance between the current assignment and a^*. Initially, the walk starts at state j. We now observe that at each step of the algorithm, the probability of moving one step closer to the state 0 is at least $\frac{1}{2k-2}$. This is easy to see, as we know that if a is not extensible, then either there is a clause $C, var(C) \subseteq V - I$, such that $C(a) = 0$, or there is an $x \in I$, and two clauses C_1, C_2 such that $C_1(a) = x$ and $C_2(a) = \neg x$. In the former case, the algorithm moves to an assignment with a lesser Hamming distance from a^* with probability at least $\frac{1}{k}$, and in the latter, with probability at least $\frac{1}{2k-2}$. The reasoning is that the values assigned to the variables in C by a and a^* have to differ at at least one variable. Similarly, the values assigned to variables in $C_1 \cup C_2 - x$ by a and a^* must differ at at least one variable. Consequently, the size of C and $C_1 \cup C_2 - x$ give the claimed probabilities. The probability of finding a^* from the chosen a (Hamming distance between a and a^* is j) in one iteration of the outer loop is at least the probability that the process moves from state j to state 0. This probability, denoted by q_j, is at least $(\frac{1}{2k-3})^j$. See [Sch99] for the derivation of this probability. Further, the success probability for one iteration of the outer loop is

$$p \geq (\tfrac{1}{2})^{n-i} \sum_{j=0}^{n-i} \binom{n-i}{j} (\tfrac{1}{2k-3})^j = (\tfrac{1}{2}(1 + \tfrac{1}{2k-3}))^{n-i}$$

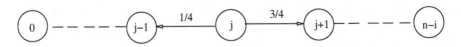

Fig. 1. Random Walk: Analysis of Local Search with Independent Set for 3-SAT

For $k = 3$, if the size of the independent set is high ($i \geq 0.3n$), then the algorithm works better than Schöning's randomized algorithm.

6   Discussion

In this paper, we have introduced the notion of an independent set of variables and use maximum independent sets in algorithms to solve k-SAT. The problem of finding a maximum independent set is a terribly hard problem. Even to find a good approximate solution in polynomial time is equally hard. However, when we permit exponential running time, finding a maximum independent set in an undirected graph has a provably better running time than the best known algorithms for k-SAT. The algorithm to find a maximum independent set due to [Bei99] runs in time $2^{.290n}$ which is approximately 1.2226^n. On the other hand, one of the best algorithms for 3-SAT is randomized and runs in time 1.3302^n [SSWH02]. Based on this observation, our approach spends some of the exponential time finding a maximum independent set, and then uses it to find a satisfying assignment. This approach is faster than [PPZ97, Sch99, SSWH02] only if the maximum independent set is sufficiently large. While there are formulae with very small independent sets, as we show below, an important direction of research is to explore the size of independent sets in random satisfiable formulae.

6.1   Formulae with Small Independent Sets

Here we construct formulae which have a small maximum independent set, and the number of clauses is also small, contradicting the intuition that small number of clauses mean large independent sets. Consider the following construction for a formula with n variables, and a parameter $1 \leq b \leq n$:

Step 1: Partition the variables into sets of b variables. There are n/b such sets.
Step 2: For each set of b variables, construct $\binom{b}{3}$ clauses made up of variables of same parity.

This formula is trivially satisfiable. The formula has $\binom{b}{3}\frac{n}{b}$ clauses, and the size of any independent set is $\frac{n}{b}$. The following table shows the sample values for different values of b.

b	no. of clauses	ind. set size
9	$9.3n$	$\frac{n}{9}$
8	$7n$	$\frac{n}{8}$

Acknowledgments. The second author would like to thank Jan Johannsen for discussions on SAT algorithms.

References

[Bei99]   R. Beigel, *Finding Maximum Independent Sets in Sparse and General Graphs*, Proceedings of the 10th Annual ACM-SIAM Symposium on Discrete Algorithms, 1999.

[DHIV01]   E. Dantsin, E.A. Hirsch, S. Ivanov, M. Vsemirnov. *Algorithms for SAT and Upper Bounds on their Complexity*. Electronic Colloquium on Computational Complexity, Report No.12, 2001.

[DLL62]   M.Davis, G. Logemann, D. Loveland, *A machine program for theorem-proving*, Communications of the ACM **5(7)** (1962), 394-397.

[DP60]   M.Davis, H. Putnam, *A computing procedure for quantification theory*, Journal of the ACM **7(3)** (1960), 201-215.

[KP92]   E. Koutsoupias, C.H. Papadimitriou, *On the Greedy algorithm for Satisfiability*,Information Processing Letters **43(1)** (1992), 53-55.

[MBFM00]   A. Z. Broder, M. Charikar, A. Frieze, and M. Mitzenmacher. *Min-wise independent permutations*, In Proceedings of the Thirtieth Annual ACM Symposium on the Theory of Computing, 1998, pages 327–336, 1998.

[MS85]   B. Monien, E. Speckenmeyer, *Solving Satisfiability in less than 2^n steps*, Discrete Applied Mathematics **10** (1985), 287-295.

[Pap91]   C.H. Papadimitriou, *On selecting a satisfying truth assignment*, Proceedings of FOCS'91, 1991, 163-169.

[PPSZ98]   R. Paturi, P. Pudlák, M.E. Saks, F. Zane, *An improved exponential-time algorithm for k-SAT*, Proceedings of FOCS'98, 1998, 628-637.

[PPZ97]   R. Paturi, P. Pudlák, F. Zane, *Satisfiability Coding Lemma*, Proceedings of FOCS'97, 1997, 566-574.

[Sch99]   U. Schöning, *A probabilistic algorithm for k-SAT and constraint satisfaction problems*, Proceedings of FOCS'99, 1999, 410-414.

[SSWH02]   T. Hofmeister, U. Schöning, R. Schuler, O. Watanabe, *A Probabilistic 3-SAT Algorithm Further Improved*, Proceedings of STACS'02, 2002, LNCS 2285:193-202.

Aligning CNF- and Equivalence-Reasoning

Marijn Heule* and Hans van Maaren

Department of Information Systems and Algorithms,
Faculty of Electrical Engineering, Mathematics and Computer Sciences,
Delft University of Technology
marijn@heule.nl, h.vanmaaren@its.tudelft.nl

Abstract. Structural logical formulas sometimes yield a substantial fraction of so called equivalence clauses after translation to CNF. Probably the best known example of this is the **parity**-family. Large instances of such CNF formulas cannot be solved in reasonable time if no detection of, and extra reasoning with, these clauses is incorporated. That is, in solving these formulas, there is a more or less separate algorithmic device dealing with the equivalence clauses, called equivalence reasoning, and another dealing with the remaining clauses. In this paper we propose a way to align these two reasoning devices by introducing parameters for which we establish optimal values over a variety of existing benchmarks. We obtain a truly convincing speed-up in solving such formulas with respect to the best solving methods existing so far.

1   Introduction

The notorious **parity-32** benchmarks [3] remained unsolved by general purpose SAT solvers for a considerable time. In [12] a method was proposed which, for the first time, could solve these instances in a few minutes. The key to this method was to detect the clauses that represented so called *equivalences* $l_1 \leftrightarrow l_2 \leftrightarrow \cdots \leftrightarrow l_n$ (where the l_i are literals, or their negations, appearing in the formula at hand) and to pre-process the set of these equivalences in such a way that dependent and independent variables became visible. The remaining clauses then were tackled with a rather straightforward DPLL procedure but in such a way that kept track of the role of these dependent and independent variables. It was developed as a two-phase method, where the equivalence part was established and transformed in a pre-processing phase.

The next important step was made by Li [6]. He incorporated a form of equivalence reasoning in every node of an emerging search tree. His approach did not incorporate a pre-processing phase (at least not regarding the equivalence clauses) and thus he established the first one-phase SAT solver **eqsatz** which could tackle these instances in reasonable time.

---

* Supported by the Dutch Organization for Scientific Research (NWO) under grant 617.023.306.

H.H. Hoos and D.G. Mitchell (Eds.): SAT 2004, LNCS 3542, pp. 145–156, 2005.

A disadvantage of his method is the fact that he uses a full look-ahead approach: all variables enter the look-ahead phase, contrary to partial look-ahead, which runs only on a pre-selected number of variables. Full look-ahead is costly for larger size formulas. In addition, his look-ahead evaluation function to measure the reduction of the formula during the look-ahead phase is - in our opinion - not optimal. Also, his equivalence reasoning is restricted to equivalences of length at most three.

Some years later Ostrowski *et al.* [10] extended the above pre-processing ideas from [12] to logical gates other than equivalences, resulting in the lsat solver. However, their DPLL approach to deal with the remaining CNF-part uses a Jeroslow-Wang branching rule and they do not perform a look-ahead phase, which is - again in our opinion - not an optimal alignment.

In this paper we propose an alignment of equivalence reasoning and DPLL reasoning which does not assume a *full* look-ahead approach. This will enforce us to introduce adequate pre-selection heuristics for selecting those variables which are allowed to enter an Iterative Unit Propagation phase. Further, we will evaluate the progress in enrolling the formula at hand in a more detailed manner as was done in eqsatz. We are forced to introduce parameters to guide this search. These parameters are needed to aggregate the reduction of the equivalence part of the formula and that of the remaining CNF part. Further, our method is able to deal with equivalences of arbitrary size. This in turn leads us to an investigation of the relative importance of equivalences of different size. Surprisingly, this relative importance turns out to be rather differently measured as would be expected when taking similar relative importance of ordinary clause-lengths as a guiding principle.

We optimise the various parameters to ensure a convincing speed-up in solving formulas with a substantial equivalence part, both with respect to the various alternative solvers available and with respect to a variety of benchmarks known of this type.

2   Equivalence Reasoning in Pre-processor

After initialisation, the first goal of the pre-processor is to simplify the formula as much as possible. This is achieved by iterative propagation of all unary clauses and binary equivalences. After this procedure, the equivalence clauses are detected using a simple syntactical search procedure, extracted from the formula and placed in a separate data-structure. We refer to this data-structure as the Conjunction of Equivalences (CoE). The aim of this equivalence reasoning enhanced pre-processor is to solve the extracted CoE sub-formula.

A solution is obtained by performing the first phase of the algorithm by Warners and Van Maaren [12]: We initialise set $I = \{x_1, \cdots, x_m\}$, the set of *independent* variables, with m referring to the initial number of variables. We loop through the equivalency clauses once, selecting variable x_i in each one to eliminate from all other equivalence clauses. Subsequently we remove x_i from I, and call it a *dependent* variable. Thus we end up with a set of equivalence clauses for which all satisfiable assignments can be constructed by assigning all possible combinations of truth values to the independent variables. The values of the dependent variables are uniquely determined by an assignment of the independent variables. Note that during the elimination process a contradiction might be derived, which implies unsatisfiability of the original problem.

Numerous of such independent sets could be obtained by this algorithm. The performance of a solver might vary significantly under different choices of the independent set, as we have observed using the march solver (developed by Joris van Zwieten, Mark Dufour, Marijn Heule and Hans van Maaren; it participated in the SAT 2002 [7], SAT 2003 [8], and SAT 2004 [9] competitions). Therefore, two enhancements are added to the original algorithm to find an independent set that would result in relatively fast performance: the first addition involves an explicit prescription for the selection of the variables to eliminate: for every equivalence clause the variable is selected that occurs (in a weighted fashion) least frequently in the CNF. Occurrences in binary clauses are counted twice as important as occurrences in n-ary clauses.

The motivation for selecting the least occurring variable is twofold: first, if the selected variable x_i does not occur in the CNF at all, the equivalence clause in which x_i occurs, becomes a *tautological clause* after elimination, because x_i could always be taken as such to satisfy it. Neglecting tautological clauses during the solving phase could result in a considerable speed-up. Second, faster reduction of the formula is expected when the independent variables occur frequently in the CNF-part: independent variables will be forced earlier to a certain truth value by constraints from both the CoE- and the CNF-part.

The second addition is a procedure that reduces the sum of the lengths of all non-tautological equivalences in the CoE. This procedure consists of two steps: the first searches for pairs of equivalence clauses that could be combined to created a binary equivalence. Note that binary equivalence clauses can always be made tautological, since one of its literals could be removed from the CNF by replacing it by the other. The second step loops through all equivalence clauses and checks whether a different choice for the dependent variable in that clause would result in a smaller sum of lenghts of non-tautological equivalences. Both methods are iteratively repeated until both yield no further reduction.

Several benchmark families in the SAT 2003, SAT 2002 and DIMACS benchmark suites[1] can be solved by merely applying the pre-processing presented above. One of these families is xor-chain which contains the smallest unsolved unsatisfiable instances from the SAT 2002 competition. Table 1 shows the required time to solve these families for various solvers. Notice that march uses the proposed pre-processing. In the table, the numbers behind the family names refer to the number of instances in a family. The last five columns show the total time required to solve a family. In these columns, numbers between braces express the number of benchmarks that could not be solved within a 120 seconds time limit. Judging from the data in the table, lsat is the only solver which can compete with the march pre-processor since it solves all but one families in comparable time.

---

[1] All three suites are available at www.satlib.org

Table 1. Performances of the solvers march, eqsatz, satzoo, lsat and zchaff in seconds on several families that could be solved by merely pre-processing

family		contributer	suite	march	eqsatz	satzoo	lsat	zchaff
bevhcube	(4)	Bevan	SAT '03	0.02	2.64 (2)	2.74 (2)	0.02	0.01 (3)
dodecahedron	(1)	Bevan	SAT '03	0.01	0.01	0.08	0.01	0.01
hcb	(4)	Bevan	SAT '03	0.16	0.01 (3)	0.01 (3)	2.53	0.01 (3)
hypercube	(4)	Bevan	SAT '03	0.09	0.40 (3)	0.08 (3)	0.33	2.56 (3)
icos	(2)	Bevan	SAT '03	0.01	2.29 (1)	4.12 (1)	0.02	—— (2)
marg	(17)	Bevan	SAT '03	0.12	195.83 (5)	52.93 (5)	0.13	0.08 (11)
urqh	(26)	Bevan	SAT '03	0.19	102.58 (20)	16.76 (20)	0.47	1.19 (22)
hardmn	(18)	Moore	SAT '03	0.87	0.75	—— (18)	0.30	—— (18)
genurq	(10)	Ostrowski	SAT '03	0.95	0.07 (7)	0.68 (1)	0.57	0.39 (6)
Urquhart	(30)	Simon	SAT '02	0.27	—— (30)	58.09 (25)	0.01	0.13 (29)
urquhart	(6)	Chu Min Li	SAT '02	0.03	0.29 (4)	95.78 (3)	0.01	0.04 (5)
xor-chain	(27)	Zhang-Lintao	SAT '02	0.16	0.17 (25)	0.77 (25)	0.02	2.25 (24)
barrel	(9)	Biere	DIMACS	2.27	2.95	99.50 (1)	2.13 (4)	46.10 (1)
dubois	(13)	Dubois	DIMACS	0.02	0.1	0.75	0.01	0.06
pret	(8)	Pretolani	DIMACS	0.03	0.08	20.81	0.01	—— (8)

3   Combined Look-Ahead Evaluation

Look-ahead appears to be a powerful technique to solve a wide range of problems. The pseudo-code of an elementary look-ahead procedure is presented in Algorithm 1. The look-ahead procedure in march closely approximates this elementary procedure. Notice that it does not perform any equivalence reasoning during this phase.

Algorithm 1. LOOK-AHEAD( )

Let $\mathcal{F}'$ and $\mathcal{F}''$ be two copies of $\mathcal{F}$
for each variable x_i in $\mathcal{P}$ **do**
 $\mathcal{F}' :=$ ITERATIVEUNITPROPAGATION($\mathcal{F} \cup \{x_i\}$)
 $\mathcal{F}'' :=$ ITERATIVEUNITPROPAGATION($\mathcal{F} \cup \{\neg x_i\}$)
 if empty clause $\in \mathcal{F}'$ **and** empty clause $\in \mathcal{F}''$ **then**
  return "unsatisfiable"
 else if empty clause $\in \mathcal{F}'$ **then**
  $\mathcal{F} := \mathcal{F}''$
 else if empty clause $\in \mathcal{F}''$ **then**
  $\mathcal{F} := \mathcal{F}'$
 else
  $H(x_i) = 1024 \times \text{DIFF}(\mathcal{F}, \mathcal{F}') \times \text{DIFF}(\mathcal{F}, \mathcal{F}'') + \text{DIFF}(\mathcal{F}, \mathcal{F}') + \text{DIFF}(\mathcal{F}, \mathcal{F}'')$
 end if
end for
return x_i with greatest $H(x_i)$ to branch on

An effective look-ahead evaluation function (DIFF in short) is critical to the effectiveness of the branching variable the look-ahead returns. Experiments on random 3-SAT instances showed that using a DIFF that counts newly created binary clauses results in fast performances on these benchmarks and many other families. Addition of new clauses of length > 2 to the DIFF requires weights that express the relative importance of clauses of various length. Weights that result in optimal performance on random k-SAT formulas could be described by linear regression: e.g. Kullmann [5] uses weights in his OKsolver that could be approximated by 0.22^{n-2}. In this equation n refers to the length of a clause, with $n \geq 2$.

Little is known about effective evaluation functions to measure the importance of a new equivalence clause. In eqsatz by Li [6] only new binary equivalences are counted. These are weighed twice as important as a new binary clause. The importance of the new equivalence clauses of various length could be obtained by measuring the reduction of its translation into CNF. Applying the approximation of the weights by Kullmann [5] results in a weight function of $2^{n-1} \times 0.22^{n-2} \approx 10.33 \times 0.44^n$ for a new equivalence of length n. However, this reference should be labelled as vague, since the weights are optimised with respect to random formulas.

Although we have indications that other models might be more appropriate when equivalence clauses are involved, we take this regression model as a first start. Performances were measured for various parameter settings of equation (1). In this equation, n refers to the reduced length of an equivalence clause. Parameter q_{base} denotes the factor that describes the decreasing importance of equivalence clauses of various length and parameter q_{const} expresses the relation between the reduction of the CNF-clauses and the equivalence clauses. Since march uses a 3-SAT translator, only new binary clauses are created. The evaluation of the look-ahead is calibrated by defining the importance of a new binary clause to value 1. The result of eq_n then defines the relative importance of a new equivalence clause of length n in relation to a new binary clause.

$$eq_n = q_{const} \times q_{base}{}^n \tag{1}$$

Wide scale experiments were troubled by the lack of useful benchmarks: many benchmark families that contain a significant part of equivalence clauses are easily solved with mere pre-processing procedures: either the solving procedure for the CoE results in a contradiction, or the propagation of the unary clauses and the binary equivalences found during pre-processing are sufficient to solve the formula. Many benchmarks families with a significant CoE-part that require a sophisticated solving procedure after pre-processing are neither useful for these experiments, because most or all of their equivalence clauses have length 3. For comparison: The SAT 2003 [8] competition suite consisted of 11 families which are solved in pre-processing while only five needed further search. Of those five only two had a large number of long equivalences after the pre-processing.

These two families are the `parity32` and the `hwb`. The first family consists of the SAT-encoding of minimal disagreement parity problems contributed by Crawford *et al.* [3]. The second consists of equivalence checking problems that arise by combining two circuits computed by the hidden weighted bit function. These latter are contributed by Stanion [8]. Both families have been used to determine

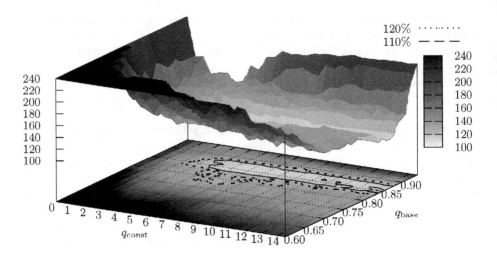

Fig. 1. Performances achieved by march on various settings of q_{base} and q_{const}. The values on the z-axis are the cumulated performances on the whole parity-32 and hwb-n20 families in seconds. Contour lines are drawn at 110% and 120% of optimal performance

Table 2. Weights to measure the reduction of equivalence clauses of various lengths

Reduced length (n):	2	3	4	5	6	7	8	9	10
CNF-reference:	2.00	0.88	0.39	0.17	0.07	0.03	0.01	0.01	0.00
Found optimum:	3.97	3.38	2.87	2.44	2.07	1.76	1.50	1.27	1.08

the parameter setting for equation (1) that results in optimal performance. The results of these experiments are shown in Fig. 1. During our experiments, the values $q_{const} = 5.5$ and $q_{base} = 0.85$ appeared optimal. Two conclusions can be derived regarding the results: (1) parameter q_{base} has a much larger influence on the performance than q_{const}. (2) Using optimal settings, the reduction of equivalences is considered far more important than the reduction of the equivalent CNF-translations would suggest: table 2 shows the weights used for both settings.

4   Pre-selection Heuristics

Although look-ahead is a powerful technique, it pays off to restrict the number of variables which enter this procedure. In Algorithm 1 this partial behaviour is achieved by performing only look-ahead on variables in set $\mathcal{P}$. At the beginning of each node, this set is filled by pre-selection heuristics that are based on an approximation function of the combined evaluation of the look-ahead (ACE). The ranking of variable x is calculated by multiplying ACE(x) and ACE($\neg x$).

$\mathcal{E}(x)$, used to obtain $\text{ACE}(x)$, refers to the set of all equivalence clauses in which x occurs and $occ_3(x)$ refers to the number of occurrences of x in ternary clauses.

$$\text{ACE}(x) = occ_3(\neg x) + \sum_{Q_i \epsilon \mathcal{E}(x)} eq_{|Q_i|-1} + \sum_{\neg x \vee y \epsilon \mathcal{F}} \left(occ_3(\neg y) + \sum_{Q_i \epsilon \mathcal{E}(y)} eq_{|Q_i|-1} \right) \tag{2}$$

In the versions of march without equivalence reasoning fast, performance is achieved on average by performing look-ahead only on the "best" 10 % of the variables. This constant percentage is not always optimal. It is not even optimal for the benchmarks used in this paper, but since it provided optimal performance on a wide scale of experiments, we restricted ourselves to this 10 %. To illustrate the diversity of partial look-ahead optima, march requires 1120 secondes to solve a benchmark provided by Philips using the 10 % setting (see table 4), while it requires only 761 seconds at the optimal setting of 8 %00. [4] provides more insight in the behaviour of march using different percentages of variables entering the look-ahead phase.

5   Additional Equivalence Reasoning

Various additional forms of equivalence reasoning are tested. These include:

- Removal of equivalence clauses that have became tautological during the solving phase. This results in a speed-up due to faster propagation.
- Propagation of binary equivalences in the CoE: replacing one of its literals by the other. This increases the chance that a variable occurs twice in an equivalence clause, so both could be removed.
- Prevention of equivalent variables to enter the look-ahead procedure, since equivalent variables will yield an equivalent DIFF.

Only the last adjustment realised a noticeable speed-up of about 10 %. The gain that other procedures accomplished were comparable to their cost, resulting in a status quo in terms of time.

6   Results

Six solvers are used to compare the results of march: eqsatz[2], lsat[3], satzoo[4], zchaff[5], limmat[6], and OKsolver[7]. The choice for eqsatz and lsat is obvious since

---

[2] version 2.0, available at http://www.laria.u-picardie.fr/~cli/EnglishPage.html

[3] version 1.1, provided by authors

[4] version 1.02, available at http://www.math.chalmers.se/~een/Satzoo/

[5] version 2003.07.01, available at http://www.ee.princeton.edu/~chaff/zchaff.php

[6] version 1.3, available at http://www2.inf.ethz.ch/personal/biere/projects/limmat/

[7] version 1.2, available at http://cs-svr1.swan.ac.uk/ csoliver/OKsolver.html

they are the only other SAT solvers performing equivalence reasoning. Since equivalence clauses merely occur in handmade and industrial problems, we added some solvers that are considered state-of-the-art in these categories: satzoo and zchaff, respectively. For extended reference we added two winners of the SAT 2002 competition: limmat and OKsolver. The last is also a look-ahead SAT solver.

All solvers were tested on an AMD 2000+ with 128Mb memory running on Mandrake 9.1. Besides the parity32 and the hwb benchmarks, we experimented on the longmult family that arises from bounded model checking [2], five unsolved benchmarks (pyhala-braun-x (pb-x in short) and lisa21-99-a) from the SAT 2002 competition contributed by Pyhala and Aloul, respectively [7], and three factoring problems (2000009987x) contributed by Purdom [11]. Except from both bounded model checking families and the benchmark provided by Philips, all benchmarks were used in the SAT 2003 competition. To enable a comparison with the SAT 2003 results[8], we used the shuffled benchmarks generated for this competition during our experiments. However, these shuffled benchmarks caused a slowdown in performance of eqsatz: e.g. eqsatz solves most original parity32 benchmarks within the 2000 seconds time limit.

Two versions of our solver are used to evaluate performance: the first, march° uses the equation $eq_n = 5.5 \times 0.85^n$ to measure the reduction of the CoE during the look-ahead, and applies it to the calculation of ACE. The second variant, march* does not use the CoE-part during the look-ahead but operates using the original CNF instead. Both march variants use a 10% partial look-ahead.

In table 3, six properties of experimented benchmarks are presented:

#Cls refers to the initial number of clauses

#Var refers to the initial number of variables

#Ind refers to the number of variables in the independent set

#Eq refers to the number of detected equivalence clauses.

#Nt refers to the number of non-tautological equivalences after pre-processing.

|Nt| refers to the average length of the non-tautological equivalences after the pre-processing.

Table 4 shows the performances of the various solvers during our experiments. Of all properties listed above, the average length (|Nt|) appears to be the most useful indicator for when to use march° instead of march*: both families that profit clearly from the equivalence reasoning have a high average length. The slowdown on the longmult family could be explained by the small number of equivalence clauses compared to the number of independent variables: the relatively costly equivalence reasoning is performed during the look-ahead while the differences in the branch decision between march° and march* are small.

---

[8] results of the SAT 2003 competition are available at
www.lri.fr/~simon/contest03/results/

Table 3. Properties of several benchmarks containing equivalence clauses

instance	#Cls	#Var	#Ind	#Eq	#Nt	|Nt|
par32-1	10227	3176	157	1158	218	7.79
par32-2	10253	3176	157	1146	218	7.83
par32-3	10297	3176	157	1168	218	7.84
par32-4	10313	3176	157	1176	218	8.12
par32-5	10325	3176	157	1182	218	7.91
par32-1-c	5254	1315	157	1158	218	7.49
par32-2-c	5206	1303	157	1146	218	8.02
par32-3-c	5294	1325	157	1168	218	7.69
par32-4-c	5326	1333	157	1176	218	7.73
par32-5-c	5350	1339	157	1182	218	8.13
hwb-n20-1	630	134	96	36	35	4.91
hwb-n20-2	630	134	96	36	35	4.88
hwb-n20-3	630	134	96	36	35	4.80
hwb-n22-1	688	144	104	38	37	4.84
hwb-n22-2	688	144	104	38	37	4.56
hwb-n22-3	688	144	104	38	37	4.62
hwb-n24-1	774	162	116	44	43	5.83
hwb-n24-2	774	162	116	44	43	4.86
hwb-n24-3	774	162	116	44	43	4.86
hwb-n26-1	832	172	124	46	45	5.04
hwb-n26-2	832	172	124	46	45	5.24
hwb-n26-3	832	172	124	46	45	5.56
longmult-6	8853	2848	1037	174	90	3.93
longmult-7	10335	3319	1276	203	105	3.93
longmult-8	11877	3810	1534	232	120	3.93
longmult-9	13479	4321	1762	261	135	3.93
longmult-10	15141	4852	2014	290	150	3.93
longmult-11	16863	5403	2310	319	165	3.93
longmult-12	18645	5974	2620	348	180	3.93
longmult-13	20487	6565	2598	377	195	3.93
longmult-14	22389	7176	2761	406	210	3.93
longmult-15	24351	7807	2784	435	225	3.93
pb-s-40-4-03	31795	9638	2860	3002	3001	3.00
pb-s-40-4-04	31795	9638	2860	2936	2935	3.00
pb-u-35-4-03	24320	7383	2132	2220	2219	3.00
pb-u-35-4-04	24320	7383	2131	2277	2276	3.00
lisa21-99-a	7967	1453	1310	460	459	3.87
2000009987fw	12719	3214	1615	1358	1319	3.54
2000009987nc	10516	2710	1303	1286	1262	3.46
2000009987nw	11191	2827	1342	1322	1299	3.38
philips	4456	3642	1005	342	224	3.50

Table 4. Performances of the solvers march, eqsatz, satzoo, lsat, zchaff, limmat, and OKsolver in seconds on various benchmarks with equivalence clauses

instance	march°	march*	eqsatz	satzoo	lsat	zchaff	limmat	OKsolver
par32-1	**0.55**	>2000	568.31	>2000	90.85	>2000	>2000	>2000
par32-2	**0.3**	>2000	>2000	>2000	88.43	>2000	>2000	>2000
par32-3	**1.08**	>2000	>2000	>2000	7.54	>2000	>2000	>2000
par32-4	**7.93**	>2000	>2000	>2000	79.87	>2000	>2000	>2000
par32-5	**8.82**	>2000	>2000	>2000	34.41	>2000	>2000	>2000
par32-1-c	**0.47**	>2000	>2000	>2000	3.91	>2000	>2000	>2000
par32-2-c	7.82	>2000	>2000	>2000	**4.45**	>2000	>2000	>2000
par32-3-c	**5.06**	>2000	>2000	>2000	33.59	>2000	>2000	>2000
par32-4-c	**0.39**	>2000	>2000	>2000	52.39	>2000	>2000	>2000
par32-5-c	**6.77**	>2000	>2000	>2000	71.98	>2000	>2000	>2000
hwb-n20-1	**18.05**	39.43	78.51	47.04	771.48	300.72	>2000	80.88
hwb-n20-2	**23.24**	50.82	83.25	73.38	738.16	461.49	>2000	69.58
hwb-n20-3	**16.16**	38.56	75.09	24.37	564.81	257.07	>2000	67.57
hwb-n22-1	**68.27**	164.65	299.45	108.	>2000	785.89	>2000	286.38
hwb-n22-2	**53.67**	145.51	297.3	85.79	>2000	1097.33	>2000	280.53
hwb-n22-3	**58.6**	148.29	306.71	60.6	>2000	1710.10	>2000	318.75
hwb-n24-1	**556.25**	796.56	>2000	624.17	>2000	>2000	>2000	>2000
hwb-n24-2	**463.46**	832.48	>2000	862.86	>2000	>2000	>2000	>2000
hwb-n24-3	**332.**	670.21	>2000	471.73	>2000	>2000	>2000	>2000
hwb-n26-1	**1203.99**	>2000	>2000	>2000	>2000	>2000	>2000	>2000
hwb-n26-2	**1777.78**	>2000	>2000	>2000	>2000	>2000	>2000	>2000
hwb-n26-3	**1703.12**	>2000	>2000	>2000	>2000	>2000	>2000	>2000
longmult-6	7.13	3.98	11.96	5.1	66.32	**1.59**	2.20	12.38
longmult-7	35.79	21.37	55.15	21.74	109.19	**13.32**	28.32	63.17
longmult-8	100.92	70.8	185.85	**66.22**	192.56	68.17	178.42	260.88
longmult-9	202.03	**130.46**	347.75	138.8	289.53	131.46	465.87	526.72
longmult-10	260.06	**168.21**	520.1	232.96	404.27	252.95	764.78	898.33
longmult-11	226.35	**151.36**	662.19	307.62	542.81	344.64	784.88	1167.59
longmult-12	128.57	**89.37**	741.19	338.48	709.15	305.07	784.72	1256.35
longmult-13	67.92	**52.69**	855.14	272.67	901.13	255.96	644.33	1516.62
longmult-14	44.33	**31.78**	985.37	383.08	1112.49	266.54	685.48	1830.4
longmult-15	25.26	**24.54**	1108.80	215.12	1236.6	207.72	630.25	>2000
pb-s-40-4-03	>2000	510.51	>2000	**184.21**	>2000	357.98	>2000	>2000
pb-s-40-4-04	>2000	**600.41**	>2000	>2000	>2000	>2000	>2000	>2000
pb-u-35-4-03	771.24	**698.22**	>2000	>2000	>2000	>2000	>2000	>2000
pb-u-35-4-04	821.43	**736.31**	>2000	>2000	>2000	>2000	>2000	>2000
lisa21-99-a	**21.26**	1170.12	>2000	>2000	>2000	>2000	>2000	>2000
2000009987fw	175.68	**115.89**	521.47	267.81	181.86	116.4	257.1	649.59
2000009987nc	137.41	**84.16**	197.27	167.4	159.55	94.61	206.75	610.84
2000009987nw	135.25	**87.9**	157.03	218.09	166.55	104.84	228.38	486.86
philips	1120.61	**1032.82**	3390.59	2114.64	1277.68	>3600	>3600	1589.59

7 Conclusions

In this paper, we presented a new alignment between equivalence reasoning and look-ahead in a DPLL Sat solver. The resulted solver outperforms existing techniques on benchmarks that contain a significant part of equivalence clauses. Two main features appeared sufficient for effective equivalence reasoning:

- an effective solving procedure for the CoE during the pre-processing.
- an effective evaluation function to measure the reduction of equivalence clauses during the look-ahead procedure.

Additional features of integration may further increase the performance, but substantial gains have not been noticed yet. However, two procedures are worth mentioning: first, integration of the effective evaluation function (ACE) into the pre-selection heuristics of the look-ahead, resulted in a speed-up up to 30%. Second, a small performance gain on practically all benchmarks, with and without equivalence clauses, was achieved by preventing equivalent variables to enter the look-ahead phase.

We conclude that aligning Equivalence- and CNF- reasoning as carried out pays off convincingly. Although some instances are *not* solved without incorporating the CoE reductions during the look-ahead phase (march°), others suffer from this additional overhead and are easier solved by updating and investigating the CoE part at the chosen path only (march*).

References

1. D. Le Berre and L. Simon, *The essentials of the SAT'03 Competition.* Springer Verlag, Lecture Notes in Comput. Sci. **2919** (2004), 452–467.
2. A. Biere, A. Cimatti, E.M. Clarke, Y. Zhu, *Symbolic model checking without BDDs.* in Proc. Int. Conf. Tools and Algorithms for the Construction and Analysis of Systems, Springer Verlag, Lecture Notes in Comput. Sci. **1579** (1999), 193–207.
3. J.M. Crawford, M.J. Kearns, R.E. Schapire, *The Minimal Disagreement parity problem as a hard satisfiability problem.* Draft version (1995).
4. M.J.H. Heule, J.E. van Zwieten, M. Dufour and H. van Maaren, *March_eq, Implementing Efficiency and Additional Reasoning in a Look-ahead SAT Solver.* Appearing in the same volume.
5. O. Kullmann, *Investigating the behaviour of a SAT solver on random formulas.* Submitted to Annals of Mathematics and Artificial Intelligence (2002).
6. C.M. Li, *Equivalent literal propagation in the DLL procedure.* The Renesse issue on satisfiability (2000). Discrete Appl. Math. **130** (2003), no. 2, 251–276.
7. L. Simon, D. Le Berre, and E. Hirsch, *The SAT 2002 competition.* Accepted for publication in Annals of Mathematics and Artificial Intelligence (AMAI) **43** (2005), 343–378.
8. L. Simon, *Sat'03 competition homepage.*
   http://www.lri.fr/~simon/contest03/results/
9. L. Simon, *Sat'04 competition homepage.*
   http://www.lri.fr/~simon/contest/results/

10. R. Ostrowski, E. Gregoire, B. Mazure, L. Sais, *Recovering and exploiting structural knowledge from CNF formulas*, in Proc. of the Eighth International Conference on Principles and Practice of Constraint Programming, Springer Verlag, Lecture Notes in Comput. Sci. **2470** (2002), 185–199.

11. P. Purdom and A. Sabry, *CNF Generator for Factoring Problems.* http://www.cs.indiana.edu/cgi-pub/sabry/cnf.htm

12. J.P. Warners, H. van Maaren, *A two phase algorithm for solving a class of hard satisfiability problems.* Oper. Res. Lett. **23** (1998), no. 3-5, 81–88.

Using DPLL for Efficient OBDD Construction

Jinbo Huang and Adnan Darwiche

Computer Science Department,
University of California, Los Angeles
{jinbo, darwiche}@cs.ucla.edu

Abstract. The DPLL procedure has found great success in SAT, where search terminates on the first solution discovered. We show that this procedure is equally promising in a problem where exhaustive search is used, given that it is augmented with appropriate caching. Specifically, we propose two DPLL-based algorithms that construct OBDDs for CNF formulas. These algorithms have a worst-case complexity that is linear in the number of variables and size of the CNF, and exponential only in the *cutwidth* or *pathwidth* of the variable ordering. We show how modern SAT techniques can be harnessed by implementing the algorithms on top of an existing SAT solver. We discuss the advantage of this new construction method over the traditional approach, where OBDDs for subsets of the CNF formula are built and conjoined. Our experiments indicate that on many CNF benchmarks, the new method runs orders of magnitude faster than a comparable implementation of the traditional method.

1 Introduction

The DPLL procedure [1] has found great success in the Propositional Satisfiability problem (SAT), attested by a series of SAT solvers that have excelled in the annual SAT competitions [2]. These solvers are, by nature, geared toward finding the first solution quickly, and not particularly concerned with any search space beyond that (solvers for *Quantified Boolean Formulas* are an exception). There has been evidence, however, that DPLL can also be useful in problems requiring exhaustive search, such as model counting [3,4]. In this paper we explore another such problem, and show that DPLL, coupled with appropriate caching, can be the basis for an efficient program that compiles propositional theories into Ordered Binary Decision Diagrams (OBDDs) [5]. Once theories are expressed as OBDDs, many important queries can be answered in constant or polynomial time, including satisfiability, equivalence, model counting, model enumeration, and clausal entailment [5,6].

Compiling propositional theories into OBDD has remained a nontrivial task. Traditionally, one enlists software packages which build OBDDs in a bottom-up fashion. For theories in Conjunctive Normal Form (CNF), this means that OBDDs are constructed for individual clauses, and conjoined to produce the OBDD for the whole theory. Although the complexity of OBDD conjunction is

H.H. Hoos and D.G. Mitchell (Eds.): SAT 2004, LNCS 3542, pp. 157–172, 2005.

only quadratic in the sizes of the conjuncts [5], this operation has to be repeatedly carried out until the final OBDD is produced. Moreover, experience has shown that the conjuncts involved in the operation—intermediate OBDDs—are often much larger than the OBDD to be finally built, leading to an accumulation of intermediate OBDD nodes that unduly exacerbates the time and space complexities of the construction.

Consider for example uf100-08.cnf, one of the standard benchmarks from the Satisfiability Library [7]. This CNF has 100 variables and its (reduced) OBDD has 176 nodes under the MINCE variable ordering [8]. Yet, to build this OBDD using the popular CUDD package and same variable ordering, a total of 30,640,582 intermediate nodes are generated, taking 25 minutes on our 2.4GHz processor.

In this paper we propose two DPLL-based algorithms that, unlike the traditional method, build OBDDs for CNFs in a top-down fashion. By using a novel caching scheme, these algorithms have a complexity that is linear in the number of variables and size of the CNF, and exponential only in the *cutwidth* and *pathwidth*, respectively, of the variable ordering with respect to a hypergraph abstraction of the CNF. As a bonus of this theoretical analysis, we provide an upper bound on the OBDD size for arbitrary CNFs. We relate these complexity results to those of some previous work that use the notions of cutwidth and pathwidth. Our upper bound on OBDD size also offers a formal explanation for the effectiveness of a recent class of variable ordering heuristics, which has hiterto been explained only intuitively.

We show next how these algorithms can be implemented on top of a SAT engine, thus harnessing the power of modern techniques, including carefully implemented *Unit Propagation* and *Nonchronological Backtracking*, that underly the success of many SAT solvers. Using multiple sets of experiments, we demonstrate the efficiency of this program and discuss a few related issues.

The rest of the paper is organized as follows. In Section 2 we describe our proposed algorithms for compilation of CNFs into OBDD, followed by a theoretical analysis of their complexities in Section 3. Section 4 is a description of our implementation of these algorithms on top of an existing SAT engine. Section 5 contains experimental results that demonstrate the efficiency of this new program and support the discussion of a few related issues. Section 6 concludes the paper.

2   Algorithms

We present in this section two DPLL-based algorithms for compiling CNF formulas into OBDD. Fig. 1 depicts a CNF Δ and its OBDD under variable order x, y, z. Recall that an OBDD is a *Directed Acyclic Graph (DAG)* where there are at most two sinks, labeled with 0 and 1 respectively, and every internal node is labeled with a variable and has exactly two children *low* and *high*; it is further required that variables appear in the same order on all paths from the root to a sink. The semantics of this graph is as follows. Given an instantiation I of the

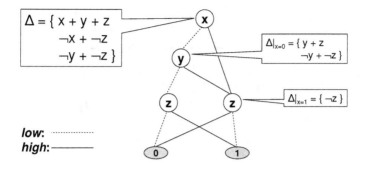

Fig. 1. A CNF and its OBDD

variables, one picks a path from the root to a sink while always choosing the low (high) child of a node if the variable associated with that node is set to 0 (1) by I. If the path ends with the 0-sink (1-sink), the theory evaluates to 0 (1) for this variable instantiation.

In this work we consider *reduced* OBDDs, where there is no node whose two children are identical, and no isomorphic sub-graphs exist. It is known that there is a unique reduced OBDD for any propositional formula under a given variable order [5]. As in SAT, the variable order plays an important role in complexity. In the rest of the paper we assume that a variable order $v_1, \ldots, v_n$ has been identified in a preprocessing step to be used for the OBDD. As we point out later, efficient tools exist that generate good variable orders.

Algorithm 1 describes a naive DPLL-style procedure that converts a CNF Δ into an OBDD by recursively converting its two restrictions, $\Delta|_{v_i=0}$ and $\Delta|_{v_i=1}$, and combining the results using *get_node* (Line 5). This is also illustrated in Fig. 1, where $\Delta|_{x=0}$ and $\Delta|_{x=1}$ are obtained by setting x to 0 and 1, respectively, in CNF Δ. Note that a common technique known as *unique nodes* is used so that the final result will be a DAG—a reduced OBDD, not a tree. Specifically, *get_node* will not construct a new node in these two cases: 1) if its last two arguments are identical, either one of them is returned immediately; 2) if there already exists a node that is labeled with the first argument and has the last two arguments as children (in the right order), that node is returned.

Note that this algorithm can have an exponential complexity even when the final OBDD has a tractable size. The reason is that when different settings of a subset of the variables lead to sub-theories that are logically equivalent, they

---

Algorithm 1. $obdd$(CNF Δ, int i): should be initially called with $i = 1$

1: **if** there is an inconsistent clause in Δ **then**
2:     return 0-sink
3: **if** there is no uninstantiated variable in Δ **then**
4:     return 1-sink
5: return $get\_node(i, obdd(\Delta|_{v_i=0}, i+1), obdd(\Delta|_{v_i=1}, i+1))$

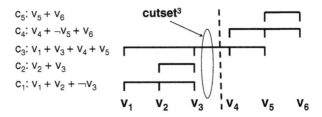

Fig. 2. Cutset-based caching

will be represented by the same OBDD node, while Algorithm 1 will convert each of these sub-theories into OBDDs only to realize that they are all the same.

Consider, for example, variable order $\pi = v_1, v_2, v_3, v_4, v_5, v_6$ for CNF $\Delta = \{c_1, c_2, c_3, c_4, c_5\}$ shown in Fig. 2. When Algorithm 1 is run on this CNF, it will spawn two recursive calls on $i = 2$ (Line 5), because there are two instantiations for variable v_1. Note that the number of recursive calls on $i = 3$ will be three, not four, because one of the four instantiations for variables v_1, v_2 results in an empty clause, terminating the recursion (Lines 1 and 2). By the same token, five recursive calls on $i = 4$ will be generated.

We will now show that three of these five recursive calls on $i = 4$ are in fact redundant and could have been avoided by caching. Specifically, let S be the set of nontrivial CNFs Δ' that can be obtained by instantiating the first three variables v_1, v_2, v_3 in Δ, we will show that $|S| \le 2$.

Note that with respect to instantiation of variables v_1, v_2, v_3, one can think of the CNF as partitioned into three sets of clauses: clauses over v_1, v_2, v_3 only, clauses over v_4, v_5, v_6 only, and the rest. Denote these three sets by $left^3 = \{c_1, c_2\}$, $right^3 = \{c_4, c_5\}$, and $cutset^3 = \{c_3\}$, respectively. After the instantiation of variables v_1, v_2, v_3, clauses $left^3$ will evaluate to a Boolean constant because their variables have all been set. If this constant is 0, we know that Δ' is a trivial CNF equal to 0 and hence not in S. Otherwise Δ' will consist of clauses $right^3$, which have not been altered because none of their variables have been set, and clauses in $cutset^3$ which must have been altered by setting variables v_1, v_2, v_3. This cutset, however, contains only one clause $v_1 + v_3 + v_4 + v_5$. After *any* instantiation of variables v_1, v_2, v_3, this clause can only be in one of two states: either satisfied, or simplified to $v_4 + v_5$. Hence, although there are eight

---

get_node(int i, BDD *low*, BDD *high*)

1: **if** $low == high$ **then**
2:     return *low*
3: **if** $(lookup = unique[(i, low, high)]) \ne$ nil **then**
4:     return *lookup*
5: $result =$ new BDD$(i, low, high)$
6: $unique[(i, low, high)] = result$
7: return *result*

Algorithm 2. *obdd*(CNF Δ, int i): *value*(C) returns a bit vector representing the states (satisfied or not) of clauses C in some fixed order

1: **if** there is an inconsistent clause in Δ **then**
2:     return 0-sink
3: **if** there is no uninstantiated variable in Δ **then**
4:     return 1-sink
5: **if** ($lookup = cache_{i-1}[value(cutset^{i-1})]) \neq$ nil **then**
6:     return $lookup$
7: $result = get\_node(i, obdd(\Delta|_{v_i=0}, i+1), obdd(\Delta|_{v_i=1}, i+1))$
8: $cache_{i-1}[value(cutset^{i-1})] = result$
9: return $result$

Algorithm 3. *obdd*(CNF Δ, int i): *value*(S) returns a bit vector representing the values of variables S in some fixed order

1: **if** there is an inconsistent clause in Δ **then**
2:     return 0-sink
3: **if** there is no uninstantiated variable in Δ **then**
4:     return 1-sink
5: **if** ($lookup = cache_{i-1}[value(separator^{i-1})]) \neq$ nil **then**
6:     return $lookup$
7: $result = get\_node(i, obdd(\Delta|_{\overline{v_i}}, i+1), obdd(\Delta|_{v_i}, i+1))$
8: $cache_{i-1}[value(separator^{i-1})] = result$
9: return $result$

different instantiations of v_1, v_2, v_3 and five that result in nontrivial CNFs, we have $|S| \leq 2 = 2^{|cutset^3|}$.

In general, the ith cutset of a variable order for a CNF is all clauses mentioning a variable at position $\leq i$ and one at position $> i$:

Definition 1. *The ith cutset of variable order $\pi = v_1, \ldots, v_n$ for CNF $\Delta = \{c_1, \ldots, c_m\}$, denoted $cutset^i_\Delta(\pi)$ or $cutset^i$ for short, is defined as $\{c \in \Delta : \exists j \leq i < k$ such that clause c mentions variables v_j and $v_k\}$.*

As we have seen, after instantiating the first i variables, each clause in $cutset^i$ can only be in one of two states. The states of clauses $cutset^i$ can therefore be represented by some bit vector $value(cutset^i)$, whose evaluation provides us with a sound equivalence test: two sub-theories, which result from two instantiations of the first i variables, must be equivalent if $cutset^i$ evaluates to the same value for both variable instantiations.

This equivalence test is used by Algorithm 2 to index a cache that stores OBDDs for all sub-theories. Specifically, when two or more sub-theories are found to have the same cutset value, only one of them will be compiled, its OBDD cached (Line 8), and others will simply generate a cache hit and have their OBDD immediately returned (Line 6). By virtue of this caching the complexity of the algorithm is only exponential in the size of the largest cutset. We discuss this complexity result in more detail in Section 3.

We now turn to Algorithm 3, which replicates Algorithm 2 except it uses a slightly different caching scheme. For position i in the variable order, let the ith separator be the subset of the first i variables that appear in clauses of the ith cutset:

Definition 2. *The ith separator of variable order $\pi = v_1, \ldots, v_n$ for CNF $\Delta = \{c_1, \ldots, c_m\}$, denoted $separator^i_\Delta(\pi)$ or $separator^i$ for short, is defined as $\{j \leq i : \exists c \in cutset^i_\Delta(\pi)$ such that clause c mentions variable $v_j\}$.*

Given an instantiation of $v_1, \ldots, v_i$, it is clear that the values of variables $separator^i$ alone determine the states of clauses $cutset^i$, and hence the sub-theory Δ'. One can represent the values of these variables, again, by some bit vector and use it to index the cache. Similarly, the complexity of this algorithm is only exponential in the size of the largest separator.

It can be seen that the value of $cutset^i$ does not determine that of $separator^i$, although the reverse, as we have just pointed out, is true. Separator caching can thus be regarded as an approximation of cutset caching, in that it may redundantly process some sub-theories that would have generated a cache hit with cutset caching. As we discuss later, though, separators may sometimes be preferable in practice as their evaluation can be less costly.

3 Complexity Results

The nature of the caching method used by our algorithms allows us to provide formal guarantees on their complexities. Our results are given in three theorems whose proofs can be found in the technical report version of this paper [9]. In stating these theorems we will refer to the size of the largest cutset as the *cutwidth*, and the size of the largest separator as the *pathwidth*, of the variable ordering with respect to the underlying CNF:[1]

Definition 3. *The cutwidth of variable order π for CNF Δ, denoted $cw_\Delta(\pi)$, is $\max_i |cutset^i_\Delta(\pi)|$.*

Definition 4. *The pathwidth of variable order π for CNF Δ, denoted $pw_\Delta(\pi)$, is $\max_i |separator^i_\Delta(\pi)|$.*

We now present the following two bounds on the time and space complexities of Algorithms 2 and 3 respectively. These results assume that *get_node* runs

---

[1] These definitions of cutwidth and pathwidth correspond precisely to those found in graph theory, given that one considers a hypergraph abstraction of the CNF formula, where each variable becomes a vertex and each clause a hyperedge enclosing its variables, and defines the cutwidth (pathwidth) of the hypergraph as the maximum cutwidth (pathwidth) among all vertex orderings. Specifically, when restricted to graphs, this notion of cutwidth is equivalent to that identified and studied in [10, 11]; this notion of pathwidth is equivalent, as proven in [12], to that originally introduced by Robertson and Seymour [13] based on the notion of *path decompositions*.

in constant time, but hold even when *unique nodes* is not used and *get_node* constructs a new node each time it is called.

Theorem 1. *For CNF Δ and variable order π, Algorithm 2 takes $O(sn2^w)$ time and space, where s is the size of Δ, n is the number of variables, and $w = cw_\Delta(\pi)$.*

Theorem 2. *For CNF Δ and variable order π, Algorithm 3 takes $O(sn2^w)$ time and space, where s is the size of Δ, n is the number of variables, and $w = pw_\Delta(\pi)$.*

As we pointed out earlier, for any given position i in the variable ordering π, the value of *separator^i* determines that of *cutset^i*. In other words, the number of possible values for *cutset^i* can never be larger than that for *separator^i*. Therefore, the result of Theorem 1 can in fact be strengthened by defining w to be $\max_i \min(|cutset^i|, |separator^i|)$. Note that this quantity is guaranteed to be a lower bound on both cutwidth and pathwidth. We will now use it in the following theorem that bounds the OBDD size for arbitrary CNF formulas, where we write $OBDD_\Delta^\pi$ to denote the OBDD for CNF Δ under variable ordering π.

Theorem 3. *For CNF Δ and variable order π, $size(OBDD_\Delta^\pi) \leq n2^w + 2$, where n is the number of variables and $w = \max_i \min(|cutset^i|, |separator^i|)$.*

We will now relate these complexity bounds to two previous results that involve similar parameters. The first of these concerns monotone 2-CNFs, which are CNFs where all clauses have length two and contain only positive literals. It has been proved in [14] that the size of any OBDD for a monotone 2-CNF is bounded by $n(2^w + 1)$, where n is the number of variables and w is the pathwidth of the *reverse* of its variable ordering.[2] This bound may look similar to that of Theorem 3, but is in fact a different result, because a variable ordering and its reverse may have quite different pathwidths. Also, the proof [14] of this result hinges on properties specific to monotone 2-CNFs and does not seem to generalize to arbitrary CNFs.

The second related result involves a SAT algorithm presented in [15], based on a static variable ordering π for a CNF. The time complexity of this algorithm is claimed to be $O(m2^w)$ where m is the number of clauses and w is the cutwidth of π. This bound is comparable to our complexity bound for Algorithm 2. However, OBDD construction is much more difficult than, and in fact subsumes, SAT solving for any given CNF. We are hence offering an algorithm that constructs an OBDD for a CNF with roughly the same time complexity as the algorithm of [15] that only solves SAT for the same CNF.

Finally, we point out that Theorem 3 offers a formal explanation for the effectiveness of a class of variable ordering techniques based on *Min-Cut Linear Arrangement* that have been recently proposed [8, 16, 15]. The MINCE variable

---

[2] The word *reverse* is not used in [14] for this result, but their definition of the pathwidth of π corresponds to the pathwidth, in our definition, of the reverse of π.

ordering [8], for example, has been shown to result in relatively small OBDDs for various benchmarks. According to its authors, MINCE minimizes the "average" cutset size of the ordering. The observed effectiveness of this technique, however, was only explained intuitively by its tendency toward grouping "connected variables together." According to Theorem 3, variable orderings that minimize cutset sizes are directly optimizing the upper bounds on OBDD size—a fundamental explanation for their effectiveness in practice.

4   Implementation

It is possible to implement Algorithms 2 and 3 in their original recursive form. One should then consider adding their own implementation of some efficient mechanism for unit propagation, nonchronological backtracking, and other important components of DPLL search. Since the zChaff SAT solver from Princeton University [17] is known to boast a highly optimized DPLL engine in these respects [2], we have decided to implement our algorithms on top of it instead. Like most modern SAT solvers, however, zChaff is based on an iterative version of DPLL, and thus not immediately adaptable for Algorithms 2 and 3. The following is pseudocode for the DPLL engine of zChaff, reproduced from [17].

```
while(1)
  if(decide_next_branch())              // branching
    while(deduce() == conflict)         // deducing
      blevel = analyze_conflicts();     // learning
      if(blevel == 0) return UNSATISFIABLE;
      else back_track(blevel);          // backtracking
  else return SATISFIABLE;              // all variables have been set
```

Implementation of (an iterative equivalent of) Algorithms 2 and 3 on top of such a SAT solver can generally be achieved in four steps, all of which are done in our case by modifying only the *decide_next_branch* function. First, make sure the program uses the variable order intended for the OBDD. Second, instruct the program to find all solutions instead of one. This can be done by adding a fake conflict clause, also known as a *blocking clause*, whenever a solution is found so that the solver will backtrack and continue to search. Third, maintain a trace of the search in the form of an OBDD (generally incomplete and nonreduced until search terminates; see Fig. 3). That is, keep an OBDD on the side and augment it during search so that it has a root-to-sink path corresponding to each solution found (see Fig. 3); all other paths should end with the zero sink. When search finishes the standard reduction algorithm [5] can be applied to obtain a reduced OBDD, which works by iteratively merging nodes that share the same label and children, and deleting nodes whose two children are identical.

Now that the program constructs OBDDs instead of just finding a solution, the fourth and key step is to put caching in place, which consists of cache insertion and cache lookup. According to Algorithms 2 and 3, every node cached represents the result of compiling some sub-theory of the original CNF into OBDD. Back to our implementation, this implies that we should only cache nodes whose construction is complete, as there are also nodes that are partially constructed. Consider, for example, the left half of the figure below, which depicts the decision stack of the program when the first solution $\bar{v}_1 v_2 \bar{v}_3 \bar{v}_4 v_5 v_6$ has just been found for the CNF from Fig. 2:

decision level 1: $\boxed{\bar{v}_1 v_2}$ (backtrack) decision level 1: $\boxed{\bar{v}_1 v_2}$

2: $\boxed{\bar{v}_3}$ =========>   2: $\boxed{\bar{v}_3 v_4}$

3: $\boxed{\bar{v}_4 v_5 v_6}$

At this point six OBDD nodes (excluding the sinks) are constructed, as shown in the first picture of Fig. 3, to form a path representing the solution. Among these, however, only the last two nodes (labeled with v_5 and v_6) are complete: their other child, although not yet drawn, must be the zero sink, because instantiations $v_5 = 1$ and $v_6 = 1$ have been implied. The other four nodes all have a child that has not been determined or has not been completely constructed. The nodes labeled with v_5 and v_6 should therefore be the only nodes to insert into the cache.

In general, whenever a solution is found by the SAT solver and the OBDD is augmented so that it contains a path corresponding to the solution, we may store in the cache all nodes on this path that come after the node labeled with the current decision variable, indexed by their corresponding separator (or cutset) value.

We now continue the example to illustrate the operation of cache lookup. After a blocking clause $v_1 + v_3 + v_4$ is added, the program will backtrack to decision level 2 and insert $v_4 = 1$ as an implication, as shown in the right half of the figure above. Before making the next decision by instantiating v_5, the program now has an opportunity to check the cache, both at position 3 (corresponding to partial assignment $\bar{v}_1 v_2 \bar{v}_3$) and position 4 (corresponding to partial assignment $\bar{v}_1 v_2 \bar{v}_3 v_4$). Note that cache lookup at preceding positions have been performed at earlier decision levels, and thus need not be repeated. As it turns out, no cache hits occur at this point.

In general, whenever the SAT solver is about to instantiate variable v_k, it may check the cache at every position i, where $j \leq i < k$ and v_j is the previous decision variable. The key used in the lookup will then be the value of $separator^i$ (or $cutset^i$). In case of a cache miss the program proceeds as usual by instantiating v_k; otherwise the OBDD is augmented so that a partial path corresponding to the current instantiation of variables $v_1, \ldots, v_{k-1}$ exists and is connected directly to the OBDD node returned from the cache (see Cache Hits in Fig. 3); again a blocking clause is added so that the solver will backtrack and continue to search.

To conclude our example with the CNF from Fig. 2, Fig. 3 shows snapshots of the OBDD maintained by the compiler at successive solution findings. Specif-

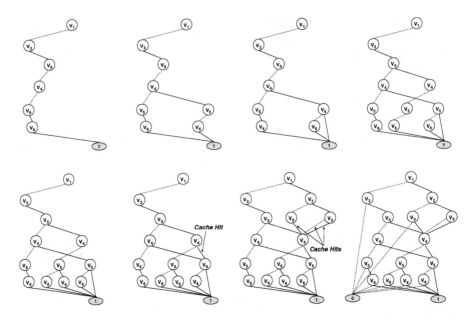

Fig. 3. Partially constructed OBDDs at various stages of DPLL, before reduction

ically, the first five shots are taken when the program has just found the first, second, third, fourth, and fifth solution, respectively. The next two pictures depict the rest of the solutions found, all through cache hits. The final picture completes the OBDD by supplying the pointers to the zero sink, which have been implicit.

The last component of the compiler is a method to properly compute the values of separators and cutsets. The former is straightforward: the value of $separator^i$ is simply the current instantiation of variables $separator^i$. The latter demands more care. Recall that the correctness of cutset caching hinges on the fact that other variables remain free when variables $v_1, \ldots, v_i$ are instantiated. This does not hold, however, in a real-world SAT solver where unit propagation constantly takes place: an instantiation of variables $v_1, \ldots, v_i$ may well have caused variables at position $> i$ to be set, which in turn alters the states of clauses $cutset^i$, obscuring their true values. To overcome this complication, the states of clauses $cutset^i$ should be determined purely on the instantiation of variables $v_1, \ldots, v_i$, pretending that other variables were all free. This process usually incurs overhead, because one can no longer rely on a quick check of some flag that may have been set by the SAT solver to indicate whether a clause has been satisfied. In our implementation, we simply walk through the literals of each clause in the cutset to determine its state.

Finally, we note that except for dynamic variable ordering, which we have turned off, all features of the original SAT solver remain in effect. In particular, we retain the benefits of unit propagation using watched literals, conflict-directed backtracking, and no-good learning.

5   Experimental Results

The purpose of our experiments is threefold. First, we demonstrate the efficiency of our program by running it against an implementation of the traditional bottom-up OBDD construction method. Second, we study the effect of caching used by our program by turning it off and observing the change in performance. Third, we investigate the intermediate explosion encountered in bottom-up construction using random CNFs with varying clauses-to-variables ratios. All our compilations use variable orders generated by *MINCE* [18], which implements the heuristic proposed in [8] for minimizing OBDD sizes. Our experiments were run on a 2.4GHz processor with 3.7GB of RAM.

Our first set of experiments are on ten groups of benchmarks taken from the Satisfiability Library [7] plus CNFs based on the first 18 of the ISCAS89 circuits [19]. Our DPLL-based compiler can be set to use either cutset (Algorithm 2) or

Table 1. Performance of DPLL vs CUDD

Benchmark	#CNFs	OBDD Size	DPLL Time (sec)	CUDD Time (sec)	CUDD Nodes
aim50	16	52	0.00	0.19	11178
aim100	16	102	0.00	21.86	2413645
ais	2	1770	0.45	15.51	613200
blocksworld	5	559	0.04	234.28	1759884
flat75	10	8610	0.29	1.37	99645
flat100	10	18515	1.61	15.41	639159
parity8	10	212	0.00	0.21	22280
parity16	8	674	4.20	800.29	38148066
uf75	10	1733	0.11	15.36	605228
uf100	10	1411	1.33	526.88	14154496
iscas89	18	85462	13.22	4.66	342313

Table 2. Effect of Caching on Performance of DPLL

CNF	OBDD Size	#Cache Hits / Entries	Time (sec)	Time without Caching (sec)
flat75-1	3966	387 / 17155	0.16	1.1
flat75-2	2231	1281 / 28180	0.28	320.83
flat75-3	14057	723 / 24208	0.29	1.5
flat100-1	10385	642 / 27232	0.78	166.13
flat100-2	14806	1902 / 71336	1.57	38.83
flat100-3	2583	464 / 17006	0.15	out of memory
iscas89-s208.1	1056	190 / 2863	0.01	0.87
iscas89-s344	10073	1154 / 20225	0.07	out of memory
iscas89-s386	14078	1399 / 180620	0.65	1.09
iscas89-s510	17366	991 / 21893	0.14	64.94
iscas89-s953	438246	56394 / 2935247	38.81	out of memory

separator (Algorithm 3) caching. For these experiments the latter has been used, as it runs slightly faster thanks to the less expensive computation of separator values. For the bottom-up method, we rely on the CUDD package from the University of Colorado [20] to build OBDDs for individual clauses and conjoin them for the final result. Since the order in which these OBDDs are built and conjoined affects the complexity of construction, we have adopted a clause ordering heuristic that was proposed in [21] exactly for use with this method of OBDD construction. This heuristic calls for clauses with higher-indexed variables to be processed first, allowing the OBDD nodes themselves to be constructed in a bottom-up fashion.

The results of these experiments are summarized in Table 1, where the two programs are referred to as DPLL and CUDD, respectively. The second column indicates the number of instances in each group of CNFs. All other figures represent group averages. The time to generate the MINCE variable order is not included as this is a preprocessing step shared by both programs. We observe that DPLL runs faster than CUDD by generally many orders of magnitude, except for the last group which we discuss more toward the end of the section. The last column gives for each group the (average) number of intermediate OBDD nodes generated by CUDD, which, compared with the OBDD size, affords an intuitive explanation for the inefficiency of the bottom-up construction method on these instances. Some instances are not included in this table because CUDD did not successfully compile them. On two of the parity16 instances, for example, DPLL finished in 6.67 and 9.53 seconds, generating 351 and 1017 OBDD nodes, respectively, but CUDD ran out of memory.

To ascertain the effect of caching on the performance of DPLL, we reran it on the same instances with caching turned off. This version of DPLL would then correspond to the original zChaff with the only change being an enforced static variable ordering and the adding of blocking clauses for enumeration of all solutions. We note that on some instances, no cache hits had occurred before and, consequently, disabling caching did not cause any noticeable change in performance. However, on other instances, especially the flat75, flat100, and iscas89 families, performance dropped significantly after caching was turned off. See the results in Table 2. Note that CUDD does better than DPLL on the iscas89 family overall, but not on all the 18 instances. In particular, all of those included in Table 2 are instances where DPLL outperforms CUDD.

We investigate next the performance of DPLL and CUDD on randomly generated 3-CNFs with varying clauses-to-variables ratios. We use *mkcnf* written by Allen van Gelder with the *forced satisfiable* option to generate the CNFs. Our suite of random 3-CNFs consists of those with n variables and m clauses, where $n = 40$, 45, and 50, and m ranges from 10 to $5n$ at intervals of 5. For each n-m combination we generate 20 instances. The OBDD sizes as well as running times we report next represent averages over these 20 instances of each ratio. Our first observation is illustrated in Fig. 4, which plots the OBDD size as a function of the clauses-to-variables ratio. It can be seen that for all three groups, the

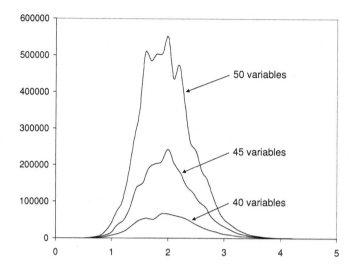

Fig. 4. OBDD size as a function of clauses-to-variables ratio

OBDD size peaks around the ratio of 2, and generally decreases toward either direction.

It is interesting to note, as shown in Fig. 5 (left) for the group of CNFs with 50 variables, that the running time of CUDD also peaks around the ratio of 2, and generally decreases toward either direction.

We now turn to an important issue with the bottom-up construction method used by CUDD—the explosion of intermediate BDD nodes. Fig. 5 (right) shows, for 50 variables, the ratio of the total number of nodes generated by CUDD over the final OBDD size, again as a function of the clauses-to-variables ratio. We observe that over the middle part of the spectrum, between the ratios of 0.6 and 3.6 for example, CUDD produces a low explosion rate. As a result, one may expect CUDD to be generally efficient on CNFs with these ratios, because relatively few dead nodes will be generated.

In fact, our next set of data indicates that it is precisely over this central portion of the gamut that CUDD outperforms DPLL.[3] To view the transition points to a higher precision, we magnify the two end portions of the spectrum and leave out the middle part, as shown in Fig. 6. It can be seen that for ratios < 0.6 and those > 3.6, DPLL is more efficient than CUDD. This corresponds roughly to what one may have predicted from Fig. 5 (right) based on the rationale that CUDD will tend to be efficient when few dead nodes are generated, which we

---

[3] For these experiments we have used a different implementation of DPLL that does not build on zChaff. Instead, it is written recursively and hence follows more closely the pseudocode of Algorithm 2. For reasons we are yet to identify, this recursive implementation runs faster than the one based on zChaff on this set of random 3-CNFs.

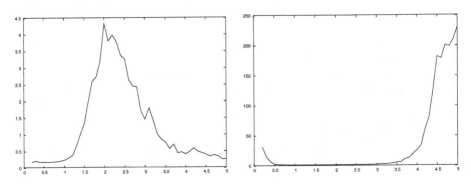

Fig. 5. CUDD time (left) and explosion rate (right) as a function of clauses-to-variables ratio, 50 variables

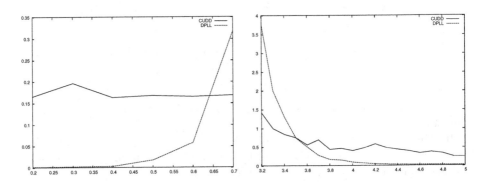

Fig. 6. Running time of CUDD vs DPLL on the two extremes of the spectrum

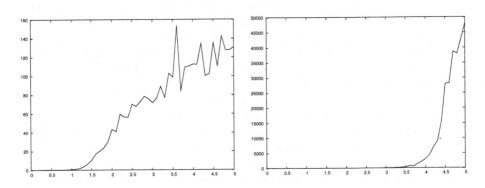

Fig. 7. CUDD time (left) and explosion rate (right) without clause ordering

have alluded to in the previous paragraph. The iscas89 family may be another example to this effect. According to Table 1, the total number of nodes generated by CUDD on this group of CNFs is only about four times the final OBDD size, and CUDD outruns DPLL by about a factor of three.

For these two extremes of the spectrum, we observed the effect of caching on DPLL by turning it off and noting the change in performance. We noticed that for the low ratios the running time increased dramatically (e.g., from 0.06 to 1252 seconds for ratio 0.6), and for the high ratios it slighly decreased (e.g., from 0.50 to 0.34 for ratio 3.6). We abscribe this phenomenon to the fact that at the low ratios there are an extremely large number of models for the CNF and hence many opportunities for cache hits, whereas at the high ratios models are sparse and one does not expect many cache hits, if at all, and the overhead of caching can slow the program down.

Finally, we offer a few more words on the two OBDD construction methods that we have been comparing. DPLL represents a top-down approach, where global properties of the CNF formula are exploited throughout the construction. The traditional bottom-up method using CUDD, on the other hand, works locally on subsets of the CNF at any given time. However, the particular implementation we have reported on does not correspond to the pure bottom-up approach, because the clause ordering heuristic we have used effectively gives it also a global view of the CNF structure, and hence some benefits of the top-down approach. In fact, in an additional set of experiments on the random 3-CNFs we turned off clause ordering and noted that the performance of the bottom-up method was now much worse. Fig. 7 (left) plots the running time of CUDD without clause ordering on the 50-variable 3-CNFs, which, instead of having a bell shape, now increases with the number of clauses. Note that the maximum value of the curve has increased from less than 4 seconds (Fig. 5) to over 150 seconds. The explosion rates have also increased significantly; see Fig. 7 (right) compared with Fig. 5 (right).

6   Conclusion

We have proposed two DPLL-based algorithms that compile CNF formulas into OBDDs. Theoretical guarantees have been provided on the complexities of these algorithms, and in the process an upper bound has been proved on the OBDD size for arbitrary CNF formulas. We have related these results to some previous complexity bounds that use similar structural parameters. We have described an implementation of these algorithms on top of an existing SAT engine, and demonstrated its efficiency in practice over the traditional bottom-up OBDD construction method on many standard benchmarks. Using randomly generated 3-CNFs, we study the relationships between the OBDD size, CUDD explosion rate, the performance of CUDD versus DPLL, and the effect of caching for varying clauses-to-variables ratios.

Acknowledgment

We wish to thank the anonymous reviewers for the SAT 2004 conference for commenting on an earlier version of this paper. This work has been partially supported by NSF grant IIS-9988543 and MURI grant N00014-00-1-0617.

References

1. Davis, M., Logemann, G., Loveland, D.: A machine program for theorem proving. Journal of the ACM **(5)7** (1962) 394–397
2. SAT Competitions: http://www.satlive.org/SATCompetition/.
3. Birnbaum, E., Lozinskii, E.: The good old Davis-Putnam procedure helps counting models. Journal of Artificial Intelligence Research **10** (1999) 457–477
4. Bayardo, R., Pehoushek, J.: Counting models using connected components. In: AAAI. (2000) 157–162
5. Bryant, R.E.: Graph-based algorithms for Boolean function manipulation. IEEE Transactions on Computers **C-35** (1986) 677–691
6. Darwiche, A., Marquis, P.: A knowledge compilation map. Journal of Artificial Intelligence Research **17** (2002) 229–264
7. Hoos, H.H., Sttzle, T.: SATLIB: An Online Resource for Research on SAT. In: I.P.Gent, H.v.Maaren, T.Walsh, editors, SAT 2000, IOS Press (2000) 283–292 SATLIB is available online at *www.satlib.org*.
8. Aloul, F., Markov, I., Sakallah, K.: Faster SAT and smaller BDDs via common function structure. In: International Conference on Computer Aided Design (IC-CAD), University of Michigan. (2001)
9. Huang, J., Darwiche, A.: Using DPLL for Efficient OBDD Construction. Technical Report D-140, Computer Science Department, UCLA, Los Angeles, CA 90095 (2004)
10. Gavril, F.: Some NP-complete problems on graphs. In: 11th conference on information sciences and systems. (1977) 91–95
11. Thilikos, D., Serna, M., Bodlaender, H.: A Polynomial Algorithm for the cutwidth of bounded degree graphs with small treewidth. Lecture Notes in Computer Science (2001)
12. Kinnersley, N.G.: The vertex separation number of a graph equals its path-width. Information Processing Letters **42** (1992) 345–350
13. Robertson, N., Seymour, P.D.: Graph minors I: Excluding a forest. Journal of Combinatorial Theory, Series B **35** (1983) 39–61
14. Langberg, M., Pnueli, A., Rodeh, Y.: The ROBDD size of simple CNF formulas. In: 12th Advanced Research Working Conference on Correct Hardware Design and Verification Methods. (2003)
15. Wang, D., Clarke, E.M., Zhu, Y., Kukula, J.: Using cutwidth to improve symbolic simulation and Boolean satisfiability. In: IEEE International High Level Design Validation and Test Workshop. (2001)
16. Aloul, F., Markov, I., Sakallah, K.: FORCE: A Fast and Easy-To-Implement Variable-Ordering Heuristic. In: Great Lakes Symposium on VLSI (GLSVLSI), Washington D.C. (2003) 116–119
17. Zhang, L., Madigan, C., Moskewicz, M., Malik, S.: Efficient conflict driven learning in a Boolean satisfiability solver. In: Proceedings of ICCAD 2001, San Jose, CA. (2001)
18. MINCE for download: http://www.eecs.umich.edu/~faloul/Tools/mince/.
19. ISCAS89 Benchmark Circuits, http://www.cbl.ncsu.edu/CBL_Docs/iscas89.html.
20. Somenzi, F.: CUDD: CU Decision Diagram Package. (Release 2.4.0)
21. Aloul, F., Markov, I., Sakallah, K.: Faster SAT and smaller BDDs via common function structure. Technical Report CSE-TR-445-01, University of Michigan (2001)

Approximation Algorithm for Random MAX-kSAT

Yannet Interian

Center for Applied Mathematics,
Cornell University, Ithaca, NY 14853, USA
interian@cam.cornell.edu

Abstract. We provide a rigorous analysis of a greedy approximation algorithm for the maximum random k-SAT (MAX-R-kSAT) problem. The algorithm assigns variables one at a time in a predefined order. A variable is assigned TRUE if it occurs more often positively than negatively; otherwise, it is assigned FALSE. After each variable assignment, problem instance is simplified and a new variable is selected. We show that this algorithm gives a 10/9.5-approximation, improving over the 9/8-approximation given by de la Vega and Karpinski [7]. The new approximation ratio is achieved by using a different algorithm than the one proposed in [7], along with a new upper bound on the maximum number of clauses that can be satisfied in a random k-SAT formula [2].

1   Introduction

In the MAX k-SAT problem we are given a Boolean formula in conjunctive normal form, with k literals in each clause, and we ask for a truth assignment that maximizes the number of satisfied clauses. The random version of this optimization problem considers inputs drawn from a predefined probability distribution. An α-approximation algorithm for the Maximum random k-SAT problem (MAX-R-kSAT) finds with high probability (w.h.p.)[1] an assignment satisfying at least α times the maximum number of possible satisfiable clauses.

The most popular model for generating random SAT problems is the uniform k-SAT model, formed by uniformly and independently selecting m clauses from the set of all $2^k \binom{n}{k}$ k-clauses on a given set of n variables. Interesting problems for this model arise when the ratio α of clauses to variables remains constant as the number of variables increases. The most famous conjecture is that such formulas exhibit a "phase transition" as a function of α [11]. There exists a constant c_k such that, uniform k-SAT problem instances with values of α below the threshold α_k, typically have one or more satisfying assignment,

---

[1] The events $\mathcal{E}_n$ hold with high probability (w.h.p.) if $\mathbf{Pr}(\mathcal{E}_n) \to 1$ when $n \to \infty$.

H.H. Hoos and D.G. Mitchell (Eds.): SAT 2004, LNCS 3542, pp. 173–182, 2005.

whereas problems with α larger than α_k have too many constraints and become unsatisfiable.

We propose the analysis of a simple greedy algorithm for approximating MAX-R-kSAT. Previous work on this problem was done by de la Vega and Karpinski [7], where a 9/8-approximation algorithm for MAX-R-3SAT is analyzed. We improve upon this ratio by analyzing a different algorithm, and using recent results Achlioptas *et al* [2] giving upper bounds on the maximum number of clauses that can be satisfied on a random k-SAT formula.

Our analysis relies on the method of differential equations studied by Wormald [13]. This method has been used extensively in the approximation (lower bounds) of the satisfiability threshold (see [1, 8, 3]). To use this method for MAX-R-kSAT, we have to be able to compute not the probability of finding an assignment using a certain algorithm, as done for the random 3-SAT problem [1, 8, 3], but the expected number of clauses that such an assignment satisfies.

2   Outline of the Results

For a k-CNF formula F, let $max(F)$ be the maximum number of clauses that can be satisfied, and $m_A(F)$ be the number of clauses satisfied by the assignment A. Let $r = r(F)$ be the ratio between the number of clauses and the number of variables for the k-CNF formula F. Denote by n, the number of variables in F, and by m the number of clauses.

We propose an algorithm which, given a k-CNF formula F, outputs an assignment A, leading to $m_A(F)$ as our approximation to $max(F)$. We prove that

$$\lim_{n\to\infty} \mathbf{Pr}\left\{\frac{max(F)}{m_A(F)} \le \alpha\right\} = 1 \tag{1}$$

To obtain (1) we prove that for a fixed k, there exists a function $g(r)$ such that $m_A(F) = g(r)m + o(m)$ w.h.p., *i.e.*, $\mathbf{Pr}\{m_A(F) = g(r)m + o(m)\}$ goes to 1 as m goes to infinity. Then we show that $\mathbf{Pr}\{max(F) \ge \alpha(g(r)m + o(m))\}$ goes to zero as m goes to infinity, where α is the approximation constant.

In part of the analysis we use an upper bound for the value of $max(F)$, as given in the results of Achlioptas *et al.* [2]. We prove the following result for random MAX-R-3SAT.

Theorem 1. *There is a polynomial time algorithm for approximating MAX-R-3SAT with approximation ratio $\alpha = 10/9.1$.*

Remarks: In section 5 we discuss how to extend the proof of theorem 1 to obtain a 10/9.5 ratio. The proof can be further generalized to obtain results for any fixed k.

3   The Algorithm

The main loop of our greedy algorithm is as follows:

Algorithm
begin
   if $r(F) > r_k$
       output a random assignment A
   Otherwise
       output the assignment A given by
       the **Majority** algorithm
  end

where r_k, is a constant that depends on k. For $k = 3$, we use $r_3 = 183$, as will be explained in the the proof of theorem 1 in section 5.

If $r(F) > r_k$, the output is a random assignment. We can change this part of the algorithm to output any fixed assignment, turning the algorithm into a deterministic algorithm. We will show below that any fixed assignment satisfies $\frac{7}{8}m + o(m)$ clauses w.h.p., and moreover we prove that $\frac{7}{8}m + o(m)$ is a good approximation of $max(F)$ when r sufficiently large.

If $r(F) \leq r_k$, the algorithm proceeds as follows: while there are unassigned variables, select an unassigned variable x. If x appears positive (*i.e.*, appears as x as opposed to $\bar{x}$) in at least half of the clauses that contain x, x is assigned to TRUE; otherwise, it is assigned to FALSE. The formula is simplified after each assignment.

Majority algorithm
begin
  While unset variables exist do
      Pick an unset variable x
      If x appears positively in at least half of the remaining
      clauses (in which x appears)
         Set x =TRUE
      Otherwise
         Set x =FALSE
      Del&Shrink
  end do
  output the current assignment
end

Chao and Franco [5] proposed a unit clause with the majority rule algorithm for the study of the satisfiability threshold for random 3-SAT formulas. The majority rule used by Chao and Franco [5] attempts to minimize the number of 3-clauses that become 2-clauses, and the unit clause rule attempts to satisfy every unit clause that is produced while running the algorithm. Such a strategy aims at finding satisfying assignments.

It's an interesting question for future research whether adding unit-clause propagation (*i.e.*, selecting variables occurring in unit clauses first) is helpful in the MAX-SAT problem when the problem instances are over-constrained. For finding a satisfying assignment (assuming such an assignment exists), unit propagation, and its generalization "the shortest clause first" heuristic, have been shown to be very effective both empirically and in formal analysis. However, in standard satisfiability testing one has in some sense "no choice" — once a unit clause is obtained, the variable in the clause has to be set such that the unit clause is satisfied. In over-constrained MAX-SAT instances, the situation is quite different. Once a unit clause is obtained, the question is whether one should proceed to satisfy that clause or instead work on satisfying other clauses and leave the unit clause unsatisfied. In order to satisfy the maximum number of clauses, it may be beneficial to not treat unit clauses any different from other clauses. Our intuition is that after a certain number of variable settings, a sufficient number of unit clauses appears, such that from then on, a procedure with unit propagation will only set variables in unit clauses (note that setting a few unit clauses will generally produce new unit clauses). It would be interesting to analyze the production of unit clauses as a branching process and check whether at some point the expected number of offspring (new unit clauses) is greater than one.

Our algorithm is also similar to the one proposed in de la Vega and Karpinski [7]. Their assignment strategy is static. The algorithm assigns every variable to its majority value, *i.e.*, x is assigned to TRUE if x appears positively more often than negatively in the original formula; otherwise, it is assigned to FALSE. Our algorithm is similar but proceeds dynamically. It assigns one variable at a time and simplifies the formula before considering a new variable to be assigned. Therefore, clauses are not taken into consideration by the algorithm if they are already satisfied.

4   Analysis

In this section we prove that if A is the assignment given by the algorithm, then we know with an $o(m)$ error, the number of clauses that A satisfies. More precisely, $m_A(F) = g(r)m + o(m)$ for some function $g(r)$ that depends only on the parameter r.

4.1   For $r > r_k$

This part of the analysis shows that for large values of the parameter r, a random assignment, or any fixed assignment, will yield a good approximation.

Any fixed assignment A will satisfy $(1 - \frac{1}{2^k})m + o(m)$ clauses w.h.p., which can be seen from the following argument. Let A be a fixed assignment and F a random k-SAT formula. The number of clauses satisfied by A is the sum of m $\{0, 1\}$-independent random variables. That is, if $X_i = 1$ if the ith clause is satisfied by A, and 0 otherwise for $1 \leq i \leq m$, then $m_A(F) = \sum_{i=1}^{m} X_i$ where X_i are independent identically distributed, and $\mathbf{Pr}(X_i = 1) = 1 - \frac{1}{2^k}$. So $\mathbf{E}(m_A) = (1 - \frac{1}{2^k})m$. Moreover, $\mathbf{Var}(m_A) = m\mathbf{Var}(X_1) = m\frac{1}{2^k}(1 - \frac{1}{2^k})$, using Chebyshev's inequality we obtain

$$\mathbf{Pr}\{|m_A - \mathbf{E}(m_A)| \geq m^{2/3}) \to 0 \text{ as } m \to \infty$$

therefore, $m_A = (1 - \frac{1}{2^k})m + o(m)$ w.h.p.

The next result says that $max(F)$ is very close to $(1 - \frac{1}{2^k})m$ for large values of the parameter r and so very close to $m_A(F)$, for any fixed A.

Lemma 1. *[7] For every ϵ there exists $r_{\epsilon,k}$ such that for $r \geq r_{\epsilon,k}$ and F a random k-SAT formula*

$$\mathbf{Pr}\{max(F) \geq (1 - \frac{1}{2^k})m(1 + \epsilon)\}$$

goes to zero as n goes to infinity.

Proof. Let $q = 1 - \frac{1}{2^k}$. Note that, the random variable $max(F) = \mathbf{max}_{A \in \{0,1\}^n} m_A(F)$ and that for any fixed A, $m_A(F)$ has distribution binomial with parameter m and q ($\mathbf{Bin}(m,q)$).

$$\begin{aligned}
\mathbf{Pr}\{max(F) \geq q\,m(1+\epsilon)\} &= \mathbf{Pr}\{|A : m_A(F) \geq q\,m(1+\epsilon)| > 0\} \\
&\leq \mathbf{E}\{|A : m_A(F) \geq q\,m(1+\epsilon)|\} \\
&= 2^n \,\mathbf{Pr}\{\,\mathbf{Bin}(m,q) \geq q\,m(1+\epsilon)\} \\
&\leq 2^n \exp(-\frac{qm\epsilon^2}{2})
\end{aligned}$$

Last inequality follows by Chernoff bound, and goes to zero for $r \geq r_{\epsilon,k} = \frac{2^{k+1}\log 2}{(2^k-1)\epsilon^2}$.

For instance, in order to obtain a 10/9.1-approximation for MAX-R-3SAT, we take $\epsilon = 1/0.91 - 1$ and obtain that a random assignment (or any fixed assignment A) gives the desired approximation ratio for $r_3 \geq 183$. To achieve the 10/9.5-approximation, we set $r_3 \geq 643.5$.

4.2 For $r < r_k$

We now analyze the majority rule algorithm. With this analysis we aim to compute how many clauses are satisfied by the assignment chosen by the algorithm, or equivalently, how many empty clauses are generated during the assignment process. To do that, we trace certain parameters during the execution of the algorithm. One of those parameters is the number o fempty clauses generated up to time t. (An empty clause is generated when all literals in the clause are assigned FALSE, so the clause is not satisfied by the current assignment.) An important aspect of the algorithm is that the order in which the variables are assigned has to be chosen in advance, *i.e.*, without looking at the particular formula. Each variable is assigned using the majority rule, *i.e.*, if the variable occurs more positively than negatively in the remaining clauses, set the variable to TRUE, else to FALSE. Using this approach, we are assured that the remaining formula is still a uniformly random formula, in the following sense.

Our algorithm sets variables one at a time. If we start for example with a 3-SAT formula, after assigning a variable we end up with a mix of 2-clauses and 3-clauses. After the next assignment, we may have some unit clauses, and even some empty clauses. Starting with a k-SAT formula, at any time t, the remaining formula has $n - t$ variables and a mix of $\{1, \ldots, k\}$-clauses and empty clauses. We define a random model that includes clauses of different length, and that includes random k-SAT as the special case in which all clauses have length k. The model is very simple. If n is the number of variables, and there are C_i, i-clauses for $1 \leq i \leq k$, we generate for each i, a i-SAT random formula with n variables and C_i clauses, and we consider the formula F to be the conjunction of all the clauses. Denote Φ_C a random formula generated in this way with $C = (C_1, \ldots, C_k)$.

Fix k, denote by $C_i(t)$ the number of clauses with i literals remaining at time t, $1 \leq i \leq k$, and $C_0(t)$ the number of empty clauses at time t. The next lemma establishes that at the end of each step t of the algorithm the remaining formula is random on the space of formulas $\Phi_{C(t)}$ with $C(t) = (C_1(t), \ldots, C_k(t))$.

Lemma 2. *[5, 6] For every time $1 \leq t \leq n$, conditional on the values $C_i(t)$, $1 \leq i \leq k$, the number of clauses of length i, the remaining formula is a random formula with parameters $C_i(t)$, $1 \leq i \leq k$ and $n' = n - t$ variables.*

Lemma 2 can be used to compute parameters of the formula conditioned on the values of $C_i(t)$ $0 \leq i \leq k$. For example we can compute expected value of $C_i(t + 1) - C_i(t)$ given the values of C_i at time t.

The analysis we propose here relies on the method of differential equations described in [13]. The sketch of the analysis is as follows: suppose $C(t) = (C_0(t), \ldots, C_k(t))$ are stochastic parameters related to a formula, in our case are the number of i-clauses at time t. We want to estimate the trajectory of $C(t)$ through the duration of our algorithm. In a restricted version, the theorem states that if

(a) $\mathbf{Pr}(|C_i(t + 1) - C_i(t)| > n^{1/5}) = o(n^{-3})$
(b) $\mathbf{E}(C_i(t + 1) - C_i(t)|\, C(t)) = f_i(t/n, C(t)/n) + o(1)$
(c) the functions f_i are continuous and satisfies a Lipschitz condition on some set D

then

$$C_i(t) = nc_i(t/n) + o(n),$$

where $c_i(x)$ is the solution of the system of differential equations

$$\frac{dc}{dt} = f(x, c) \qquad c(0) = (\frac{C_0(0)}{n}, \ldots, \frac{C_k(0)}{n}) = (0, \ldots, \alpha),$$

and $c = (c_0, c_1, \ldots, c_k)$.

In our case, the equation for the conditional expectation of $C_i(t + 1) - C_i(t)$ is as follows:

$$\mathbf{E}[C_i(t + 1) - C_i(t)|\, C_0(t), \ldots, C_k(t)] = -\frac{iC_i}{n - t}\delta_{i \neq 0} + \mu_\lambda \frac{(i + 1)C_{i+1}}{\rho}\delta_{i \neq k} \quad (2)$$

where $i = 0, 1 \ldots k$ and $\delta_{i \neq j}$ is 0 if $i = j$ and 1 otherwise. Note we have an equation for each value of i.

To better understand these difference equations, let's consider a specific case. For example, with $k = 3$ and $i = 3$, we obtain

$$\mathbf{E}[C_3(t+1) - C_3(t) \mid C_0(t), \ldots, C_3(t)] = -\frac{3C_3}{n-t}$$

The terms on the right, measure the expected reduction of ternary clauses at time t. Note that, at time t there are $n - t$ variables, so the probability that a 3-clause has a fixed variable x is $\frac{\binom{n-1}{2}}{\binom{n}{3}}$. Therefore, $\frac{3C_3}{n-t}$ is the expected number of 3-clauses with a variable x.

The definition of ρ and μ_λ are as follows. Let $C(t) = (C_1(t), \ldots, C_k(t))$, $F \in \Phi_{C(t)}$ and X be the random variable defined as the number of clauses in F where the random literal l appears. The distribution of X can be approximated by a Poisson random variable with parameter $\lambda = \frac{\rho}{2(n-t)}$, where $\rho = C_1(t) + 2C_2(t) + \cdots + kC_k(t)$. The algorithm takes a variable x and satisfies the literal that appears the most among $\{x, \bar{x}\}$. Denote by Z the number of clauses in which the falsified literal appears (that is, the literal that appears the least number of times from among $\{x, \bar{x}\}$). Z has the distribution of $\min(X, X')$, where X' is independent to X and have both distribution Poisson with parameter λ. Let $\mu_\lambda = \mathbf{E}(Z)$, be the expected value of Z.

Wormald's theorem says that we can approximate the values of $C_i(t)$ by the solutions $c_i(x)$ of the following system of differential equations.

$$\frac{dc_i}{dx} = -\frac{ic_2}{1-x}\delta_{i \neq 0} + \mu_\lambda \frac{(i+1)c_{i+1}}{\rho}\delta_{i \neq k} \tag{3}$$

For $i = 0, 1, \ldots, k$ and with initial conditions $c_i(0) = \frac{C_i(0)}{n}$. Here $\rho = c_1 + 2c_2 + \cdots + kc_k$, the scaled number of literals in the formula and $\lambda = \frac{\rho}{2(1-x)} \cdot \mu_\lambda$ has the same definition as before.

At any time $t < (1 - \epsilon)n$, $c_i(t/n)$ gives a good approximation of the scaled values of $C_i(t)$. More precisely,

$$C_i(nx) = c_i(x)n + o(n) \tag{4}$$

with high probability when n goes to infinity.

Wormald's theorem can be applied for $0 \leq x \leq 1 - \epsilon$, for any $\epsilon > 0$. That is because $x = 1$ is a singularity point for our function f. In our analysis, we take $\epsilon = 10^{-5}$. To get around the problem of not having equation (4) for all the values of t, we analyze the algorithm for $0 \leq t \leq n(1 - \epsilon)$ and then count all the remaining clauses plus the empty clauses as not satisfied by the assignment. Let $t_\epsilon = n(1 - \epsilon)$, we use the following bound $C_0(n) \leq C_0(t_\epsilon) + C_1(t_\epsilon) + \cdots + C_k(t_\epsilon)$ for the number of empty clauses generated by the algorithm. A precise statement of the theorem is given on the Appendix.

5   Proof of the Theorem. Results for MAX-R-3SAT

We solve the differential equations (3) numerically using the *ode45* function of matlab. The values of μ_λ are approximated numerically. The results are in agreement with simulations of the algorithm on randomly generated 3-SAT formulas.

In figure 1 we give the results of approximating $1 - \frac{C_0(n)}{m}$, the fraction of clauses that are satisfied by the algorithm, with a lower bound $1 - \frac{c_0(x_\epsilon)+c_1(x_\epsilon)+c_2(x_\epsilon)+c_3(x_\epsilon)}{r}$, where $x_\epsilon = 1 - \epsilon = t_\epsilon/n$ and $c_i(x)$, $1 \le i \le 3$ are the solutions of the differential equations.

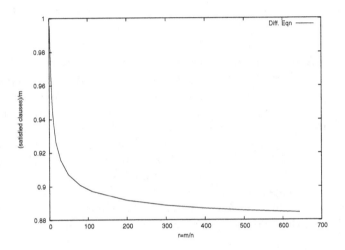

Fig. 1. Fraction of satisfied clauses ($g(r)$) as the function of r. Results from the solution of the differential equations

We will use the following result in the proof of theorem 1.

Theorem 2. *[2] Let F be a k-CNF random formula, if*

$$r(F) > \tau(p) = 2^k \ln 2/(p + (1-p)\ln(1-p)),$$

then $\mathbf{Pr}(max(F) > (1 - 2^k(1-p))m)$ *goes to zero as n goes to infinity.*

The result in theorem 2 provides an upper bound on the maximum number of clauses that can be satisfied in a typical random k-CNF formula.

Proof. of Theorem 1.
To prove the 10/9.1-approximation for MAX-R-3SAT, we choose $r_3 = 183$ as the parameter for the algorithm. The result for $r = m/n > r_3$ holds just by the lemma in the subsection 4.1.

For $r = m/n \le r_3$ we split our proof into two parts, for $r \le 12$ and $12 \le r \le 183$. For $r = 12$, the function $g(r) \ge 0.9357$, as $g(r)$ is a decreasing function of r

then $g(r) \geq 0.9357$ for $r \leq 12$. Then $\frac{max(F)}{m_A(F)} \leq \frac{m}{g(r)m} < \frac{10}{9.1}$ w.h.p. Here we take m as an approximation to the optimal value of $max(F)$.

For $12 < r \leq 183$ using theorem 2 for $p = 0.8$, $k = 3$ we get that for $r > 11.6$ the probability that $0.975m$ clauses can be satisfied goes to zero as n goes to infinity. Therefore, we can use that $max(F) \leq 0.975m$ w.h.p. and the fact that for $r \leq 183$ $g(r) \geq 0.8922$ to obtain $\frac{max(F)}{m_A(F)} \leq \frac{0.975m}{g(r)m} < \frac{10}{9.1}$ w.h.p.

The $10/9.5$-approximation result can be obtained by carefully dividing the interval $r \in (0, 643.5)$ in several pieces. For each piece, using theorem 2, we get an upper-bound for $max(F)$, and our function $g(r)$ for the approximation of $m_A(F)$. The analysis for $r \geq 643.5$ comes from the results in subsection 4.2.

References

1. D. Achlioptas. Lower Bounds for Random 3-SAT via Differential Equations. Theoretical Computer Science, 265 (1-2), p.159-185 (2001).
2. D. Achlioptas, A. Naor, and Y. Peres. On the Fraction of Satisfiable Clauses in Typical Formulas. Extended Abstract in FOCS'03, p. 362-370.
3. D. Achlioptas and G. B. Sorkin. Optimal Myopic Algorithms for Random 3-SAT. In Proceedings of FOCS 00, p. 590-600.
4. A. Z. Broder, A. M. Frieze, and E. Upfal. On the satisfiability and maximum satisfiability of random 3-CNF formulas. In Proc. 4th Annual ACM-SIAM Symposium on Discrete Algorithms, p. 322–330, (1993).
5. M-T. Chao and J. Franco. Probability analysis of two heuristics for the 3-satisfiability problem. SIAM J. Comput., 15(4) p.1106-1118, (1986).
6. M. T. Chao, and J. Franco. Probabilistic analysis of a generalization of the unit clause selection heuristic for the k-satisfiability problem. Information Sciences 51 p. 289-314, (1990).
7. W. Fernandez de la Vega, and M. Karpinski. 9/8-Approximation Algorithm for Random MAX-3SAT. Electronic Colloquium on Computational Complexity (ECCC)(070) (2002).
8. A. C. Kaporis, L. M. Kirousis, and E. G. Lalas. The probabilistic analysis of a greedy satisfiability algorithm. In 10th Annual European Symposium on Algorithms (Rome, Italy, 2002).
9. A. C. Kaporis, L. M. Kirousis, and E. Lalas. Selecting complementary pairs of literals. Electronic Notes in Discrete Mathematics, Vol. 16 (2003).
10. A.C. Kaporis, L.M. Kirousis, and Y.C. Stamatiou. How to prove conditional randomness using the Principle of Deferred Decisions. Technical Report, Computer Technology Institute, Greece, 2002. Available at: www.ceid.upatras.gr/faculty/kirousis/kks-pdd02.ps.
11. D. Mitchell, B. Selman, and H. Levesque. Hard and easy distributions of sat problems. In Proc. 10-th National Conf. on Artificial Intelligence (AAAI-92), p. 459–465.
12. B. Selman, D. Mitchell, and H. Levesque. Generating Hard Satisfiability Problems. Artificial Intelligence, Vol. 81, p. 17–29, (1996).
13. N. C. Wormald. Differential equations for random processes and random graphs. Ann. Appl. Probab. 5 (4) p. 1217–1235. 36, (1995).

A   Differential Equations

We consider here a sequence of random process $Y_t = Y_t(n), n = 1, 2, \ldots$. For simplicity the dependence on n is dropped from the notation. Let $\mathcal{F}_t$ be the the σ-algebra generated by the process up to time t, i.e $\mathcal{F}_t = \sigma(Y_0, Y_1, \ldots, Y_t)$. Our process $Y_t = (Y_t^{(1)}, \ldots, Y_t^{(j)})$ is a vector of dimension j. Let $\|Y\| = \max(|Y^{(1)}|, \ldots, |Y^{(j)}|)$. Suppose that $Y_0 = z_0 n$ the value of the process at time 0.

We say that $X = o(f(n))$ always if $\max\{x : \mathbf{Pr}(X = x) \neq 0\} = o(f(n))$. The term *uniformly* means that the convergence implicit in the $o()$ is uniform on t.

Theorem 3. *[13] Let $f : \Re^{j+1} \to \Re^j$. Suppose there exists a constant C such that the process Y_t is bounded by Cn, i.e $\|Y_t\| < Cn$. Suppose also that for some function $m = m(n)$:*

(i) *uniformly over all $t < m$*

$$\mathbf{Pr}(\|Y_{t+1} - Y_t\| > n^{1/5} | \mathcal{F}_t) = o(n^{-3})$$

always;

(ii) *for all l and uniformly over all $t < m$,*

$$\mathbf{E}(Y_{t+1} - Y_t | \mathcal{F}_t) = f(t/n, Y_t/n) + o(1)$$

always;

(iii) *The function f is continuous and satisfies a Lipschitz condition on D, where D is some bounded open set containing $(0, z_0^{(1)}, \ldots, z_0^{(j)})$.*

then:

(a) *The system of differential equations*

$$\frac{dz}{ds} = f(s, z)$$

has a unique solution in D for $z : \Re \to \Re^j$ with initial conditions $z(0) = z_0$ and which extends to points arbitrarily closed to the boundary of D.

(b)

$$Y_t = nz(t/n) + o(n) \ w.h.p$$

uniformly for $0 \leq t \leq \min\{\sigma n, m\}$, where σ is the supremum of those s to which the solution can be extended.

Clause Form Conversions for Boolean Circuits

Paul Jackson and Daniel Sheridan

School of Informatics,
University of Edinburgh, Edinburgh, UK
pbj@inf.ed.ac.uk
d.j.sheridan@sms.ed.ac.uk

Abstract. The Boolean circuits is well established as a data structure for building propositional encodings of problems in preparation for satisfiability solving. The standard method for converting Boolean circuits to clause form (naming every vertex) has a number of shortcomings.

In this paper we give a projection of several well-known clause form conversions to a simplified form of Boolean circuit. We introduce a new conversion which we show is equivalent to that of Boy de la Tour in certain circumstances and is hence optimal in the number of clauses that it produces. We extend the algorithm to cover reduced Boolean circuits, a data structure used by the model checker NuSMV.

We present experimental results for this and other conversion procedures on BMC problems demonstrating its superiority, and conclude that the CNF conversion has a significant role in reducing the overall solving time.

1   Introduction

SAT solvers based on the DPLL procedure typically require their input to be in conjunctive normal form (CNF). Earlier papers dealing with encoding to SAT, particularly much of the planning literature, encode directly from the input representation to clause form. More recent encoding work makes little mention of CNF conversion. Biere et al., proposing BMC [3], give an encoding to propositional logic — we assume from their space complexity claim that a DAG representation is in use — but they make no mention of the final conversion to CNF.

The microprocessor verification work of Velev includes a thorough analysis of improving the clause form generated [10], but the work is not immediately applicable to general propositional logic. Nevertheless, Velev is able to claim a speed up by a factor of 32 by altering the clause form conversion.

There is other evidence to motivate the study of clause form conversions for SAT. While focussing on CNF representations of cardinality constraints, Bailleux and Boufkhad [2] give a reformulation of the parity problems which have been standard SAT benchmarks for a number of years. They argue that the problems are made harder than they should be by a poor clause form representation, and demonstrate a dramatic speedup on the par32 problem with modern solvers on the reformulated problem.

H.H. Hoos and D.G. Mitchell (Eds.): SAT 2004, LNCS 3542, pp. 183–198, 2005.
© Springer-Verlag Berlin Heidelberg 2005

In the first-order logic domain, the CNF conversion problem was handled comprehensively by Boy de la Tour [4]. The algorithm given is impractical without the improvements by Nonnengart et al. [8], and the resulting algorithm is fiddly to implement making it hard to be confident of a correct implementation.

In this paper we introduce a simple and easy to understand CNF conversion algorithm for propositional logic constructed as a hybrid between the structure-preserving conversion [9] and the standard distributivity law application. We prove that it produces the minimum number of clauses for certain classes of formula. As its time complexity is linear in the size of the input formula, it represents a significant improvement over the (quadratic time) Boy de la Tour algorithm. Of course, it is well known that problem size does not necessarily correspond to solving time in SAT, so we present some experimental results demonstrating the effect that our algorithm has on some BMC [3] problems.

1.1   Notation Conventions

In an attempt to improve the clarity of the presentation, we use a number of conventions in our notation. Much of the work is concerned with both graphs and propositional logic, so we distinguish between *graph variables* ranging over vertices and edges given in italic capitals (X, Y) and *propositional variables* given in italic lower case (x, y); vertices are typically denoted V and edges E and this notation is significant in determining the type of a function. We will use the shorthand of referring to a subgraph by a single edge; the subgraph thus identified includes all of the descendents of the edge given, and such an edge is called the *root* of the subgraph and denoted T. Sets of vertices or edges are given in bold type $(\mathbf{X}, \mathbf{Y})$.

Where a function creates new propositional variables, these are given the name x_i where i is some identifier (typically a graph vertex). These variables are assumed to be unused in any other context.

2   Boolean Circuits

In contrast to the formulaic representation of propositional logic normally used, Boolean circuits are much closer to an electronics view of logic. Labelled input *wires* take the place of variables and together with (possibly unlabelled) internal wires they are connected by logic *gates* which compute various logic functions. This makes it very natural for the results of sub-circuits to be shared amongst other parts of the circuit, as would be expected in the physical world.

Boolean circuits may be efficiently represented as directed acyclic graphs (DAGs). Vertices having outgoing edges correspond to gates, with the edges pointing to the inputs to the gate. Vertices without outgoing edges (which we will call *leaf* vertices) are the inputs and outputs for the circuit, corresponding to variables in a propositional formula.

Abdulla, Bjesse, and Eén proposed *reduced* Boolean circuits (RBCs) [1] as a DAG representation of a propositional formula with additional restrictions on

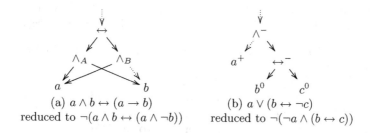

(a) $a \wedge b \leftrightarrow (a \rightarrow b)$
reduced to $\neg(a \wedge b \leftrightarrow (a \wedge \neg b))$

(b) $a \vee (b \leftrightarrow \neg c)$
reduced to $\neg(\neg a \wedge (b \leftrightarrow c))$

Fig. 1. Example RBCs showing vertex labelling

the type and relationships of the gates which place RBCs somewhere between being a normal form and a canonical form for propositional formulæ. One of the key strengths of Boolean circuits is the ability to use one circuit to represent a formula both positively and negatively. To preserve this property, Abdulla et al. eschew NNF in favour of restricting gates to conjunctions and equivalences (bi-implications), marking negation on the edges of the graph.

Definition 1. *An RBC is a DAG consisting of edges* **E** *and vertices* **V** $=$ **V$_I$** $\cup$ **V$_L$** *where internal vertices* **V$_I$** *represent operators, and leaf vertices* **V$_L$** *represent variables. The following properties are required to hold and form the encoding of Boolean circuits as DAGs:*

- *Each* $V \in$ **V$_I$** *consists of an operator* $op(v) \in \{\wedge, \leftrightarrow\}$ *and a left and right edge (left(V), right(V)* $\in$ **E***).*
- *Each* $V \in$ **V$_L$** *contains a variable var(V).*
- *Each* $E \in$ **E** *has a sign sign(E)* $\in \{+, -\}$ *and a target vertex target(E)* $\in$ **V***.*

The sign attribute encodes negation, where $sign(E) = +$ *indicates an unnegated edge and* $sign(E) = -$ *indicates a negated edge. The following additional properties serve to reduce the number of representations possible for equivalent formulæ:*

- *All common subformulæ are shared:*
  $\forall V, V' \in$ **V$_I$**$, left(V) = left(V') \wedge right(V) = right(V') \rightarrow V = V'$.
- *The constant* $\top$ *only occurs in single-vertex RBCs.*
- *For all vertices, left(V)* $\neq$ *right(V).*
- *If* $op(V) = \leftrightarrow$ *then left(V) and right(V) are unsigned.*
- *There is a total order* $\prec$ *such that for all* $V \in$ **V***, left(V)* $\prec$ *right(V).*

For example, Figure 1a shows the RBC representing the formula $a \wedge b \leftrightarrow \neg(a \rightarrow b)$, with some internal vertices annotated by a subscript capital. The annotations allow us to refer to the subformula $a \wedge b$ by the vertex A, for example, and also allows us to depict RBC fragments by identifying a vertex without giving any further details.

To simplify the definitions in this paper we extend the set of properties on RBC vertices and edges with the inverse function of *target*:

$$source(E) = \begin{cases} V & \text{if } E = left(V) \vee E = right(V) \\ \text{undefined} & \text{otherwise} \end{cases}$$

RBC Operations. Two RBCs rooted at edges L and R, are composed given an operation $o \in \{\wedge, \leftrightarrow\}$ and a sign $s \in \{+, -\}$ to give the RBC $rbc(L, R, o, s)$:

- If o may be trivially evaluated using identity and other properties, return the result of doing so.
- Otherwise, check $L \prec R$ and swap if not.
- If $o = \leftrightarrow$ then s becomes $s \oplus sign(L) \oplus sign(R)$, and $sign(L)$ and $sign(R)$ become $+$ ($\oplus$ is the exclusive-or operation).
- The new vertex $V = \langle o, L, R \rangle$ is inserted into the DAG.
- The result is the edge $\langle sign, V \rangle$.

3   CNF Conversions on Linear Trees

We begin by examining CNF conversions for a restriction of RBCs, which will become a building block for the CNF conversions of full RBCs. Linear trees represent linear formulæ (those without equivalence operators) without taking into account the possibility for sharing.

Definition 2. *A* linear tree *is an RBC with the following changes to its structure:*

- *The only internal vertices are conjunction vertices*
- *No vertices are shared: the graph is a tree*

Given a linear tree, we define the following additional properties over vertices

$$inedge(V) = E \qquad \text{where } target(E) = V$$

$$sib(V) = \begin{cases} target(left(V')) & \text{if } inedge(V) = right(V') \\ target(right(V')) & \text{if } inedge(V) = left(V') \end{cases}$$

We give the various well-known CNF conversions informally and as depth-first procedures on linear trees. Each conversion produces a set of clauses; we write $|C|$ for the number of clauses in C, and use the union ($\cup$) operator to combine sets of clauses, and the cross-multiply operator ($\times$), to form the set of clauses corresponding to the disjunction of two sets, obtained by

$$A \times B = \{x \cup y \mid x \in A, y \in B\}$$

The *standard* CNF conversion is that obtained by exploiting the distributive properties of $\wedge$ and $\vee$ on a formula already in NNF to push disjunctions in towards the literals. This produces an equivalent (rather than equisatisfiable) formula at the expense of a potentially exponential number of clauses. Nevertheless, the conversion is optimal for some input formulæ. We define the conversion for linear trees as a recursive descent. $CNF(T)$ given in Figure 2 denotes the standard CNF conversion of the subtree beginning at a root edge T.

$$\mathrm{CNF}(E) = \begin{cases} \mathrm{CNF}(target(V)) & \text{if } sign(E) = + \\ \mathrm{CNF}^-(target(V)) & \text{if } sign(E) = - \end{cases}$$

$$\mathrm{CNF}^-(E) = \begin{cases} \mathrm{CNF}^-(target(V)) & \text{if } sign(E) = + \\ \mathrm{CNF}(target(V)) & \text{if } sign(E) = - \end{cases}$$

$$\mathrm{CNF}(V) = \begin{cases} var(V) & \text{if } V \in \mathbf{V_L} \\ \mathrm{CNF}(left(V)) \cup \mathrm{CNF}(right(V)) & \text{if } op(V) = \wedge \end{cases}$$

$$\mathrm{CNF}^-(V) = \begin{cases} \neg\, var(V) & \text{if } V \in \mathbf{V_L} \\ \mathrm{CNF}^-(left(V)) \times \mathrm{CNF}^-(right(V)) & \text{if } op(V) = \wedge \end{cases}$$

Fig. 2. The standard clause form conversion for linear trees

3.1   Clause Form Conversions with Renaming

Renaming subformulæ is a strategy for reducing the number of clauses produced by a formula. The observation is made that a subformula may be replaced by a single variable if clauses are given to constrain that variable such that the satisfiability of the overall formula is unaffected. Such a conversion is said to be *equisatisfiable*: the introduced variables break equivalency. For example, the formula $(a \wedge b \wedge c) \vee (d \wedge e \wedge f)$ produces nine clauses in the standard conversion; introducing a new variable for the left-hand disjunct to produce the formula

$$x_{a \wedge b \wedge c} \vee (d \wedge e \wedge f) \quad \wedge \quad x_{a \wedge b \wedge c} \leftrightarrow (a \wedge b \wedge c)$$

with $x_{a \wedge b \wedge c}$ constrained by the equivalence on the right hand side results in only seven clauses. Nevertheless, it is satisfiable by precisely those assignments that satisfy the original formula.

$$\mathrm{DEF}(E) = \begin{cases} \mathrm{DEF}(target(V)) & \text{if } sign(E) = + \\ \mathrm{DEF}^-(target(V)) & \text{if } sign(E) = - \end{cases}$$

$$\mathrm{DEF}(V) = \begin{cases} var(V) & \text{if } v \in \mathbf{V_L} \\ \{\{\neg x_V, x_{target(left(V))}\}, \{\neg x_V, x_{target(right(V))}\}\} \\ \quad \cup \{\{x_V, \neg x_{target(left(V))}, \neg x_{target(right(V))}\}\} \\ \quad \cup \mathrm{DEF}(left(V)) \cup \mathrm{DEF}(right(V)) & \text{if } op(V) = \wedge \end{cases}$$

$$\mathrm{DEF}^-(V) = \begin{cases} \neg\, var(V) & \text{if } v \in \mathbf{V_L} \\ \{\{x_V, x_{target(left(V))}\}, \{x_V, x_{target(right(V))}\}\} \\ \quad \cup \{\neg x_V, \neg x_{target(left(V))}, \neg x_{target(right(V))}\}\} \\ \quad \cup \mathrm{DEF}(left(V)) \cup \mathrm{DEF}(right(V)) & \text{if } op(V) = \wedge \end{cases}$$

Fig. 3. The definitional clause form conversion

$$\mathrm{ren}(T, \mathbf{R}) = rbc(\mathrm{def}(T, T, \mathbf{R}), \mathrm{sub}(T, \mathbf{R}), \wedge, +)$$

$$\mathrm{def}(T, E, \mathbf{R}) = \mathrm{def}(T, target(E), \mathbf{R})$$

$$\mathrm{def}(T, V, \mathbf{R}) = \begin{cases} V & \text{if } V \in \mathbf{V_L} \\ rbc(\begin{cases} \top & \text{if } V \notin \mathbf{R} \\ rbc(x_V, \mathrm{sub}^-(V, \mathbf{R} \setminus \{V\}), \wedge, -) & \text{if } \mathrm{pol}(T, V) = 1 \\ rbc(\neg x_V, \mathrm{sub}^+(V, \mathbf{R} \setminus \{V\}), \wedge, -) & \text{if } \mathrm{pol}(T, V) = -1 \end{cases}, \\ \quad rbc(\mathrm{def}(T, left(V), \mathbf{R}), \mathrm{def}(T, right(V), \mathbf{R}), \wedge, +), \\ \quad \wedge, +) & \text{if } V \in \mathbf{V_I} \end{cases}$$

$$\mathrm{sub}(E, \mathbf{R}) = \mathrm{sub}^{sign(E)}(target(E), \mathbf{R})$$

$$\mathrm{sub}^s(V, \mathbf{R}) = \begin{cases} V & \text{if } V \in \mathbf{V_L} \\ x_V & \text{if } V \in \mathbf{R} \\ rbc(\mathrm{sub}(left(V), \mathbf{R}), \mathrm{sub}(right(V), \mathbf{R}), op(V), s) & \text{otherwise} \end{cases}$$

Fig. 4. The vertex-based renaming construction $\mathrm{ren}(T, \mathbf{R})$. $\mathbf{R}$ identifies the subgraphs to be renamed; $\mathrm{sub}(T, \mathbf{R})$ returns a copy of the graph with root edge T, replacing renamed subgraphs by new variables; $\mathrm{def}(T, T', \mathbf{R})$ returns the graph which is the conjunction of the definition of all of the introduced variables below T' (T is used to establish the polarity of the subgraph)

The most straightforward algorithm of this type gives a new name to every internal vertex of the tree and is known as the *definitional* clause form conversion, given in Figure 3. In fact, as observed by Plaisted and Greenbaum [9], if a subformula occurs with positive or negative polarity — if it appears under an even or odd number of negations — then only an implication is required to constrain the new variable, with the direction of the implication corresponding to the polarity of the subformula. We define the polarity function $\mathrm{pol}(T, V)$ for a vertex V in a linear trees T as

$$\mathrm{pol}(T, T) = 1$$

$$\mathrm{pol}(T, E) = \begin{cases} \mathrm{pol}(T, source(E)) & \text{if } sign(E) = + \\ -\mathrm{pol}(T, source(E)) & \text{if } sign(E) = - \end{cases}$$

$$\mathrm{pol}(T, V) = \mathrm{pol}(inedge(V))$$

In the example above, the subformula $a \wedge b \wedge c$ appears positively, so the renaming can be shortened to

$$x_{a \wedge b \wedge c} \vee (d \wedge e \wedge f) \quad \wedge \quad x_{a \wedge b \wedge c} \rightarrow (a \wedge b \wedge c)$$

producing only six clauses.

For linear trees, we consider only renamings of *vertices* (other analyses place an equivalent restriction on renaming subfomulæ with negation as the main connective). The order in which renamings are made does not affect the final result due to the commutivity of $\wedge$, so we are able to give renaming-based clause form conversions in terms of the sets of vertices that they rename. The general transformation in Figure 4 constructs the renamed formula in two parts: a copy of the original graph with renamed subgraphs replaced by the appropriate variables; and a graph giving the definitions of all of the introduced variables. The case split by polarity in $\text{def}(T, T', \mathbf{R})$ constructs the RBCs for $x_V \to V$ or $V \to x_V$ as appropriate; the remainder of $\text{def}(T, T', \mathbf{R})$ simply constructs a tree of conjunctions while traversing the RBC recursively. The correctness of $\text{ren}(T, \mathbf{R})$ and hence of the conversions based on it follows directly from the correctness of renaming in general. We can now write the structure-preserving clause form conversion due to Plaisted and Greenbaum [9] as simply

$$SP(T) = \text{CNF}(\text{ren}(T, \mathbf{V_I}))$$

It is easy to construct cases where the definitional and structure-preserving conversions perform significantly worse than the standard conversion, despite the difference in asymptotic complexity — the worst case is a formula already in CNF, the SP conversion doubling the size of the clause form while the standard conversion leaves the formula unchanged. By carefully selecting the vertices to rename we can obtain a blend of the two algorithms.

3.2   The Conversion Due to Boy de la Tour

Boy de la Tour [4] presents a comprehensive solution to the problem of choosing the subformulæ to rename. The approach taken is to compute the impact of renaming any given subformula and to perform the renaming only if it will not increase the number of clauses produced by the formula as a whole. We give a very terse presentation below as our main interest is in making use of its optimality for formulæ without equivalences.

Boy de la Tour defines the functions $p^+(T) = |\text{CNF}(T)|$ and $p^-(T) = |\text{CNF}(\neg T)|$ using a simple lookup table (Table 1) which enables these values to be computed without constructing the clauses themselves. The *benefit* (the reduction in the total number of clauses) of renaming a vertex V in a tree T is given by

$$B(T, V) = p^+(T) - p^+(\text{ren}(T, \{V\}))$$

Table 1. The clause counting functions $p^+(V)$ and $p^-(V)$

	$p^+(E)$	$p^-(E)$
$sign(E) = +$	$p^+(target(E))$	$p^-(target(E))$
$sign(E) = -$	$p^-(target(E))$	$p^+(target(E))$
	$p^+(V)$	$p^-(V)$
$v \in \mathbf{V_L}$	1	1
$op(V) = \wedge$	$p^+(left(V)) + p^+(right(V))$	$p^-(left(V))p^-(right(V))$

Table 2. Computation of the coefficients a_V^T and b_V^T

	a_E^T	b_E^T
$E = T$	1	0
$sign(E) = +$	$a_{source(E)}^T$	$b_{source(E)}^T$
$sign(E) = -$	$b_{source(E)}^T$	$a_{source(E)}^T$
	a_V^T	b_V^T
$op(V) = \wedge$	$a_{inedge(V)}^T$	$b_{inedge(V)}^T p^-(sib\ V)$

$$\mathrm{BDLT}(T, E) = \mathrm{BDLT}(T, target(V))$$

$$\mathrm{BDLT}(T, V) = \begin{cases} \emptyset & \text{if } v \in \mathbf{V_L}, \text{ or} \\ \mathrm{BDLT}(T, left(V)) \cup \mathrm{BDLT}(T, right(V)) & \text{if } B(T, V) < 0, \text{ or} \\ \{V\} \cup \mathrm{BDLT}(\mathrm{ren}(T, \{V\}), left(v)) & \\ \quad \cup \mathrm{BDLT}(\mathrm{ren}(T, \{V\}), right(V)) & \text{if } B(T, V) \geq 0 \end{cases}$$

Fig. 5. Renaming sets construction for the Boy de la Tour conversion

In order to make a decision about renaming at a particular vertex without needing to analyse the whole tree, $p^+(T)$ is rewritten in terms of $p^+(V)$ and $p^-(V)$:

$$p^+(T) = a_V^T p^+(V) + b_V^T p^-(V) + c_V^T$$

Where the coefficients a, b may be considered as the number of occurrences of the clauses representing V and $\neg V$ respectively, such that the first sum counts the total number of clauses including subformulæ of V; the coefficient c represents the number of clauses due to the rest of the tree. a and b are computed from the context of V as in Table 2. Note that the values are related to the polarity of the vertices: $a_V^T = 0$ if $\mathrm{pol}(T, V) = -1$ and $b_V^T = 0$ if $\mathrm{pol}(T, V) = 1$. When computing the benefit, the coefficient c is cancelled, so we do not need to give its construction. The benefit function can now be given in terms of polarity as

$$a_V^T p^+(V) - (a_V^T + p^+(V)) \qquad \text{if } \mathrm{pol}(T, V) = 1$$
$$b_V^T p^-(V) - (b_V^T + p^-(V)) \qquad \text{if } \mathrm{pol}(T, V) = -1$$

The algorithm given by Boy de la Tour is a top-down computation of the benefit of a renaming given those that have gone before. The construction of the renaming set in Table 5 allows us to write the algorithm as

$$\mathrm{BDLT}(T) = \mathrm{CNF}(\mathrm{ren}(T, \mathrm{BDLT}^+(T, T) \cup \mathrm{BDLT}^-(T, T)))$$

A dynamic programming implementation of $B(T, V)$ as given by Boy de la Tour [4] requires $O(1)$ computations at each vertex but the arithmetic is on $|\mathbf{V}|$-bit words which leads to a per-vertex complexity of $O(|\mathbf{V}|)$. The resulting algorithm is $O(|\mathbf{V}|^2)$ in contrast to DEF and SP which are both linear.

The presentation by Nonnengart et al. [8] removes the need for arbitrary-length arithmetic by reducing $B(T, V) \geq 0$ to an elaborate series of syntactic conditions on the formula.

4   The Compact Conversion

We present a new clause form conversion, the *compact* conversion, which computes the sets of renamed vertices locally and bottom-up. For each vertex we consider the number of clauses it will generate based on whether a child vertex is renamed. Consider a disjunction $\phi \vee \psi$, with all subformulæ of ϕ and ψ already renamed as appropriate. The disjunction is converted by either renaming an argument, eg ϕ to x_ϕ, which produces a definition $x_\phi \rightarrow \phi$ and replaces the disjunction by the renamed form $x_\phi \vee \psi$; or alternatively computing $\mathrm{CNF}(\phi) \times \mathrm{CNF}(\psi)$ — the standard conversion of the disjunction. The decision is made based on which generates the most clauses, determined by the sum or the product, respectively, of the number of clauses in ϕ and ψ.

More precisely, we define the function $\mathrm{COMP}(T, V)$ in Figure 6 to give the set of renamings on the tree beginning at V. The auxiliary function $\mathrm{dis}(V)$ chooses the best child of V, if any, to rename by using the sum-vs-product decision. The renaming condition is computed on the tree after all vertices below that considered have been renamed. We define a new pair of clause-counting functions $p_r^+(V, \mathbf{R})$ and $p_r^-(V, \mathbf{R})$ giving the number of clauses produced by the graph beginning at vertex V after the application of renaming $\mathbf{R}$ (Table 3). That is, $p_r^s(V, \mathbf{R}) = |\mathrm{sub}^s(V, \mathbf{R})|$ (the clauses in $\mathrm{def}^s(V, \mathbf{R})$ are disregarded as they play no further part in determining the size of the result).

Since we are targeting a SAT solver with this conversion, with its (assumed) exponential complexity in the number of variables, we choose to rename only if it *reduces* the number of clauses produced; the analysis of the Boy de la Tour conversion is simplified by allowing renamings which result in the same number of clauses.

$$\mathrm{COMP}(T, E) = \mathrm{COMP}(T, target(V))$$

$$\mathrm{COMP}(T, V) = \begin{cases} \emptyset & \text{if } V \in \mathbf{V_L}, \text{ or} \\ \mathrm{COMP}(T, left(V)) \cup \mathrm{COMP}(T, right(V)) & \text{if } \mathrm{pol}(T, V) = 1, \text{ or} \\ \mathrm{dis}(V) \cup \mathrm{COMP}(T, left(V)) \cup \mathrm{COMP}(T, right(V)) & \text{if } \mathrm{pol}(T, V) = -1 \end{cases}$$

$$\mathrm{dis}(V) = \left\{ \begin{array}{ll} \emptyset & \text{if } n_l n_r \leq n_l + n_r, \text{ or} \\ \{left(V)\} & \text{if } n_l > n_r \\ \{right(V)\} & \text{if } n_l \leq n_r \end{array} \right\} \text{ where } \left\{ \begin{array}{l} n_l = p_r^-(left(V), \\ \qquad \mathrm{COMP}(T, left(V))) \\ n_r = p_r^-(right(V), \\ \qquad \mathrm{COMP}(T, right(V))) \end{array} \right\}$$

Fig. 6. Renaming sets construction for the compact conversion

Table 3. The renaming-compensated clause counting functions $p_r^+(T, \mathbf{R})$ and $p_r^-(T, \mathbf{R})$

	$p_r^+(E, \mathbf{R})$	$p_r^-(E, \mathbf{R})$
$sign(E) = +$	$p_r^+(target(E), \mathbf{R})$	$p_r^-(target(E), \mathbf{R})$
$sign(E) = -$	$p_r^-(target(E), \mathbf{R})$	$p_r^+(target(E), \mathbf{R})$
	$p_r^+(V, \mathbf{R})$	$p_r^-(V, \mathbf{R})$
$V \in \mathbf{V_L}$	1	1
$V \in \mathbf{R}$	1	1
$op(V) = \wedge$	$p_r^+(left(V), \mathbf{R}) + p_r^+(right(V), \mathbf{R})$	$p_r^-(left(V), \mathbf{R}) \cdot p_r^-(right(V), \mathbf{R})$

4.1   Optimality of the Compact Conversion for Linear Trees

The main result of this paper is to show the optimality of the compact conversion which we do by a comparison with the Boy de la Tour conversion. We establish which vertices appear in the renaming sets of one conversion and not the other, and then analyse the impact that the differences make.

When comparing the decision taken to include a vertex in the renaming sets by the two algorithms we take into account the different contexts: in the Boy de la Tour algorithm, the superformulæ have already been renamed; in the compact conversion the subformulæ have been renamed. Writing $\mathbf{R}$ for a set of renamings, we have $\mathbf{R}_{\neg V}$ for the subset of renamings involving the superformulæ of V and $\mathbf{R}_{\sqsubset V}$ for the subset involving the subformulæ of V. The compact conversion depends only on p_r^+ and p_r^- but these are computed after subformula renaming. That is, the decision to rename the vertex V_1 in $V_1 \wedge V_2$ is based on the values $p_r^+(V_1, \mathbf{R}_{\sqsubset V_1})$, $p_r^-(V_2, \mathbf{R}_{\sqsubset V_2})$ and their complements. In contrast, for the Boy de la Tour algorithm the decision is based on the values $a_{V_1}^{\text{ren}(T, \mathbf{R}_{\neg V_1})}$, $b_{V_1}^{\text{ren}(T, \mathbf{R}_{\neg V_1})}$, $p^+(V_1)$, $p^-(V_1)$

We begin by establishing some basic lemmas about the Boy de la Tour coefficients and the clause counting functions; in each case we refer to a vertex V and renamings $\mathbf{R}$ and $\mathbf{R}'$ on a tree T.

Lemma 1. $a_V^{\text{ren}(T, \mathbf{R})} = 1$ if $\text{pol}(T, V) = 1$, and $b_V^{\text{ren}(T, \mathbf{R})} = 1$ if $\text{pol}(T, V) = -1$

Proof. After renaming, a vertex V becomes part of the definition of the replacement variable x_V. According to Figure 4, the definition is attached by a tree of positive conjunctions to the root with the sign of the inedge of V reflecting its original polarity. By the definition of a_V^T and b_V^T on conjunctions, the lemma holds.

Lemma 2. If $\mathbf{R}' \subseteq \mathbf{R}$, $p_r^s(V, \mathbf{R}) \leq p_r^s(V, \mathbf{R}') \leq p^s(V)$

Proof. This follows from the definitions of p_r^s and p_r. Both increase monotonically with tree depth. As renaming effectively prunes part of the tree, it can only reduce the values of the functions.

4.2   Positive Polarity

Lemma 3. *Neither conversion renames the children of positive polarity conjunctions. That is, for* $\mathbf{pc} = \{V \in \mathbf{V}_I | \, \mathrm{pol}(T, source(inedge(V))) = 1\}$, $\mathbf{pc} \cap \mathrm{BDLT}(T, V) = \emptyset$ *and* $\mathbf{pc} \cap \mathrm{COMP}(T, V) = \emptyset$

Proof. The argument for the compact conversion follows trivially from its definition. For the Boy de la Tour conversion, consider the vertex X in Figure 7a. The benefit of renaming, $B(T, X)$, is evaluated in the context $\mathrm{ren}(T, \mathbf{R}_{\neg X})$. From Figure 2, $a_X^{\mathrm{ren}(T,\mathbf{R}_{\neg X})} = a_B^{\mathrm{ren}(T,\mathbf{R}_{\neg X})}$, hence the benefit is

$$a_B^{\mathrm{ren}(T,\mathbf{R}_{\neg X})} p^+(X) \; - \; (a_B^{\mathrm{ren}(T,\mathbf{R}_{\neg X})} + p^+(X))$$

The condition $B(T, X) \geq 0$ reduces to $a_B^{\mathrm{ren}(T,\mathbf{R}_{\neg X})} \geq 2$ and $p^+(X) \geq 2$. From Lemma 1, in order to obtain the former vertex B must not be renamed. From $B \notin \mathbf{R}$, we deduce $\mathbf{R}_{\neg B} = \mathbf{R}_{\neg X}$ and hence write the condition $B(T, B) < 0$ as

$$a_B^{\mathrm{ren}(T,\mathbf{R}_{\neg X})} p^+(B) \; - \; (a_B^{\mathrm{ren}(T,\mathbf{R}_{\neg X})} + p^+(B)) < 0$$

which together with the earlier conditions constrains $p^+(B) = 1$. Since B is a conjunction it produces $p^+(X) + p^+(Y)$ clauses and the condition on $p^+(X)$ is thus in conflict with the condition that B is not renamed.

The argument for Y follows similarly, as do the cases of BX or BY being signed.

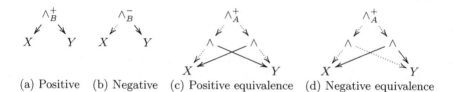

(a) Positive   (b) Negative   (c) Positive equivalence   (d) Negative equivalence

Fig. 7. RBC subgraphs for the optimality proofs and equivalence discussion

4.3   Negative Polarity

We break the negative polarity argument into several pieces, firstly simplifying the Boy de la Tour benefit function. Consider vertex X in Figure 7b. From Figure 2, $b_X^{\mathrm{ren}(T,\mathbf{R}_{\neg X})} = b_B^{\mathrm{ren}(T,\mathbf{R}_{\neg X})} p^-(Y)$, hence the benefit of renaming $B(T, X)$, in the context $\mathrm{ren}(T, \mathbf{R}_{\neg X})$ is

$$b_B^{\mathrm{ren}(T,\mathbf{R}_{\neg X})} p^-(Y) p^+(X) \; - \; (b_B^{\mathrm{ren}(T,\mathbf{R}_{\neg X})} p^-(Y) + p^+(X))$$

We consider two cases for $B(T, X) \geq 0$. If $b_B^{\mathrm{ren}(T,\mathbf{R}_{\neg X})} = 1$ then the renaming decision is localised: it is based only on $p^+(X)$ and $p^-(Y)$:

$$B'(T, X) = p^-(Y) p^+(X) - (p^-(Y) + p^+(X))$$

If $b_B^{\mathrm{ren}(T,\mathbf{R}_\sqsupset X)} \geq 2$, we must consider the same situation as for the positive case: the condition that $B \notin \mathbf{R}^-$. The inequality $B(T,B) < 0$ reduces to

$$b_B^{\mathrm{ren}(T,\mathbf{R}_\sqsupset X)} p^-(B) < b_B^{\mathrm{ren}(T,\mathbf{R}_\sqsupset X))} + p^-(B)$$

This holds only when $p^-(B) = 1$. Given $p^-(B) = p^+(X)p^-(Y)$ we also have $p^+(X) = p^-(Y) = 1$ and hence the vertex X is not renamed. This configuration is covered by the reduced condition $B'(T,X)$ which is thus sufficient condition for making the renaming decision. That is, the renaming decision is made independently of the value of $b_B^{\mathrm{ren}(T,\mathbf{R}_\sqsupset X)}$.

Lemma 4. *For linear trees, the renaming given by the Boy de la Tour algorithm with benefit function $B'(T,V)$ is the same as with the original function $B(T,V)$.*

Proof. The argument for the children of negative polarity vertices is given above (the arguments for Y and different edge signs follow similarly). For children of positive polarity vertices, it is easy to see that Lemma 3 still holds. The remaining case is the root vertex, which is not renamed under either condition.

Using this reduced condition, the Boy de la Tour conversion has no restriction on the order of evaluation: we can compare it more directly with the compact conversion. We define the conversion $\mathrm{BDLT}'(T,V)$ to be a bottom-up conversion using the benefit function $B'(T,V)$. From Lemmas 3 and 4 we know that $\mathrm{BDLT}(T,T) = \mathrm{BDLT}'(T,T)$ for all linear trees T. All remaining lemmas are on this bottom-up conversion.

Lemma 5. *For all linear trees T, $\mathrm{COMP}(T,T) \subseteq \mathrm{BDLT}'(T,T)$*

Proof. We argue in the negative as it is more convenient. Consider vertex X in Figure 7a, with $X \notin \mathrm{BDLT}'(T,T)$. From the definition of the Boy de la Tour conversion, $B'(T,X) < 0$ which reduces to the two possibilities $p(X) = 1$ or $p(Y) = 1$. By Lemma 2, this means that either $p_r^+(X,\mathbf{R}_{\sqsubset B}) = 1$ or $p_r^+(Y,\mathbf{R}_{\sqsubset B}) = 1$ and hence the renaming condition for the compact conversion, $p_r^+(X,\mathbf{R}_{\sqsubset B})p_r^+(Y,\mathbf{R}_{\sqsubset B}) > p_r^+(X,\mathbf{R}_{\sqsubset B}) + p_r^+(Y,\mathbf{R}_{\sqsubset B})$, is violated.
    The argument follows similarly for Y.

Lemma 6. *For all linear trees T, with a renaming $\mathbf{R} = \mathrm{COMP}(T,T)$, for all $V \notin \mathbf{R}$, $p_r^s(V,\mathbf{R}) = 1 \rightarrow p^s(V) = 1$.*

Proof. We show this by induction on the structure of the tree. The base case $V \in \mathbf{V_L}$ (V is a leaf) is trivial from the definition of p. For the step case, if V is a disjunction, then $p_r^s(left(V),\mathbf{R}) = p_r^s(right(V),\mathbf{R}) = 1$. This means, if $X = target(left(V))$ and $Y = target(right(V))$,

- $X \notin \mathbf{R}, Y \notin \mathbf{R}$: proof follows from the inductive hypothesis
- $X \notin \mathbf{R}, Y \in \mathbf{R}$: the condition necessary to rename Y is violated because, by Lemma 2, $p_r^s(X,\mathbf{R}_{\sqsubset V}) = p^s(X) = 1$.

- $X \in \mathbf{R}, Y \notin \mathbf{R}$: as above, by symmetry
- $X \in \mathbf{R}, Y \in \mathbf{R}$: prohibited by the definition of the compact conversion

V cannot be a conjunction as $p_r^s(V, \mathbf{R}) \geq 2$ is in contradiction with the induction hypothesis.

We can now fix the precise difference between the two conversions. Consider vertex X in Figure 7a, with $X \notin \text{COMP}(T, T)$. By the definition of the compact conversion, $p_r^+(X, \mathbf{R}_{\sqsubset B})p_r^+(Y, \mathbf{R}_{\sqsubset B}) \leq p_r^+(X, \mathbf{R}_{\sqsubset B}) + p_r^+(Y, \mathbf{R}_{\sqsubset B})$ which reduces to the three possibilities $p_r^+(X, \mathbf{R}_{\sqsubset B}) = 1$ or $p_r^+(Y, \mathbf{R}_{\sqsubset B}) = 1$ or $p_r^+(X, \mathbf{R}_{\sqsubset B}) = p_r^+(Y, \mathbf{R}_{\sqsubset B}) = 2$. In the first case, X may be a leaf vertex, in which case $X \notin \text{BDLT}'(T, T)$, or a disjunction, in which case by Lemma 6, $p^+(X) = 1$ and hence[1] $X \notin \text{BDLT}'(T, T)$. A conjunction is ruled out by the restriction on the number of clauses. The cases for Y and for signed edges follow similarly. For the final case, by Lemma 2, the Boy de la Tour conversion always renames either X or Y: this defines the set of vertices renamed by Boy de la Tour but not by compact.

Lemma 7. *For all linear trees T, $\text{COMP}(T, T) \cup \mathbf{Z} = \text{BDLT}'(T, T)$ where $\mathbf{Z}$ is the set of vertices such that for all $V \in \mathbf{Z}$, $p_r^+(V, \text{COMP}(T, V)) = 2$ and $p_r^+(sib(V), \text{COMP}(T, V)) = 2$*

Proof. From above and by Lemma 5, no other vertex is in $\text{BDLT}'(T, T)$ that is not in $\text{COMP}(T, T)$.

Theorem 1. *The size of the clause form generated by the compact and Boy de la Tour conversions is the same: $p_r^+(T, \text{COMP}(T, T)) = p_r^+(T, \text{COMP}(T, T))$*

Proof. Since renamings may be applied in any order, we show that after applying those in $\text{COMP}(T, T)$, the benefit of applying any of those in $\mathbf{Z}$ is zero. By Boy de la Tour's *fundamental theorem of monotonicity* [4], the members of $\mathbf{Z}$ may be considered in any order for this proof.

Consider a vertex $X \in \mathbf{Z}$ as depicted in Figure 7b. The benefit $B'(T, X)$ of renaming X after $\text{COMP}(T, T)$ is $p_r^+(X, \text{COMP}(T, T))p_r^-(Y, \text{COMP}(T, T)) - (p_r^+(X, \text{COMP}(T, T)) + p_r^-(Y, \text{COMP}(T, T)))$. However, by the definition of $\mathbf{Z}$ in Lemma 7, and by Lemma 2, $p_r^+(X, \text{COMP}(T, T)) = 2$ and $p_r^+(Y, \text{COMP}(T, T)) = 2$, and hence $B'(T, V) = 0$.

5   Extension to RBCs

We have shown that the compact conversion produces an optimal number of clauses for linear trees, so we now extend the algorithm to general RBCs. The extension is heuristic: like Boy de la Tour, we do not claim optimality for the resulting clause form conversion.

---

[1] The case split for BDLT′ is given in the proof of Lemma 5

Removal of Equivalences. An RBC with equivalence vertices can be transformed into a linear RBC with only a linear increase in size by replacing equivalences with the subgraphs given in Figures 7c and d. The different treatments for
positive and negative polarity equivalences reduce the number of clauses generated [9]. Note that a negative equivalence is replaced by a positive subgraph so
the incoming edge has its sign changed. The conversion remains optimal provided
equivalences are not nested.

Polarity Zero Vertices. The children of equivalence nodes are referenced both
positively and negatively (as can be seen from the replacement subgraphs), sometimes referred to as *zero* polarity. Similarly, the sharing used in RBCs encourages
a single vertex to be referenced with both polarities. We can convert an RBC
with zero polarity vertices to one without by splitting every zero polarity vertex
into a pair, one of each polarity, and suitably treating the incoming edges. Such
treatment results in at most a doubling of the size of the RBC.

The substitution and subsequent splitting of equivalences differs significantly
from the direct treatment of Boy de la Tour. In particular, Boy de la Tour's
algorithm renames a descendant vertex of an equivalence both positively and
negatively, simultaneously. This sometimes results in a tradeoff: the renaming
of one polarity must have sufficient benefit to outweigh any negative benefit
of renaming the other polarity. By splitting the polarities and treating them
independently we improve the flexibility of the conversion and reduce the number
of clauses in some circumstances, as compared to Boy de la Tour.

Shared Subgraphs. Having removed equivalences and zero polarity vertices
we are close to a linear tree structure. In fact, we can see the resulting structure
as a collection of trees joined at the shared vertices. We can incorporate treatment of shared vertices into the bottom-up compact conversion algorithm by
renaming any shared vertex which generates more than one clause and repeating the subgraph otherwise. The resulting algorithm is locally optimal as each
constituent tree is optimally converted and the shared subgraphs are renamed
only when renaming does not increase the resulting size. We believe that a truly
optimal handling of shared vertices is impossible without a significant increase
in conversion complexity. This solution is a good compromise for RBCs with a
small proportion of shared vertices.

6   Implementation and Evaluation

We have implemented the RBC extension of the compact conversion as part
of the NuSMV model checker [5]. The implementation performs the substitutions and duplications described above implicitly rather than constructing the
resulting graph explicitly. Each vertex is considered simultaneously as both a
positive and a negative polarity vertex; a depth-first traversal marks each vertex with the number of incoming edges in each polarity. A second depth-first
traversal produces the clause form. Bottom-up, each vertex is annotated with

Table 4. Benchmark results for three clause form conversions

Problem	Conv.	Clauses	Vars	Total literals	zChaff [6] Decisions	Time (s)	Jerusat [7] Time (s)
	Def	89150	31328	229882	40332	24.2	155.3
DME (Access)	SP	53285	22866	129840	39283	25.8	104.9
	Comp	22979	4986	70278	48232	10.6	32.1
	Def	234515	79577	569387	28798	52.5	149.8
DME (Priority)	SP	109637	51965	273339	21894	8.1	47.1
	Comp	52312	7587	456576	34936	5.2	3.53
	Def	737157	247079	1741365	25991	181.3	1084
DME (OT)	SP	280979	140302	700484	32023	50.4	150.9
	Comp	141604	12779	3322302	34808	10.4	38.9
	Def	234397	78483	548461	52450	68.3	369.2
Elevator	SP	109677	39373	274751	147791	74.4	338.3
	Comp	83901	23157	343673	168902	190	15.1

the clauses produced after renaming (ie, $\text{CNF}(\text{sub}(V, \text{COMP}(T, V))))$, the definitional clauses being saved in a global variable (ie, $\text{CNF}(\text{def}(V, \text{COMP}(T, V))))$. Whenever a shared vertex is encountered, it is renamed according to the strategy described above. No explicit computation of $p_r^s(V)$ is required: they correspond to the sizes of the sets of clauses — a constant time operation.

In Table 4 we compare the behaviour of the built-in CNF conversion in NuSMV (the definitional conversion) against the structure-preserving conversion and the compact conversion using two leading satisfiability solvers. We report results for the standard DME benchmark[2] and a deadlock problem[3] (Elevator), as they were found to be representative of other hardware and deadlock problems. The compact conversion consistently generates fewer clauses and the solving times are also better in most cases, sometimes dramatically so; the conversion time was similar in all cases, and negligible compared to the solving time. More surprising is the increase in the number of decisions made by zChaff in every case: for the DME example, decisions are made more quickly, while for the Elevator, the rise in the number of decisions is more dramatic and the time taken by zChaff is increased. Interestingly, the time taken by Jerusat in this case is dramatically better than the best case for zChaff; it is outperformed by zChaff in most cases.

The results also illustrate the effect of the compact conversion preferring to repeat small sets of clauses rather than renaming them: the total number of literals is, in the worst case, double that for the definitional conversion; this is contrasted with the order of magnitude reductions in the number of variables.

---

[2] See the NuSMV distribution

[3] Thanks to Toni Jussila for providing the files for this example

7   Conclusions and Future Work

Despite optimising a problem attribute that is not directly connected to the solving time — the number of clauses — the compact conversion algorithm produces a set of clauses that are in most cases more quickly solved. With the compact conversion, in contrast to the Boy de la Tour conversion or the use of preprocessing procedures to obtain similar results, this is achieved without changing the complexity class (or the observed time taken) of the conversion as compared to the more well-known clause form conversions.

The empirical study above is limited in its scope; the next step for this work must be a more thorough experimental analysis including not only much large BMC problems, but also a wider variety of leading SAT solvers.

References

1. Parosh Aziz Abdulla, Per Bjesse, and Niklas Eén. Symbolic reachability analysis based on SAT-solvers. In S. Graf and M. Schwartzbach, editors, *Tools and Algorithms for the Construction and Analysis of Systems, 6th International Conference, TACAS'00*, volume 1785 of *Lecture Notes in Computer Science*, pages 411–425. Springer-Verlag, March 2000.
2. Olivier Bailleux and Yacine Boufkhad. Efficient CNF encoding of Boolean cardinality constraints. In *Principles and Practice of Constraint Programming — 9th International Conference, CP 2003*, Lecture Notes in Computer Science, 2003.
3. Armin Biere, Alessandro Cimatti, Edmund Clarke, and Yunshan Zhu. Symbolic model checking without BDDs. In W.R. Cleaveland, editor, *Tools and Algorithms for the Construction and Analysis of Systems. 5th International Conference, TACAS'99*, volume 1579 of *Lecture Notes in Computer Science*, pages 193–207. Springer-Verlag, July 1999.
4. Thierry Boy de la Tour. An optimality result for clause form translation. *Journal of Symbolic Computation*, 14:283–301, 1992.
5. A. Cimatti, E.M. Clarke, F. Giunchiglia, and M. Roveri. NuSMV: a new Symbolic Model Verifier. In N. Halbwachs and D. Peled, editors, *Proceedings of the Eleventh Conference on Computer-Aided Verification (CAV'99)*, number 1633 in Lecture Notes in Computer Science, pages 495–499, Trento, Italy, July 1999. Springer-Verlag.
6. M. Moskewicz, C. Madigan, Y. Zhao, L. Zhang, and S. Malik. Chaff: Engineering an efficient SAT solver. In *39th Design Automation Conference*, pages 530–535, Las Vegas, June 2001.
7. Alexander Nadel. Backtrack search algorithms for propositional logic satisfiability: Review and innovations. Master's thesis, Tel-Aviv University, November 2002.
8. Andreas Nonnengart, Georg Rock, and Christoph Weidenbach. On generating small clause normal forms. In Claude Kirchner and Hélène Kirchner, editors, *Fifteenth International Conference on Automated Deduction*, volume 1421 of *Lecture Notes in Artificial Intelligence*, pages 397–411. Springer-Verlag, 1998.
9. David A. Plaisted and Steven Greenbaum. A structure-preserving clause form translation. *Journal of Symbolic Computation*, 2(3):293–304, September 1986.
10. M. N. Velev. Efficient translation of Boolean formulas to CNF in formal verification of microprocessors. In *Asia and South Pacific Design Automation Convference (ASP-DAC '04)*, January 2004.

From Spin Glasses to Hard Satisfiable Formulas

Haixia Jia[1], Cris Moore[1,2], and Bart Selman[3]

[1] Computer Science Department, University of New Mexico,
Albuquerque NM 87131
{hjia, moore}@cs.unm.edu
[2] Department of Physics and Astronomy, University of New Mexico,
Albuquerque NM 87131
[3] Computer Science Department, Cornell University, Ithaca NY
selman@cs.cornell.edu

Abstract. We introduce a highly structured family of hard satisfiable 3-SAT formulas corresponding to an ordered spin-glass model from statistical physics. This model has provably "glassy" behavior; that is, it has many local optima with large energy barriers between them, so that local search algorithms get stuck and have difficulty finding the true "ground state," i.e., the unique satisfying assignment. We test the hardness of our formulas with two Davis-Putnam solvers, Satz and zChaff, the recently introduced Survey Propagation (SP), and two local search algorithms, WalkSAT and Record-to-Record Travel (RRT). We compare our formulas to random 3-XOR-SAT formulas and to two other generators of hard satisfiable instances, the minimum disagreement parity formulas of Crawford et al., and Hirsch's hgen2. For the complete solvers the running time of our formulas grows exponentially in $\sqrt{n}$, and exceeds that of random 3-XOR-SAT formulas for small problem sizes. SP is unable to solve our formulas with as few as 25 variables. For WalkSAT, our formulas appear to be harder than any other known generator of satisfiable instances. Finally, our formulas can be solved efficiently by RRT but only if the parameter d is tuned to the height of the barriers between local minima, and we use this parameter to measure the barrier heights in random 3-XOR-SAT formulas as well.

1   Introduction

3-SAT, the problem of deciding whether a given CNF formula with three literals per clause is satisfiable, is one of the canonical NP-complete problems. Although it is believed that it requires exponential time in the worst case, many heuristic algorithms have been proposed and some of them seem to be quite efficient on average. To test these algorithms, we need families of hard benchmark instances; in particular, to test incomplete solvers we need hard but *satisfiable* instances. Several families of such instances have been proposed, including quasigroup completion [21, 15, 1] and random problems with one or more "hidden" satisfying assignments [3, 24, 2].

H.H. Hoos and D.G. Mitchell (Eds.): SAT 2004, LNCS 3542, pp. 199–210, 2005.

In this paper we introduce a new family of hard satisfiable 3-SAT formulas, based on a model from statistical physics which is known to have "glassy" behavior. Physically, this means that its energy function has exponentially many local minima, i.e., states in which any local change increases the energy, and which moreover are separated by energy barriers of increasing height. In terms of SAT, the energy is the number of dissatisfied clauses and the global minimum, or "ground state," is the unique satisfying assignment. In other words, there are exponentially many truth assignments which satisfy all but a few clauses, which are separated from each other and from the satisfying assignment by assignments which dissatisfy many clauses. Therefore, we expect local search algorithms like WalkSAT to get stuck in the local minima, and to have a difficult time finding the satisfying assignment.

We start with a spin-glass model introduced by Newman and Moore [18] and also studied by Garrahan and Newman [11]. It is like the Ising model, except that each interaction corresponds to the product of three spins rather than two; thus it corresponds to a family of 3-XOR-SAT formulas. Random 3-XOR-SAT formulas, which correspond to a similar three-spin interaction on a random hypergraph and which are also known to be glassy, have been studied by Franz, Mézard, Ricci-Tersenghi, Weigt, and Zecchina [10, 22, 16], Barthel et al. [4], and Cocco, Dubois, Mandler, and Monasson [6]. In contrast, the Newman-Moore model is defined on a simple periodic lattice, so it has no disorder in its topology.

We test our formulas against five leading SAT solvers: two complete solvers, zChaff and Satz, and three incomplete ones, WalkSAT, RRT and the recently introduced SP. We compare them with random 3-XOR-SAT formulas, and also with two other hard satisfiable generators, the minimum disagreement parity formulas of Crawford et al. [7] and Hirsch's hgen2 [12]. For Davis-Putnam solvers, our formulas are easier than random 3-XOR-SAT formulas of the same density in the limit of large size, although they are harder below a certain crossover at about 900 variables. For SP, both our formulas and random 3-XOR-SAT formulas appear to be impossible to solve beyond very small sizes. For WalkSAT, our formulas appear to be harder than any other known generator of satisfiable instances. We believe this is because our formulas' lattice structure gives them a very high "configurational entropy," i.e., a very large number of local minima, in which local search algorithms like WalkSAT get stuck for long periods of time.

The RRT algorithm solves our formulas efficiently only if its parameter d is set to the barrier height between local minima, which for our formulas we know exactly to be $\log_2 L + 1$. Although the barrier height in random 3-XOR-SAT formulas seems to grow more quickly with n than in our glassy formulas, when $\sqrt{n} = L \leq 13$ our formulas are harder for RRT than random 3-XOR-SAT formulas of the same density, even when we use the value of d optimized for each type of formula. We propose using RRT to measure barrier heights in other families of instances as well.

2   The Model and Our Formulas

The Newman-Moore model [18] consists of spins $\sigma_{i,j} = \pm 1$ on a triangular lattice. Each spin interacts only with its nearest neighbors, and only in groups of three lying at the vertices of a downward-pointing triangle. If we encode points in the triangular lattice as (i, j), where the neighbors of each point are $(i \pm 1, j)$, $(i, j \pm 1)$, and $(i \pm 1, j \mp 1)$, the model's Hamiltonian (energy function) is

$$H = \frac{1}{2} \sum_{i,j} \sigma_{i,j} \sigma_{i,j+1} \sigma_{i+1,j}$$

Let us re-define our variables so that they take Boolean values, $s_{i,j} \in \{0, 1\}$. Then, up to a constant, the energy can be re-written

$$H = \sum_{i,j} (s_{i,j} + s_{i,j+1} + s_{i+1,j}) \bmod 2$$

In particular, we will focus on the case where the lattice is an $L \times L$ rhombus with cyclic boundary conditions; then

$$H = \sum_{i,j=0}^{L-1} (s_{i,j} + s_{i,j+1 \bmod L} + s_{i+1 \bmod L,j}) \bmod 2 .$$

Clearly we can think of this as a set of L^2 3-XOR-SAT clauses of the form

$$s_{i,j} \oplus s_{i,j+1 \bmod L} \oplus s_{i+1 \bmod L,j} = 0$$

in which case H is simply the number of dissatisfied clauses. Each one of these can then be written as a conjuction of four 3-SAT clauses,

$$(\overline{s}_{i,j} \vee s_{i,j+1 \bmod L} \vee s_{i+1 \bmod L,j}) \wedge (s_{i,j} \vee \overline{s}_{i,j+1 \bmod L} \vee s_{i+1 \bmod L,j})$$
$$\wedge (s_{i,j} \vee s_{i,j+1 \bmod L} \vee \overline{s}_{i+1 \bmod L,j}) \wedge (\overline{s}_{i,j} \vee \overline{s}_{i,j+1 \bmod L} \vee \overline{s}_{i+1 \bmod L,j})$$

producing a 3-SAT formula with L^2 variables and $4L^2$ clauses for a total of $12L^2$ literals.

There is always at least one satisfying assignment, i.e., where $s_{i,j} = 0$ for all i, j. However, using algebraic arguments [18] one can show that this satisfying assignment is unique whenever L has no factors of the form $2^m - 1$, and in particular when L is a power of 2.

To "hide" this assignment, we flip the variables randomly; that is, we choose a random assignment $A = (a_{i,j}) \in \{0, 1\}^{L^2}$ and define a new formula in terms of the variables $x_{i,j} = s_{i,j} \oplus a_{i,j}$. While some other schemes for hiding a random satisfying assignment in a 3-SAT formula create an "attraction" that allows simple algorithms to find it quickly, Barthel et al. [4] pointed out that for XOR-SAT formulas these attractions cancel and make the hidden assignment quite difficult to find. (Another approach pursued by Achlioptas, Jia, and Moore is to simultaneously hide two complementary assignments [2].) Of course, XOR-SAT is solvable in polynomial time by Gaussian elimination, but Davis-Putnam and local search algorithms can still take exponential time on random XOR-SAT formulas [4, 22].

In general, XOR-SAT formulas have local minima because flipping any variable will dissatisfy all the currently satisfied clauses it appears in. However, the lattice structure of the Newman-Moore model allows us to say much more. In particular, let us call an unsatisfied XOR-clause a "defect." Then if L is a power of 2, there is exactly one state of the lattice for any choice of defect locations [18]. To see this, consider the state shown in Figure 1. Here there is a single defect (the three cells outlined in black) in which just one XOR-SAT clause (in fact, just one 3-SAT clause) is dissatisfied. However, since satisfying the XOR-SAT clause at i, j implies that

$$s_{i,j+1} = s_{i,j} \oplus s_{i+1,j} ,$$

the truth values below the defect are given by a mod-2 Pascal's triangle. If L is a power of 2 the L'th row of this Pascal's triangle consists of all 0's, so wrapping around the torus matches its first row except for the defect.

This gives a truth assignment which satisfies all but one clause. Moreover, this assignment has a large Hamming distance from the satisfying assignment; namely, the number of 1's in the Pascal's triangle, which is $H(L) = L^{\log_2 3}$ since it obeys the recurrence $H(2L) = 3H(L)$. It also has a large energy barrier separating it from the satisfying assignment: to fix the defect with local moves it is necessary to first introduce $\log_2 L$ additional defects [18].

Now, by taking linear combinations (mod 2) of single-defect assignments we can construct truth assignments with arbitrary sets of defects, and whenever these defects form an independent set on the triangular lattice, the corresponding state is a local energy minimum. Thus the number of local minima equals the number of independent sets of the triangular lattice, which grows exponentially as κ^{L^2} where $\kappa \approx 1.395$ is the *hard hexagon constant* [11, 5].

To recap, when $L = 2^k$, there is a unique satisfying assignment. The system is glassy in that there are many truth assignments which are far from the satisfying assignment, but which satisfy all but a small number of clauses. Escaping these local minima requires us to first increase the number of unsatisfied clauses by roughly $\log_2 L$. Newman and Moore [18] studied the behavior of this model under simulated annealing, and found that the system is unable to find its ground state unless the cooling rate is exponentially slow; similarly, we expect the running time of local search algorithms like WalkSAT to be exponentially large.

Below, we compare our formulas to random satisfiable 3-XOR-SAT formulas, which were proposed in [22] (and also in [4] as the special case $p_0 = 1/4$). These are formed with a random hidden assignment in the following way: given variables $x_1, \ldots, x_n$, select a random truth assignment $A \in \{0,1\}^n$. Then, m times, select a triple x_i, x_j, x_k uniformly without replacement, and add the 3-XOR clause consistent with A, i.e. $x_i \oplus x_j \oplus x_k = a_i \oplus a_j \oplus a_k$. To compare with our formulas, we set $n = m = L^2$ so the resulting 3-XOR-SAT formula has a density of one clause per variable.

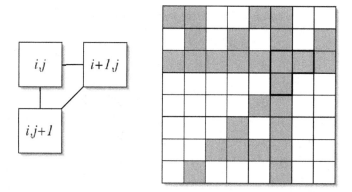

Fig. 1. A local minimum with a single defect. Grey and white cells correspond to $s_{i,j} = 1$ and 0 respectively; the XOR-SAT clause corresponding to the three cells outlined in black is dissatisfied, and all the others are satisfied. The Hamming distance from the satisfying assignment is the number of grey cells, $L^{\log_2 3} = 27$ since $L = 8$

3   Experimental Results

3.1   Davis-Putnam Solvers: zChaff and Satz

We obtained zChaff from the Princeton web site [25] and Satz from the SATLIB web site [14]. Figure 2 shows a log-log plot of the median number of decisions or branches that zChaff and Satz took as a function of the lattice size L. For both algorithms the slope for our glassy formulas is roughly 1, indicating that the running time for zChaff and Satz to solve our formulas grows as $2^L = 2^{\sqrt{n}}$. The reason for this is that, due to a process similar to bootstrap percolation [13], when a sufficient number of variables are set by the algorithm (for instance, the variables in a single row) the remainder of the variables in the lattice are determined by unit propagation. For random 3-XOR-SAT formulas, the running time is exponential in $n = L^2$, but with a smaller constant, so that for $L \lesssim 30$ (i.e., $n \lesssim 900$) our formulas are harder than random 3-XOR-SAT formulas of the same size.

3.2   SP

SP is an incomplete solver recently introduced by Mézard and Zecchina [17] based on a generalization of belief propagation called *survey propagation*. For random 3-SAT formulas it is extremely successful; it can find a satisfiable assignment efficiently for random 3-SAT formulas up to size $n = 10^7$ near the satisfiability threshold $m/n \approx 4.25$ where random 3-SAT appears to be hardest.

We found that SP cannot solve our formulas for $L \geq 5$, i.e., with $n = 25$ variables. The cavity biases continue to change, and never converge to a fixed point, so no variables are ever set by the decimation process. There are several possible reasons for this. One is the large number of local minima; another is that

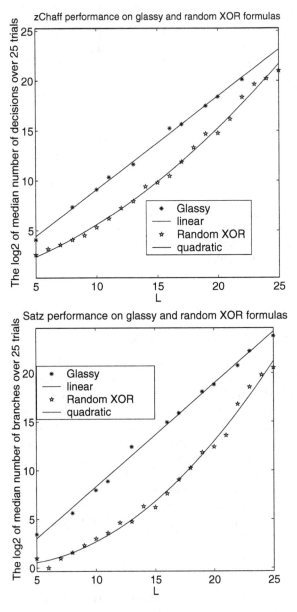

Fig. 2. The number of branches made by zChaff and Satz on our formulas and on random 3-XOR-SAT formulas of the same size and density, as a function of the lattice size L. The running time for random 3-XOR-SAT is exponential in $L^2 = n$, while for our formulas it is exponential in $L = \sqrt{n}$. Nevertheless, for small values of n our formulas are harder. Each point is the median of 25 trials; for our formulas, only values of L for which the satisfying assignment is unique are shown

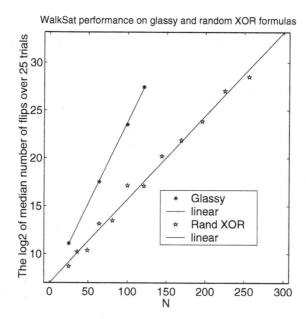

Fig. 3. The median number of flips made by WalkSAT on our formulas and random 3-XOR-SAT formulas of the same size. For our formulas, only values of L for which the satisfying assignment is unique are shown. Each point is the median of 25 trials

the symmetry in XOR clauses may produce conflicting messages; another is that our formulas have small loops which violate SP's assumption that the formula is locally treelike and that neighbors are statistically independent. (Random 3-XOR-SAT formulas are also quite hard for SP, although we found that SP solved about 25% of them with $n = m = 25$.)

3.3   Local Algorithms: WalkSAT

WalkSAT [20] is an algorithm which combines a random walk search strategy with a greedy bias towards assignments with more satisfied clauses. WalkSAT has been shown to be highly effective on a range of problems, such as hard random k-SAT problems, graph coloring, and the circuit synthesis problem. We performed trials of up to 10^9 flips for each formula, without random restarts, where each step does a random or greedy flip with equal probability. Figure 3 shows a semi-log plot of the median number of flips as a function of $n = L^2$. We only choose four different values of L, namely 5, 8, 10 and 11, because WalkSAT was unable to solve the majority of formulas with larger values of L (for which the satisfying assignment is unique) within 10^9 flips.

For both our formulas and random 3-XOR-SAT formulas, the median running time of WalkSAT grows exponentially in n. However, the slope of the exponential is considerably larger for our formulas, making them much harder than the random ones. We believe this is due to a larger number of local minima.

3.4   Local Algorithms: RRT

Record-to-Record Travel (RRT) [9, 19] is a variant of WalkSAT which works as follows:

1. Start from a random truth assignment;
2. Randomly choose a variable from an unsatisfied clause;
3. Flip it if this leads to a configuration that has at most d more unsatisfied clauses than the best configuration found so far (the "record"). Otherwise, do nothing;
4. Repeat steps 2 and 3 until it finds the satisfying truth assignment.

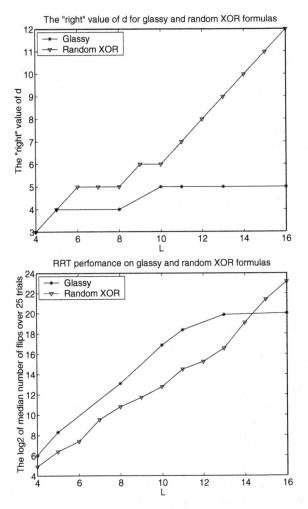

Fig. 4. The optimal value of d and the running time for our formulas and random 3-XOR-SAT formulas of the same size. Shown is the number of flips per variable

Since d stays fixed throughout the algorithm, if we wish to solve a formula of a given size and type, we need to find the optimal value of d. If d is too small, RRT fails because it cannot escape the local minima; and if d is too big, it escapes the local minima but takes a long time to find the solution since it is not greedy enough to move toward it. Our goal is to find a vale of d for which RRT escapes the local minima and finds the solution quickly.

We tested RRT on our formulas with $L = 4, 5, 8, 10, 11, 13$ and 16 and we performed trials of up to 10^7 flips for each formula. Newman and Moore [18] showed that the largest barrier height is $\log_2 L + 1$, and in fact a typical path from an initial state to the solution must cross multiple barriers of this height. In fact, it turns out that RRT solved our formulas efficiently only for this value of d (see Figure 4). With $L = 16$ and $d = 5$, RRT solved our formulas in all of 50 trials with a median number of flips 1.10×10^6; but when we set $d = 4$ or 6, RRT failed to solve any of the formulas with $L = 16$ within 10^7 flips. Thus RRT solves our glassy formulas only if we set d equal to the barrier height.

We also tested RRT on random 3-XOR-SAT formulas with $n = m = L^2$ ranging from 16 to 256 so the resulting 3-XOR-SAT formula has same density as our glassy formulas. Since we don't know the barrier height between local minima in these formulas, we tried RRT with different values of d to find the optimal d for each n. As a rough measurement of the barrier heights, we measured the value v for which RRT solved more than half the formulas with $d = v$ but failed to solve half of them with $d = v - 1$. We set the maximum running time to 10^7 flips.

Figure 4 shows the optimal value of d and the running time for each value of n. We see that the barrier height in random 3-XOR-SAT formulas seems to grow more quickly with n than in our glassy formulas. However, when $\sqrt{n} = L \leq 13$, our formulas are harder for RRT than random 3-XOR-SAT formulas of the same density, even when we use the value of d optimized for each type of formula.

We find it interesting that RRT can be used to measure the barrier heights between local minima, and we propose to do this for other families of formulas as well.

3.5 Comparison with Other Hard SAT Formulas

To further demonstrate the hardness of our glassy formulas, we compare them to other two generators of hard instances: the parity formulas introduced by Crawford et al. [7] and the hgen2 formulas introduced by E.A. Hirsch [12]. The parity formulas of [7] are translated from minimal disagreement parity problems and are considered very hard. While hgen2 does not generate parity formulas, we include it because it produced the winner of the SAT 2003 competition for the hardest satisfiable formula [23].

We compared our glassy formulas with 10 formulas of Crawford et al., obtained from [7], and with 25 hgen2 formulas using the generator obtained from [12]. We ran zChaff, Satz, WalkSAT and SP; we did not test RRT on these formulas.

For WalkSAT, we ran 25 trials of up to 10^9 flips each, and labeled the formula "not solved" if none of these trials succeeded. Comparing our glassy formulas with those of Crawford et al., taking similar numbers of variables and clauses

Table 1. Comparison of our glassy lattice formulas with the parity formulas of Crawford et al., Hirsch's **hgen2**, and random 3-XOR-SAT formulas

Formulas	Literals	Variables	Decisions (zChaff)	Branches (Satz)	Flips (WalkSAT)
par8-1-c.cnf	732	64	17	3	1494
par8-2-c.cnf	780	68	9	1	2371
par8-3-c.cnf	864	75	18	4	5638
par8-4-c.cnf	768	67	7	1	2811
par8-5-c.cnf	864	75	12	3	4828
par16-1-c.cnf	3670	317	2073	1591	2.5×10^8
par16-2-c.cnf	4054	349	11117	499	1.3×10^8
par16-3-c.cnf	3874	334	7505	1489	1.0×10^8
par16-4-c.cnf	3754	324	2181	4415	1.4×10^8
par16-5-c.cnf	3958	341	2758	1296	4.1×10^8
Glassy 8×8	768	64	167	50	219455
Glassy 16×16	3072	256	39293	32219	not solved
Random XOR	768	64	23	3	9167
Random XOR	3072	256	1427	198	3.9×10^8
hgen2	3096	295	not solved	1478340	751723

(e.g. comparing our $L = 16$ formulas, which have 256 variables and 3072 clauses, with theirs with roughly 300 variables and 4000 clauses) we see from Table 1 that our formulas are significantly harder than theirs for zChaff, Satz, and WalkSAT. (SP didn't solve any of these formulas, so it doesn't provide a basis for comparison.) Compared to **hgen2** formulas with 195 variables and 3096 clauses, our formulas are not as hard for the complete solvers, but appear to be harder for WalkSAT, again perhaps due to their large number of local minima.

4   Conclusion

We have introduced a new generator of hard satisfiable SAT formulas derived from a two-dimensional spin-glass model. We tested our formulas against five leading SAT solvers, and compared them with random 3-XOR-SAT formulas, the minimal disagreement parity formulas of Crawford et al., and Hirsch's **hgen2** generator. For complete solvers, the running time of our formulas grows exponentially only in $L = \sqrt{n}$, but they are harder than random 3-XOR-SAT formulas when n is small. For SP our formulas appear to be impossible for $n \geq 25$ variables. For WalkSAT our formulas appear to be harder than any other known generator of satisfiable instances. Finally, the RRT algorithm solves our formulas only if d is set to the barrier height between local minima, which we know exactly to be $\log_2 L + 1$. We propose that RRT can be used to measure the barrier heights between local minima in other families of instances, and we have done this for random 3-XOR-SAT formulas.

Since XOR-SAT is solvable in polynomial time, it would be interesting to have a provably glassy set of formulas which would be NP-complete to solve.

One approach would be a construction along the lines of [7], where "noise" is introduced to the underlying parity problem so that it is no longer polynomial-time solvable.

Finally, we feel that the highly structured nature of our formulas, which makes it possible to prove the existence of exponentially many local optima with large barriers between them, suggests an interesting direction for future work. For instance, are there families of formulas based on spin-glass models in three or more dimensions which would be even harder to solve?

Acknowledgments. We are grateful to Pekka Orponen for helpful discussions on the RRT algorithm and we are also grateful to the anonymous referees for proposing several directions for further work. C.M. and H.J. are supported by NSF grant PHY-0200909 and H.J. is supported by an NSF Graduate Research Fellowship. C.M. is also grateful to Tracy Conrad for helpful discussions.

References

1. D. Achlioptas, C. Gomes, H. Kautz, and B. Selman, Generating satisfiable problem instances. *Proc. AAAI '00* 256-261.
2. D. Achlioptas, H. Jia, and C. Moore, Hiding satisfying assignments: two are better than one. Submitted.
3. Y. Asahiro, K. Iwama, and E. Miyano, Random generation of test instances with controlled attributes. In [8], op. cit.
4. W. Barthel, A.K. Hartmann, M. Leone, F. Ricci-Tersenghi, M. Weigt, and R. Zecchina, Hiding solutions in random satisfiability problems: a statistical mechanics approach. *Phys. Rev. Lett.* 88 (2002) 188701.
5. R.J. Baxter, "Hard hexagons: exact solution." *J. Physics A* 13 (1980) 1023-1030.
6. S. Cocco, O. Dubois, J. Mandler, and R. Monasson, Rigorous decimation-based construction of ground pure states for spin glass models on random lattices. *Phys. Rev. Lett.* 90(4) (2003) 047205.
7. J.M. Crawford and M.J. Kearns, The Minimal Disagreement Parity Problem as a Hard Satisfiability Problem, ftp://dimacs.rutgers.edu/pub/challenge/ satisfiability/benchmarks/cnf/ and J.M. Crawford, M.J. Kearns, and R.E. Schapire, The minimal disagreement parity problem as a hard satisfiability problem. Technical report, CIRL, 1994.
8. Second DIMACS Implementation Challenge, 1993. Published as *DIMACS Series in Disc. Math. and Theor. Comp. Sci.* vol. 26, D. Johnson and M. Trick, Eds. AMS, 1996.
9. G. Dueck, New optimization heuristics: the great deluge algorithm and the record-to-record travel. *J. Comp. Phys.* 104 (1993) 86–92.
10. S. Franz, M. Mézard, F. Ricci-Tersenghi, M. Weigt, and R. Zecchina, A ferromagnet with a glass transition. *Europhys. Lett.* 55 (2001) 465.
11. J.P. Garrahan and M.E.J. Newman, Glassiness and constrained dynamics of short-range non-disordered spin model. *Phys. Rev. E* 62 (2000) 7670–7678.
12. E.A. Hirsch, hgen2 formula generator source site. http://logic.pdmi.ras.ru/ hirsch/
13. A. E. Holroyd, Sharp metastability threshold for two-dimensional bootstrap percolation. *Prob. Theory and Related Fields* 125 (2003) 195–224.

14. H.H. Hoos, SATLIB, A collection of SAT tools and data. www.informatik.tu-darmstadt.de/AI/SATLIB
15. H. Kautz, Y. Ruan, D. Achlioptas, C. Gomes, B. Selman, and M. Stickel, Balance and Filtering in Structured Satisfiable Problems. *Proc. IJCAI '01* 351–358.
16. M. Mézard, F. Ricci-Tersenghi, and R. Zecchina, Alternative solutions to diluted p-spin models and XORSAT problems. *J. Stat. Phys.* 111 (2003) 505.
17. M. Mézard and R. Zecchina, Random K-satisfiability: from an analytic solution to a new efficient algorithm. *Phys. Rev. E* 66 (2002). See also A. Braunstein, M. Mézard, M., and R. Zecchina, Survey propagation: an algorithm for satisfiability. Preprint, 2002, http://www.ictp.trieste.it/~zecchina/SP/.
18. M.E.J. Newman and C. Moore, Glassy Dynamics in an Exactly Solvable Spin Model. *Phys. Rev. E* 60 (1999) 5068–5072.
19. S. Seitz and P. Orponen, An efficient local search method for random 3-satisfiability. *LICS'03*, workshop on Typical Case Complexity and Phase Transitions. *Electronic Notes in Discrete Mathematics* 16 (2003).
20. B. Selman, H.A. Kautz, and B. Cohen, Noise strategies for improving local search. *Proc. AAAI* (1994).
21. P. Shaw, K. Stergiou, and T. Walsh, Arc consistency and quasigroup completion. *ECAI '98*, workshop on binary constraints.
22. F. Ricci-Tersenghi, M. Weigt, and R. Zecchina, Simplest random K-satisfiability problem. *Phys. Rev. E* 63 (2001) 026702.
23. SAT2003 Competition result site, http://www.satlive.org/SATCompetition/2003/results.html
24. A. Van Gelder, Problem generator mkcnf.c contributed to the DIMACS 1993 Challenge archive.
25. L. Zhang, zChaff source site, http://ee.princeton.edu/~chaff/zChaff.php

CirCUs: A Hybrid Satisfiability Solver

HoonSang Jin and Fabio Somenzi

University of Colorado at Boulder
{Jinh, Fabio}@colorado.EDU

Abstract. CirCUs is a satisfiability solver that works on a combination of an And-Inverter-Graph (AIG), Conjunctive Normal Form (CNF) clauses, and Binary Decision Diagrams (BDDs). We show how BDDs are used by CirCUs to help in the solution of SAT instances given in CNF. Specifically, the clauses are sorted by solving a hypergraph linear arrangement problem. Then they are clustered by an algorithm that strives to avoid explosion in the resulting BDD sizes. If clustering results in a single diagram, the SAT instance is solved directly. Otherwise, search for a satisfying assignment is conducted on the original clauses, enhanced with information extracted from the BDDs. We also describe a new decision variable selection heuristic that is based on recognizing that the variables involved in a conflict clause are often best treated as a related group. We present experimental results that demonstrate CirCUs's efficiency especially for medium-size SAT instances that are hard to solve by traditional solvers based on DPLL.

1 Introduction

Different representations of Boolean functions have peculiar strengths in regard to satisfiability (SAT) problems. Conjunctive Normal Form (CNF) is often used because it can be manipulated efficiently and because constraints of various provenance are easily translated into it. Boolean circuits, especially semi-canonical ones like the And-Inverter Graph (AIG) [24], allow a variety of simplification techniques that may significantly speed up subsequent analyses. For other representations, like the Disjunctive Normal Form (DNF) and Binary Decision Diagrams (BDDs) [6], the hurdle lies in converting the problem specification into the required form; if this can be accomplished, satisfiability is then trivial. In particular, with BDDs, determining whether a function is satisfiable requires constant time, while a satisfying assignment, if it exists, can be found in $O(n)$ time, where n is the number of variables. Since converting a Boolean circuit into a BDD may incur an exponential blow-up, naive application of BDDs to SAT lacks robustness. On the other hand, there exist numerous cases in which a proper mix of canonical (e.g., BDDs) and non-canonical representations (e.g., CNF or AIG) is very beneficial [25, 8]. This is true, in particular, of SAT solvers based on search, and applied to instances for which compact search trees do not exist or are hard to find.

CirCUs is a SAT solver that accepts as input a combination of an AIG, CNF clauses, and BDDs. Rather than converting all into one form as a preprocessing step, CirCUs

---

* This work was supported in part by SRC contract 2003-TJ-920.

H.H. Hoos and D.G. Mitchell (Eds.): SAT 2004, LNCS 3542, pp. 211–223, 2005.

operates on all three representations, transforming, when appropriate, parts of the input from one of them to another. For instance, in Bounded Model Checking (BMC) [4] applications, CirCUs reads the input as an AIG with additional constraints given as clauses, and transforms part of the AIG into BDDs, so that it may apply powerful implication and conflict analysis algorithms [23, 21]. The conflict clauses, on the other hand, are recorded in CNF form as suggested in [13]. Because of this ability to operate on multiple representations, we call CirCUs a *hybrid* SAT solver.

In this paper we discuss how CirCUs handles SAT instances given in CNF. After a review of related work in Sect. 2, in Sect. 4, we show how the clauses may be "conditioned" with the help of BDDs so as to allow the solution of some hard, though not very large, problems. The conditioning consists of building BDDs from the clauses in such a way that resource limits are not exceeded. This implies that more than one BDD may be built. If that is the case, CNF clauses are extracted from the BDDs to replace the original ones.

Section 5 presents a new decision variable selection heuristic, which is based on the observation that variables appearing in one conflict clause should be treated as a related group. In Sect. 6 we present empirical evidence that for mid-size hard instances, CNF conditioning is very effective, and that our decision variable heuristic consistently improves over the VSIDS rule of [32]. Finally, we draw conclusions in Sect.7.

2   Related Work

Considerable work has been done in which constraints are represented by a collection of BDDs. In symbolic model checking, the transition relation is often represented in such an implicitly conjoined form [39, 7, 19, 30, 17, 22]. The partitioned representation was also applied to the problem of minimum-cost satisfiability in [20]. In our work we leverage several techniques from this body of literature, especially from [22].

More recently, there has been considerable interest in BDD-based techniques for the SAT problem. Gupta el al. [15] proposed BDD-based learning while solving Bounded Model Checking (BMC [4]) instances with a circuit SAT solver. The BDDs are used to supplement conflict-learned clauses. They are created from portions of the circuit that defines the BMC instance. Their approach is similar to our approach in the sense that they use BDD to extract helpful CNF from it. On the other hand, we do no assume the existence of a circuit, and our algorithms are different.

Damiano and Kukula [9] replace clauses with BDDs in a classical DPLL solver, while in [12], the authors propose the method that uses BDDs to precompute complete lookahead information to drive the search. This is done by converting each BDD into a finite state machine that reads assignments to the BDD inputs and outputs implied values. During a preprocessing phase, Franco et al. use *strengthening* to infer additional literals and equivalences, since their BDD is highly localized because of BDD blow-up. The search is then conducted on the modified BDDs. By contrast, the technique we discuss in this paper either solves the SAT instance without search, or eventually operates on CNF that has been possibly enhanced using the extracted BDDs.

3   Preliminaries

We consider three ways of representing a Boolean function. The first is a Boolean circuit, that is, a directed acyclic graph whose nodes correspond to input variables and Boolean gates. Specifically, we use a form of Boolean circuit called And-Inverter Graph (AIG) in which each node's function is one of $x \wedge y$, $x \wedge \neg y$, $\neg x \wedge y$, and $\neg x \wedge \neg y$. An AIG contains no isomorphic subgraphs; for this reason, it is called *semicanonical*.

The second representation is Conjunctive Normal Form (CNF). A CNF formula is a set of *clauses*; each clause is a set of *literals*; each literal is either a variable or its complement. The function of a clause is the disjunction of its literals, and the function of a CNF formula is the conjunction of its clauses.

The last representation of Boolean functions is Binary Decision Diagrams (BDDs). A BDD is a Boolean circuit such that each node is labeled by either a Boolean constant (terminal node) or a variable (internal node). Each internal node has two children, T and E. The function of an internal node labeled by v is defined recursively by $(v \wedge f(T)) \vee (\neg v \wedge f(E))$, where $f(T)$ and $f(E)$ are the functions of T and E. A reduced BDD is one in which there are no isomorphic subgraphs, and no node has identical children. (Such nodes are redundant.) A BDD is ordered if the variables encountered along all paths from root to leaves respect a fixed order. Reduced, ordered BDDs are canonical: for a given variable order, two functions are the same if and only if they have the same BDD [6]. We shall refer to reduced, ordered BDDs simply as BDDs. Another form of diagrams that are useful in manipulating Boolean functions are Zero-suppressed BDDs (ZDDs). The difference between BDDs and ZDDs is that in the former, nodes with identical children are removed, while in the latter nodes whose T child is the constant 0 are removed. ZDDs are usually more concise than BDDs when representing sets of clauses (each clause corresponding to a path in the diagram). BDDs, on the other hand, are usually better when representing the functions themselves.

CirCUs is a SAT solver based on the DPLL procedure [11, 10] and conflict clause recording [36, 42, 32, 14]. It is built on top of VIS [5, 41], and uses the CUDD package [37] for BDD and ZDD manipulations. Figure 1 describes the core of the decision procedure, whose input is an AIG, a set of CNF clauses, and a set of BDDs.

The pseudo-code of DPLL procedure is presented in Fig. 1. Procedure CHOOSENEXTASSIGNMENT checks the implication queue. If the queue is empty, the procedure

```
1   DPLL() {
2       while  (CHOOSENEXTASSIGNMENT() == FOUND)
3           while  (DEDUCE() == CONFLICT) {
4               blevel = ANALYZECONFLICT();
5               if (blevel ≤ 0) return UNSATISFIABLE;
6               else BACKTRACK(blevel);
7           }
8       return SATISFIABLE;
9   }
```

Fig. 1. DPLL algorithm

makes a *decision*: it chooses one unassigned variable and a value for it, and adds the assignment to the implication queue. If none can be found, it returns `false`. This causes DPLL to return an affirmative answer, because the assignment to the variables is complete and no conflict is detected. If a new assignment has been chosen, its implications are added by DEDUCE to the queue. If the implications yield a conflict, this is analyzed to produce two important results. The first is a clause implied by the given circuit and objectives. This *conflict clause* is added to the clauses of the circuit. Termination relies on conflict clauses, because they prevent the same variable assignment from being tried more than once. The second result of conflict analysis is the *backtracking level*: Each assignment to a variable has a *level* that starts from 0 and increases with each new decision. When a conflict is detected, the algorithm determines the lowest level at which a decision was made that eventually caused the conflict. The search for a satisfying assignment resumes from this level by deleting all assignments made at higher levels. This *non-chronological backtracking* allows the decision procedure to ignore inconsequential decisions that have provably no part in the conflict being analyzed.

The pseudo-code of Fig. 1 is essentially the same used to describe CNF SAT solvers like GRASP and Zchaff. However, in CirCUs all operations are carried out on the three Boolean function representations at once. CNF clauses and BDDs are connected to the AIG so that propagation of implications and conflict analysis proceed seamlessly on all of them. The algorithm uses a common assignment stack and implication queue. The decision variable selection is also common. In particular, the DVH heuristic of Sect. 5 is used by CirCUs regardless of the mix of function representations. The specific implication and conflict analysis algorithms for AIG, clauses, and BDDs are described in [24, 32, 23].

When the input is in the form of an AIG, replacing parts of it by BDDs allows CirCUs to reduce the number of decisions and conflicts without slowing down implication too much. In this paper, we consider the case in which the input is a set of clauses. The strategy of [23], which replaces *fanout-free* subcircuits of the AIG with BDDs, is not applicable. Instead, we try to improve the given CNF as described in Sect. 4.

4   CNF Conditioning

For hard CNF SAT instances with moderate numbers of variables and clauses, it is often advantageous to *condition* the given set of clauses. In the following, we describe the approach implemented in CirCUs.

A *hypergraph* $G = (V, H)$ consists of a set of vertices V and a multiset of hyperedges H. Each hyperedge is a subset of V. A *linear arrangement* of G is a bijection $\alpha : V \to \{1, \ldots, |V|\}$.

A set of Boolean functions can be regarded as a hypergraph by associating variables to vertices and functions to hyperedges. A hyperedge connects all the variables appearing in the function to which it is associated. Linear arrangement has been used in [1, 2] to derive variables orders for both BDD construction and SAT. Our use is closer in spirit to the one of [22], in which the objective is to derive a good order for the conjunction of the functions.

We compute a linear arrangement by *force-directed* (or *quadratic*) placement [33], as done in [2]. Given a linear arrangement α_i, the algorithm computes the *center of mass* of hyperedge $h \in H$ thus:

$$COM(h) = \frac{\sum_{v \in h} \alpha_i(v)}{|h|} \ . \tag{1}$$

The center of mass of a vertex is computed as the average of the centers of mass of all hyperedges incident on the vertex. Finally, α_{i+1} is obtained by sorting vertices according to their centers of mass. The process is iterated starting from an initial given arrangement α_0 until the cost function stops decreasing, or until the alloted computational resources are exhausted. The cost function is the sum of the hyperedge spans, where the span of hyperedge h under arrangement α is

$$span(h) = \max_{v \in h}\{\alpha(v)\} - \min_{v \in h}\{\alpha(v)\} \ . \tag{2}$$

Once the final vertex arrangement is determined, the order of the hyperedges is given by their centers of mass.

Once the clauses of the given CNF are sorted, if the numbers of variables and clauses do not exceed specified thresholds, the clustering algorithm of [22] is invoked to try to conjoin all clauses into one BDD. The algorithm works on a list of Boolean functions initialized to the sorted list of clauses. It selects a set of adjacent functions to be conjoined, and tries to construct a BDD for them. If the BDD can be built without exceeding a threshold on the number of nodes, it replaces the functions that were conjoined in the list. The candidates are chosen so as to favor the confinement of as many variables as possible to one cluster only. A detailed description of the algorithm can be found in [22]. The thresholds on the numbers of variables and clauses are chosen so that it is likely that all clauses will be conjoined into one BDD. When this happens, the SAT instance is solved directly.

For this purpose, the clustering algorithm iterates until no new clusters are created in one pass. At each pass, it creates a list of candidates. Each candidate is a pair of clusters. The list is ordered in decreasing order of the number of isolated variable to favor candidates that allow many variables to be quantified. (This is beneficial when trying to build one BDD from all clauses.) As a tie-breaker, the upper bound on the number of variables in the resulting cluster is used. This policy favors the creation of small clusters that may be merged in subsequent passes. If a given instance is unsatisfiable, it will result in the constant zero BDD; otherwise it will result in the constant one BDD because all variables are quantified while clustering.

To get a satisfying assignment without saving all the BDDs produced during clustering, we save the last two BDDs, so that a partial assignments can be extracted from them. We then use this partial assignment as a constraint for the CNF SAT solver. This results in a quick solution of the CNF instance because the clustering process is such that the last two BDDs tend to contain the global variables of the function.

If, on the other hand, the initial CNF is too large, or the conjunction of all clauses cannot be carried out without exceeding the resource limits, several BDDs are built, each to be used in conditioning a subset of the CNF formula. The clauses are divided into *short* (one or two literals) and *long* (more than two literals). The long clauses are conjoined in the order determined by the linear arrangement until the BDD for the resulting cluster exceeds a given size, at which point a new cluster is started. Let f be the function for such a cluster. The next step consists of conjoining all the short clauses that share at least one variable with f into a function g. Since g is implied by the original set of clauses, any function f_g such that $f_g \wedge g = f$ can replace f. Therefore, we are interested in a simple CNF representing a function from the interval $[f, f \vee \neg g]$. This is computed by the Morreale-Minato algorithm for prime and irredundant covering of a Boolean function [31, 27]. The algorithm is called on the interval $[\neg f \wedge g, \neg f]$, and DeMorgan's Laws are applied to the resulting DNF.

The result of the Minato-Morreale algorithm is computed as a Zero-Suppressed BDD (ZDD) [28]. The clauses are then obtained by enumeration of the paths of the ZDD. Since the computed CNF is not guaranteed to have the minimum number of clauses, it is possible that more clauses be extracted than were used to produce f. If this happens, the process is abandoned, and the original clauses are used instead. Even in such a case, the construction of the BDD may be helpful: If a variable occurring in the clauses conjoined to obtain f does not occur either in f or in the other clauses, then it can be universally quantified from the original clauses.

The final step of conditioning consists of extracting all short clauses from the function in the interval $[f, f \vee \neg g]$ chosen by the Morreale-Minato algorithm. This is accomplished by a single traversal of each BDD, during which the short clauses of a BDD with top node   are obtained from the short clauses of the children of   [38]. The procedure extends the one for unit clauses of [20]. Both procedures, as well as the Morreale-Minato algorithm, are implemented in CUDD [37].

5   Decision Variable Selection

The choice of the decision variables has a large impact on the run time of the DPLL procedure. Hence, considerable attention has been devoted to the problem. (See, for instance, [35, 26, 18].) Many rules have been proposed that are based on the frequency of literals in unresolved clauses; for instance, the Dynamic Largest Individual Sum (DLIS) heuristic of GRASP [36]. Chaff's VSIDS rule [32] disregards whether a clause is resolved in the interest of speed. It also introduces the notion that recently generated conflict clauses should be given more weight in the choice of the next variable. The VSIDS rule increases the score of a literal whenever a clause containing that literal is added to the database. Scores are periodically halved to give more weight to recently added conflict clauses. The literal with the highest score is chosen whenever a decision must be made.

Though non-chronological backtracking helps the DPLL procedure to recover from poorly chosen decision variables, it is only effective once a conflict has been detected. Suppose a conflict clause γ involves variables at decision levels $d_0 \ldots, d_k$. Ideally, one would have $d_{i+1} = d_i + 1$ for $0 < i < k$. Otherwise, the work done in propagating the effects of the irrelevant intervening decisions is wasted. Increasing the scores of the variables in γ as done in VSIDS helps because the variables at the higher decision levels will be chosen earlier in the sequel of the search. However, the variables in the conflict clause at the lower decision levels will also be chosen earlier. More importantly, it may take several conflicts for a group of related variables to have similar scores if their initial scores are sufficiently different. In BerkMin [14] this problem is addressed by choosing the decision literal from the unassigned variables in the most recent conflict clause that is unsatisfied. The limitation of this approach is that a conflict clause's ability to cause its literals to be treated like a related group is lost as soon as it is no longer the most recent unsatisfied clause.

By contrast, the approach followed in CirCUs is the following. Suppose a new conflict clause $\gamma = \{l_0, \ldots, l_k\}$ is generated. Suppose that d_i is the decision level of l_i, and, w.l.o.g., that $d_i < d_{i+1}$ for $0 < i < k$. The scores of all literals in the clause are incremented by one with the exception of the literal l_k at the current decision level, whose score is set equal to one less than the score of l_{k-1}. Boosting the score of the most recent decision variable causes the relation between l_{k-1} and l_k to be recorded in the scores, producing a longer lasting effect than in the BerkMin case. We call the new heuristic Deepest Variable Hiking (DVH).

Figure 2 shows two series of decisions to illustrate the advantages of the DVH heuristic. Each circle represents a decision made by a score-based heuristic and the dark circles represent decisions whose implications are involved in the conflict-learned clause. We assume that the conflict occurs in both cases at decision level d_{i+4}.

If all the decisions are relevant to the current conflict, then the conflict-learned clause will contain literals implied by all previous decisions as shown in Fig. 2 (a). In this case we backtrack to the decision level d_{i+3}. If, however, irrelevant intervening decisions were made, such as those at levels d_{i+2} and d_{i+3} in Fig. 2 (b), then backtracking will be to a lower decision level like d_{i+2} in the example. Since the decisions made at level d_{i+2} and d_{i+3} are not related to the conflict-learned clause, the cost of BCP for those decisions is wasted. Even though the current scores of the decision variables at levels d_{i+2} and d_{i+3} are higher than the one at level d_{i+4}, the variable of d_{i+4} is a better choice. Thanks to the DVH decision heuristic, we can avoid the waste of effort even when exploring subspaces in which the clause derived from the current conflict is satisfied.

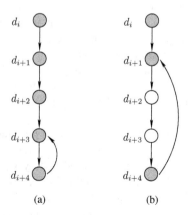

(a)   (b)

Fig. 2. Two examples of decision

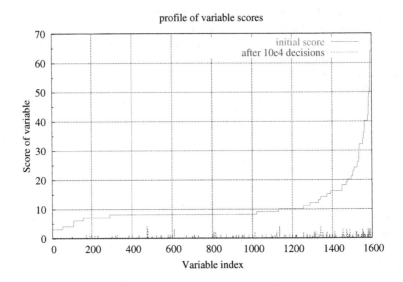

Fig. 3. Scores of variables while solving C880.cnf using VSIDS

Since we increase the score of the variable of level d_{i+4} to one less than the score of the variable at level d_{i+2}, the score-based decision heuristic treats them as a related group. If they are not relevant variables for the rest of the search, then the periodic decay will reduce their scores thereby decreasing their importance automatically.

Suppose the data inputs of a multiplexer are driven by two subcircuits having disjoint supports and that the *sel* signal selects which circuitry is connected to the output of multiplexer. Once *sel* is decided then the variables in the unselected circuitry can be ignored since they no longer affect the value of the circuit. Silva et al. address this problem in [35] and a related approach is presented in [3]. Gupta et al. [16] use circuit SAT to identify the unobservable gates and disable the corresponding clauses in the CNF database. In [40], the author proposes an efficient translation of CNF from circuits that considers unobservable gates. Even though the DVH heuristic does not explicitly address unobservable gates in a circuit, it does help when such gates are present thanks to its ability to increase the dynamics of decision heuristics. For instance, once the *sel* signal is assigned and we find a conflict from the circuitry feeding one of the inputs to the multiplexer, the DVH heuristic helps the decision procedure focus on the part where the conflict was found.

The VSIDS rule as implemented in Zchaff halves the literal scores once every so many decisions. If the ratio of decisions to conflicts is large, most scores decay to 0. In Fig. 3 We show the profile of variable scores produced by VSIDS for C880.cnf, which is one of SAT 2003 industrial benchmark. In the figure, one can find two lines. They are the profiles of initial scores and the scores after 10000 decisions are made. The variables are sorted according to their initial scores. One can see from the figure that not only most variables have scores of zero, but also the few non-null scores take only a very limited number of values.

When this is the case, variables are chosen on the basis on insufficient information. The DVH heuristic of CirCUs tries to overcome this problem by reducing the halving frequency if the ratio of decisions to conflicts is too high.

6   Experimental Results

We performed two sets of experiments to assess the impact of the techniques described in Sections 4 and 5. The first set studies the effects of CNF conditioning on the speed of the SAT solver for 89 examples from the hand-made category of the SAT2003 benchmark set. These examples are not very large   up to 2,000 variables and 60,000 clauses   but some of them are hard for many solvers. The experiments were run on a 2.4 GHz Pentium IV with 500 MB of RAM running Linux. Runs longer than 2,000 s were terminated.

Table 1 shows the examples that were used for the CNF conditioning experiments. The columns comparing CPU time show that CirCUs achieves huge improvements over Zchaff. We also show the numbers of completed instances with in parenthesis.

Figure 4 shows a log-log scatterplot that compares CirCUs runtimes with and without CNF conditioning. One can easily identify two groups of instances. Those for which reshaping is effective, including those for which a monolithic BDD can be built, and those near or above the main diagonal, for which conditioning does not appreciably change the CNF. In the latter group, the overhead of constructing the BDDs is not recovered.

It should be pointed out that sorting the clauses by linear arrangement and applying the clustering algorithm of [22] are fundamental for efficiency. Many of the examples that terminate in a few seconds with the algorithm of Sect. 4 cannot be completed otherwise.

The second set of experiments compares the DVH variable selection heuristic of Sect. 5 to the popular VSIDS heuristic used in Chaff. We compared three sets of 50 runs: Zchaff [32], CirCUs with VSIDS, and CirCUs with DVH. CNF conditioning was not used in these experiments that were performed on a 1.7 GHz Pentium IV with 2 GB

Table 1. Examples from hand-made category of the SAT2003 benchmark set

Benchmark name	Number of instances	CPU time	
		Zchaff	CirCUs
bevan/marg*	14	4330.36(12)	1.33(14)
bevan/urqh1c*	13	14638.84(8)	6.77(13)
bevan/urqh*	12	20028.83(2)	4.81(12)
markstrom/mm*	8	1047.22(8)	1262.46(8)
purdom/	4	3160.07(3)	1720.21(4)
simon/sat02/x1*	19	38000.00(0)	6.97(19)
simon/sat02/x2*	9	18000.00(0)	3.38(9)
simon/sat02/Urquhart*	10	20000.00(0)	4.59(10)

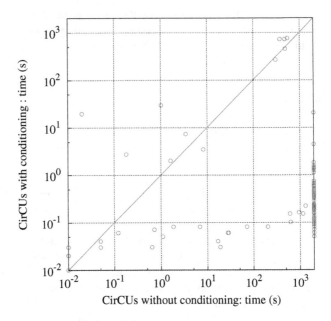

Fig. 4. Effects of CNF conditioning

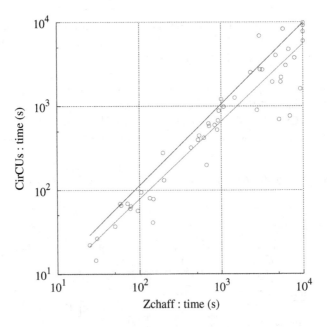

Fig. 5. DVH versus VSIDS

of RAM running Linux. The timeout was set at 10,000 s. The SAT instances are derived from BMC experiments on models from the VIS Verilog benchmark collection [41].

The results are summarized in Fig. 5. The log-log scatterplot shows the points comparing CirCUs with DVH to Zchaff. The two straight lines are regression curves of the form $y = \kappa \cdot x$, where κ and   are obtained by least-square fitting. The upper line is for the comparison of CirCUs with VSIDS to Zchaff; it is provided for calibration. It shows that the two solvers are quite comparable in performance when using the same decision heuristic. The lower line is for CirCUs with DVH vs. Zchaff. The separation of the two lines indicates that DVH provides a speedup of almost 2 over VSIDS.

CirCUs was in second stage of industrial category of SAT'04 solver competition. One can find more experimental results in SAT'04 competition wep page [34].

7   Conclusions

We have presented CirCUs, a hybrid SAT solver that operates on an And-Inverter Graph, a set of CNF clauses, and a set of BDDs. We have described the approach used to speed up the solver when the input is in CNF form. By converting the clauses into one or more BDDs, we are often able to either solve the problem directly, or extract an improved CNF formula. We have shown the effectiveness of this strategy on small-but-hard examples from the SAT2003 benchmark set.

We have also presented an improved decision variable selection heuristic, and shown its effectiveness by comparing it to the popular VSIDS heuristic of Zchaff.

Our results demonstrate the usefulness of allowing the SAT solver to operate on multiple representations of the input problem. We intend to explore more applications of this principle, and we are busy improving the efficiency of the current implementation. For instance, we plan to compare the current force-directed placement approach to the MLP algorithm [29].

References

[1] F. A. Aloul, I. L. Markov, and K. A. Sakallah. Mince: A static global variable-ordering for SAT and BDD. Presented at IWLS01, June 2001.

[2] F. A. Aloul, I. L. Markov, and K. A. Sakallah. FORCE: A fast and easy-to-implement variable-ordering heuristic. In *Proceedings of the Great Lakes Symposium on VLSI*, pages 116–119, Washington, DC, Apr. 2003.

[3] C. W. Barrett, D. L. Dill, and A. Stump. Checking satisfiability of first-order formulas by incremental translation to SAT. In E. Brinksma and K. G. Larsen, editors, *Fourteenth Conference on Computer Aided Verification (CAV'02)*, pages 236–249. Springer-Verlag, Berlin, July 2002. LNCS 2404.

[4] A. Biere, A. Cimatti, E. Clarke, and Y. Zhu. Symbolic model checking without BDDs. In *Fifth International Conference on Tools and Algorithms for Construction and Analysis of Systems (TACAS'99)*, pages 193–207, Amsterdam, The Netherlands, Mar. 1999. LNCS 1579.

[5] R. K. Brayton et al. VIS: A system for verification and synthesis. In T. Henzinger and R. Alur, editors, *Eighth Conference on Computer Aided Verification (CAV'96)*, pages 428–432. Springer-Verlag, Rutgers University, 1996. LNCS 1102.

[6] R. E. Bryant. Graph-based algorithms for Boolean function manipulation. *IEEE Transactions on Computers*, C-35(8):677–691, Aug. 1986.

[7] J. R. Burch, E. M. Clarke, and D. E. Long. Representing circuits more efficiently in symbolic model checking. In *Proceedings of the Design Automation Conference*, pages 403–407, San Francisco, CA, June 1991.

[8] J. R. Burch and V. Singhal. Tight integration of combinational verification methods. In *Proceedings of the International Conference on Computer-Aided Design*, pages 570–576, San Jose, CA, Nov. 1998.

[9] R. Damiano and J. Kukula. Checking satisfiability of a conjunction of BDDs. In *Proceedings of the Design Automation Conference*, pages 818–823, June 2003.

[10] M. Davis, G. Logemann, and D. Loveland. A machine program for theorem proving. *Communications of the ACM*, 5:394–397, 1962.

[11] M. Davis and H. Putnam. A computing procedure for quantification theory. *Journal of the Association for Computing Machinery*, 7(3):201–215, July 1960.

[12] J. Franco, M. Kouril, J. Schlipf, J. Ward, S. Weaver, M. Dransfield, and W. M. Vanfleet. SBSAT: A state-based, BDD-based satisfiability solver. In *International Conference on Theory and Applications of Satisfiability Testing (SAT 2003)*, Portofino, Italy, May 2003.

[13] M. K. Ganai, P. Ashar, A. Gupta, L. Zhang, and S. Malik. Combining strengths of circuit-based and CNF-based algorithms for a high-performance SAT solver. In *Proceedings of the Design Automation Conference*, pages 747–750, New Orleans, LA, June 2002.

[14] E. Goldberg and Y. Novikov. BerkMin: A fast and robust SAT-solver. In *Proceedings of the Conference on Design, Automation and Test in Europe*, pages 142–149, Paris, France, Mar. 2002.

[15] A. Gupta, M. Ganai, C. Wang, Z. Yang, and P. Ashar. Learning from BDDs in SAT-based bounded model checking. In *Proceedings of the Design Automation Conference*, pages 824–829, June 2003.

[16] A. Gupta, A. Gupta, Z. Yang, and P. Ashar. Dynamic detection and removal of inactive clauses in SAT with application in image computation. In *Proceedings of the Design Automation Conference*, pages 536–541, Las Vegas, NV, June 2001.

[17] A. Gupta, Z. Yang, P. Ashar, and A. Gupta. SAT-based image computation with application in reachability analysis. In W. A. Hunt, Jr. and S. D. Johnson, editors, *Formal Methods in Computer Aided Design*, pages 354–271. Springer-Verlag, Nov. 2000. LNCS 1954.

[18] M. Herbstritt and B. Becker. Conflict-based selection of branching rules. In *Sixth International Conference on Theory and Application in Satisfiability Testing (SAT2003)*, pages 441–451, Portofino, Italy, May 2003. Springer. LNCS 2919.

[19] A. J. Hu and D. Dill. Efficient verification with BDDs using implicitly conjoined invariants. In C. Courcoubetis, editor, *Fifth Conference on Computer Aided Verification (CAV '93)*, pages 3–14. Springer-Verlag, Berlin, 1993. LNCS 697.

[20] S.-W. Jeong and F. Somenzi. A new algorithm for 0-1 programming based on binary decision diagrams. In T. Sasao, editor, *Logic Synthesis and Optimization*, chapter 7, pages 145–165. Kluwer Academic Publishers, Boston, MA, 1993.

[21] H. Jin, M. Awedh, and F. Somenzi. CirCUs: A satisfiability solver geared towards bounded model checking. In R. Alur and D. Peled, editors, *Sixteenth Conference on Computer Aided Verification (CAV'04)*. Springer-Verlag, Berlin, July 2004. To appear.

[22] H. Jin, A. Kuehlmann, and F. Somenzi. Fine-grain conjunction scheduling for symbolic reachability analysis. In *International Conference on Tools and Algorithms for Construction and Analysis of Systems (TACAS'02)*, pages 312–326, Grenoble, France, Apr. 2002. LNCS 2280.

[23] H. Jin and F. Somenzi. CirCUs: Speeding up circuit SAT with BDD-based implications. Submitted for publication, Apr. 2004.

[24] A. Kuehlmann, M. K. Ganai, and V. Paruthi. Circuit-based Boolean reasoning. In *Proceedings of the Design Automation Conference*, pages 232–237, Las Vegas, NV, June 2001.

[25] A. Kuehlmann and F. Krohm. Equivalence checking using cuts and heaps. In *Proceedings of the Design Automation Conference*, pages 263–268, Anaheim, CA, June 1997.

[26] P. Liberatore. On the complexity of choosing the branching literal in DPLL. *Artificial Intelligence*, 116(1–2):315–326, 2000.

[27] S.-I. Minato. Fast generation of irredundant sums-of-products forms from binary decision diagrams. In *SASIMI '92*, pages 64–73, Kyoto, Japan, Apr. 1992.

[28] S.-I. Minato. Zero-suppressed BDDs for set manipulation in combinatorial problems. In *Proceedings of the Design Automation Conference*, pages 272–277, Dallas, TX, June 1993.

[29] I.-H. Moon, G. D. Hachtel, and F. Somenzi. Border-block triangular form and conjunction schedule in image computation. In W. A. Hunt, Jr. and S. D. Johnson, editors, *Formal Methods in Computer Aided Design*, pages 73–90. Springer-Verlag, Nov. 2000. LNCS 1954.

[30] I.-H. Moon, J. H. Kukula, K. Ravi, and F. Somenzi. To split or to conjoin: The question in image computation. In *Proceedings of the Design Automation Conference*, pages 23–28, Los Angeles, CA, June 2000.

[31] E. Morreale. Recursive operators for prime implicant and irredundant normal form determination. *IEEE Transactions on Computers*, C-19(6):504–509, June 1970.

[32] M. Moskewicz, C. F. Madigan, Y. Zhao, L. Zhang, and S. Malik. Chaff: Engineering an efficient SAT solver. In *Proceedings of the Design Automation Conference*, pages 530–535, Las Vegas, NV, June 2001.

[33] N. R. Quinn. The placement problem as viewed from the physics of classical mechanics. In *Proceedings of the Design Automation Conference*, pages 173–178, Boston, MA, June 1975.

[34] URL: http://www.lri.fr/~simon/contest/results.

[35] J. P. M. Silva. The impact of branching heuristics in propositional satisfiability algorithms. In *Proceedings of the 9th Portuguese Conference on Artificial Intelligence (EPIA)*, Sept. 1999.

[36] J. P. M. Silva and K. A. Sakallah. Grasp   a new search algorithm for satisfiability. In *Proceedings of the International Conference on Computer-Aided Design*, pages 220–227, San Jose, CA, Nov. 1996.

[37] F. Somenzi. *CUDD: CU Decision Diagram Package*. University of Colorado at Boulder, ftp://vlsi.colorado.edu/pub/.

[38] F. Somenzi and K. Ravi. Extracting simple invariants from BDDs. Unpublished manuscript, May 2002.

[39] H. Touati, H. Savoj, B. Lin, R. K. Brayton, and A. Sangiovanni-Vincentelli. Implicit enumeration of finite state machines using BDD's. In *Proceedings of the IEEE International Conference on Computer Aided Design*, pages 130–133, Nov. 1990.

[40] M. N. Velev. Exploiting signal unobservability for efficient translation to CNF in formal verification of microprocessor. In *In the Proceedings of the IEEE/ACM Design, Automation and Test in Europe Conference (DATE)*, pages 10266–10271, Feb. 2004.

[41] URL: http://vlsi.colorado.edu/~vis.

[42] H. Zhang. SATO: An efficient propositional prover. In *Proceedings of the International Conference on Automated Deduction*, pages 272–275, July 1997. LNAI 1249.

Equivalence Models for Quantified Boolean Formulas

Hans Kleine Büning[1] and Xishun Zhao[2,*]

[1] Department of Computer Science,
Universität Paderborn, 33095 Paderborn (Germany)
kbcsl@upb.de
[2] Institute of Logic and Cognition,
Sun Yat-sen University, 510275 Guangzhou, (P.R. China)
hsdp08@zsu.edu.cn

Abstract. In this paper, the notion of equivalence models for quantified Boolean formulas with free variables is introduced. The computational complexity of the equivalence model checking problem is investigated in the general case and in some restricted cases. We also establish a connection between the structure of some quantified Boolean formulas and the structure of models.

Keywords: Quantified Boolean formula, equivalence model, model checking, complexity, equivalence, satisfiability.

1 Introduction

For several applications the satisfiability and the equivalence problems for quantified Boolean formulas (QBF) (with free variables) play a central role. We not only want to decide whether the satisfiability or equivalence holds, but we want to know for which assignments, that means, for which Boolean functions, the desired properties are fulfilled. That leads to the question of whether certain satisfiability or equivalence models exist. The paper will introduce the notion of equivalence models and presents some results for some subclasses of QBF and some restricted sets of Boolean functions, where Boolean functions are considered as propositional formulas.

The notion of (satisfiability) models for formulas in *QBF* (i.e., the class of quantified Boolean formulas in prenex normal form without free variables) has been introduced in [8, 9]. Generally speaking, an assignment for a formula in *QBF* is a mapping which maps each existential variable to a propositional formula over universal variables whose quantifiers precede the quantifier of the existential variable. An assignment M is a model for a quantified formula Φ (with existential variables $\boldsymbol{y} = y_1, \cdots, y_m$) if the resulting formula $\Phi[\boldsymbol{y}/M]$ after

---

* The author was supported partially by the SSFC project 02BZX046 and the NSFC project "Algorithmic Foundation of Computability and Complexity".

H.H. Hoos and D.G. Mitchell (Eds.): SAT 2004, LNCS 3542, pp. 224–234, 2005.

replacing each existential variable by its corresponding formula (and removing existential quantifiers from the prefix) is true.

The notion of models for closed QBF can be easily extended to formulas in $QBF\,^*$, the class of quantified Boolean formulas with (or without) free variables, by just allowing free variables occurring in the propositional formulas of assignments.

In this paper, we often write $\Phi = Q\phi(x, y)$ and $\Phi(z) = Q\phi(x, y, z)$ for formulas in QBF and $QBF\,^*$, respectively, with universal variables $x = x_1, \cdots, x_n$, existential variables $y = y_1, \cdots, y_m$, and free variables $z = z_1, \cdots, z_r$.

Please note that if M is a model for a formula $\Phi \in QBF$ then Φ and $\Phi[y/M]$ are equivalent. However, this is generally invalid for formulas in $QBF\,^*$ and their models. For example, the formula $\Phi(z_1, z_2) = \exists y((z_1 \vee y) \wedge (\neg y \vee z_2))$ is equivalent to the formula $(z_1 \vee z_2)$. For $f_y(z_1, z_2) = 1$, $M = (f_y)$ is a model, since $\Phi[y/f_y] = (z_1 \vee 1) \wedge (0 \vee z_2) \approx z_2$ is satisfiable. But the resulting formula is not equivalent to the input formula $\Phi(z_1, z_2)$.

This motivates us to introduce and investigate equivalence models for formulas in $QBF\,^*$, which deserve attention because quantified Boolean formulas can be used to represent Boolean functions with essentially small size. There are Boolean functions which can be represented by a formula in $QBF\,^*$ with quadratic size while every propositional formula representing the same function has super-polynomial size [7]. An assignment M is an equivalence model for a formula $\Phi \in QBF\,^*$ if Φ and $\Phi[y/M]$ are equivalent. We are interested in the equivalence model checking problem of determining whether an assignment is an equivalence model of a formula $\Phi \in QBF\,^*$, and the equivalence model problem of deciding whether a formula in $QBF\,^*$ has an equivalence model. Since the equivalence model checking problem involves testing the equivalence of two quantified formulas, the problem is $PSPACE$–complete. Without any restriction we will see that any $QBF\,^*$ formula has an equivalence model. In this paper, we restrict the two problems to some subclasses of $QBF\,^*$ and some models consisting of propositional formulas with special structures.

We are also interested in discovering some connections between the structure of formulas in $QBF\,^*$ and that of models. We will show that $Q2\text{-}CNF\,^*$ formulas always have an equivalence model consisting of formulas in $1\text{-}CNF \cup 1\text{-}DNF \cup \{0, 1\}$.

In the remainder of this section we will recall and introduce some notations and terminologies.

The classes of propositional formulas such as CNF, DNF, $k\text{-}CNF$, $HORN$ and so on, are defined as usual.

QBF is the class of closed quantified Boolean formula (i.e., without free variables). The formula Φ is in prenex normal form, if $\Phi = Q_1 v_1 \cdots Q_n v_n \phi$, where $Q_i \in \{\forall, \exists\}$ and ϕ is a propositional formula over variables $v_1, \cdots, v_n$. $Q_1 v_1 \cdots Q_n v_n$ is called the prefix and ϕ the matrix or kernel of Φ. Usually, we simply write $\Phi = Q\phi$. A literal x or $\neg x$ is called a universal resp. existential literal, if the variable x is bounded by a universal quantifier resp. by an existential quantifier. A closed formula $\Phi \in QBF$ in prenex normal form is called satisfiable or true, if there exists an assignment of truth values to the existential variables depending on the truth values for the preceding universal variables, for which the propositional kernel of the formula is true. $QCNF$ denotes the class of QBF formulas in prenex normal form with matrix in CNF, likewise for Qk-CNF , $QHORN$.

The classes $QCNF^*$, $Q2$-CNF *, $QHORN$ * are defined in the same way as $QCNF$, $Q2$-CNF , $QHORN$, respectively, except allowing free variables. A formula in $QCNF^*$ is satisfiable if and only if there is a truth assignment for the free variables, such that the closed QBF formula resulting from the partial evaluation is true. The class $\exists^* CNF^*$ is the subset of $QCNF^*$ in which any formula has a purely existential prefix and a CNF kernel.

In our investigations we will make use of substitutions of existential variables by propositional formulas. For a quantified Boolean formula Φ with or without free variables $\Phi[y_1/f_1, \cdots, y_m/f_m]$ denotes the formula obtained by simultaneously substituting the occurrences of each variables y_i by the formula f_i and removing quantifiers of y_i. For $\Phi[y_1/f_1, \cdots, y_m/f_m]$, $\boldsymbol{y} = y_1, \cdots, y_m$, and $M = (f_1, \cdots, f_m)$ we write $\Phi[\boldsymbol{y}/M]$.

2   Models

In this section we present two definitions of models for quantified Boolean formulas and prove some basic results. The first definition is based on satisfiability and has been investigated in [8,9] for closed formulas.

Definition 1. *(Satisfiability Model) [8] Let* $\Phi(\boldsymbol{z}) = Q\phi(\boldsymbol{x}, \boldsymbol{y}, \boldsymbol{z})$ *be a formula in* $QCNF^*$, *where* $\boldsymbol{x} = x_1, \cdots, x_n$ *are the universal variables,* $\boldsymbol{y} = y_1, \cdots, y_m$ *the existential variables, and* $\boldsymbol{z} = z_1, \cdots, z_r$ *the free variables. For propositional formulas* f_{y_i} *over* $\boldsymbol{z}$ *and universal variables whose quantifiers precede* $\exists y_i$, *we say* $M = (f_{y_1}, \cdots, f_{y_m})$ *is a* **(satisfiability) model** *for* $\Phi(\boldsymbol{z})$ *if and only if* $\forall x_1 \cdots \forall x_n \ \phi(\boldsymbol{x}, \boldsymbol{y}, \boldsymbol{z})[\boldsymbol{y}/M]$ *is satisfiable.*

If the propositional formulas f_{y_i} *belong to a class K of propositional formulas, then M is called a* K−**model** *for* $\Phi(\boldsymbol{z})$.

For example, the formula $\Phi(z) = \forall x \exists y ((x \vee y) \wedge (\neg x \vee \neg y) \wedge z)$ is satisfiable and for $f_y(x, z) = \neg x$, $M = (f_y)$ is a model for $\Phi(z)$, because

$$\forall x((x \vee y) \wedge (\neg x \vee \neg y) \wedge z)[y/f_y] = \forall x((x \vee \neg x) \wedge (\neg x \vee x) \wedge z)$$

is satisfiable (set $z = 1$).

The formula $\Phi(z_1, z_2) = \exists y ((z_1 \vee y) \wedge (\neg y \vee z_2))$ is logically equivalent to the formula $(z_1 \vee z_2)$. For $f_y(z_1, z_2) = 1$, $M = (f_y)$ is a model, since $\Phi(z_1, z_2)[y/f_y] = (z_1 \vee 1) \wedge (0 \vee z_2) \approx z_2$ is satisfiable. But the formula $\Phi(z_1, z_2)[y/f_y]$ is not equivalent to the input formula $\Phi(z_1, z_2)$. For $g_y(z_1, z_2) = z_2$, however, we get the equivalence $\Phi(z_1, z_2)[y/g_y] = (z_1 \vee z_2) \wedge (\neg z_2 \vee z_2) \approx (z_1 \vee z_2) \approx \Phi(z_1, z_2)$. That means, the substitution of the existential variable y by the associated propositional formula g_y preserves the image of the formula. This simple observation motivates a second definition of models for quantified Boolean formulas.

Definition 2. *(Equivalence Model) Let* $\Phi(z) = Q\phi(x, y, z)$ *be a formula in* $QCNF^*$, *where* $x = x_1, \cdots, x_n$ *are the universal variables,* $y = y_1, \cdots, y_m$ *the existential variables, and* $z = z_1, \cdots, z_r$ *the free variables. For propositional formulas* f_{y_i} *over* z *and universal variables whose quantifiers precede* $\exists y_i$, *we say* $M = (f_{y_1}, \cdots, f_{y_m})$ *is an* **equivalence model** *for* $\Phi(z)$ *if and only if* $\Phi(z) \approx \forall x_1 \cdots \forall x_n \phi(x, y, z)[y/M]$.

If the propositional formulas f_{y_i} *belong to a class* K *of propositional formulas, then* M *is called a* K–**equivalence model** *for* $\Phi(z)$.

The formula $\Phi(z) = \forall x_1 \forall x_2 \exists y((x_1 \vee y) \wedge (x_2 \vee \neg y \vee z))$ is equivalent to z. For $f_y(x_1, x_2, z) = \neg x_1$, $M = (f_y)$ is an equivalence model, since $\Phi(z)[y/f_y] = \forall x_1 \forall x_2((x_1 \vee \neg x_1) \wedge (x_2 \vee x_1 \vee z)) \approx \forall x_1 \forall x_2(x_2 \vee x_1 \vee z) \approx z$.

Obviously, any unsatisfiable formula has an equivalence model. In that case any propositional formula over the corresponding variables is an equivalence model.

Lemma 1. *Any formula in* $QCNF^*$ *has an equivalence model.*

Proof. Suppose, we have a formula $\Phi(z) = Q\phi(z, x, y) \in QCNF^*$ equivalent to the Boolean function $F(z)$ with free variables $z = z_1, \cdots, z_m$, universal variables $x = x_1, \cdots, x_t$, and existential variables $y = y_1, \cdots, y_n$. For fixed tuples of truth values $a \in \{0, 1\}^m$ the formula $\Phi(a)$ is a closed formula.

If the formula $\Phi(a)$ is true, then there is a satisfiability model $M^a = (f_{y_1}^a, \cdots, f_{y_n}^a)$. That means $\Phi(a)[y/M^a]$ is true.

If the formula $\Phi(a)$ is false, then for $M^a = (0, \cdots, 0)$ the formula $\Phi(a)[y/M^a]$ is false.

Now, we combine these 2^m cases to an equivalence model as follows: Let $x^i = x_1, \cdots, x_{r_i}$ be the preceding universal variables for y_i. We define a Boolean function $f_{y_i}(z, x^i) = f_{y_i}^a(x^i)$, if $z = a$. Since for any Boolean function there is an equivalent propositional formula, for $M = (f_{y_1}, \cdots, f_{y_n})$ we have $\Phi(z)[y/M] \approx F(z) \approx \Phi(z)$. Hence, M is an equivalence model for $\Phi(z)$. ∎

The next proposition states some simple observations for which we omit the proof.

Proposition 1. *Let* $\Phi(z) = Q\phi(x, y, z)$ *be an arbitrary formula in* $QCNF^*$, $M = (f_{y_1}, \cdots, f_{y_m})$ *any sequence of propositional formulas.*

1. $\Phi(z)[y/M] \models \Phi(z)$. *Moreover,* M *is an equivalence model for* $\Phi(z)$ *if and only if* $\Phi(z) \models \Phi(z)[y/M]$
2. *If* Φ *is closed (i.e.,* z *is empty) and true, then* M *is a (satisfiability) model for* Φ *if and only if* M *is an equivalence model for* Φ.

Let K be a class of propositional formulas and $X \subseteq QCNF^*$. We are mainly interested in the following problems:

K–**Equivalence Model Checking Problem for** X:

    Instance: A formula $\Phi \in X$ and $M = (f_1, \cdots, f_n)$ a sequence of propositional formulas $f_i \in K$.

    Query: Is M a K–equivalence model for Φ?

K–Equivalence Model Problem for X:
Instance: A formula $\Phi \in X$.
Query: Does there exist a K–equivalence model M for Φ?

A procedure for deciding whether M is an equivalence model for $\Phi(z)$ is as follows:

1. Substitute the existential variables by the associated propositional formulas and remove from the prefix the existential quantifiers.
2. Test whether the resulting formula is equivalent to the input formula.

Because the equivalence problem between quantified formulas is *PSPACE–* complete even if one of them is very simple, we obtain the following lemma.

Lemma 2. *The equivalence model checking problem for $QCNF^*$ is PSPACE–complete.*

Proof. Obviously, the equivalence model checking problem is in *PSPACE*, since the satisfiability and the equivalence problem for quantified Boolean formulas are in *PSPACE*. We prove the *PSPACE*–hardness from a reduction of the *PSPACE*–complete evaluation problem for $QCNF$ [7]. We associate to a closed formula $\Phi = Q\phi \in QCNF$ for a new variable y the formula $\Phi' = \exists y Q(\phi \wedge y)$. Then, Φ is true if and only if Φ' is true. Suppose, Φ has the existential variables $\boldsymbol{y} = y_1, \cdots, y_n$. For $M = (f_y, f_{y_1}, \cdots f_{y_n})$, where all Boolean functions are the constant 0, that means $f_y = 0, f_{y_i} = 0$, $\Phi'[\boldsymbol{y}/M]$ is false. Hence, M is an equivalence model for Φ' if and only if Φ is false. ∎

That means, equivalence model checking is much harder than satisfiability model checking which has been shown in [8] to be *coNP*–complete for $QCNF$.

The next theorem states that the upper bound for the complexity of the equivalence checking problem for classes X depends on the complexity of the satisfiability problem for X. We say a class $X \subseteq QCNF^*$ is *closed under constant–substitutions* if and only if for every formula $\Phi(z) \in X$ and for all combinations of constants $\boldsymbol{a}$ in $\{0,1\}^r$ the formula $\Phi(z)[z/a] = \Phi(a)$ is in X.

We recall the notion of the polynomial-time hierarchy. Σ_2^P is a class of problems defined as ($k \geq 0$): $\Sigma_0^P = \Pi_0^P = P$ the class of polynomial-time solvable problems, $\Sigma_{k+1}^P = NP^{\Sigma_k^P}, \Pi_{k+1}^P = co - \Sigma_{k+1}^P$. Thus, $NP = \Sigma_1^P$ and $coNP = \Pi_1^P$. Relationships between prefix classes of QBF^* and classes of the polynomial-time hierarchy have been shown for example in [15].

Theorem 1. *For every class $X \subseteq QCNF^*$ which is closed under constant-substitutions we have*

1. *If $X \cap QSAT$ is polynomial-time solvable (i.e., in $\Sigma_0^P = \Pi_0^P$), then the equivalence model checking problem for X is in $coNP = \Pi_1^P$.*
2. *For $k \geq 1$, if $X \cap QSAT$ is in Σ_k^P, then the equivalence model checking problem for X is in Π_k^P.*
3. *For $k \geq 1$, if $X \cap QSAT$ is in Π_k^P, then the equivalence model checking problem for X is in Π_{k+1}^P.*

Here, QSAT is the class of all satisfiable formulas in QCNF*.

Proof. For any $\Phi(z) = Q\phi(x, y, z)$ and any sequence $M = (f_{y_1}, \cdots, f_{y_m})$ of propositional formulas. we have the following equivalence relations.

M is not an equivalence model for $\Phi(z)$
$\Leftrightarrow \Phi(z) \not\approx \Phi(z)[y/M] \Leftrightarrow \Phi(z) \not\models \Phi(z)[y/M]$
$\Leftrightarrow \exists z : (\Phi(z)$ is true and $\Phi(z)[y/M]$ is false$)$
$\Leftrightarrow \exists z : (\Phi(z)$ is true and $\forall x_1 \cdots \forall x_n \phi(x, z, y)[y/M]$ is false $)$
$\Leftrightarrow \exists z : (\Phi(z)$ is true and $\exists x_1 \cdots \exists x_n \neg\phi(x, y, z)[y/M]$ is true $)$
$\Leftrightarrow \exists z \exists x' : (\Phi(z)$ is true and $\neg\phi(x', y, z)[y/M']$ is true $)$.

The propositional formula $\neg\phi(x', y, z)[y/M']$ contains the variables x' and z, where $x' := x'_1, \cdots, x'_n$ and M' is the result of renaming x_i by x'_i in M for every i. Whether for fixed values for x' and z the formula evaluates to true, can be decided in linear time. Please note that for fixed values for z the formula $\Phi(z)$ is in X.

If the satisfiability of formulas in X is solvable in polynomial time, then the non-equivalence model checking problem is in $NP = \Sigma_1^P$. Hence, the complementary problem — the equivalence model checking problem — is in $coNP = \Pi_1^P$.

If the satisfiability problem for formulas in X is in Σ_k^P then the problem whether $\exists x \Phi(x)$ is satisfiable remains in Σ_k^P. Hence, the equivalence model checking problem is in Π_k^P.

If the satisfiability problem for formulas in X is in Π_k^P then the problem whether $\exists x \Phi(x)$ is satisfiable is in Σ_{k+1}^P. Hence, the equivalence model checking problem is in Π_{k+1}^P.   ∎

Please note, that Theorem 1 establishes only an upper bound for the complexities. Classes with a tractable satisfiability problem are $Q2$-CNF * and $QHORN^*$ [1, 4], whereas the satisfiability problem for $\exists^* CNF^*$ is obviuosly NP–complete. With respect to the completeness of the various problems we can prove

Lemma 3. *The equivalence model checking problems for $\exists^* CNF^*$, $Q2$-CNF * and $QHORN^*$ are coNP–complete.*

Proof. The $coNP$–hardness of the equivalence checking problem for $\exists^* CNF^*$ follows from a reduction from the $coNP$–complete unsatisfiability problem for CNF. We associate to every propositional CNF formula $\phi(x)$ the formula $\Phi(x) = \exists y(y \wedge \phi(x))$ and $M = (f_y(x))$ with $f_y(x) = 0$. The formula $\Phi(x)$ is equivalent to $\phi(x)$. Substituting the existential variable y by the model function f_y we obtain $\Phi(x)[y/f_y(x)] = 0 \wedge \phi(x) \approx 0$. Hence, M is an equivalence model for $\Phi(x)$ if and only if $\phi(x)$ is unsatisfiable.

That the problem is in $coNP = \Pi_1^P$ follows from Theorem 1, because $\exists^* CNF$ is closed under constant–substitutions and the satisfiability problem is in NP.

The other classes $Q2$-CNF * and $QHORN^*$ are closed under constant–substitution and their satisfiability problems are solvable in polynomial time [7]. Hence, by Theorem 1 the problems are in $coNP$. The $coNP$–hardness follows from a reduction from the *coNP*–complete tautology problem for *DNF*. We associate to a formula $\psi \in DNF$ over the variables $x_1, \cdots, x_n$ the quantified Boolean formula $\Phi = \forall x_1 \cdots \forall x_n \exists y : \neg y$ and $M = (f_y(x_1, \cdots, x_n))$, where $f_y(x_1, \cdots, x_n) = \neg \psi$. Then, M is an equivalence model for Φ if and only if $\neg f_y(x_1, \cdots, x_n) = \psi \in DNF$ is a tautology.   ∎

3   Special Classes of Models

In this section we investigate the problems for various classes of model formulas and input formulas. Some of the results are depicted in figure 1. Two classes K_0, the set of constants 0 and 1, and K_2, the set of monomials, are defined as $K_0 := \{f \mid f \text{ is } 0 \text{ or } 1\}$ and $K_2 := \{f \mid \exists I \subseteq \{1, \cdots, n\} : f(x_1, ..., x_n) = \bigwedge_{i \in I} x_i, \; n \in \mathbf{N}\} \cup K_0$

Fig. 1

Boolean functions	$QCNF^*$–class	equivalence model checking
$K_0 = \{0, 1\}$	$QCNF$	$PSPACE$-complete
1-*CNF* ∪ 1-*DNF* ∪ $\{0, 1\}$	$Q2$-CNF *	polytime
1-*DNF*	$QHORN^*$	*coNP*-complete
K_2	$QHORN^*$	polytime

For a formula $\Phi = \exists^* \phi \in \exists^* CNF^*$, if the kernel ϕ is satisfiable then Φ has a K_0-model. However, this is not true for equivalence models. The formula $\Phi(z_1, z_2) = \exists y((z_1 \vee y) \wedge (\neg y \vee z_2))(\approx (z_1 \vee z_2))$ has no K_0-equivalence model, since

$$\Phi(z_1, z_2)[y/0] = (z_1 \vee 0) \wedge (1 \vee z_2) \approx (z_1) \not\approx (z_1 \vee z_2)$$
$$\Phi(z_1, z_2)[y/1] = (z_1 \vee 1) \wedge (0 \vee z_2) \approx (z_2) \not\approx (z_1 \vee z_2)$$

Since the equivalence problem for $QCNF$ formulas is $PSPACE$-complete even if one of the formulas is very simple, we have the following lemma.

Lemma 4.

1. *The K_0–equivalence model checking problem for QCNF is PSPACE–complete.*
2. *The K_0–equivalence model problem for QCNF is PSPACE–complete.*

Proof. Ad 1: (see proof of Lemma 2)

Ad 2: Obviously, the problem is in *PSPACE*. For the *PSPACE*–hardness, we associate to *QCNF* formulas $\Phi = Q\phi(\boldsymbol{x}, \boldsymbol{y})$ the *QCNF* formula $\Phi' := \forall x_0 \exists y_0 Q(\phi(\boldsymbol{x}, \boldsymbol{y}) \wedge (x_0 \vee y_0) \wedge (\neg x_0 \vee \neg y_0))$ with new variables x_0 and y_0. Φ is true if and only if Φ' is true, since $\Psi := \forall x_0 \exists y_0((x_0 \vee y_0) \wedge (\neg x_0 \vee \neg y_0))$ is true for example with $f_{y_0}(x_0) = \neg x_0$. But Ψ has no K_0–equivalence model. That can be seen by a case distinction $y_0 = 0$ and $y_0 = 1$. Therefore, if Φ is true, then Φ' has no K_0–equivalence model. Suppose Φ is false, then Φ' is false. Thus, Φ' has a K_0–equivalence model. Altogether, Φ is false if and only if Φ' has a K_0–equivalence model. Since the evaluation problem for $QCNF$ is $PSPACE$–complete, we have shown our desired result.   ■

Satisfiable $Q2$-CNF formulas have always a satisfiability model consisting of formulas of the form $f_y(x) = (\neg)x_i$ for some i, $f_y(x) = 0$, or $f_y(x) = 1$. For $Q2$-CNF^* these model formulas are not sufficient as the following example shows $\Phi(z_1, z_2) = \exists y((z_1 \vee y) \wedge (z_2 \vee y) \wedge (\neg y \vee z_3) \wedge (\neg y \vee z_4))$. The proof is straight forward by a case distinction. We will see that the class of models defined as $\mathcal{B} = 1$-$CNF \cup 1$-$DNF \cup \{0, 1\}$ characterizes in a certain sense equivalence models for $Q2$-CNF^*.

Theorem 2.

1. *Any formula in $Q2$-CNF^* has a $\mathcal{B}$–equivalence model.*
2. *The $\mathcal{B}$–equivalence model checking for $Q2$-CNF^* is solvable in polynomial time.*

Proof. Ad 1: Suppose, we have a formula $\Phi(z) = Q\phi(x, y, z) \in Q2$-$CNF^*$. If $\Phi(z)$ is unsatisfiable, then there is a $\{0, 1\}$–equivalence model, and therefore a $\mathcal{B}$–equivalence model. Now, we assume the satisfiability of the input formula. In a first step we apply the Q-resolution as long as possible with $\Phi(z)$ [7]. The resulting formula, called $\Psi(z)$, is again in $Q2$-CNF^* and for any truth assignment for z, $\Phi(z)$ is true if and only if $\Psi(z)$ is true. Next we will define f_{y_j} for each y_j by means of the derived unit clauses.

Case 1. y_j or $\neg y_j$ occurs in $\Psi(z)$ as a unit clause. Then define $f_{y_j} = 1$ or $f_{y_j} = 0$ accordingly.

Case 2. y_j occurs in a $\exists$-unit clause (i.e., a clause with one existential literal and the other literal is universal), but $\neg y_i$ does not occur in any $\exists$-unit clause. Let $w_1 \vee y_j, \cdots, w_k \vee y_j$ be all the $\exists$-unit clause containing y_j. Then define $f_{y_j} = \neg w_1 \vee \cdots \vee \neg w_k$.

Case 3. $\neg y_j$ occurs in a $\exists$-unit clause, but y_i does not occur in any $\exists$-unit clause. Let $w_1 \vee \neg y_j, \cdots, w_k \vee \neg y_j$ be all the $\exists$-unit clause containing $\neg y_j$. Then define $f_{y_j} = w_1 \wedge \cdots \wedge w_k$.

Case 4. Both y_j and $\neg y_j$ occur in some $\exists$-unit clauses. Since Ψ is satisfiable, there are exactly two clauses containing y_j or $\neg y_j$, and they must be of the form $w \vee y_j$ and $\neg w \vee \neg y_j$. Then define $f_{y_j} = \neg w$.

Case 5. y_j or $\neg y_j$ is derivable from free-unit clauses (by a free-unit clause we mean a clause with at most one existential literal and the other literals are literals over free variables). Then define $f_{y_j} = 1$ or $f_{y_j} = 0$ accordingly.

Case 6. y_j or $\neg y_j$ is a pure literal. Then define $f_{y_j} = 1$ or $f_{y_j} = 0$ accordingly.

Case 7. Note cases 1–6. Let $y_j \vee v_1, \cdots, y_j \vee v_k$ and $\neg y_j \vee u_1, \cdots, \neg y_j \vee u_r$ be all free-unit clauses containing y_j or $\neg y_j$. Then define f_{y_j} either to be $\neg v_1 \vee \cdots \vee \neg v_k$ or to be $u_1 \wedge \cdots \wedge u_r$.

Case 8. There are no free–unit clauses containing y_j or $\neg y_j$. That is, y_j has nothing to do with free variables and existential variables. Thus, in this case f_{y_j} is either 0 or 1.

It is not hard to see that in any case $\Psi(z)[y_j/f_{y_j}]$ is true if and only if $\Psi(z)$ is true for any truth assignment for z. Consequently, $(f_{y_1}, \cdots, f_{y_m})$ is an equivalence model for $\Phi(z)$.

Ad 2: Let $\Phi(z) = Q\phi(x, y, z)$ be in $Q2\text{-}CNF^*$. For a sequence of propositional formulas $M = (f_{y_1}, \cdots, f_{y_m})$, where $f_{y_i} \in \mathcal{B}$, we want to decide whether M is an equivalence model for $\Phi(z)$.

At first we can calculate by applying a polynomial-time algorithm to $\Phi(z)$ a logically equivalent propositional formula $F(z) \in 2\text{-}CNF$. The length of $F(z)$ is bound by $O(|\phi|^2)$ (see Theorem 7.4.6 and Theorem 7.6.1 in [7]). In the next step we substitute in the initial formula the existential variables y_i by the model–functions f_{y_i}. That means, we have $\Phi(z)[y/M] = \forall x_1 \cdots \forall x_n \phi(x, y, z)[y/M]$. Please note that $\Phi(z)[y/M]$ may not be in $QCNF^*$. However, it can be transformed in polynomial time into an equivalent formula with CNF kernel by applying the distributivity law. The result is denoted as $\Psi(z)$ (which still contains only universal quantifiers). Further, we can calculate in polynomial time an equivalent propositional formula $G(z)$ of length less or equal than the length of $\Psi(z)$. If $\Psi(z)$ contains a $\forall$-clause then $G(z)$ is *false*. Otherwise, $G(z)$ is obtained by deleting all universal literals and removing the quantifiers. It is not difficult to see that $G(z)$ and $\Psi(z)$ are equivalent. Finally it remains to decide whether $F(z) \models G(z)$. Since $F(z)$ is a propositional $2\text{-}CNF$ formula that can be done in polynomial time.

Altogether, we have an polynomial-time procedure for the $\mathcal{B}$–equivalence model checking problem for $2\text{-}CNF^*$.   ∎

For $QHORN^*$ the regular equivalence problem — the problem whether two quantified Horn formulas are equivalent — is *coNP*–complete. Further, any $QHORN^*$ formula is equivalent to a $HORN$ formula, but sometimes of length essentially different [7]. The next lemma shows that for very simple model formulas the *coNP*–completeness persist.

Lemma 5. *The 1-DNF–equivalence model checking problem for $QHORN^*$ is coNP–complete.*

Proof. By Theorem 1 the problem is in *coNP*. We show the *coNP*–hardness by a reduction from the *coNP*–complete tautology problem for $3\text{-}DNF$ formulas. We associate to the DNF formula $\psi = \bigvee_{1 \leq i \leq m}(L_{i,1} \wedge L_{i,2} \wedge L_{i,3})$ with literals $L_{i,j}$ over the variables $z_1 \cdots, z_r$ the quantified Horn formula $\Phi(z) = \exists y \exists y_1 \cdots \exists y_m((\neg y_1 \vee$

$\cdots \vee \neg y_m) \wedge (\neg z_1 \vee \cdots \vee \neg z_r \vee \neg y))$ and $M = (f_y, f_{y_1}, \cdots, f_{y_m})$, where $f_y(z) = (\neg z_1 \vee \cdots \vee \neg z_r)$, $f_{y_i}(z) = (\neg L_{i,1} \vee \neg L_{i,2} \vee \neg L_{i,3}) \in$ *1-DNF*. $\Phi(z)$ is always true, that means equivalent to the constant 1.

We have $\Phi(z)[y/M] = (\neg f_{y_1}(z) \vee \cdots \vee \neg f_{y_m}(z)) \wedge (\neg z_1 \vee \cdots \vee \neg z_r \vee (z_1 \wedge \cdots \wedge z_n)) \approx (\neg f_{y_1}(z) \vee \cdots \vee \neg f_{y_m}(z))$ and this formula is equivalent to $\Phi(z)$ if and only if the propositional *DNF* formula $\psi = (\neg f_{y_1}(z) \vee \cdots \vee \neg f_{y_m}(z))$ is a tautology.

Hence, our *1-DNF*–equivalence model checking problem is *coNP*–complete. ∎

With respect to the satisfiability models, we know that every satisfiable *QHORN* has a K_2–model [8]. That does not hold for the equivalence model and *QHORN**. The formula

$$\Phi_n(z_{1,1}, \cdots, z_{n,n}) := \exists y_1 \cdots \exists y_n ((\neg y_1 \vee \cdots \vee \neg y_n) \wedge \bigwedge_{1 \le i,j \le n} (y_i \vee \neg z_{i,j}))$$

is equivalent to $\bigvee_{1 \le i \le n} (\neg z_{i,1} \wedge \cdots \wedge \neg z_{i,n})$. $M = (f_{y_1}, \cdots, f_{y_n})$ is an equivalence model for Φ_n if $f_{y_i}(z_{1,1}, \cdots, z_{n,n}) = (z_{i,1} \vee \cdots \vee z_{i,n})$. But $\Phi_n(z_{1,1}, \cdots, z_{n,n})$ has no K_2–equiv–model.

Lemma 6. *The K_2–equivalence model checking for QHORN* is solvable in polynomial time.*

Proof. Suppose $\Phi(z) = Q\phi(x, y, z) \in$ *QHORN** and $M = (f_{y_1}, \cdots, f_{y_m})$, where $f_{y_i} \in K_2$. That means, if $x^i = x_1, \cdots, x_{r_i}$ are the preceding universal variables for y_i, then we have $f_{y_i}(z, x^i) = \bigwedge_{j \in J_i} v_j$, $v_j \in \{x_1, \cdots, x_n, z_1, \cdots, z_r\}$ for some J_i, $f_{y_i} = 0$, or $f_{y_i} = 1$.

If the formula $\Phi(z) \in$ *QHORN** is unsatisfiable, which can be decided in polynomial time, then M is an equivalence model for the formula. We continue assuming that the formula $\Phi(z)$ is satisfiable.

The substitution $\Phi(z)[y/M]$ can lead to a non-Horn kernel. Since every clause in the kernel of the input formula $\Phi(z)$ contains at most one positive literal, by the distributivity law we can transform in polynomial time the formula $\Phi(z)[y/M]$ into a universally quantified *QHORN** formula, say $\Psi(z) = \forall x \bigwedge \psi_j(x, z)$. We can simplify the formula to obtain an equivalent propositional Horn formula $\bigwedge \psi'_j(z)$ by removing the universal literals and all quantifiers. To test whether M is an equivalence model, it suffices to decide whether $\Phi(z) \models \psi'_j(z)$, that means $\Phi(z) \wedge \neg \psi'_j(z)$ is unsatisfiable. But that is the problem of deciding whether a *QHORN** formula is satisfiable and this problem is solvable in polynomial time. ∎

4   Conclusion and Outlook

The results presented in the paper are a first step in understanding the structure of equivalence models and the complexity of the problems. There are various open problems. Take *QHORN** as an example, try to establish a class of propositional formulas $K \subseteq$ *CNF* with the following properties:

1. Any formula in $QHORN^*$ has a K–equivalence model.
2. A K–equivalence model for $\Phi(z) \in QHORN^*$ can be constructed in polynomial time.
3. The K–equivalence model checking problem for $QHORN^*$ is solvable in polynomial time.

References

1. B. Aspvall, B., M. F. Plass, M. F., and Tarjan, R. E.: A Linear-Time Algorithm for Testing the Truth of Certain Quantified Boolean Formulas, *Information Processing Letters*, **8** (1979), 121-123
2. Cadoli, M., Schaerf, M., Giovanardi, A., and Giovanardi, M.: An Algorithm to Evaluate Quantified Boolean Formulas and its Evaluation, In: *highlights of Satisfiability Research in the Year 2000*, IOS Press, 2000.
3. Cook, S., Soltys, M.: Boolean Programs and Quantified Propositional Proof Systems, *Bulletin of the Section of Logic*, **28** (1999), 119-129.
4. Flögel, A., Karpinski, M., and Kleine Büning, H.: Resolution for Quantified Boolean Formulas, *Information and Computation* **117** (1995), 12-18
5. Garey, M.R., Johnson, D.S.: *Computers and Intractability: A Guide to the Theory of NP-Completeness*. W.H. Freeman Company, San Francisco, 1979.
6. Giunchiglia, E., Narizzano, M., and Tacchella, A.: QuBE: A System for Deciding Quantified Boolean Formulas, In: *Proceedings of IJCAR*, Siena, 2001.
7. Kleine Büning, H., Lettmann, T.: *Propositional Logic: Deduction and Algorithms*, Cambridge University Press, 1999.
8. Kleine Büning, H., Subramani, K., and Zhao, X.: On Boolean Models for Quantified Boolean Horn Formulas, *SAT 2003*, Italy. *Lecture Notes in Computer Science* **2919**, 93-104, 2004.
9. Kleine Büning, H., Zhao, X.: On Models for Quantified Boolean Formulas, to appear in LNCS, 2004.
10. Letz, R.: Advances in Decision Procedure for Quantified Boolean Formulas, In: *Proceedings of IJCAR*, Siena, 2001.
11. Meyer, A. R., Stockmeyer, L. J.: Word Problems Requiring Exponential Time, In: *Preliminary Report, Proc. 5th Ann. Symp. on Theory of Computing*, (1973), pp 1-9
12. Papadimitriou, C. H.: *Computational Complexity*, Addison-Wesley, New York, 1994.
13. Rintanen, J.T.: Improvements to the Evaluation of Quantified Boolean Formulae, In: *Proceedings of IJCAI*, 1999.
14. Schaefer, T.J.: The Complexity of Satisfiability Problem, In: *Proceedings of the 10th Annual ACM Symposium on Theory of Computing (ed. A. Aho)*, 216-226, New York City, ACM Press, 1978.
15. Stockmeyer, L. J.: The Polynomial-Time Hierarchy, In: *Theoretical Computer Science*, **3**(1977), 1-22.

Search vs. Symbolic Techniques in Satisfiability Solving

Guoqiang Pan* and Moshe Y. Vardi*

Department of Computer Science, Rice University, Houston, TX
{gqpan, vardi}@cs.rice.edu

Abstract. Recent work has shown how to use OBDDs for satisfiability solving. The idea of this approach, which we call *symbolic quantifier elimination*, is to view an instance of propositional satisfiability as an existentially quantified propositional formula. Satisfiability solving then amounts to quantifier elimination; once all quantifiers have been eliminated we are left with either **1** or **0**. Our goal in this work is to study the effectiveness of symbolic quantifier elimination as an approach to satisfiability solving. To that end, we conduct a direct comparison with the DPLL-based ZChaff, as well as evaluate a variety of optimization techniques for the symbolic approach. In comparing the symbolic approach to ZChaff, we evaluate scalability across a variety of classes of formulas. We find that no approach dominates across all classes. While ZChaff dominates for many classes of formulas, the symbolic approach is superior for other classes of formulas.

Once we have demonstrated the viability of the symbolic approach, we focus on optimization techniques for this approach. We study techniques from constraint satisfaction for finding a good plan for performing the symbolic operations of conjunction and of existential quantification. We also study various variable-ordering heuristics, finding that while no heuristic seems to dominate across all classes of formulas, the maximum-cardinality search heuristic seems to offer the best overall performance.

1   Introduction

Propositional-satisfiability solving has been an active area of research through out the last 40 years, starting from the resolution-based algorithm in [21] and the search-based algorithm in [20]. The latter approach, referred to as the DPLL approach, has since been the method of choice for satisfiability solving. In the last ten years, much progress have been made in developing highly optimized DPLL solvers, leading to efficient solvers such as ZChaff [49] and BerkMin [30], all of which use advanced heuristics in choosing variable splitting order, in performing efficient Boolean constraint propagation, and in conflict-driven learning to prune unnecessary search branches. These solvers are so effective that they are used as generic problem solvers, where problems such as bounded model checking [6], planning [34], scheduling [17], and many others are typically solved by reducing them to satisfiability problems.

Another successful approach to propositional reasoning is that of decision diagrams, which are used to represent propositional functions. An instance of the approach is that

---

* Supported in part by NSF grants CCR-9988322, CCR-0124077, CCR-0311326, IIS-9908435, IIS-9978135, EIA-0086264, ANI-0216467, and by BSF grant 9800096.

H.H. Hoos and D.G. Mitchell (Eds.): SAT 2004, LNCS 3542, pp. 235–250, 2005.

of ordered Boolean decision diagrams (OBDDs) [9], which are used successfully in model checking [11] and planning [14]. The zero-suppressed variant (ZDDs) is used in prime implicants enumeration [39]. A decision-diagram representation also enables easy satisfiability checking, which amounts to deciding whether it is different than the empty OBDD [9]. Since decision diagrams usually represent the set of all satisfying truth assignments, they incur a significant overhead over search techniques that focus on finding a single satisfying assignment [16]. Thus, published comparisons between search and OBDD techniques [47, 35] used search to enumerate all satisfying assignments. The conclusion of that comparison is that no approach dominates; for certain classes of formulas search is superior, and for other classes of formulas OBDDs are superior.

Recent work has shown how to use OBDDs for satisfiability solving rather for enumeration [42]. The idea of this approach, which we call *symbolic quantifier elimination*, is to view an instance of propositional satisfiability as an existentially quantified propositional formula. Satisfiability solving then amounts to quantifier elimination; once all quantifiers have been eliminated we are left with either **1** or **0**. This enables us to apply ideas about existential quantifier elimination from model checking [41] and constraint satisfaction [23]. The focus in [42] is on expected behavior on random instances of 3-SAT rather than on efficiency. In particular, only a minimal effort is made to optimize the approach and no comparison to search methods is reported. Nevertheless, the results in [42] show that OBDD-based algorithms behave quite differently than search-based algorithms, which makes them worthy of further investigation. (Other recent approaches reported using decision diagrams in satisfiability solving [12, 26, 40]. We discuss these works in our concluding remarks).

Our goal in this paper is to study the effectiveness of symbolic quantifier elimination as an approach to satisfiability solving. To that end, we conduct a direct comparison with the DPLL-based ZChaff, as well as evaluate a variety of optimization techniques for the symbolic approach. In comparing the symbolic approach to ZChaff we use a variety of classes of formulas. Unlike, however, the standard practice of comparing solver performance on benchmark suites [37], we focus here on *scalability*. That is, we focus on scalable classes of formulas and evaluate how performance scales with formula size. As in [47] we find that no approach dominates across all classes. While ZChaff dominates for many classes of formulas, the symbolic approach is superior for other classes of formulas.

Once we have demonstrated the viability of the symbolic approach, we focus on optimization techniques. The key idea underlying [42] is that evaluating an existentially quantified propositional formula in conjunctive-normal form requires performing several instances of conjunction and of existential quantification. The goal is to find a good plan for these operations. We study two approaches to this problem. The first is Bouquet's method (BM) of [42] and the second is the *bucket-elimination* (BE) approach of [23]. BE aims at reducing the size of the support set of the generated OBDDs through quantifier elimination and it has the theoretical advantage of being, in principle, able to attain optimal support set size, which is the *treewidth* of the input formula [25]. Nevertheless, we find that for certain classes of formulas BM is superior to BE.

The key to good performance in both BM and BE is in choosing a good variable order for quantification and OBDD order. Finding an optimal order is by itself a difficult problem (computing the treewidth of a given graph is NP-hard [2]), so one has to resort to various heuristics, cf. [36]. No heuristic seems to dominate across all classes of formulas, but the maximal-cardinality-search heuristic seems to offer the best overall performance.

We start the paper with a description of symbolic quantifier elimination as well as the BM approach in Section 2. We then describe the experimental setup in Section 3. In Section 4 we compare ZChaff with BM and show that no approach dominates across all classes of formulas. In Section 5 we compare BM with BE and study the impact of various variable-ordering heuristics. We conclude with a discussion in Section 6.

2   Background

An binary decision diagram (BDD) is a rooted directed acyclic graph that has only two terminal nodes labeled **0** and **1**. Every non-terminal node is labeled with a Boolean variable and has two outgoing edges labeled 0 and 1. An ordered binary decision diagram (OBDD) is a BDD with the constraint that the input variables are ordered and every path in the OBDD visits the variables in ascending order. We assume that all OBDDs are *reduced*, which means that where every node represents a distinct logic function. OBDDs constitute an efficient way to represent and manipulate Boolean functions [9], in particular, for a given variable order, OBDDs offer a canonical representation. Checking whether an OBDD is satisfiable is also easy; it requires checking that it differs from the predefined constant **0** (the empty OBDD). We used the CUDD package for managing OBDDs [45]. The *support set* of an OBDD is the set of variables labeling its internal nodes.

In [47, 16], OBDDs are used to construct a compact representation of the set of all satisfying truth assignments of CNF formulas. The input formula φ is a conjunction $c_1 \wedge \ldots \wedge c_m$ of clauses. The algorithm constructs an OBDD A_i for each clause c_i. (Since a clause excludes only one assignments to its variables, A_i is of linear size.) An OBDD for the set of satisfying truth assignment is then constructed incrementally; B_1 is A_1, while B_{i+1} is the result of APPLY$(B_i, A_i, \wedge)$, where APPLY$(A, B, \circ)$ is the result of applying a Boolean operator $\circ$ to two OBDDs A and B. Finally, the resulting OBDD B_m represents all satisfying assignments of the input formula.

We can apply existential quantification to an OBDD B:

$$(\exists x)B = \text{APPLY}(B|_{x \leftarrow 1}, B|_{x \leftarrow 0}, \vee),$$

where $B|_{x \leftarrow c}$ restricts B to truth assignments that assign the value c to the variable x. Note that quantifying x existentially eliminates it from the support set of B. The satisfiability problem is to determine whether a given formula $c_1 \wedge \ldots \wedge c_m$ is satisfiable. In other words, the problem is to determine whether the existential formula $(\exists x_1) \ldots (\exists x_n)(c_1 \wedge \ldots \wedge c_m)$ is true. Since checking whether the final OBDD B_m is equal to **0** can be done by CUDD in constant time, it makes little sense, however, to apply existential quantification to B_m. Suppose, however, that a variable x_j does not occur in the clauses $c_{i+1}, \ldots, c_m$. Then the existential formula can be rewritten as

$$(\exists x_1) \ldots (\exists x_{j-1})(\exists x_{j+1}) \ldots (\exists x_n)((\exists x_j)(c_1 \wedge \ldots \wedge c_i) \wedge (c_{i+1} \wedge \ldots \wedge c_m)).$$

This means that after constructing the OBDD B_i, we can existentially quantify x_j before conjuncting B_i with $A_{i+1}, \ldots, A_m$.

This motivates the following change in the earlier OBDD-based satisfying-solving algorithm [42]: after constructing the OBDD B_i, quantify existentially variables that do not occur in the clauses $c_{i+1}, \ldots, c_m$. In this case we say that the quantifier $\exists x$ has been *eliminated*. The computational advantage of quantifier elimination stems from the fact that reducing the size of the support set of an OBDD typically (though not necessarily) results in a reduction of its size; that is, the size of $(\exists x)B$ is typically smaller than that of B. In a nutshell, this method, which we describe as *symbolic quantifier elimination*, eliminates all quantifiers until we are left with the constant OBDD 1 or 0. Symbolic quantifier elimination was first applied to SAT solving in [31] (under the name of *hiding functions*) and tried on random 3-SAT instances. The work in [42] studied this method further, and considered various optimizations. The main interest there, however, is in the behavior of the method on random 3-SAT instances, rather in its comparison to DPLL-based methods.[1]

So far we processed the clauses of the input formula in a linear fashion. Since the main point of quantifier elimination is to eliminate variables as early as possible, re-ordering the clauses may enable us to do more aggressive quantification. That is, instead of processing the clauses in the order $c_1, \ldots, c_m$, we can apply a permutation π and process the clauses in the order $c_{\pi(1)}, \ldots, c_{\pi(m)}$. The permutation π should be chosen so as to minimize the number of variables in the support sets of the intermediates OBDDs. This observation was first made in the context of symbolic model checking, cf. [10, 29, 33, 7]. Unfortunately, finding an optimal permutation π is by itself a difficult optimization problem, motivating heuristic approaches.

A particular heuristic that was proposed in the context of symbolic model checking in [41] is that of *clustering*. In this approach, the clauses are not processed one at a time, but several clauses are first partitioned into several clusters. For each cluster C we first apply conjunction to all the OBDDs of the clauses in C to obtain an OBDD B_C. The clusters are then combined, together with quantifier elimination, as described earlier. Heuristics are required both for clustering the clauses and ordering the clusters. Bouquet proposed the following heuristic in [8] (the focus there is on enumerating prime implicants). Consider some order of the variables. Let the *rank* (from 1 to n) of a variable x be $rank(x)$, let the rank $rank(\ell)$ of a literal ℓ be the rank of is underlying variable, and let the rank $rank(c)$ of a clause c be the maximum rank of its literals. The clusters are the equivalence classes of the relation $\sim$ defined by: $c \sim c'$ iff $rank(c) = rank(c')$. The rank of a cluster is the rank of its clauses. The clusters are then ordered according to increasing rank. Satisfiability solving using symbolic quantifier elimination combined with Bouquet's clustering is referred to in [42] as *Bouquet's Method*, which we abbreviate here is as BM.

---

[1] Note that symbolic quantifier elimination provides *pure* satisfiability solving; the algorithm returns 0 or 1. To find a satisfying truth assignment when the formula is satisfiable, the technique of self-reducibility can be used, cf. [3].

We still have to chose a variable order. An order that is often used in constraint satisfaction [22] is the "maximum cardinality search" (MCS) order of [46], which is based on the graph-theoretic structure of the formula. The graph associated with a CNF formula $\varphi = \bigwedge_i c_i$ is $G_\varphi = (V, E)$, where V is the set of variables in φ and an edge $\{x_i, x_j\}$ is in E if there exists a clause c_k such that x_i and x_j occur in c_k. We refer to G_φ as the *Gaifman graph* of φ. MCS ranks the vertices from 1 to n in the following way: as the next vertex to number, select the vertex adjacent to the largest number of previously numbered vertices (ties can be broken in various ways). Our first experiment is a performance comparison of MCS-based BM to ZChaff.

3   Experimental Setup

We compare symbolic quantifier elimination to ZChaff across a variety of classes of formulas. Unlike the standard practice of comparing solver performance on benchmark suites [37], our focus here is not on simple time comparison, but rather on *scalability*. That is, we focus on scalable classes of formulas and evaluate how performance *scales* with formula size. We are interested in seeing which method scales better, i.e., polynomial vs. exponential scalability, or different degrees of exponential scalability. Our test suite includes both random and nonrandom formulas (for random formulas we took 60 samples per case and reported median time). Experiments were performed using x86 emulation on the Rice Terascale Cluster[2], which is a large Linux cluster of Itanium II processors with 4GB of memory each.

Our test suite includes the following classes of formulas:

- Random 3-CNF: We chose uniformly k 3-clauses over n variables. The *density* of an instance is defined as k/n. We generate instances at densities 1.5, 6, 10, and 15, with up to 200 variables, to allow comparison for both under-constrained and over-constrained cases. (It is known that the satisfiability threshold of such formulas is around 4.25 [44]).

- Random affine 3-CNF: Affine 3-CNF formulas are generated in the same way as random 3-CNF formulas, except that the constraints are not 3-clauses, but parity equations in the form of $l_1 \oplus l_2 \oplus l_3 = 1$. Each constraint is then converted into four clauses, yielding CNF formulas. The satisfiability threshold of such formula is found empirically to be around density (number of equations divided by number of variables) 0.95. We generate instances of density 0.5 and 1.5, with up to 400 variables.

- Random biconditionals: Biconditional formulas, also known as Urquhart formulas, form a class of affine formulas that have provably exponential resolution proofs. A biconditional formula has the form $l_1 \leftrightarrow (l_2 \leftrightarrow (\ldots (l_{k-1} \leftrightarrow l_k) \ldots))$, where each l_i is a positive literal. Such a formula is valid if either all variables occur an even number of times or all variables occur an odd number of times [48]. We generate valid formulas with up to 100 variables, where each variable occurs 3 times on average.

---

[2] http://www.citi.rice.edu/rtc/

– **Random chains:** The classes described so far all have an essentially uniform random Gaifman graph, with no underlying structure. To extend our comparison to structured formulas, we generate random chains [24]. In a random chain, we form a long chain of random 3-CNF formulas, called *subtheories*. (The chain structure is reminiscent to the structure typically seen in satisfiability instances obtained from bounded model checking [6] and planning [34].) We use a similar generation parameters as in [24], where there are 5 variables per sub-theory and 5-23 clauses per sub-theory, but that we generate instances with a much bigger number of sub-theories, scaling up to > 20000 variables and > 4000 sub-theories.

– **Nonrandom formulas:** As in [47], we considered a variety of formulas with very specific scalable structure:
  - The n-Rooks problem (satisfiable).
  - The n-Queens problem (satisfiable for $n > 3$).
  - The pigeon-hole problem with $n + 1$ pigeons and n holes (unsatisfiable).
  - The mutilated-checkerboard problem, where an $n \times n$ board with two diagonal corner tiles removed is to be tiled with 1×2 tiles (unsatisfiable).

4   Symbolic vs. Search Approaches

Our goal in this section is to address the viability of symbolic quantifier elimination. To that end we compare the performance of BM against ZChaff, a leading DPLL-based solver across the classes of formulas described above, with a focus on scalability. For now, we use the MCS variable order.

In Figure 1, we can see that BM is not very competitive for random 3-CNF formulas. At density 1.5, ZChaff scales polynomially, while BM scales exponentially. At density 6.0 and at higher densities, both methods scale exponentially, but ZChaff scales exponentially better. (Note that above density 6.0 both methods scale better as the density increases. This is consistent with the experimental results in [16] and [42].) A similar pattern emerges for random affine formulas, see Figure 2. Again, ZChaff scales exponentially better than BM. (Note that both methods scale exponentially at the higher density, while it is known that affine satisfiability can be determined in polytime using Gaussian elimination [43].)

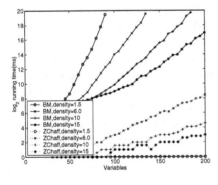

Fig. 1. Random 3-CNF

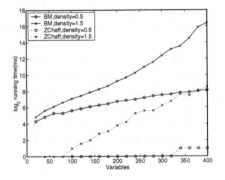

Fig. 2. Random 3-Affine

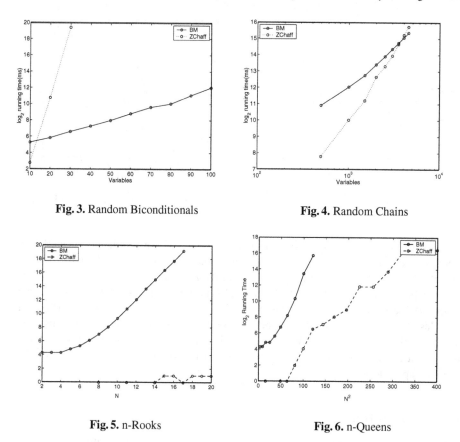

Fig. 3. Random Biconditionals

Fig. 4. Random Chains

Fig. 5. n-Rooks

Fig. 6. n-Queens

The picture changes for biconditional formulas, as shown in Figure 3. Again, both methods are exponential, but BM scales exponentially better than ZChaff. (This result is consistent with the finding in [12], which compares search-based methods to ZDD-based multi-resolution.)

For random chains, see Figure 4, which uses a log-log scale. Both methods scale polynomially on random chains. (Because density for the most difficult problems change as the size of the chains scales, we selected here the hardest density for each problem size.) Here BM scales polynomially better than than ZChaff. Note that for smaller instances ZChaff outperforms BM, which justifies our focus on scalability rather than on straightforward benchmarking.

Finally, we compare BM with ZChaff on the non-random formulas of [47]. The n-Rooks problem is a simpler version of n-Queens problem, where the diagonal constraints are not used. For n-Rooks, the results are as in Figure 5. This problem have the property of being *globally consistent*, i.e., any consistent partial solution can be extended to a solution [22]. Thus, the problem is trivial for search-based solvers, as no backtracking is need. In contrast BM scales exponentially on this problem. For n-Queens, see Figure 6, both methods scale exponentially (in fact, they scale exponentially in n^2), but ZChaff scales exponentially better than BM. Again, a different picture

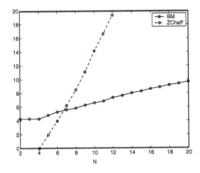

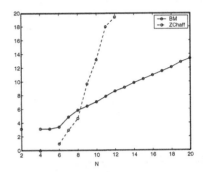

Fig. 7. Pigeon Hole   **Fig. 8.** Mutilated Checkerboard

emerges when we consider the pigeon-hole problem and the mutilated-checkerboard problem, see Figure 7 and Figure 8. On both problems both BM and ZChaff scale exponentially, but BM scales exponentially better than ZChaff.

As in [47], who compared OBDDs and DPLL for solution enumeration, we find that no approach dominates across all classes. While ZChaff dominates for many classes of formulas, the symbolic approach is superior for other classes of formulas. This suggests that the symbolic quantifier elimination is a viable approach and deserves further study. In the next section of this work we focus on various optimization strategies for the symbolic approach.

5   Optimizations

So far we have described one approach to symbolic quantifier elimination. There are, however, many choices one needs to make to guide an implementation. The order of variables is both used to guide clustering and quantifier elimination, as well as to order the variables in the underlying OBDDs. Both clustering and cluster processing can be performed in several ways. In this section, we investigate the impact of choices in clustering, variable order, and quantifier elimination in the implementation of symbolic algorithms. Our focus here is on measuring the impact of variable order on BDD-based SAT solving; thus, the running time for variable ordering, which is polynomial for all algorithms, is not counted in our figures.

5.1   Cluster Ordering

As argued earlier, the purpose of quantifier elimination is to reduce support-set size of intermediate OBDDs. What is the best reduction one can hope for? This question has been studied in the context of constraint satisfaction. It turns out that the optimal schedule of conjunctions and quantifier eliminations reduces the support-set size to one plus the *treewidth* of the Gaifman graph of the input formula [18]. The treewidth of a graph is a measure of how close this graph is to being a tree [25]. Computing the treewidth of a graph is known to be NP-hard, which is why heuristic approaches are employed [36]. It turns out that by processing clusters in a different order we can

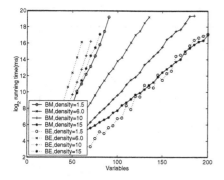
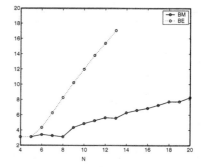

Fig. 9. Clustering Algorithms - Random 3-CNF   **Fig. 10.** Clustering Algorithms - Pigeon Hole

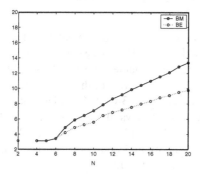

Fig. 11. Clustering Algorithms - Mutilated Checkerboard

attain the optimal support-set size. Recall that BM processes the clusters in order of increasing ranks. *Bucket elimination* (BE), on the other hand, processes clusters in order of decreasing ranks [23]. Maximal support-size set of BE with respect to optimal variable order is defined as the *induced width* of the input instance, and the induced width is known to be equal to the treewidth [23, 27]. Thus, BE with respect to optimal variable order is guaranteed to have polynomial running time for input instances of logarithmic treewidth, since this guarantees a polynomial upper bound on OBDD size. We now compare BM and BE with respect to MCS variable order (MCS is the preferred variable order also for BE).

The results for the comparison on random 3-CNF formulas is plotted in Figure 9. We see that the difference between BM and BE is density dependent, where BE excels in the low-density case, which have low treewidth, and BM excels in the high-density cases, which has high treewidth. Across our other classes of random formulas, BM is typically better, except for a slight edge that BE sometimes has for low-density instances. A similar picture can be seen on most constructed formulas, where BM dominates, except for mutilated-checkerboard formulas, where BE has a slight edge. (Note that treewidth for mutilated checkerboard problems grows only at $O(n)$ compared to $O(n^2)$ for other constructed problems.) We plot the performance comparison for pigeon-hole formulas in Figure 10 and mutilated checkerboard problems in Figure 11.

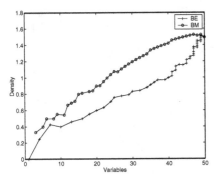
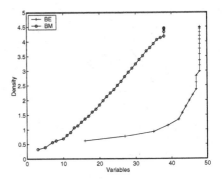

Fig. 12. Clustering Algorithms, Density=1.5   **Fig. 13.** Clustering Algorithms, Density=6.0

To understand the difference in performance between BM and BE, we study their effect on intermediate OBDD size. OBDD size for a random 3-CNF instance depends crucially on both the number of variables and the density of the instance. Thus, we compare the effect of BM and BE in terms of these measures for the intermediate OB-DDs. We apply BM and BE to random 3-CNF formulas with 50 variables and densities 1.5 and 6.0. We then plot the density vs. the number of variables for the intermediate OBDDs generated by the two cluster-processing schemes. The results are plotted in in Figure 12 and Figure 13. Each plotted point corresponds to an intermediate OBDD, which reflects the clusters processed so far.

As can be noted from the figures, BM increases the density of intermediate results much faster than BE. This difference is quite dramatic for high-density formulas. The relation between density of random 3-CNF instance and OBDD size has been studied in [16], where it is shown that OBDD size peaks at around density 2.0, and is lowest when the density is close to 0 or the satisfiability threshold. This enables us to offer an possible explanation to the superiority of BE for low-density instances and the superiority of BM for high-density instances. For formulas of density 1.5, the density of intermediate results is smaller than 2.0 and BM's increased density results in larger OBDDs. For formulas of density 6.0, BM crosses the threshold density 2.0 using a smaller number of variables, and then BM's increased density results in smaller OBDDs.

The general superiority of BM over BE suggests that minimizing support-set size ought not to be the dominant concern. OBDD size is correlated with, but not dependent on, support-set size. More work is required in order to understand the good performance of BM. Our explanation argues that, as in [1], BM first deals with the most constrained subproblems, therefore reducing OBDD-size of intermediate results. While the performance of BE can be understood in terms of treewidth, we still lack, however, a fundamental theory to explain the performance of BM.

5.2   Variable Ordering

As mentioned earlier, when selecting variables, MCS has to break ties, which happens quite often. One can break ties by minimizing degree to unselected variables [42] or

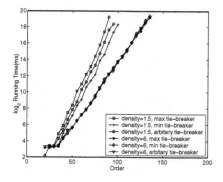

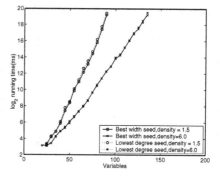

Fig. 14. Variable Ordering Tie-breakers   **Fig. 15.** Initial Variable Choice

by maximizing it [4]. (Another choice to to break ties uniformly at random, but this choice is expensive to implement, since it is difficult to choose an element uniformly at random from a heap.) We compare these two heuristic with an arbitrary tie-breaking heuristic, in which we simply select the top variable in the heap. The results are shown in Figure 14 for random 3-CNF formulas. For high density formulas, tie breaking made no significant difference, but least-degree tie breaking is markedly better for the low density formulas. This seems to be applicable across a variety of class of formulas and even for different orders and algorithms.

MCS typically has many choices for the lowest-rank variable. In Koster et. al. [36], it is recommended to start from every vertex in the graph and choose the variable order that leads to the lowest treewidth. This is easily done for instances of small size, i.e. random 3-CNF or affine problems; but for structured problems, which could be much larger, the overhead is too expensive. Since min-degree tie-breaking worked quite well, we used the same idea for initial variable choice. In Figure 15, we see that our assumption is well-founded, that is, the benefit of choosing the best initial variable compared to choosing a min-degree variable is negligible.

Algorithms for BDD variable ordering in the model checking area are often based on circuit structures, for example some form of traversal [38, 28] or graph evaluation [13]. Since we only have the graph structure based on the CNF clauses, we do not have the depth or direction information that circuit structure can provide. As the circuits in question become more complex, the effectiveness of simple traversals would also reduce. So, we use the graph-theoretic approaches used in constraint satisfaction instead of those from model checking.

MCS is just one possible vertex-ordering heuristics. Other heuristics have been studied in the context of treewidth approximation. In [36] two other vertex-ordering heuristics that based on local search are studied: LEXP and LEXM.[3] Both LEXP and LEXM are based on *lexicographic breadth-first search*, where candidate variables are lexicographically ordered with a set of labels, and the labels are either the set of already cho-

---

[3] The other heuristic mentioned in [36] is MSVS, which constructes a tree-decomposition instead of a variable order.

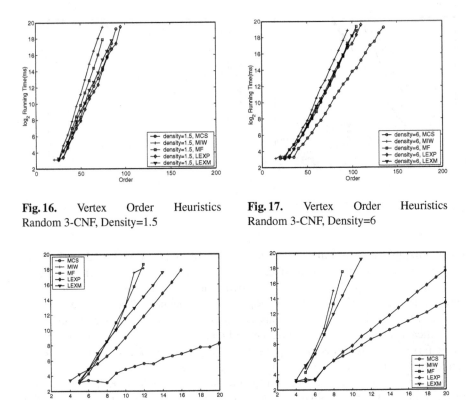

Fig. 16.   Vertex   Order   Heuristics
Random 3-CNF, Density=1.5

Fig. 17.   Vertex   Order   Heuristics
Random 3-CNF, Density=6

Fig. 18. Vertex Order Heuristics - Pigeon Hole

Fig. 19. Vertex Order Heuristics - Mutilated
Checkerboard

sen neighbors (LEXP), or the set of already chosen vertices reachable through lower-ordered vertices (LEXM). Both algorithms try to generate vertex orders where a triangulation would add a small amount of edges, thus reducing treewidth. In [22], Dechter also studied heuristics like Min-Induced-Width (MIW) or Min-Fill (MF), which are greedy heuristics based on choosing the vertex that have the least number of induced neighbors (MIW) or the vertex that would add the least number of induced edges (MF).

In Figure 16 and 17, we compare variable orders constructed from MCS, LEXP, LEXM, MIW, and MF for random 3-CNF formulas. For high-density cases, MCS is clearly superior. For low-density formulas, LEXP has a small edge, although the difference is quite minimal. Across the other problem classes (for example, pigeon-hole formulas as in Figure 18 and mutilated checkerboard as in Figure 19), MCS uniformly appears to be the best order, generally being the top performer. Interestingly, while other heuristics like MF often yield better treewidth, MCS still yields better runtime performance. This indicates that minimizing treewidth need not be the dominant concern.

5.3   Quantifier Elimination

So far we argued that quantifier elimination is the key to the performance of the symbolic approach. In general, reducing support-set size does result in smaller OBDDs. It is known, however, that quantifier elimination may incur non-negligible overhead and may not always reduce OBDD size [9]. To understand the role of quantifier elimination in the symbolic approach, we reimplemented BM and BE without quantifier elimination. Thus, we do construct an OBDD that represent all satisfying truth assignments, but we do that according to the clustering and cluster processing order of BM and BE.

In Figure 20, we plotted the running time of both BM and BE, with and without quantifier elimination on random 3-CNF formulas. We see that for BM there is a trade-off between the cost and benefit of quantifier elimination. For low-density instances, where there are many solutions, the improvement from quantifier elimination is clear, but for high-density instances, quantifier elimination results in no improvement (while not reducing OBDD size). For BE, where the overhead of quantifier elimination is lower, quantifier elimination improves performance, although the improvement is less significant for high densities. On the other hand, quantifier elimination is important for the constructed formulas, for example, for the pigeon-hole formulas in Figure 21 and the mutilated checkerboard formulas in Figure 22.

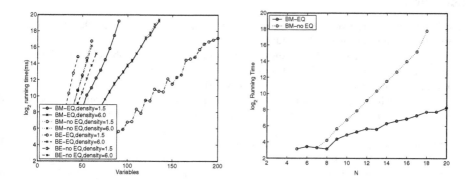

Fig. 20. Quantifier Elimination-Random 3-CNF   **Fig. 21.** Quantifier Elimination-Pigeon Hole

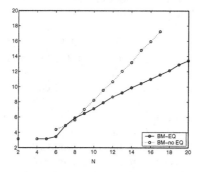

Fig. 22. Quantifier Elimination - Mutilated Checkerboard

6   Discussion

Satisfiability solvers have made tremendous progress over the last few years, partly driven by frequent competitions, cf. [37]. At the same time, our understanding of why extant solvers perform so well is lagging. Our goal in this paper is not to present a new competitive solver, but rather to call for a broader research agenda. We showed that a symbolic approach can outperform a search-based approach, but much research is needed before we can have robust implementations of the symbolic approach. Recent works have suggested other symbolic approaches to satisfiability solving, e.g., ZDD-based multi-resolution in [12], compressed BFS search in [40], and BDD representation for non-CNF constraint in the framework of DPLL search in [19, 26]. These works bolster our call for a broader research agenda in satisfiability solving. Such an agenda should build connections with two other successful areas of automated reasoning, namely model checking [15] and constraint satisfaction [22]. Furthermore, such an agenda should explore *hybrid* approaches, combining search and symbolic techniques, cf. [32, 40, 19, 26]. One hybrid approach that have shown promise is that of the QBF solver Quantor [5], where quantifier elimination is applied until the formula become propositional, then a search-based solver takes over.

References

1. E. Amir and S. McIlraith. Solving satisfiability using decomposition and the most constrained subproblem. In *SAT 2001*, June 2001.
2. S. Arnborg, D. Corneil, and A. Proskurowski. Complexity of finding embeddings in a k-tree. *SIAM J. Alg. Disc. Math*, 8:277–284, 1987.
3. J. Balcazar. Self-reducibility. *J. Comput. Syst. Sci.*, 41(3):367–388, 1990.
4. D. Beatty and R. Bryant. Formally verifying a microprocessor using a simulation methodology. In *Proc. 31st Design Automation Conference*, pages 596–602. IEEE Computer Society, 1994.
5. A. Biere. Resolve and expand. In *SAT 2004*, 2004.
6. A. Biere, C. A, E. Clarke, M. Fujita, and Y. Zhu. Symbolic model checking using SAT procedures instead of BDD. In *Proc. 36th Conf. on Design Automation*, pages 317–320, 1999.
7. M. Block, C. Gröpl, H. Preuß, H. L. Proömel, and A. Srivastav. Efficient ordering of state variables and transition relation partitions in symbolic model checking. Technical report, Institute of Informatics, Humboldt University of Berlin, 1997.
8. F. Bouquet. *Gestion de la dynamicite et enumeration d'implicants premiers, une approche fondee sur les Diagrammes de Decision Binaire*. PhD thesis, 1999.
9. R. Bryant. Graph-based algorithms for Boolean function manipulation. *IEEE Trans. on Comp.*, Vol. C-35(8):677–691, August 1986.
10. J. Burch, E. Clarke, and D. Long. Symbolic model checking with partitioned transition relations. In *Int. Conf. on Very Large Scale Integration*, 1991.
11. J. Burch, E. Clarke, K. McMillan, D. Dill, and L. Hwang. Symbolic model checking: 10^{20} states and beyond. *Infomation and Computation*, 98(2):142–170, 1992.
12. P. Chatalic and L. Simon. Multi-Resolution on compressed sets of clauses. In *Twelfth International Conference on Tools with Artificial Intelligence (ICTAI'00)*, pages 2–10, 2000.
13. P. Chung, I. Hajj, and J. Patel. Efficient variable ordering heuristics for shared robdd. In *Proc. Int. Symp. on Circuits and Systems*, 1993.

14. A. Cimatti and M. Roveri. Conformant planning via symbolic model checking. *J. of AI Research*, 13:305–338, 2000.
15. E. Clarke, O. Grumberg, and D. Peled. *Model Checking*. MIT Press, 1999.
16. C. Coarfa, D. D. Demopoulos, A. San Miguel Aguirre, D. Subramanian, and M. Vardi. Random 3-SAT: The plot thickens. *Constraints*, pages 243–261, 2003.
17. J. Crawford and A. Baker. Experimental results on the application of satisfiability algorithms to scheduling problems. In *AAAI*, volume 2, pages 1092–1097, 1994.
18. V. Dalmau, P. Kolaitis, and M. Vardi. Constraint satisfaction, bounded treewidth, and finite-variable logics. In *CP'02*, pages 310–326, 2002.
19. R. F. Damiano and J. H. Kukula. Checking satisfiability of a conjunction of BDDs. In *DAC 2003*, 2003.
20. M. Davis, G. Logemann, and D. Loveland. A machine program for theorem proving. *Journal of the ACM*, 5:394–397, 1962.
21. S. Davis and M. Putnam. A computing procedure for quantification theory. *Journal of ACM*, 7:201–215, 1960.
22. R. Dechter. *Constraint Processing*. Morgan Kaufmann, 2003.
23. R. Dechter and J. Pearl. Network-based heuristics for constraint-satisfaction problems. *Artificial Intelligence*, 34:1–38, 1987.
24. R. Dechter and I. Rish. Directional resolution: The Davis-Putnam procedure, revisited. In *KR'94: Principles of Knowledge Representation and Reasoning*, pages 134–145. 1994.
25. R. Downey and M. Fellows. *Parametrized Complexity*. Springer-Verlag, 1999.
26. J. Franco, M. Kouril, J. Schlipf, J. Ward, S. Weaver, M. Dransfield, and W. Vanfleet. SBSAT: a state-based, BDD-based satisfiability solver. In *SAT 2003*, 2003.
27. E. Freuder. Complexity of *k*-tree structured constraint satisfaction problems. In *Proc. AAAI-90*, pages 4–9, 1990.
28. M. Fujita, H. Fujisawa, and N. Kawato. Evaluation and improvements of Boolean comparison method based on binary decision disgrams. In *ICCAD*, 1988.
29. D. Geist and H. Beer. Efficient model checking by automated ordering of transition relation partitions. In *CAV 1994*, pages 299–310, 1994.
30. E. Goldberg and Y. Novikov. BerkMin: A fast and robust SAT solver, 2002.
31. J. F. Groote. Hiding propositional constants in BDDs. *FMSD*, 8:91–96, 1996.
32. A. Gupta, Z. Yang, P. Ashar, L. Zhang, and S. Malik. Partition-based decision heuristics for image computation using SAT and BDDs. In *ICCAD*, 2001.
33. R. Hojati, S. C. Krishnan, and R. K. Brayton. Early quantification and partitioned transition relations. pages 12–19, 1996.
34. H. Kautz and B. Selman. Planning as satisfiability. In *Proc. Eur. Conf. on AI*, pages 359–379, 1992.
35. S. Khurshid, D. Marinov, I. Shlyyakhter, and D. Jackson. A case for efficient solution enumeration. In *SAT 2003*, 2003.
36. A. Koster, H. Bodlaender, and S. van Hoesel. Treewidth: Computational experiments. Technical report, 2001.
37. D. Le Berre and L. Simon. The essentials of the SAT'03 competition. In *SAT 2003*, 2003.
38. S. Malik, A. Wang, R. Brayton, and A. Sangiovanni Vincentelli. Logic verification using binary decision diagrams in a logic synthesis environment. In *ICCAD*, 1988.
39. S. Minato. *Binary Decision Diagrams and Applications to VLSI CAD*. Kluwer, 1996.
40. D. B. Motter and I. L. Markov. A compressed breadth-first search for satisfiability. In *LNCS 2409*, pages 29–42, 2002.
41. R. Ranjan, A. Aziz, R. Brayton, B. Plessier, and C. Pixley. Efficient BDD algorithms for FSM synthesis and verification. In *Proc. of IEEE/ACM Int. Workshop on Logic Synthesis*, 1995.

42. A. San Miguel Aguirre and M. Y. Vardi. Random 3-SAT and BDDs: The plot thickens further. In *Principles and Practice of Constraint Programming*, pages 121–136, 2001.

43. T. Schaefer. The complexity of satisfiability problems. In *STOC'78*, pages 216–226, 1978.

44. B. Selman, D. G. Mitchell, and H. J. Levesque. Generating hard satisfiability problems. 81(1-2):17–29, 1996.

45. F. Somenzi. CUDD: CU decision diagram package, 1998.

46. R. E. Tarjan and M. Yannakakis. Simple linear-time algorithms to tests chordality of graphs, tests acyclicity of hypergraphs, and selectively reduce acyclic hypergraphs. *SIAM J. Comput.*, 13(3):566–579, 1984.

47. T. E. Uribe and M. E. Stickel. Ordered binary decision diagrams and the Davis-Putnam procedure. In *1st Int. Conf. on Constraints in Computational Logics*, pages 34–49, 1994.

48. A. Urquhart. The complexity of propositional proofs. *the Bulletin of Symbolic Logic*, 1:425–467, 1995.

49. L. Zhang and S. Malik. The quest for efficient Boolean satisfiability solvers. In *CAV 2002*, pages 17–36, 2002.

Worst Case Bounds for Some NP-Complete Modified Horn-SAT Problems

Stefan Porschen and Ewald Speckenmeyer

Institut für Informatik, Universität zu Köln, D-50969 Köln, Germany
{porschen, esp}@informatik.uni-koeln.de

Abstract. We consider the satisfiability problem for CNF formulas that contain a (hidden) Horn and a 2-CNF (also called quadratic) part, called *mixed (hidden) Horn formulas*. We show that SAT remains NP-complete for such instances and also that any SAT instance can be encoded in terms of a mixed Horn formula in polynomial time. Further, we provide an exact deterministic algorithm showing that SAT for mixed (hidden) Horn formulas containing n variables is solvable in time $O(2^{0.5284n})$. A strong argument showing that it is hard to improve a time bound of $O(2^{n/2})$ for mixed Horn formulas is provided. We also obtain a fixed-parameter tractability classification for SAT restricted to mixed Horn formulas C of at most k variables in its positive 2-CNF part providing the bound $O(\|C\|2^{0.5284k})$. Motivating examples for mixed Horn formulas are level graph formulas [14] and graph colorability formulas.

Keywords: (hidden) Horn formula, quadratic formula, satisfiability, NP-completeness, minimal vertex cover, fixed-parameter tractability.

1 Introduction and Motivation

In recent time the interest in designing exact algorithms providing better upper time bounds than the trivial ones for NP-complete problems and their NP-hard optimization versions has increased. Of particular interest in this context is the investigation of exact algorithms for testing the satisfiability (SAT) of propositional formulas in conjunctive normal form (CNF). This interest stems from the fact that SAT is well known to be a fundamental NP-complete problem appearing naturally or via reduction as the abstract core of many application-relevant problems. Not only the whole class CNF is of interest in this context. In several applications subclasses of CNF are of importance for which SAT unfortunately remains NP-complete. Nevertheless, it is often possible by exploiting the specific structure of such formulas to design fast exact algorithms for them. Such subclasses can be obtained by composing or *mixing* formulas of two different parts each of which separately is SAT-testable in polynomial time (see also [10]).

In this paper we introduce and study so-called *mixed Horn formulas* which roughly speaking are formulas composed of a quadratic part and a Horn part. More precisely, for a positive monotone 2-CNF formula P (containing only 2-clauses) and a Horn formula H, we call the formula $M = H \wedge P$ a *mixed Horn*

H.H. Hoos and D.G. Mitchell (Eds.): SAT 2004, LNCS 3542, pp. 251–262, 2005.
© Springer-Verlag Berlin Heidelberg 2005

formula (MHF). It is well known that 2-SAT and Horn-SAT are solvable in linear time [1, 12], but SAT for MHF's (shortly MHF-SAT) remains NP-complete. The main purpose of this paper is to prove a non-trivial worst case upper time bound for solving MHF-SAT, namely $O(2^{0.5284n})$ where n is the number of variables in the input instance. Moreover, we obtain a fixed-parameter tractability classification (cf. e.g. [3]) of SAT restricted to MHF's $M = P \wedge H$ where P has a fixed number k of different variables, provided by the polynomial bound $O(\|M\|2^{0.5284k})$, where $\|M\|$ is the length of M.

We also analyse the connection of MHF-SAT to unrestricted SAT. Specifically we show that each CNF formula C with n different variables can be transformed in polynomial time into a MHF $M = P \wedge H$, such that P has $k \leq 2n$ different variables. Then C is satisfiable if and only if M is satisfiable, and the question, whether $M \in$ SAT, can be answered in time $O(\|C\|2^{k/2})$. Hence, if there is an $\alpha < \frac{1}{2}$ such that every MHF $M = P \wedge H$ can be solved in time $O(\|C\|2^{\alpha k})$, then there is $\beta \leq 2\alpha < 1$ such that SAT for an arbitrary CNF-formula C can be decided in time $O(\|C\|2^{\beta n})$. The MHF-formulation of a CNF-formula C yields a partition of all variables in C into the *essential* variables (variables occuring in P) and the remaining ones.

The introduction and investigation of MHF's is by no means artificial. Well known problems for level graphs, like level-planarity test or the NP-hard crossing-minimization problem, can be formulated conveniently in terms of MHF's (for more details see [14]). This was our motivation for considering MHF's. Also graph colorability naturally leads to MHF's. To see this, consider a simple graph $G = (V, E)$ and a set of r colors $[r] := \{1, \ldots, r\}$. The decision whether G is r-colorable, i.e., whether at most r colors can be assigned to all vertices in V such that no two adjacent vertices are colored equally, can be encoded into MHF-SAT as follows: For every vertex $x \in V$ introduce r variables $x_i, i \in [r]$, and one clause $x_1 \vee x_2 \vee \cdots \vee x_r$. For every edge $x - y \in E$, we have to ensure that x and y are colored differently. So we introduce for each color $i \in [r]$ the clause $\neg(x_i \wedge y_i) \equiv (\overline{x_i} \vee \overline{y_i})$ yielding r 2-clauses for every edge. In summary, we obtain a CNF formula $C(G)$ consisting of $|V| + r|E|$ clauses and containing $r|V|$ different variables. Finally complementing all variables in $C(G)$ turns all its r-clauses into negative monotone clauses and its 2-clauses into positive monotone clauses, hence yields a MHF $\tilde{C}(G)$. It is easy to verify that G is r-colorable if and only if the MHF $\tilde{C}(G)$ is satisfiable via the interpretation that setting variable x_i to FALSE means that the corresponding vertex x is colored by i. Notice that introducing only one r-clause for every vertex ensuring at least (instead of exactly) one color for every vertex suffices for deciding r-colorability. Another source supplying the interest in Horn clauses contained in CNF formulas stems from recent observations of hidden threshold phenomena [17] according to a fixed fraction of Horn clauses in CNF formulas.

The paper is organized as follows. Section 2 introduces definitions and notations used throughout the paper. In Section 3, three versions of MHF's are introduced. Each of these classes is NP-complete w.r.t. SAT as follows from the above described transformation of the well known NP-complete graph coloring problem [4]. We provide another polynomial time transformation of CNF-SAT to MHF-SAT on which some investigations in this paper rely. In Section 4, a vertex cover based algorithm for determining SAT of a MHF M is presented having running time $O(2^{0.5284n})$, with n the number of variables in M. The approach also yields a classification of MHF's allowing for a fixed-parameter tractability result. Section 5 provides a strong argument telling that it is hard to improve an $O(2^{n/2})$ time bound for solving MHF-SAT. Section 6 finally, describes a further vertex cover based technique for speeding up the MHF-SAT algorithm. Some experimental results illustrating the usefulness of this approach are presented.

2   Basic Definitions and Notation

Let CNF denote the set of formulas (free of duplicate clauses) in conjunctive normal form over a set $V = \{x_1, \ldots, x_n\}$ of propositional variables ($n \in \mathbb{N}$). Each formula $C \in$ CNF is considered as a clause set $C = \{c_1, \ldots, c_{|C|}\}$. Each clause $c \in C$ is a disjunction of different literals, and is also represented as a set $c = \{l_1, \ldots, l_{|c|}\}$. The length of a formula C is denoted by $\|C\|$ whereas $|C|$ denotes the number of its clauses. Each variable x induces a positive literal (variable x) or a negative literal (negated variable: $\overline{x}$). Clauses containing positive (negative) literals only are called *positive (negative) monotone*. We denote by $V(C)$ the set of variables occuring in formula C. The satisfiability problem (SAT) asks in its *decision* version, whether a given CNF instance C is *satisfiable*, i.e., whether C has a *model*, which is a truth assignment $\tau : V(C) \rightarrow \{0,1\}$ setting at least one literal in each clause of C to 1 (TRUE). For convenience we allow the empty set to be a formula: $\varnothing \in$ CNF which is always satisfiable. In its *search* version SAT means to find a model τ if the input formula is satisfiable.

For $X \subseteq V(C)$, we denote by C^X the formula obtained from C by flipping all variables in X; in case $X = V(C)$ we write $C^\gamma := C^{V(C)}$. Given a formula $C \in$ CNF and a partial truth assignment $\tau : V(C) \rightarrow \{0,1\}$, we denote by $C[\tau]$ the formula obtained from C by removing all clauses satisfied by τ and removing all literals from the remaining clauses which are set to 0 (FALSE) by τ. Obviously, if τ is a model of C then $C[\tau] = \varnothing$. For two partial truth assignments τ, τ_1 of a formula C, satisfying $\tau_1 \subseteq \tau$, i.e., $D(\tau_1) \subseteq D(\tau)$ (for their domains) and $\tau_1^{-1}(1) \subseteq \tau^{-1}(1)$, obviously holds: If τ satisfies C, then $C[\tau_1]$ is satisfiable.

For $k \in \mathbb{N}$, let k-CNF (resp. CNF($= k$)) denote the subset of formulas C such that each clause has length at most (resp. exactly) k. Moreover $\mathcal{M}_\varepsilon, \varepsilon \in \{+, -\}$, denotes the set of ε-monotone (CNF-)formulas, i.e., for $\varepsilon = + (-)$ all clauses are positive (negative) monotone. Let $\mathcal{H}$ denote the set of all *Horn* formulas; each clause of which has at most one positive literal. For a *hidden* Horn formula H, by definition there exists a subset $X \subseteq V(H)$ such that H^X is a Horn formula. The set of all hidden Horn formulas is denoted by $\hat{\mathcal{H}}$.

For a monotone formula $C \in \mathcal{M}_\varepsilon$ ($\varepsilon \in \{+, -\}$), we can construct its *formula graph* G_C with vertex set $V(C)$ in linear time. Two vertices are joined by an edge if there is a clause in C containing the corresponding variables. Clearly, for each $c \in C$ the subgraph $G_C|c$ of G_C is isomorphic to the complete graph $K_{|c|}$. In the particular case of $C \in \mathcal{M}_\varepsilon(= 2)$, i.e., C is a monotone formula containing 2-clauses only, G_C contains exactly one edge for every clause in C. Note that a monotone formula $C \in \text{CNF}(= 2)$ with each variable occuring only once corresponds to a graph consisting of isolated edges only, and whose number of edges is half the number of vertices.

3   Mixed Horn Formulas

Let $\mathcal{C}_1, \mathcal{C}_2 \subset \text{CNF}$ be two classes of formulas over the same variable set V. A formula $C \in \text{CNF}$ such that there are formulas $C_i \in \mathcal{C}_i, i = 1, 2$, with $C = C_1 \wedge C_2$, is called *mixed (over $\mathcal{C}_1, \mathcal{C}_2$)*. The collection of formulas mixed over $\mathcal{C}_1, \mathcal{C}_2$ is denoted as $\mathcal{C}_1 \wedge \mathcal{C}_2$. In this paper we are interested in specific mixed formulas containing Horn subformulas:

Definition 1. *We define the class of* mixed Horn formulas *by* $\text{MHF} := \mathcal{H} \wedge$ 2-CNF *and the class of* mixed hidden Horn formulas *by* $\text{MHHF} := \hat{\mathcal{H}} \wedge$ 2-CNF. *The set of mixed Horn formulas based on the negative monotone formulas, called* negative mixed Horn formulas *is denoted by* $\text{MHF}_- := \mathcal{M}_- \wedge$ 2-CNF.

Because all 2-clauses which are not positive monotone are Horn, every formula $M \in \text{MHF}$ has the unique representation $M = H \wedge P$, where P is the collection of all positive monotone 2-clauses in M and H is the remaining Horn subformula. Given $M \in \text{MHF}$ we thus write $P(M), H(M)$ for these subformulas, respectively.

The question arises whether the mixed formulas introduced in the Definition above can be recognized fast. It is obvious that membership of MHF and MHF_- for an instance $C \in \text{CNF}$ can be recognized in time $O(\|C\|)$. The next lemma gives a positive answer also for recognizing mixed hidden Horn formulas.

Lemma 1. *For $C \in \text{CNF}$ with $n = |V(C)|$, it can be decided in time $O(\|C\|)$ whether $C \in \text{MHHF}$.*

PROOF. Let $C \in \text{CNF}$ such that $C \notin \text{MHF}$, otherwise we are done. Let $T \in \text{CNF}(= 2)$ be the collection of all 2-clauses in C. Observe that T is Horn except for its positive monotone part. This means that it suffices to check whether $C' := C \setminus T$ is a hidden Horn formula, since flipping variables in T has no effect regarding the mixed hidden Horn status of the input instance C. It is well known that a hidden Horn formula C' can be recognized in time $O(\|C'\|)$ [11], from which the assertion follows.   □

We therefore have for every instance $\hat{M} \in \text{MHHF}$ a unique decomposition $\hat{M} = \hat{H} \wedge T$ where $T \in \text{CNF}(= 2)$ contains all 2-clauses in $\hat{M}$ and $\hat{H} \in \hat{\mathcal{H}}$. We shall also write $T(\hat{M})$ and $\hat{H}(\hat{M})$ for given $M \in \text{MHHF}$.

It is not hard to see that the reduction from graph colorability to MHF-SAT presented in the introduction is de facto a reduction to MHF_--SAT. Using the (proper) inclusions $\text{MHF}_- \subset \text{MHF} \subset \text{MHHF}$ we readily obtain:

Proposition 1. SAT *remains NP-complete for instances from* MHF_-, MHF, *and* MHHF.   □

Next we describe a transformation of CNF-SAT to MHF-SAT, which is reconsidered in Section 5. This transformation also provides a different look at CNF-SAT solving from the point of view of MHF's.

Transformation $(*)$:
Input: $C \in$ CNF
Output: $M_C \in$ MHF$_-$, s.t. $M_C \in$ SAT iff $C \in$ SAT
begin
Let $V_+(C) \subseteq V(C)$ be the set of all variables that occur positive in at least one k-clause of C with $k \geq 3$.
For every variable $x \in V_+(C)$ introduce a new variable $y_x \notin V(C)$. Then:
1.) Replace all positive occurences of $x \in V_+(C)$ in the k-clauses $k \geq 3$ by $\overline{y}_x$, for every $x \in V_+(C)$. Let the formula obtained be C'.
2.) Add the constraints $\overline{y}_x \leftrightarrow x$ to C', for all $x \in V_+(C)$. This yields the new CNF formula

$$M_C := C' \cup \bigcup_{x \in V_+(C)} \{y_x, x\} \cup \{\overline{y}_x, \overline{x}\}$$

end

In the last step we have used the simple equivalences $\overline{y}_x \leftrightarrow x \equiv \overline{y}_x \rightarrow x \wedge y_x \rightarrow x$ and $a \rightarrow b \equiv \overline{a} \vee b$. Because all positive literals occuring in k-clauses of C with $k \geq 3$ are removed, $M_C \in$ MHF$_-$ holds.

Transformation $(*)$ obviously is polynomial time and a reduction in the sense that $C \in$ SAT if and only if $M_C \in$ SAT. It can be adapted also to obtain a MHF, which is not necessarily a member of the class MHF$_-$. For this, it is often not necessary to create for every $x \in V_+(C)$ a new variable as indicated above. A subset of $V_+(C)$, as small as possible, suffices to yield a (not necessarily negative monotone) Horn part and thus may produce a smaller positive monotone part P of 2-clauses. It turns out that the size of P is the crucial quantity regarding the running time of Algorithm MHFSAT described in the next section.

4   A SAT-Algorithm for Mixed Horn Formulas

We aim at providing a non-trivial exact deterministic algorithm solving the SAT search problem for the classes MHF$_-$, MHF, and MHHF. As it turns out it is convenient to address first the class MHF. For such an instance $M \in$ MHF, we assume that $P := P(M) \in \mathcal{M}_+(= 2)$ is not the empty formula. Since otherwise a model for $M = H(M) \in \mathcal{H}$ can be found by Horn-SAT, if existing. Recall that Horn-SAT even provides a minimal model. Since P is monotone and each of its clauses is a 2-clause, the formula graph G_P of P has exactly one edge for each clause in P, i.e. $G_P = (V(P), P)$. By monotonicity P obviously is satisfiable. Observe that for satisfying P it suffices to find a set of variables X hitting all clauses of P and to set every variable in X to 1. The remaining variables in P are free, i.e., independent of P and if possible should be assigned appropriately to satisfy the remaining Horn formula, too. In terms of the formula graph G_P, such a set X corresponds to a vertex cover of G_P. In other words running through all vertex covers of G_P means running through all models of P. For every such model of P, we can test by Horn-SAT whether it can be extended to a model of the remaining Horn formula $H(M)$ and thus to a model of the whole instance M. Due to the next lemma it is not necessary to test every vertex cover of G_P:

Lemma 2. *An instance* $M = P \wedge H \in \mathrm{MHF}$ *is satisfiable if and only if there exists a minimal vertex cover of* G_P *which can be extended to a model of* M.

PROOF. Suppose that $M = H \wedge P \in \mathrm{MHF}$ is satisfiable and let σ be a model of M. Then $H \in \mathrm{SAT}$ and $\sigma_P := \sigma | V(P)$ is a model of P. Restricting the domain of σ_P to those variables $x \in V(P)$ with $\sigma_P(x) = 1$ also yields a model τ of P with $D(\tau) = \tau^{-1}(1)$, because P is positive monotone. Clearly, the set $X := \{x \in V(P) : \sigma_P(x) = 1\}$ represents a vertex cover of G_P. If X is a minimal vertex cover of G_P we are done. Otherwise, this vertex cover contains a minimal vertex cover of G_P corresponding to a truth assignment τ' that is also a model of P. By construction $D(\tau') = \tau'^{-1}(1) \subset \sigma^{-1}(1)$ holds. Hence τ' is contained in σ yielding $M[\tau'] \in \mathrm{SAT}$ which means that τ' can be extended to a model of M proving the only-if part of the Lemma. The converse direction is obvious. $\square$

Hence, an algorithm that enumerates all minimal vertex covers of G_P and that for each cover separately checks in linear time whether the remaining Horn formula is satisfiable, definitely performs the task of solving SAT for M. It is well known that the complement of a vertex set of a minimal vertex cover of G_P is a maximal independent set in G_P. Thus, it suffices to compute all maximal independent sets in G_P. Fortunately an algorithm of computing all maximal independent sets in graphs, with polynomial delay only, has been developed by Johnson et al. see [7]. Exploiting this algorithm we use a procedure MinVC(G) to generate all minimal vertex covers of a graph G with polynomial delay. Similarly, we will use a procedure HornSat(H) that returns a minimal model τ of H if and only if H is a satisfiable Horn formula, else returns **nil**, for an appropriate Horn-SAT algorithm see e.g. [12, 9]. Now we are ready to state algorithm MHFSAT determining a model τ of $M \in \mathrm{MHF}$, if M is satisfiable, otherwise unsatisfiability (**nil**) of M is reported. For convenience we identify a vertex cover X of G_P and the corresponding partial model in $M = H \wedge P \in \mathrm{MHF}$. X becomes **nil** if all minimal vertex covers of G_P have been enumerated:

Algorithm MHFSAT(M, τ)
Input: $M \in \mathrm{MHF}$
Output: model τ for M, if $M \in \mathrm{SAT}$, **nil** otherwise
begin
compute $P := P(M)$
if $P = \varnothing$ **then return** $\tau \leftarrow$ HornSat(M)
compute graph G_P
$\tau \leftarrow$ **nil**; $X \leftarrow$ **nil**
repeat

compute by MinVC(G_P) the next minimal vertex cover X of G_P
  if $X \neq$ **nil then** $\tau \leftarrow$ HornSat($M[X]$)
until $\tau \neq$ **nil or** $X =$ **nil**
return $X \cup \tau$
end

Theorem 1. *Algorithm* MHFSAT *solves* SAT *of an input formula* $M \in$ MHF *in time* $O(2^{0.5284n})$, *where* $n = |V(M)|$.

PROOF. The correctness of the algorithm follows from the argumentation above. Moreover, it is not hard to see that $X =$ **nil** if and only if $\tau =$ **nil** in the last line of the algorithm. Hence, it is ensured that the returned value either is a model for the input instance M or is **nil**.

Addressing the running time, we can compute $P := P(M)$, and the formula graph $G_P = (V(P), P)$ of P in linear time $O(\|M\|)$. If $P = \varnothing$ we are done in linear time by Horn-SAT. If $P \neq \varnothing$, we have to execute the repeat-until loop. During each iteration we never consume more than the polynomial delay for computing the next minimal vertex cover followed by a linear time Horn-SAT computation, thus needing only polynomial time. The number of iterations is bounded by the number of all minimal vertex covers of G_P. Given a graph G, it is a long standing result by Moon and Moser [13] that the number of its maximal independent sets is bounded by $3^{\frac{1}{3}|V(G)|} \lesssim 2^{0.5284|V(G)|}$. In fact, this is a tight bound in the sense that there exist graphs achieving this number. Such graphs consist of $n/3$ copies of the K_3. For every triangle independently contributes three different minimal vertex covers, resp. maximal independent sets. Hence, we conclude that SAT for an arbitrary instance $M \in$ MHF is solvable in time $O(p(n)3^{\frac{n}{3}})$ where p denotes an appropriate polynomial, thus providing the claimed time bound of $O(2^{0.5284n})$.   □

Due to Lemma 1 we can solve the SAT search problem for mixed hidden Horn formulas $C \in$ MHHF within the same time asymptotically:

Corollary 1. *For input instances* $C \in$ MHHF, *the search version of* SAT *can be solved in time* $O(2^{0.5284n})$ *where* $n = |V(C)|$.

We shall derive another consequence from the preceding discussion. Notice that the variables of $P(M)$ are crucial for the running time of Algorithm MHFSAT only, because they form the vertex set of the graph $G_{P(M)}$ that has to be investigated:

Corollary 2. *For* $M = H \wedge P \in$ MHF *the search version of* SAT *is solvable in time* $O(\|M\|2^{0.5284|V(P)|})$.

Let MHF$_k$ denote the special class of MHF's $C = H \wedge P$ with $|V(P)| \leq k$, i.e., the positive monotone 2-clauses have at most k different variables for fixed $k \in \mathbb{N}$. W.r.t. the classes of MHF$_k$, $k \geq 0$, we have a fixed-parameter tractability classification of the SAT problem for MHF:

Corollary 3. *For an input instance* $M \in$ MHF$_k$, $k \geq 0$, SAT *can be decided in polynomial time* $O(\|M\|2^{0.5284k})$.

Similarly, denote by MHHF_k the subset of mixed hidden Horn formulas $\hat{M}$ whose maximal subformula $T(\hat{M}) \in \text{CNF}(= 2)$ has at most k different variables. By Lemma 1 and Corollary 3, we also obtain w.r.t. the classes of MHHF_k, $k \geq 0$, a fixed-parameter tractability classification of the SAT problem for MHHF:

Corollary 4. *For an input instance* $\hat{M} \in \text{MHHF}_k, k \geq 0$, *SAT can be decided in polynomial time* $O(\|\hat{M}\|2^{0.5284k})$, *where* $n := |V(\hat{M})|$.

For some subclasses of MHF we have slightly better bounds than stated in Corollary 2:

Proposition 2. *Let* $M = H \wedge P \in \text{MHF}$ *with* $k = |V(P)|$ *and formula graph* $G := G_P$ *associated to* P.

(1) There is a polynomial p such that SAT *is solvable for M in time* $O(p(k)2^{k/2})$ *in either of the following cases:*
  (i) G is triangle-free,
  (ii) G is connected and contains at most one cycle.
(2) If G contains at most $r \geq 1$ cycles and has at least $3r$ vertices, then SAT *is solvable for M in time* $O(p(k)3^r 2^{\frac{k-3r}{2}})$, *for an appropriate polynomial p.*

PROOF. It suffices to verify that Algorithm MHFSAT has the claimed running times for the special instances fulfilling the stated properties. Case (1)(i), for G triangle-free, has been solved by Hujter et al. [6], who have shown that a triangle-free graph of at least four vertices contains at most 2^s maximal independent sets if $|V(G)| = 2s$ and at most $5 \cdot 2^{s-2}$ maximal independent sets if $|V(G)| = 2s + 1$. The extremal graphs achieving these bounds consist of s copies of the K_2, respectively, $s - 2$ copies of K_2 and one copy of C_5. Case (ii) was solved by Griggs, and Jou et al. [5,8]. They have shown that a connected graph with at most one cycle admits at most $3 \cdot 2^{s-2}$ maximal independent sets for $|V(G)| = 2s$ and at most $2^s + 1$ maximal independent sets if $|V(G)| = 2s + 1$.

Assertion (2) follows by the above argumentation from the results obtained by Sagan et al. [16]. They have shown that the number of maximal independent sets in graphs containing at most $r \geq 1$ (not necessarily nonintersecting) cycles and at least $3r$ vertices is upper bounded by $3^r 2^{\frac{k-3r}{2}}$. They also have shown that this bound is tight and is achieved by graphs that consist of copies of an appropriate number of K_3 and K_2. Notice that assertion (2) implies (1),(ii), for $r = 1$. $\qquad\qquad\square$

5   Hardness of Improving Theorem 1

Next we address the question which improvements of the time bound for solving MHF-SAT presented in Theorem 1 can be expected.

Theorem 2. *Every instance $C \in \text{CNF}$ can be transformed in linear time into a corresponding instance $M_C \in \text{MHF}$ such that M_C can be tested for* SAT *in time* $O(p(n)2^{n/2})$, *where* $n := |V(M_C)| \leq 2|V(C)|$ *and p is an appropriate polynomial.*

PROOF. To an arbitrary formula $C \in \mathrm{CNF}$ we apply Transformation $(*)$ with the slight modification that also all positive monotone 2-clauses in C (if some exist) are considered. It is easy to verify that the resulting transformation changes C into an instance $M_C \in \mathrm{MHF}_-$ such that $G_{P(M_C)}$ consists of isolated edges only. Hence, we obtain the assertion by Proposition 2, (1)(i). $\qquad \square$

It seems to be very hard to improve on the bound stated in the last theorem significantly, since otherwise SAT for an arbitrary $C \in \mathrm{CNF}$ ($n := |V(C)|$) could be solved significantly faster than in 2^n steps. For suppose there is an algorithm solving SAT for MHF's $M = H \wedge P$ with $n = |V(P)|$ in $O(2^{\alpha n})$ steps for some $\alpha < 1/2$. Then we can transform an arbitrary CNF formula C into a sat-equivalent MHF $M_C = H_C \wedge P_C$ with at most $2n$ variables contained in P_C. So, by Corollary 2, SAT for M_C can be solved in $O(2^{2\alpha n})$ steps, where $2\alpha < 1$. Although, there has been made some progress recently in finding non-trivial bounds for SAT for arbitrary CNF formulas [2], it would require a significant breakthrough in our understanding of SAT to obtain upper time bounds of the form $O(2^{(1-\epsilon)n})$ for some constant $\epsilon > 0$.

6   An Approach for Reducing the Number of Essential Variables

The number of new introduced variables necessary to transform $C \in \mathrm{CNF}$ into $M_C = H_C \wedge P_C \in \mathrm{MHF}$ is crucial regarding the running time of Algorithm MHFSAT. This is due to the fact that these variables contribute vertices to the formula graph of P_C. The requirement to keep this set small leads us to the following notion:

Definition 2. *For $C \in \mathrm{CNF}$, a minimal set $X \subseteq V(C)$, for which the transformation in the proof of Theorem 2 yields a MHF formulation $M_C := H_C \wedge P_C \in \mathrm{MHF}$ of C via the corresponding set X' of new variables ($|X| = |X'|$), is called an* essential set of variables *(of C).*

Observe that there may exist many essential sets of variables of a formula C not necessary of the same cardinality. To obtain a smallest essential set of variables one can proceed as follows: For each clause $c \in C$ that is not Horn, let c' denote the positive monotone part of c. For example $c = \{x, \overline{y}, z\}$ delivers $c' = \{x, z\}$. Collecting these parts c' of all clauses c in C, yields a positive monotone formula $C' \in \mathcal{M}_+$. It remains to transform C' into a Horn formula with least effort. Resting on the formula graph $G_{C'}$ determined from $C \in \mathrm{CNF}$ we obtain:

Lemma 3. *Let $C \in \mathrm{CNF}$ and let C' be defined as above. Every essential set of variables $X \subset V(C)$ is a minimal vertex cover of the formula graph $G_{C'}$ of C' and vice versa.*

A smallest essential set of variables of a formula $C \in \mathrm{CNF}$ by Lemma 3 is a minimum vertex cover of the formula graph $G_{C'}$. A minimum vertex cover of a graph with n vertices can be computed in time $O(2^{n/4})$ by the Robson algorithm [15] determining a maximum independent set. This algorithm can be used to speed-up Algorithm MHFSAT for solving some MHF instances M by treating them as CNF formulas. It is the part $P := P(M)$ of a formula $M \in$ MHF that can be made into a Horn formula by the transformation shown above using a minimum vertex cover of G_P as essential set, if it has appropriate size. Thus instead of Algorithm MHFSAT we shall proceed in the following way for solving MHF-SAT: Let $k := |V(P)|$ and compute a minimum vertex cover X of G_P by the Robson algorithm in time $O(2^{k/4})$ (let $j := |X|$). In case of $j \geq 0.5284 \cdot k$, we proceed by the usual Algorithm MHFSAT, for the original instance M. Otherwise, i.e. $(*) : j < 0.5284 \cdot k$, we use X as an essential set of variables for a reformulation of M resulting in a new MHF M', whose positive monotone part P' contains $|V(P')| =: k' = 2j$ variables. Moreover, the formula graph $G_{P'}$ by construction consists of isolated edges only (cf. the proof of Theorem 2). Now we complete the computation by running Algorithm MHFSAT on the modified instance M'. Because of the structure of $G_{P'}$ and according to Proposition 2, (1)(i), we obtain in this branch of the modified algorithm a better running time of $O(\|M'\|2^{k'/2}) = O(\|M'\|2^j)$, where the exponential factor has decreased due to $(*)$.

To illustrate the usefulness of essential sets, consider again the graph coloring problem. Let $C(G)$ be the 3-CNF formula corresponding to the 3-colorability problem of a given graph $G = (V, E)$ as mentioned in the introduction. $C(G)$ consists of $|V|$ positive monotone 3-clauses and $3|E|$ negative monotone 2-clauses and is therefore no MHF formula. Clearly, complementing all variables yields a MHF $H \wedge P$. Unfortunately, the crucial subformula P becomes large by this operation. In order to speed up the SAT test of $H \wedge P$, an essential set of variables in $C(G)$ turning it into a MHF of a smallest P-part is required. As an example, take the triangle graph K_3 with vertex set $\{a, b, c\}$ leading to the CNF formula $C(G) = C(V) \cup C(E)$ with corresponding clause sets:

$$C(V) := \{\{a_1, a_2, a_3\}, \{b_1, b_2, b_3\}, \{c_1, c_2, c_3\}\},$$
$$C(E) := \{\{\overline{a}_1, \overline{b}_1\}, \{\overline{a}_2, \overline{b}_2\}, \{\overline{a}_3, \overline{b}_3\}\}$$
$$\cup \{\{\overline{a}_1, \overline{c}_1\}, \{\overline{a}_2, \overline{c}_2\}, \{\overline{a}_3, \overline{c}_3\}\}$$
$$\cup \{\{\overline{b}_1, \overline{c}_1\}, \{\overline{b}_2, \overline{c}_2\}, \{\overline{b}_3, \overline{c}_3\}\}$$

Turning this into a MHF by complementing all variables yields a P-subformula of 9 clauses and 18 variables. Taking instead only an essential set of 6 variables, namely two variables of each 3-clause in $C(V)$, e.g. $\{a_i, b_i, c_i : i = 1, 2\}$, yields a MHF $M(G) = H \wedge P$ with

$$P := \{\{a_1, a_1'\}, \{a_2, a_2'\}, \{b_1, b_1'\}, \{b_2, b_2'\}, \{c_1, c_1'\}, \{c_2, c_2'\}\},$$
$$H := C(E) \cup P^\gamma \cup \{\{\overline{a'}_1, \overline{a'}_2, a_3\}, \{\overline{b'}_1, \overline{b'}_2, b_3\}, \{\overline{c'}_1, \overline{c'}_2, c_3\}\}$$

Recall that C^γ means to complement all variables in formula C. The new formula P contains only 6 clauses and only 12 variables instead of 18, moreover the formula graph consists of isolated edges only. Although the example is simple, it describes the usefulness of essential sets, which becomes explicit when dealing with larger instances.

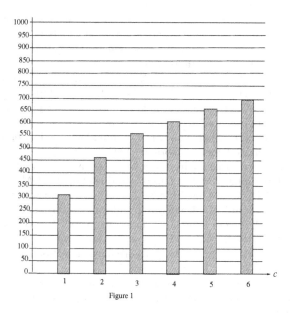

Figure 1

Fig. 1. Number of essential variables in MHF's with 1000 variables and $c \cdot$ 1000 clauses $(c = 1, \ldots, 6)$

To supply these observations we had run several experiments for $\mathrm{CNF}(= 3)$ formulas with 1000 variables and $c \cdot 1000$ clauses, for $c = 1, \ldots, 6$. Each instance $C \in \mathrm{CNF}(= 3)$ has been generated randomly and was transformed into $M_C = H \wedge P$. The new introduced variables form an essential set of variables of C. Figure 1 displays the average number of essential variables, computed by a vertex-cover heuristic, obtained from samples of 100 formulas, for each $c = 1, \ldots, 6$.

Acknowledgement. We are grateful to Hans van Maaren for pointing out the encoding of 3-colorability into MHF-SAT.

References

1. B. Aspvall, M. R. Plass, and R. E. Tarjan, A linear-time algorithm for testing the truth of certain quantified Boolean formulas, Inform. Process. Lett. 8 (1979) 121-123.
2. E. Dantsin and A. Wolpert, Algorithms for SAT based on search in Hamming balls, ECCC Report No. 17, 2004.
3. R. G. Downey and M. R. Fellows, Parameterized Complexity, Springer-Verlag, New York, 1999.
4. M. R. Garey and D. S. Johnson, Computers and Intractability: A Guide to the Theory of NP-Completeness, W. H. Freeman and Company, San Francisco, 1979.

5. J. R. Griggs, C. M. Grinstead, and D. R. Guichard, The number of maximal independent sets in a connected graph, Discrete Math. 68 (1988) 211-220.
6. M. Hujter and Z. Tuza, The number of maximal independent sets in triangle-free graphs, SIAM J. Discrete Math. 6 (1993) 284-288.
7. D. S. Johnson, M. Yannakakis, and C. H. Papadimitriou, On Generating All Maximal Independent Sets, Inform. Process. Lett. 27 (1988) 119-123.
8. M. Jou and G. J. Chang, Maximal independent sets in graphs with at most one cycle, Discrete Applied Math. 79 (1997) 67-73.
9. H. Kleine Büning and T. Lettman, Propositional logic, deduction and algorithms, Cambridge University Press, Cambridge, 1999.
10. D. E. Knuth, Nested satisfiability, Acta Informatica 28 (1990) 1-6.
11. H. R. Lewis, Renaming a Set of Clauses as a Horn Set, J. ACM 25 (1978) 134-135.
12. M. Minoux, LTUR: A Simplified Linear-Time Unit Resolution Algorithm for Horn Formulae and Computer Implementation, Inform. Process. Lett. 29 (1988) 1-12.
13. J. W. Moon and L. Moser, On cliques in graphs, Israel J. Math. 3 (1965) 23-28.
14. B. Randerath, E. Speckenmeyer, E. Boros, P. Hammer, A. Kogan, K. Makino, B. Simeone, and O. Cepek, A Satisfiability Formulation of Problems on Level Graphs, ENDM, Vol. 9, 2001.
15. J. M. Robson, Finding a maximum independent set in time $O(2^{n/4})$, Technical Report, Univ. Bordeaux, http://dept-info.labri.u-bordeaux.fr/ robson/mis/techrep.html, 2001.
16. B. E. Sagan and V. R. Vatter, Maximal and Maximum Independent Sets In Graphs With At Most r Cycles, Preprint, arXiv:math CO/0207100 v2, 2003.
17. H. van Maaren and L. van Norden, Hidden threshold phenomena for fixed-density SAT-formulae, in: "E. Giunchiglia, A. Tacchella (Eds.), Proceedings of the 6th International Conference on Theory and Applications of Satisfiability Testing (SAT'03), Santa Margherita Ligure, Italy", Lecture Notes in Computer Science, Vol. 2919, pp. 135-149, Springer-Verlag, Berlin, 2004.

Satisfiability Threshold of the Skewed Random k-SAT

Danila A. Sinopalnikov

Department of Mathematics and Mechanics,
St. Petersburg State University, Russia
dasinopalnikov@yahoo.com

Abstract. We consider the k-satisfiability problem. It is known that the random k-SAT model, in which the instance is a set of m k-clauses selected uniformly from the set of all k-clauses over n variables, has a phase transition at a certain clause density, below which most instances are satisfiable and above which most instances are unsatisfiable. The essential feature of the random k-SAT is that positive and negative literals occur with equal probability in a random formula. How does the phase transition behavior change as the relative probability of positive and negative literals changes?

In this paper we focus on a distribution in which positive and negative literals occur with different probability. We present empirical evidence for the satisfiability phase transition for this distribution. We also prove an upper bound on the satisfiability threshold and a linear lower bound on the number of literals in satisfying partial assignments of skewed random k-SAT formulas.

1 Introduction

The problem to decide whether a given propositional formula has a satisfying truth assignment (SAT) is one of the first for which NP-completeness was proven. Nowadays it attracts much attention, since many hard combinatorial problems in areas including planning [12, 13] and finite mathematics [18, 19] can be naturally encoded and studied as SAT instances.

While the SAT hardness is determined by the difficulty of solving an instance of the problem in the worst case, the scientific interest is also focused on randomly chosen SAT instances in attempt to determine the typical-case complexity. The choice of the probabilistic distribution is critical for the significance of such a study. In particular, it was proven that in some probabilistic spaces a random formula is easy-to-decide with high probability [6, 9, 15]. To date, most of the research in the field is concentrated on the random k-SAT model $RD(n, k, m)$, which appears to be more robust in this respect.

Let X be a set of n boolean variables; a proper k-clause is a disjunction of k distinct and non-contradictory literals corresponding to variables in X. Under the random k-SAT distribution $RD(n, k, m)$, a random formula $F_k(n, m)$ is built

H.H. Hoos and D.G. Mitchell (Eds.): SAT 2004, LNCS 3542, pp. 263–275, 2005.

by selecting uniformly, independently and with replacement m clauses from the set of all proper k-clauses over n variables.

Numerous empirical results suggested that $RD(n, k, m)$ exhibits a phase transition behavior as the clause density $\delta = m/n$ changes [2, 14, 16, 17]. When the number of variables tends to infinity and the clause density remains constant, the random formula $F_k(n, \delta n)$ is almost surely satisfiable for low clause densities while for higher clause densities it is almost surely unsatisfiable. The satisfiability threshold conjecture asserts that for every $k \geq 2$ there exists δ_k such that

$$
\begin{aligned}
\delta_k = \sup\{\delta|\ \lim_{n \to \infty} P(F_k(n, \delta n) \text{ is satisfiable}) = 1\} = \\
= \inf\{\delta|\ \lim_{n \to \infty} P(F_k(n, \delta n) \text{ is satisfiable}) = 0\} \ .
\end{aligned}
\tag{1}
$$

This conjecture was settled for $k = 2$ with $\delta_2 = 1$ by Chvátal and Reed [3], Goerdt [8] and Fernandez de la Vega [4]. A threshold sequence $r_k(n)$ is proven to exist by the following theorem, due to Friedgut,

Theorem 1. [7] *For every $k \geq 2$, there exists a sequence $r_k(n)$ such that for all $\epsilon > 0$*

$$
\lim_{n \to \infty} P(F_k(n, (r_k(n) - \epsilon)n) \text{ is satisfiable}) = 1 \ ,
\tag{2}
$$

$$
\lim_{n \to \infty} P(F_k(n, (r_k(n) + \epsilon)n) \text{ is satisfiable}) = 0 \ .
\tag{3}
$$

More recently the asymptotic form of the conjecture was established by Achlioptas and Peres [1].

The essential feature of the random k-SAT is that positive and negative literals occur in a formula with equal probability. In this paper we consider satisfiability of random formulas from the skewed random k-SAT distribution, in which positive and negative literals occur with different probability. To the best of our knowledge, there has not been much work on this generalization of random k-SAT. The paper answers the question whether the satisfiability phase transition manifests in the skewed distributions and presents a proof of an upper bound on the threshold location for skewed random k-SAT. We expect that this study will provide further insight into the nature of the phase transition phenomenon in the boolean satisfiability problem.

We also investigate the minimal number of literals in a satisfying partial assignment of a random formula. This study is motivated by the fact that if a random k-CNF formula is satisfiable then it has an exponential number of satisfying assignments with high probability [11]. On the other hand, it is known that k-CNF formulas with many satisfying assignments have short satisfying partial assignments [10]. This might imply that a random formula with clause density far below the satisfiability threshold is likely to have short satisfying partial assignments. In this paper we elaborate on this intuition and prove a linear lower bound on the number of literals in a satisfying partial assignment of a random formula for skewed and plain random k-SAT distributions.

The paper is organized as follows. Section 2 contains basic definitions. In Section 3 we present the results of empirical study of the phase transition behavior of the skewed distribution. We formulate and prove the main results in Sections 4 and 5. Section 6 concludes the paper.

2   Basic Definitions

Let X be a set of n boolean variables. A *literal* is a variable (positive literal) or its negation (negative literal). A variable and its negation are *contradictory* literals. A k-clause is an ordered collection of k literals. A *clause density* of a formula F is the ratio of number of clauses in F to the number of variables. A *complementary* formula for F is a formula obtained from F by replacing all literals with their negations.

A *partial assignment* σ is an arbitrary set of non-contradictory literals. The *size* of a partial assignment is the number of literals in it. A *complete* assignment on n variables is a partial assignment of size n. A *complementary* partial assignment for σ is an assignment obtained from σ by replacing of all literals with their negations. A partial assignment σ is *satisfying* for a formula F if in each clause of F there is at least one literal from σ.

We use the following notation:

$\Phi(n, k)$ denotes the set of all k-CNF formulas over n variables,
$A(n, m)$ denotes the set of all partial assignments of size m over n variables.

For a formula $F \in \Phi(n, k)$ and a partial assignment σ:

$F \in SAT$ means F is satisfiable,
$\bar{F}, \bar{\sigma}$ denote the complementary formula and assignment respectively,
$\sigma \in S(F)$ means σ satisfies F,
$pl(\sigma)$ denotes the number of positive literals in σ,
$minsat(F)$ denotes the minimum size of a satisfying partial assignment of F.

For $\lambda \in [0, 1]$ denote $H(\lambda) = \lambda^\lambda (1 - \lambda)^{1-\lambda}$.

Let $n, k, m \in N$, $p \in (0, 1)$. The skewed random k-SAT distribution $SD(n, k, m, p)$ is the distribution, where a random formula is obtained by building m k-clauses as follows: for each clause we select independently, uniformly and with replacement k variables from the set of n boolean variables; then for each selected variable we take a positive literal with probability p and a negative literal otherwise and add it to the clause.

3   The Phase Transition Behavior of the Skewed Random k-SAT

In this section we present empirical evidence that random k-SAT instances undergo a satisfiability phase transition even if positive and negative literals occur

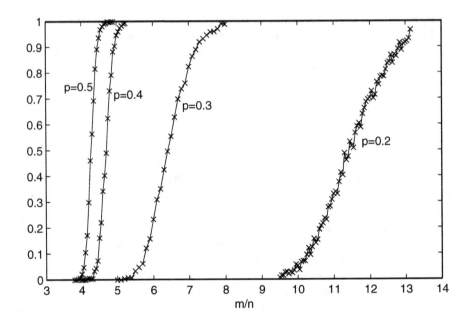

Fig. 1. The probability that a skewed random 3-SAT formula is unsatisfiable, as a function of the clause density

with different probabilities. We took skewed random 3-SAT distributions at four levels of skewness: $p = 0.5$(the plain 3-SAT); $0.4; 0.3; 0.2$.

Figure 1 shows the sample probability that a random clause is unsatisfiable at the particular clause density and skewness. There are four curves, one for each level of skewness. Along the horizontal axis is the clause density. Each sample contains 500 random 3-SAT instances on 200 variables.

We observe that all four distributions exhibit the satisfiability phase transition. The transition manifests at a greater density and becomes less distinct as the skewness of the distribution increases and the number of variables remains constant.

4   An Upper Bound for the Satisfiability Threshold for *SD*

For the skewed random k-SAT distribution with clause density σ and skewness $p \le 1/2$ we consider the following equation

$$(1 - p^k)^{-\delta} = 1 + \exp\left(-\frac{k\delta p^k}{1 - p^k} \cdot \frac{1 - 2p}{p}\right). \tag{4}$$

We first show that this equation has a single root and this root is positive. Then we prove that the root constitutes an upper bound on the satisfiability threshold location for the skewed distribution.

Lemma 1. *Let $p \in (0, 1/2]$, $k \in N$.*
Then there exists a single δ_0 that satisfies (4), $\delta_0 > 0$ and

$$\forall \delta \; \delta > \delta_0 \Leftrightarrow (1 - p^k)^{-\delta} > 1 + \exp\left(-\frac{k\delta p^k}{1 - p^k} \cdot \frac{1 - 2p}{p}\right) \tag{5}$$

Proof. Take arbitrary $p \in (0, 1/2]$, $k \in N$. Consider two functions

$$f(\delta) = (1 - p^k)^{-\delta},$$

$$g(\delta) = 1 + \exp\left(-\frac{k\delta p^k}{1 - p^k} \cdot \frac{1 - 2p}{p}\right).$$

$f(\delta)$ is a continuous, strictly increasing function of δ. $\lim_{\delta \to +\infty} f(\delta) = +\infty$, $f(0) = 1$. $g(\delta)$ is a continuous, decreasing function of δ, $g(0) = 2$. This implies the statement of the lemma.

Definition 1. *Let $p \in (0, 1/2]$, $k \in N$.*
$\Delta_s(p, k)$ *is the only root of the equation (4).*

Theorem 2. *Let $\delta > 0$, $\pi \in (0, 1)$, $p = \min(\pi, 1 - \pi)$, $k \in N$.*
Let $F_k(n, \delta, \pi)$ be a random formula from $SD(n, k, \delta n, \pi)$.
Then

$$\delta > \Delta_s(p, k) \Rightarrow \lim_{n \to \infty} P(F_k(n, \delta, \pi) \in SAT) = 0. \tag{6}$$

Proof. By the definition of the distribution SD,

$$\forall F \in \Phi(n, k) \; P(F_k(n, \delta, \pi) = F) = P(F_k(n, \delta, 1 - \pi) = \bar{F}). \tag{7}$$

Note that an assignment σ satisfies a formula F if and only if $\bar{\sigma}$ satisfies $\bar{F}$, so

$$\forall F \in \Phi(n, k) \; F \in SAT \Leftrightarrow \bar{F} \in SAT. \tag{8}$$

(7) and (8) imply that

$$\lim_{n \to \infty} P(F_k(n, \delta, \pi) \in SAT) = 0 \Leftrightarrow \lim_{n \to \infty} P(F_k(n, \delta, 1 - \pi) \in SAT) = 0 \tag{9}$$

So it is sufficient to prove the theorem for $\pi \in (0, 1/2]$, $p = \pi$.
Let X denote the number of satisfying assignments of $F_k(n, \delta, \pi)$.

$$P(F_k(n, \delta, \pi) \in SAT) = P(X \geq 1) \leq E[X] \tag{10}$$

$$E[X] = \sum_{\sigma \in A(n,n)} P(\sigma \in S(F_k(n, \delta, \pi))) = \tag{11}$$

Due to the symmetry of the distribution, the probability that σ satisfies a random formula depends only on $pl(\sigma)$.

$$= \sum_{z=0}^{n} \binom{n}{z} P(\sigma \in S(F_k(n, \delta, \pi)) | pl(\sigma) = z). \tag{12}$$

Let C and l denote random formulas from $SD(n, k, 1, \pi)$ and $SD(n, 1, 1, \pi)$ respectively (C and l can be viewed as a random clause and a random literal from $F_k(n, \delta, \pi)$). Then

$$P(\sigma \in S(F_k(n, \delta, \pi)) | pl(\sigma) = z) = P(\sigma \in S(C) | pl(\sigma) = z)^{\delta n} = \quad (13)$$

$$= (1 - P(\sigma \notin S(C) | pl(\sigma) = z))^{\delta n} = (1 - P(\sigma \notin S(l) | pl(\sigma) = z)^k)^{\delta n}. \quad (14)$$

The probability that a random literal l is unsatisfied by σ can be derived as follows

$$P(\sigma \notin S(l) | pl(\sigma) = z) = P(\sigma \notin S(l) | pl(\sigma) = z \ \& \ l \text{ - positive}) \cdot P(l \text{ - positive}) +$$

$$+ P(\sigma \notin S(l) | pl(\sigma) = z \ \& \ l \text{ - negative}) \cdot P(l \text{ - negative}) =$$

$$= \left(1 - \frac{z}{n}\right) \cdot p + \frac{z}{n} \cdot (1 - p) = p + \frac{z}{n}(1 - 2p). \quad (15)$$

Then

$$E[X] = \sum_{z=0}^{n} \binom{n}{z} \left(1 - \left(p + \frac{z}{n}(1 - 2p)\right)^k\right)^{\delta n} = \quad (16)$$

$$= \sum_{z=0}^{n} \binom{n}{z} \left(1 - p^k \left(1 + \frac{z}{n} \cdot \frac{1 - 2p}{p}\right)^k\right)^{\delta n} \leq \quad (17)$$

Now use the standard bound $(1 + x)^\alpha \geq 1 + \alpha x$, $\alpha > 1$, $x > 0$ and obtain

$$E[X] \leq \sum_{z=0}^{n} \binom{n}{z} \left(1 - p^k \left(1 + k \cdot \frac{z}{n} \cdot \frac{1 - 2p}{p}\right)\right)^{\delta n} = \quad (18)$$

$$= (1 - p^k)^{\delta n} \sum_{z=0}^{n} \binom{n}{z} \left(1 - \frac{p^k}{1 - p^k} \cdot k \cdot \frac{z}{n} \cdot \frac{1 - 2p}{p}\right)^{\delta n} \leq \quad (19)$$

Note that for $p \in (0, 1/2]$, $k \in N$ $\frac{kp^k}{1 - p^k} \cdot \frac{1 - 2p}{p} < 1$,
so we can apply $(1 - \alpha)^\beta \leq e^{-\alpha\beta}, \alpha \in [0, 1], \beta > 0$ and get

$$E[X] \leq (1 - p^k)^{\delta n} \sum_{z=0}^{n} \binom{n}{z} \exp\left(-\delta n \frac{p^k}{1 - p^k} \cdot k \cdot \frac{z}{n} \cdot \frac{1 - 2p}{p}\right) = \quad (20)$$

$$= (1 - p^k)^{\delta n} \sum_{z=0}^{n} \binom{n}{z} \exp\left(-\frac{k\delta p^k}{1 - p^k} \cdot \frac{1 - 2p}{p} \cdot z\right) = \quad (21)$$

$$= (1 - p^k)^{\delta n} \left(1 + \exp\left(-\frac{k\delta p^k}{1 - p^k} \cdot \frac{1 - 2p}{p}\right)\right)^n. \quad (22)$$

Table 1. Approximate values of Δ_s

p	0.1	0.2	0.3	0.4	0.5
k=3	95.662	22.385	10.204	6.346	5.191
k=4	783.420	94.318	29.749	14.665	10.740
k=5	6680.783	409.448	88.114	33.843	21.832
k=6	58520.315	1816.785	264.895	78.449	44.014

Finally, for every $\delta > \Delta_s(p, k)$ the Lemma 1 implies that

$$(1 - p^k)^{-\delta} > 1 + \exp\left(-\frac{k\delta p^k}{1 - p^k} \cdot \frac{1 - 2p}{p}\right). \tag{23}$$

So we obtain an upper bound on the base in (22)

$$(1 - p^k)^{\delta}\left(1 + \exp\left(-\frac{k\delta p^k}{1 - p^k} \cdot \frac{1 - 2p}{p}\right)\right) < 1. \tag{24}$$

This implies that for any clause density above $\Delta_s(p, k)$ the expected number of satisfying assignments of a random formula tends to zero as the number of variables tends to infinity. From (10) we get that the formula is satisfiable with probability approaching 0.

$$\lim_{n \to \infty} P(F_k(n, \delta, \pi) \in SAT) = 0. \tag{25}$$

This statement proves the theorem.   □

$\Delta_s(1/2, k) = -\dfrac{\ln 2}{\ln(1 - 1/2^k)}$ gives the known upper bound for the satisfiability threshold (see [5]).

Table 1 provides approximate values of $\Delta_s(p, k)$ obtained numerically for $p = 0.2; 0.3; 0.4; 0.5$ and $k = 3; 4; 5; 6$ from (4).

5   A Lower Bound on the Size of Satisfying Partial Assignments of a Skewed Random k-SAT Formula

If a formula over n variables is satisfiable, one might be interested in finding a satisfying partial assignment of the minimum size, that corresponds to the largest cluster of satisfying assignments. In this section we prove a linear lower bound on the size of a satisfying partial assignment of a random formula from a skewed random k-SAT distribution.

Consider the following equation for $k \in N$, $\lambda \in (0, 1]$, $q \in (1 - \lambda, 1 - \lambda/2]$.

$$\frac{H(\lambda)}{(1 - q^k)^{\delta}} = \left(1 + \exp\left(-\frac{k\delta q^k}{1 - q^k} \cdot \frac{2(1 - q) - \lambda}{q}\right)\right)^{\lambda} \tag{26}$$

We first prove that for all k, λ and suitable q this equation has a single root and this root is positive. Then we show that the root allows to define a linear lower bound on the size of the satisfying partial assignment of a skewed random k-SAT formula.

Lemma 2. *Let $k \in N$, $\lambda \in (0,1]$, $q \in (1-\lambda, 1-\lambda/2]$.*
Then there exists a single δ_0 that satisfies (26), $\delta_0 > 0$ and for all δ

$$\delta > \delta_0 \Leftrightarrow \frac{H(\lambda)}{(1-q^k)^\delta} > \left(1 + \exp\left(-\frac{k\delta q^k}{1-q^k} \cdot \frac{2(1-q)-\lambda}{q}\right)\right)^\lambda \qquad (27)$$

Proof. Take arbitrary $k \in N$, $\lambda \in (0,1]$, $q \in (1-\lambda, 1-\lambda/2]$.
Consider two functions

$$f(\delta) = \frac{H(\lambda)}{(1-q^k)^\delta}, \qquad (28)$$

$$g(\delta) = \left(1 + \exp\left(-\frac{k\delta q^k}{1-q^k} \cdot \frac{2(1-q)-\lambda}{q}\right)\right)^\lambda. \qquad (29)$$

$f(\delta)$ is a continuous, strictly increasing function of δ, $\lim_{\delta \to +\infty} f(\delta) = +\infty$, $f(0) = H(\lambda) \le 1$. $g(\delta)$ is a continuous, decreasing function of δ, $g(0) = 2^\lambda > 1$. This implies the statement of the lemma.

Definition 2. *Let $k \in N$, $\lambda \in (0,1]$, $q \in (1-\lambda, 1-\lambda/2]$.*
Then $\Delta_{sp}(q,k,\lambda)$ is the root of the equation (26)

Theorem 3. *Let $\delta > 0$, $\lambda \in (0,1]$, $\pi \in (0,1)$, $p = \min(\pi, 1-\pi)$, $k \in N$.*
Let $F_k(n,\delta,\pi)$ be a random formula from $SD(n,k,\delta n,\pi)$.
Then

$$\delta > \Delta_{sp}(1-\lambda(1-p),k,\lambda) \Rightarrow \lim_{n\to\infty} P(minsat(F_k(n,\delta,\pi)) \le \lambda n) = o(1) \ as \ n \to \infty. \qquad (30)$$

Proof. Take arbitrary $k \in N$, $\lambda \in (0,1]$, $m = \lambda n$. A partial assignment $\sigma \in A(n,m)$ satisfies a formula F if and only if the complementary assignment $\bar{\sigma}$ satisfies the complementary formula $\bar{F}$. So, similarly to the Theorem 2 it is sufficient to prove for $\pi \in (0,1/2]$, $p = \pi$.
The total number of partial assignment of size m is $|A(n,m)| = 2^m \binom{n}{m}$,
Let X_m denote the number of partial assignments of size m that satisfy $F_k(n,\delta,\pi)$.

$$E[X_m] = \sum_{\sigma \in A(n,m)} P(\sigma \in S(F_k(n,\delta,\pi))). \qquad (31)$$

Again due to the symmetry of the distribution, the probability that σ satisfies a random formula depends only on $pl(\sigma)$, so

$$E[X_m] = \sum_{z=0}^{m} \sum_{\substack{\sigma \in A(n,m) \\ pl(\sigma)=z}} P(\sigma \in S(F_k(n,\delta,\pi)) | pl(\sigma) = z). \qquad (32)$$

Now let's compute the probability that a partial assignment with a fixed number of positive literals satisfies a random formula. Let C and l denote random formulas from $SD(n, k, 1, \pi)$ and $SD(n, 1, 1, \pi)$ respectively (C and l can be viewed as a random clause and a random literal from $F_k(n, \delta, \pi)$). Then, since clauses in $F_k(n, \delta, \pi)$ are independent,

$$P(\sigma \in S(F_k(n, \delta, \pi))|pl(\sigma) = z) = (1 - P(\sigma \notin S(C)|pl(\sigma) = z))^{\delta n}. \quad (33)$$

Let $var(C)$ and $var(\sigma)$ denote the set of variables in C and σ respectively, $overlap(C, \sigma) = |var(C) \cap var(\sigma)|$ - the number of variables shared by C and σ. $|var(C)| \leq k$, since a clause can contain repeated or contradictory literals the distribution SD, so $overlap(C, \sigma) \leq k$.

$$P(\sigma \notin S(C)|pl(\sigma) = z) =$$

$$= \sum_{j=0}^{k} P(\sigma \notin S(C)|overlap(C, \sigma) = j \ \& \ pl(\sigma) = z)P(overlap(C, \sigma) = j). \quad (34)$$

Literals in a random clause are independent, so

$$P(overlap(C, \sigma) = j) = \binom{k}{j} \left(\frac{m}{n}\right)^j \left(\frac{n - m}{n}\right)^{k-j}, \quad (35)$$

$$P(\sigma \notin S(C)|overlap(C, \sigma) = j \ \& \ pl(\sigma) = z) = P(\sigma \notin S(l)|pl(\sigma) = z \ \& \ var(l) \in var(\sigma))^j. \quad (36)$$

Now we can apply (15),

$$P(\sigma \notin S(l)|pl(\sigma) = z \ \& \ var(l) \in var(\sigma)) = p + \frac{z}{m}(1 - 2p). \quad (37)$$

Plugging this into (34), we get

$$P(\sigma \notin S(C)|pl(\sigma) = z) = \sum_{j=0}^{k} \left(p + \frac{z}{m}(1 - 2p)\right)^j \binom{k}{j} \left(\frac{m}{n}\right)^j \left(\frac{n - m}{n}\right)^{k-j} =$$

$$= \left(\frac{n - m}{n} + \frac{m}{n} \cdot \left(p + \frac{z}{m}(1 - 2p)\right)\right)^k. \quad (38)$$

The number of partial assignments of size m containing z positive literals is $\binom{n}{m}\binom{m}{z}$, so we can return to (32),

$$E[X_m] = \sum_{z=0}^{m} \sum_{\substack{\sigma \in A(n,m) \\ pl(\sigma)=z}} \left(1 - \left(1 - \frac{m}{n} \cdot \left(1 - p - \frac{z}{m}(1 - 2p)\right)\right)^k\right)^{\delta n} = \quad (39)$$

$$= \binom{n}{m} \sum_{z=0}^{m} \binom{m}{z} \left(1 - \left(1 - \frac{m}{n} \cdot (1 - p) + \frac{m}{n} \cdot \frac{z}{m}(1 - 2p)\right)^k\right)^{\delta n} = \quad (40)$$

$$= \binom{n}{m} \sum_{z=0}^{m} \binom{m}{z} \left(1 - \left(1 - \frac{m}{n}(1-p)\right)^k \left[1 + \frac{\frac{z}{n}(1-2p)}{1 - \frac{m}{n} \cdot (1-p)}\right]^k\right)^{\delta n}. \quad (41)$$

Now we can apply $(1+x)^\alpha \geq 1 + \alpha x$, $\alpha > 1$, $x > 0$ and get

$$E[X_m] \leq \binom{n}{m} \sum_{z=0}^{m} \binom{m}{z} \left(1 - \left(1 - \frac{m}{n}(1-p)\right)^k \left[1 + k \cdot \frac{\frac{z}{n}(1-2p)}{1 - \frac{m}{n} \cdot (1-p)}\right]\right)^{\delta n}. \quad (42)$$

Recall that $m = \lambda n$, so

$$E[X_m] \leq \left(1 - (1 - \lambda(1-p))^k\right)^{\delta n} \binom{n}{\lambda n} \times$$

$$\times \sum_{z=0}^{\lambda n} \binom{\lambda n}{z} \left(1 - \frac{(1 - \lambda(1-p))^k}{1 - (1 - \lambda(1-p))^k} \cdot k \cdot \frac{z}{n} \cdot \frac{\lambda(1-2p)}{1 - \lambda(1-p)}\right)^{\delta n}. \quad (43)$$

Since for $p \in (0, 1/2]$, $k \in N$

$$\frac{(1 - \lambda(1-p))^k}{1 - (1 - \lambda(1-p))^k} \cdot k \cdot \frac{\lambda(1-2p)}{1 - \lambda(1-p)} < 1, \quad (44)$$

we can apply $(1-\alpha)^\beta \leq e^{-\alpha\beta}, \alpha \in [0,1], \beta > 0$ and get

$$E[X_m] \leq \left(1 - (1 - \lambda(1-p))^k\right)^{\delta n} \binom{n}{\lambda n} \times$$

$$\times \sum_{z=0}^{\lambda n} \binom{\lambda n}{z} \exp\left(-\frac{k\delta (1 - \lambda(1-p))^k}{1 - (1 - \lambda(1-p))^k} \cdot \frac{\lambda(1-2p)}{1 - \lambda \cdot (1-p)} \cdot z\right) \leq \quad (45)$$

Using the standard bound

$$\binom{n}{\lambda n} \leq H(\lambda)^{-n}, \quad (46)$$

we obtain

$$E[X_m] \leq H(\lambda)^{-n} \left(1 - (1 - \lambda(1-p))^k\right)^{\delta n} \times$$

$$\times \left(1 + \exp\left(-\frac{k\delta (1 - \lambda(1-p))^k}{1 - (1 - \lambda(1-p))^k} \cdot \frac{\lambda(1-2p)}{1 - \lambda \cdot (1-p)}\right)\right)^{\lambda n} \quad (47)$$

We take $p \in (0, 1/2]$, so $1 - \lambda(1-p) \in (1 - \lambda, 1 - \lambda/2]$ and we can apply the Lemma 2

Table 2. Approximate values of $\Delta_{sp}(1 - \lambda(1 - p), 3, \lambda)$

λ	0.1	0.2	0.3	0.4	0.5	0.6	0.7	0.8	0.9	1.0
p=0.2	0.256	0.670	1.266	2.108	3.314	5.088	7.789	12.000	18.263	22.385
p=0.3	0.238	0.605	1.110	1.783	2.675	3.855	5.413	7.419	9.705	10.204
p=0.4	0.220	0.545	0.976	1.526	2.216	3.068	4.095	5.269	6.406	6.346
p=0.5	0.203	0.489	0.860	1.325	1.897	2.592	3.414	4.335	5.215	5.191

$$\delta > \Delta_{sp}(1 - \lambda(1 - p), k, \lambda) \Rightarrow$$

$$\Rightarrow \frac{H(\lambda)}{\left(1 - (1 - \lambda(1 - p))^k\right)^{\delta}} > \left(1 + \exp\left(-\frac{k\delta\left(1 - \lambda(1 - p)\right)^k}{1 - (1 - \lambda(1 - p))^k} \cdot \frac{\lambda(1 - 2p)}{1 - \lambda \cdot (1 - p)}\right)\right)^{\lambda} \Rightarrow$$

$$\Rightarrow P(minsat(F_k(n, \delta, \pi)) \leq \lambda n) = P(X_m \leq 1) \leq E[X_m] = o(1). \tag{48}$$

This proves the theorem. □

For $\lambda = 1$ we get an upper bound on the satisfiability threshold for a skewed random k-SAT formula (see Theorem 2).

For $p = 1/2$ we get a lower bound on the size of satisfying partial assignments for a plain random k-SAT formula.

$$\forall \lambda \in (0, 1] \; \delta > \frac{\ln H(\lambda) - \lambda \ln 2}{\ln\left(1 - \left(1 - \frac{\lambda}{2}\right)^k\right)} \Rightarrow$$

$$\Rightarrow P(minsat(F_k(n, \delta)) \leq \lambda n) = o(1) \text{ as } n \to \infty. \tag{49}$$

Table 2 provides approximate values of $\Delta_{sp}(1 - \lambda(1 - p), k, \lambda)$ obtained numerically from (26) for $k = 3$.

6 Conclusion

In this paper we considered a skewed random k-SAT distribution and investigated the phase transition behavior in this model. Empirical evidence for the satisfiability phase transition was presented. Further experiments suggest that even for a highly skewed random k-SAT distribution the phase transition becomes sharp as the number of variables increases.

We proved an upper bound on the satisfiability threshold and a lower bound on the number of literals in satisfying partial assignments for a skewed random k-SAT formula. For the considered skewed distribution there is still a large gap between the observed threshold location and the proved upper bound, so better bounds are still to be obtained. Lower bounds on the threshold and upper bounds on the minimum number of literals in a satisfying partial assignment of a skewed random k-SAT formula are needed to complete the picture.

Another interesting direction is to evaluate the computational hardness of skewed random k-SAT formulas with respect to the skewness of the distribution for a fixed clause density. The possible candidates for the maximum hardness are the non-skewed distribution and the skewed distribution that undergoes the satisfiability phase transition at this clause density.

Acknowledgments

The author would like to thank Edward Hirsch for bringing this problem to his attention and for valuable comments.

References

1. D. Achlioptas and Y. Peres. The threshold for random k-SAT is $2^k ln2 - O(k)$. Submitted for publication.
2. P. Cheeseman, B. Kanefsky, W. Taylor. Where the really hard problems are. *12th International Joint Conference on Artificial Intelligence (IJCAI-91)*, volume 1, pages 331-337. Morgan Kaufman, 1991.
3. V. Chvátal and B. Reed. Mick gets some (the odds are on his side). *33th Annual Symposium on Foundation of Computer Science (Pittsburg, PA, 1992)*, pages 620-627. IEEE Comput. Soc. Press, Los Alamitos, CA, 1992.
4. W. Fernandez de la Vega. On random 2-SAT. Manuscript, 1992.
5. J. Franco, M. Paull. Probabilistic analysis of the Davis-Putnam procedure for solving satisfiability. *Discrete Applied Mathematics*, 5, pages 77-87, 1983.
6. J. Franco, R. Swaminathan. Average case results for satisfiability algorithms under the random clause width model. *Annals of Mathematics and Artificial Intelligence* 20(1-4), pages 357-391, 1997.
7. E. Friedgut. Necessary and sufficient conditions for sharp thresholds of graph properties, and the k-SAT problems. *J. Amer. Math. Soc.*, 12, pages 1017-1054, 1999.
8. A. Goerdt. A threshold for unsatisfiability. *J. Comput. System Sci.*, 53(3), pages 469-486, 1996.
9. A. Goldberg. Average case complexity of the satisfiability problem. *Proceedings of 4th Workshop on Automated Deduction*, 1979.
10. E. A. Hirsch. A Fast Deterministic Algorithm for Formulas That Have Many Satisfying Assignments. *Logic Journal of the IGPL*, Vol.6, No.1, Oxford University Press, pages 59-71, 1998.
11. A. Kamath, R. Motwani, K. Palem, P. Spirakis. Tail bounds for occupancy and the satisfiability threshold conjecture. *Random structures and algorithms* 7(1), pages 59-80, 1995.
12. H. Kautz and B. Selman. Planning as satisfiability. *In Proceedings of ECAI'92*, volume 2, pages 1194-1201. pages 359-363. John Wiley & Sons, 1996.
13. H. Kautz and B. Selman. Pushing the envelope: planning, propositional logic and stochastic search. *In Proceedings of AAAI'96*, volume 2, pages 1194-1201. MIT Press, 1996.
14. S. Kirkpatrick, B. Selman. Critical behavior in the satisfiability of random boolean expressions. *Science* 264, pages 1297-1301, 1994.
15. Elias Koutsoupias and Christos H. Papadimitriou. On the greedy algorithm for satisfiability. IPL, 43(1):53–55, 1992.

16. T. Larrabee, Y. Tsuji. Evidence for satisfiability threshold for random 3CNF formulas. *Proceedings of the AAAI Symposium on Artificial Intelligence and NP-hard problems*, 112. 1993.

17. D. Mitchell, B. Selman, H. Levesque. Hard and easy distributions of SAT problems. *Proceedings of 10th National Conference on Artificial Intelligence*, pages 459-465. AAAI Press, Menlo Park, CA, 1992.

18. J. Slaney, M. Fujita, M. Stickel. Automated reasoning and exhaustive search: quasigroup existence problems. *Computers and Mathematics with Applications*, 29, pages 115-132.

19. H. Zhang, M. Bonacina, J. Hsiang. PSATO: a distributed propositional prover and its application to quasigroup problems. *Journal of Symbolic Computation*, 21(4), pages 543-560.

NiVER: Non-increasing Variable Elimination Resolution for Preprocessing SAT Instances[*]

Sathiamoorthy Subbarayan[1] and Dhiraj K. Pradhan[2]

[1] Department of Innovation,
IT-University of Copenhagen, Copenhagen, Denmark
sathi@itu.dk
[2] Department of Computer Science,
University of Bristol, Bristol, UK
pradhan@cs.bris.ac.uk

Abstract. The original algorithm for the SAT problem, Variable Elimination Resolution (VER/DP) has exponential space complexity. To tackle that, the backtracking-based DPLL procedure [2] is used in SAT solvers. We present a combination of two techniques: we use NiVER, a special case of VER, to eliminate some variables in a preprocessing step, and then solve the simplified problem using a DPLL SAT solver. NiVER is a strictly formula size not increasing resolution based preprocessor. In the experiments, NiVER resulted in up to 74% decrease in N (Number of variables), 58% decrease in K (Number of clauses) and 46% decrease in L (Literal count). In many real-life instances, we observed that most of the resolvents for several variables are tautologies. Such variables are removed by NiVER. Hence, despite its simplicity, NiVER does result in easier instances. In case NiVER removable variables are not present, due to very low overhead, the cost of NiVER is insignificant. Empirical results using the state-of-the-art SAT solvers show the usefulness of NiVER. Some instances cannot be solved without NiVER preprocessing. NiVER consistently performs well and hence, can be incorporated into all general purpose SAT solvers.

1 Introduction

The Variable Elimination Resolution (VER) [1] has serious problems due to exponential space complexity. So, modern SAT solvers are based on DPLL [2]. Preprocessors (simplifiers) can be used to simplify SAT instances. The simplified formula can then be solved by using a SAT Solver. Preprocessing is worthwhile only if the overall time taken for simplification as well as for solving the simplified formula is less than the time required to solve the unsimplified formula. This paper introduces NiVER (Non-increasing VER), a new preprocessor based on VER. NiVER is a limited version of VER, which resolves away a variable only if

---

[*] Research reported supported in part by EPSRC(UK). Most of this work was done when the first author was working at University of Bristol.

H.H. Hoos and D.G. Mitchell (Eds.): SAT 2004, LNCS 3542, pp. 276–291, 2005.

there will be no resulting increase in space. For several instances, NiVER results in reducing the overall runtime. In many cases, NiVER takes less than one second CPU time. Because, NiVER consistently performs well, like clause learning and decision heuristics, NiVER can also be integrated into the DPLL framework for general purpose SAT solvers.

The important contribution of this paper is the observation that most of the real-life SAT instances have several NiVER removable variables, which can be resolved away without increase in space. The structure of the real-life instances are such that for several variables, most of the resolvents are tautologies. Hence, there will be no increase in space due to VER on them. Empirical results show that NiVER not only decreases the time taken to solve an instance, but also decreases the amount of memory needed to solve it. Some of the instances cannot be solved without NiVER preprocessing, because of more memory requirement. Importantly, empirical results also show that the benefit of NiVER preprocessing increases with the increase in the size of the problem. As the size of the problem instances increases, it is useful to have a NiVER preprocessor in general SAT solvers. Another advantage is its simplicity: NiVER can be implemented with the conventional data structures used for clause representation in modern SAT solvers. In fact, in our implementation we have used the data structures from the zChaff SAT solver [6]. Hence, there will not be a significant overhead in implementation of NiVER with SAT solvers.

The next section presents a brief overview of previous SAT preprocessing techniques. Section 3 presents the NiVER preprocessor and gives some examples. Empirical results are presented in section 4. We conclude in section 5.

2   SAT Preprocessors

Simplifiers are used to change the formula into a simpler one, which is easier to solve. Pure literal elimination and unit propagation are the two best known simplification methods used in most of the DPLL based SAT solvers. Although several preprocessors have been published [3],[4], the current state-of-the-art SAT solvers [6],[5], just use these two simplifications. The 2-simplify preprocessor by Brafman [4], applies unit clause resolution, equivalent variable substitution, and a limited form of hyper-resolution. It also generates new implications using binary clause resolution. The recent preprocessor, HyPre [3] applies all the rules in 2-simplify and also does hyper-binary resolution. For some problems HyPre preprocessor itself solves the problem. But for other instances, it takes a lot of time to preprocess, while the original problem is easily solvable by SAT solvers.

VER has already been used as a simplifier, but to a lesser extent. In [7], variables with two occurences are resolved away. For a class of random benchmarks, [7] has empirically shown that the procedure, in average case, results in polynomial time solutions. In 2clsVER [8], VER was used. They resolved away a variable rather than splitting on it, if the VER results in less than a 200 increase in L (Number of literals). It was done inside a DPLL method, not as a preprocessor. But that method was not successful when compared to the state-of-the-art DPLL algorithms. A variant of the NiVER method, which does not allow an increase in K, was used in [14] to obtain the current best worst-case upper bounds. The method in [14] was used not just as a preprocessor, but, at each node of a DPLL search. However, no implementation was found.

Algorithm 1. NiVER CNF Preprocessor

```
 1: NiVER(F)
 2: repeat
 3:    entry = FALSE
 4:    for all V ∈ Var(F) do
 5:       P_C = {C | C ∈ F, l_V ∈ C }
 6:       N_C = {C | C ∈ F, l̄_V ∈ C }
 7:       R = { }
 8:       for all P ∈ P_C do
 9:          for all N ∈ N_C do
10:             R = R ∪ Resolve(P,N)
11:          end for
12:       end for
13:       Old_Num_Lits = Number of Literals in (P_C∪N_C)
14:       New_Num_Lits = Number of Literals in R
15:       if (Old_Num_Lits ≥ New_Num_Lits) then
16:          F=F-(P_C∪N_C), F=F+R, entry = TRUE
17:       end if
18:    end for
19: until ¬entry
20: return F
```

---

3 NiVER: Non-increasing VER

Like other simplifiers, NiVER takes a CNF as input and outputs another CNF, with a lesser or equal number of variables. The VER, the original algorithm for SAT solving, has exponential space complexity, while that of DPLL is linear. Both have exponential time complexity. In NiVER, as we do not allow space increasing resolutions, we have linear space complexity. The strategy we use is to simplify the SAT instance as much as possible using NiVER, a linear-space version of VER. Then, the resulting problem is solved using a DPLL-based SAT solver. NiVER does not consider the number of occurrences of variables in the formula. In some instances, NiVER removes variables having more than 25 occurrences. For each variable, NiVER checks whether it can be removed by VER, without increasing L. If so, it eliminates the variable by VER. The NiVER procedure is shown in Algorithm 1. When VER removes a variable, all resolvents of the variable have to be added. We discard trivial resolvents (tautologies). The rest of the resolvents are added to the formula. Then, all the clauses containing the variable are deleted from the formula. In many real-life instances (Figure 1 and Figure 2) we observed that for many variables, most of the resolvents are tautologies. So, there will be no increase in space when those variables are resolved away. Apart from checking for tautologies, NiVER does not do any complex steps like subsumption checking. No other simplification is done. Variables are checked in the sequence of their numbering in the original formula. There is not much difference due to different variable orderings. Some variable removals cause other variables to become removable. NiVER iterates until no more variable can be removed. In the present implementation, NiVER does not even check whether any unit clause is present. Rarely, when a variable is removed, we observed an increase in K, although, NiVER does not allow L to increase. Unlike HyPre or 2-simplify, NiVER does not do unit propagation, neither explicitly nor implicitly.

Clauses with literal l_{24}

$(\bar{l}_{23} + l_{24})$ $(\bar{l}_{22} + l_{24})$ $(l_{24} + \bar{l}_{31})$

$(l_2 + \bar{l}_{15} + l_{24})$ $(\bar{l}_2 + l_{15} + l_{24})$

Old_Num_Lits = 22

Clauses with literal $\bar{l}_{24}$

$(l_{22} + l_{23} + \bar{l}_{24} + l_2 + l_{15})$

$(l_{22} + l_{23} + \bar{l}_{24} + \bar{l}_2 + \bar{l}_{15})$

Number of Clauses deleted = 7

Added Resolvents

$(\bar{l}_{31} + l_{22} + l_{23} + l_2 + l_{15})$ $(\bar{l}_{31} + l_{22} + l_{23} + \bar{l}_2 + \bar{l}_{15})$

Eight other resolvents are tautologies

New_Num_Lits = 10

Number of Clauses added = 2

Fig. 1. NiVER Example 1: Elimination of Variable numbered *24* of *barrel8* instance from Bounded Model Checking

Figure 1 shows an example of variable elimination by NiVER, when applied to the *barrel8* instance from Bounded Model Checking [15]. In this example, variable *24* has 10 resolvents. Among them, eight are tautologies, which can be discarded. Only the two remaining non-trivial resolvents are added. The seven old clauses containing a literal of variable *24* are deleted. The variable elimination decreases N (number of variables) by one, K (number of clauses) by 12, and L (literal count) by five.

Figure 2 shows another example of variable elimination by NiVER, when applied to the *6pipe* instance from a microprocessor verification benchmark suite[16]. In this example, variable *44* has nine resolvents. Among them, five are tautologies, which can be discarded. Only the four remaining non-trivial resolvents are added. The six old clauses containing a literal of variable *44* are deleted. The variable elimination decreases N by one, K by two, and L by two.

Table 1 shows the effect of NiVER on a few instances from [9]. For the *fifo8_400* instance, NiVER resulted in a 74% decrease in N, a 58% decrease in K and a 46% decrease in L. The benefit of these reductions is shown in the results section. In many of the real-life instances, NiVER decreases N, K and L. This might be a reason for the usefulness of NiVER on those instances.

NiVER preserves the satisfiability of the original problem. If the simplified problem is unsatisfiable, then the original is also unsatisfiable. If the simplified problem is satisfiable, the assignment for the variables in the simplified formula

Clauses with literal l_{44}

$(l_{44} + l_{6315} + \bar{l}_{15605})$
$(l_{44} + l_{6192} + \bar{l}_{6315})$
$(l_{44} + \bar{l}_{3951} + \bar{l}_{11794})$

Clauses with literal $\bar{l}_{44}$

$(\bar{l}_{44} + l_{6315} + \bar{l}_{15605})$
$(\bar{l}_{44} + \bar{l}_{6192} + \bar{l}_{6315})$
$(\bar{l}_{44} + \bar{l}_{3951} + l_{11794})$

Old_Num_Lits = 18

Number of Clauses deleted = 6

Added Resolvents

$(l_{6315} + \bar{l}_{15605} + \bar{l}_{3951} + l_{11794})$
$(l_{6192} + \bar{l}_{6315} + \bar{l}_{3951} + l_{11794})$
$(\bar{l}_{3951} + \bar{l}_{11794} + l_{6315} + l_{15605})$
$(\bar{l}_{3951} + \bar{l}_{11794} + \bar{l}_{6192} + \bar{l}_{6315})$

Discarded Resolvents(Tautologies)

$(l_{6315} + \bar{l}_{15605} + l_{6315} + l_{15605})$
$(l_{6315} + \bar{l}_{15605} + \bar{l}_{6192} + \bar{l}_{6315})$
$(l_{6192} + \bar{l}_{6315} + l_{6315} + l_{15605})$
$(l_{6192} + \bar{l}_{6315} + \bar{l}_{6192} + \bar{l}_{6315})$
$(\bar{l}_{3951} + \bar{l}_{11794} + \bar{l}_{3951} + l_{11794})$

New_Num_Lits = 16

Number of Clauses added = 4

Fig. 2. NiVER Example 2: Elimination of Variable numbered _44_ of 6_pipe_ instance from Microprocessor Verification

Table 1. Effect of NiVER preprocessing. N-org, N-pre: N (Number of variables) in original and simplified formulas. %N↓ : The percentage of variables removed by NiVER. Corresponding information about clauses are listed in consecutive columns. %K↓ : The percentage decreases in K due to NiVER. %L↓ : The percentage decreases in L (Number of Literals) due to NiVER. The last column reports the CPU time taken by NiVER preprocessor in seconds. Some good entries are in bold. A machine with AthlonXP1900+ processor and 1GB memory was used in the experiments

Benchmark	N-org	N-pre	%N↓	K-org	K-pre	%K↓	L-org	L-pre	%L↓	Time
6pipe	15800	15067	5	394739	393239	0.4	1157225	1154868	0.2	0.5
f2clk_40	27568	10408	**62**	80439	44302	**45**	234655	157761	**32.8**	1.3
ip50	66131	34393	48	214786	148477	**31**	512828	398319	**22.3**	5.2
fifo8_400	259762	68790	**74**	707913	300842	**58**	1601865	858776	**46.4**	14.3
comb2	31933	20238	37	112462	89100	21	274030	230537	**15.9**	1
cache_10	227210	129786	**43**	879754	605614	**31**	2191576	1679937	**23.3**	20.1
longmult15	7807	3629	**54**	24351	16057	**34**	58557	45899	**21.6**	0.2
barrel9	8903	4124	**54**	36606	20973	**43**	102370	66244	**35.2**	0.4
ibm-rule20_k45	90542	46231	49	373125	281252	25	939748	832479	11.4	4.5
ibm-rule03_k80	88641	55997	37	375087	307728	18	971866	887363	8.7	3.6
w08_14	120367	69151	**43**	425316	323935	24	1038230	859105	**17.3**	5.45
abp1-1-k31	14809	8183	45	48483	34118	**30**	123522	97635	**21.0**	0.44
guidance-1-k56	98746	45111	**54**	307346	193087	37	757661	553250	**27.0**	2.74

is a subset of at least one of the satisfying assignments of the original problem. For variables removed by NiVER, the satisfying assignment can be obtained by a well-known polynomial procedure, in which the way NiVER proceeds is simply reversed. The variables are added back in the reverse order in which they were eliminated. While adding each variable, assignment is made to that variable such that the formula is satisfied.

For example, let F be the original formula. Let C_x refer to a set of clauses containing literals of variable x. Let C_{xr} represent the set of clauses obtained by resolving clauses in C_x on variable x. NiVER first eliminates variable a from F, by removing C_a from F and adding C_{ar} to F, resulting in the new formula F_a. Then NiVER eliminates variable b by deleting C_b from F_a and adding C_{br} to F_a, resulting in F_{ab}. Similarly, eliminating c results in F_{abc}. Now NiVER terminates and let a SAT solver finds a satisfying assignment, A_{abc}, for F_{abc}. A_{abc} will contain satisfying values for all variables in F_{abc}. Now, add variables in the reverse order they were deleted. First, add C_c to F_{abc}, resulting in F_{ab}. Assign to c either the value one or the value zero, such that F_{ab} is satisfied. At least one among those assignments will satisfy F_{ab}. Similarly, add C_b and find a value for b and then for a. During preprocessing, just the set of clauses, C_a, C_b and C_c, should be stored, so that a satisfying assignment can be obtained if the DPLL SAT solver finds a satisfying assignment for the simplified theory.

4   Experimental Results

This section contains two subsections. The subsection 4.1 presents the effect of the NiVER Preprocessor on two state-of-the-art SAT solvers: Berkmin and Siege. As the two SAT solvers have different decision strategies, the effect of NiVER on them can be studied. The other subsection presents the effect of NiVER on time and memory requirement for solving large instances from four families. The time listed for NiVER preprocessing in the tables are just the time taken for preprocessing the instance. At present, as NiVER is a separate tool, and, if the instance is *very large*, it might take few additional seconds to read the file and write back into the disk. But, as the data structures used in NiVER implementation are those used in modern SAT solvers to represent clauses, the time taken for reading and writing back can be avoided when integrated with the SAT solver.

4.1   Effect of NiVER on SAT-Solvers: BerkMin and Siege

The SAT benchmarks used in this subsection are from [9], [10] and [11]. Benchmarks used in [3] were mostly used. The NiVER software is available at [13]. Experiments were done with, Berkmin [5], a complete deterministic SAT solver and Siege(v_4) [12] , a complete randomized SAT Solver. Two SAT solvers have different decision strategies and hence the effect of NiVER on them can be studied. In Table 2 runtimes in CPU seconds for experiments using Berkmin are shown. In Table 3 corresponding runtimes using Siege are tabulated. All experiments using Siege were done with 100 as the random seed parameter. For every benchmark, four types of experiments were done with each solver. The first type is just using the solvers to solve the instance. The second one is using the NiVER preprocessor and solving the simplified theory by the SAT solvers. The third type of experiments involves two preprocessors. First the benchmark is simplified by NiVER and then by HyPre. The output of HyPre is then solved using the SAT solvers. A fourth type of experiment uses just HyPre simplifier

Table 2. Results with Berkmin (Ber) SAT solver. CPU Time (seconds) for four types of experiments, along with class type (Cls) for each benchmark. An underlined entry in the second column indicates that NiVER+Berkmin results in better runtime than just using the solver. NSpdUp column lists the speedup due to NiVER+Berkmin over Berkmin. A machine with AthlonXP1900+ processor and 1GB memory was used in the experiments

| BenchMark | Berkmin | Berkmin with | | | Cls | (UN)SAT | NSpdUP |
		NiVER	NiVER +HyPre	HyPre			
6pipe	**210**	222	392	395	I	UNSAT	0.95
6pipe_6_ooo	276	**253**	738	771	I	UNSAT	**1.09**
7pipe	**729**	734	1165	1295	I	UNSAT	0.99
9vliw_bp_mc	**90**	100	1010	1031	I	UNSAT	0.90
comb2	305	<u>**240**</u>	271	302	II	UNSAT	**1.27**
comb3	817	<u>407</u>	**337**	368	II	UNSAT	**2**
fifo8_ 300	16822	<u>13706</u>	**244**	440	II	UNSAT	**1.23**
fifo8_400	42345	<u>1290</u>	**667**	760	II	UNSAT	**32.82**
ip38	256	<u>99</u>	**52**	105	II	UNSAT	**2.59**
ip50	341	<u>313</u>	**87**	224	II	UNSAT	**1.09**
barrel9	106	<u>39</u>	**34**	114	II	UNSAT	**2.71**
barrel8	368	<u>34</u>	**10**	38	II	UNSAT	**10.82**
ibm-rule20_k30	475	554	**116**	305	II	UNSAT	0.86
ibm-rule20_k35	1064	1527	**310**	478	II	UNSAT	0.70
ibm-rule20_k45	5806	8423	**757**	1611	II	SAT	0.69
ibm-rule03_k70	21470	<u>9438</u>	**399**	637	II	SAT	**2.28**
ibm-rule03_k75	30674	<u>29986</u>	**898**	936	II	SAT	**1.02**
ibm-rule03_k80	31206	58893	1833	**1343**	II	SAT	0.53
abp1-1-k31	1546	3282	1066	**766**	IV	UNSAT	0.47
abp4-1-k31	1640	<u>949</u>	1056	**610**	IV	UNSAT	**1.72**
avg-checker-5-34	1361	<u>1099</u>	**595**	919	II	UNSAT	**1.24**
guidance-1-k56	90755	<u>17736</u>	**14970**	22210	III	UNSAT	**5.17**
w08_14	3657	4379	**1381**	1931	III	SAT	0.84
ooo.tag14.ucl	18	<u>**8**</u>	399	1703	III	UNSAT	**2.25**
cache.inv14.ucl	36	<u>**7**</u>	396	2502	III	UNSAT	**5.14**
cache_05	3430	<u>**1390**</u>	2845	3529	III	SAT	**2.47**
cache_10	22504	55290	**12449**	15212	III	SAT	0.41
f2clk_30	100	<u>61</u>	**29**	53	IV	UNSAT	**1.64**
f2clk_40	2014	<u>1848</u>	1506	**737**	IV	UNSAT	**1.09**
longmult15	183	<u>160</u>	128	**54**	IV	UNSAT	**1.14**
longmult12	283	<u>233</u>	180	**39**	IV	UNSAT	**1.21**
cnt10	4170	<u>2799</u>	193	**134**	IV	SAT	**1.49**

and the SAT-solvers. When preprocessor(s) are used, the reported runtimes are the overall time taken to find satisfiability.

Based on the experimental results in these two tables of this subsection, we classify the SAT instances into four classes. Class-I: Instances for which pre-

Table 3. Results with Siege (Sie) SAT solver. A machine with AthlonXP1900+ processor and 1GB memory was used in the experiments

| BenchMark | Siege | Siege with | | | Cls | (UN)SAT | NSpdUP |
		NiVER	NiVER +HyPre	HyPre			
6 pipe	79	**70**	360	361	I	UNSAT	**1.13**
6pipe_6_ooo	187	**156**	743	800	I	UNSAT	**1.20**
7pipe	185	**177**	1095	1183	I	UNSAT	**1.05**
9vliw_bp_mc	52	**46**	975	1014	I	UNSAT	**1.14**
comb2	407	266	**257**	287	II	UNSAT	**1.53**
comb3	550	419	396	**366**	II	UNSAT	**1.31**
fifo8_300	519	310	**229**	281	II	UNSAT	**1.68**
fifo8_400	882	657	**404**	920	II	UNSAT	**1.34**
ip38	146	117	**85**	115	II	UNSAT	**1.25**
ip50	405	258	**131**	234	II	UNSAT	**1.57**
barrel9	59	**12**	16	54	II	UNSAT	**4.92**
barrel8	173	25	**6**	16	II	UNSAT	**6.92**
ibm-rule20_k30	216	131	**112**	315	II	UNSAT	**1.65**
ibm-rule20_k35	294	352	**267**	482	II	UNSAT	**0.84**
ibm-rule20_k45	1537	1422	1308	**827**	II	SAT	**1.08**
ibm-rule03_k70	369	360	**223**	516	II	SAT	**1.03**
ibm-rule03_k75	757	**492**	502	533	II	SAT	**1.54**
ibm-rule03_k80	946	781	**653**	883	II	SAT	**1.21**
abp1-1-k31	559	471	**281**	429	II	UNSAT	**1.19**
abp4-1-k31	455	489	**303**	346	II	UNSAT	**0.93**
avg-checker-5-34	619	621	**548**	690	II	UNSAT	1
guidance-1-k56	9972	8678	**6887**	20478	II	UNSAT	**1.15**
w08_14	1251	**901**	1365	1931	III	SAT	**1.39**
ooo.tag14.ucl	15	**6**	396	1703	III	UNSAT	**2.5**
cache.inv14.ucl	39	**13**	396	2503	III	UNSAT	**3**
cache_05	238	**124**	2805	3540	III	SAT	**1.92**
cache_10	1373	**669**	10130	13053	III	SAT	**2.05**
f2clk_30	70	48	53	**41**	IV	UNSAT	**1.46**
f2clk_40	891	988	802	**519**	IV	UNSAT	**0.90**
longmult15	325	198	169	**54**	IV	UNSAT	**1.64**
longmult12	471	256	292	**72**	IV	UNSAT	**1.84**
cnt10	236	139	193	**134**	IV	SAT	**1.70**

processing results in no significant improvement. Class-II: Instances for which NiVER+HyPre preprocessing results in best runtimes. Class-III: Instances for which NiVER preprocessing results in best runtimes. Class-IV: Instances for which HyPre preprocessing results in best runtimes. The sixth column in the tables lists the class to which each problem belongs. When using SAT solvers to solve problems from a particular domain, samples from the domain can be used to classify them into one of the four classes. After classification, the corresponding type of framework can be used to get better run times. In case of Class-I problems, NiVER results are almost same as the pure SAT solver results. But HyPre takes a lot of time for preprocessing some of the Class-I problems like *pipe* instances. There are several Class-I problems not listed in tables here, for which neither NiVER nor HyPre results in any simplification, and hence no significant overhead. In case of Class-II problems, NiVER removes many variables and results in a simplified theory F_N. HyPre further simplifies F_N and results in F_{N+H} which is easier for SAT solvers. When HyPre is alone used for Class-II problems, they simplify well, but the simplification process takes more time than for simplifying corresponding F_N. NiVER removes many variables and results in F_N. But the cost of reducing the same variables by comparatively complex procedures in HyPre is very high. Hence, for Class-II, with few exceptions, HyPre+Solver column values are more than the values in NiVER+HyPre+Solver column. For Class-III problems, HyPre takes a lot of time to preprocess instances, which increases the total time taken to solve by many magnitudes than the normal solving time. In case of *cache.inv14.ucl* [11], NiVER+Siege takes 13 seconds to solve, while HyPre+Siege takes 2503 seconds. The performance of HyPre is similar to that on other benchmarks generated by an infinite state systems verification tool [11]. Those benchmarks are trivial for DPLL SAT Solvers. The Class-IV problems are very special cases in which HyPre outperform others. When NiVER is applied to these problems, it destroys the structure of binary clauses in the formula. HyPre which relies on hyper binary resolution does not perform well on the formula simplified by NiVER. In case of *longmult15* and *cnt10*, the HyPre preprocessor itself solves the problem. When just the first two types of experiments are considered, NiVER performs better in almost all of the instances.

4.2   Effect of NiVER on Time and Memory Usage by Siege SAT Solver

In this subsection the effect of the NiVER preprocessor on four families of large SAT instances are presented. As the results in the previous subsection show, Siege SAT solver is better than Berkmin. As we are primarily interested in studying the effect of NiVER on memory and time requirements of SAT solvers, all the results in this section are done using the Siege SAT solver, alone. Again, the random seed parameter for the Siege SAT solver was fixed at 100.

In the tables in this subsection: *NiVER Time (sec)* refers to the CPU-time in seconds taken by NiVER to preprocess the benchmark. *Time (sec)-Siege* refers to the CPU-time in seconds taken by Siege to solve the original benchmark. *Time (sec)-NiVER+Siege* refers to the sum of the time taken for NiVER preprocessing and the time taken for solving the NiVER-preprocessed instance by Siege. *Time (sec)-HyPre+Siege* refers to the sum of the time taken for HyPre preprocessing and the time taken for solving the HyPre-preprocessed instance by Siege. The other three columns list the amount of memory, in MegaBytes (MB), used by Siege SAT solver to solve the original and the corresponding preprocessed instances. *MA-S* means memory-abort by Siege SAT solver. *MA-H* means memory-abort by HyPre preprocessor. *SF* mentions segmentation fault by HyPre preprocessor.

The *x-x-barrel* family of instances are generated using the tools, *BMC* and *genbarrel*, available at [15]. The *BMC* tool, bounded model checker, takes as input: a model and a parameter, k, the bound for which the model should be verified. It then creates a SAT formula, whose unsatisfiability implies the verification of a property for the specified bound. The *genbarrel* tool takes an integer parameter and generates a model of the corresponding size. The *x-x-barrel* instances are generated using the command: *genbarrel x | bmc -dimacs -k x > x-x-barrel.cnf*

The results for the *x-x-barrel* family are listed in Table 4. Three types of experiments are done. One just using the plain Siege SAT solver. Second one using the NiVER preprocessor. Third one using the HyPre preprocessor. Corresponding time and memory usage are listed. Figure 3 shows the graph obtained when the time usage for *x-x-barrel* instances are plotted. Figure 4 shows the graph obtained when the memory usage for *x-x-barrel* instances are plotted. The HyPre preprocessor was not able to handle instances larger than the *10-10-barrel*, and aborted with a segmentation fault error message. Although the Siege SAT solver was able to solve all the original *x-x-barrel* instances, it used more time and space, than those for the corresponding NiVER simplified instances. As the Figure 3 shows, with the increase in the size of the instance, the benefit of NiVER preprocessing increases. For example, the speed-up due to NiVER in case of *8-8-barrel* instance is 6.1 (Table3), while it is 13.6 in case of *14-14-barrel* instance. Similar trend is also observed in the graph (Figure 4) for memory usage comparison. More than half of the variables in *x-x-barrel* instances are resolved away by NiVER preprocessor. In several other bounded model checking generated SAT instances, we observed similar amount of decrease in the number of variables due to NiVER preprocessing. For example, the *longmult15* instance and *w08_14* instance in Table 1 are both generated by bounded model checking. In both cases, approximately half of the variables are removed by NiVER. In all the *x-x-barrel* instances, the time taken for NiVER preprocessing is very small, and insignificant, when compared with the original solution time.

The other three families of large SAT instances: *xpipe_k*, *xpipe_q0_k*, and, *xpipe_xooo_q0_T0*, are all obtained from the Microprocessor Verification bench-

Table 4. Results for *x-x-barrel* family. Experiments were done on an Intel Xeon 2.8 GHz machine with 2 GB of memory. *SF*: Segmentation fault by HyPre preprocessor

Benchmark	N	%N↓	NiVER Time (sec)	Time (sec)			Memory (MB)		
				Siege	NiVER +Siege	HyPre +Siege	Siege	NiVER +Siege	HyPre +Siege
8-8-barrel	5106	56	0.2	86	14	**9**	59	26	**13**
9-9-barrel	8903	53	0.4	29	**7**	25	20	**8**	41
10-10-barrel	11982	54	1	45	**10**	65	31	**12**	63
11-11-barrel	15699	54	1	77	**12**	SF	52	**14**	SF
12-12-barrel	20114	55	2	580	**73**	SF	147	**50**	SF
13-13-barrel	25287	55	3	429	**42**	SF	120	**34**	SF
14-14-barrel	31278	55	5	1307	**96**	SF	208	**70**	SF

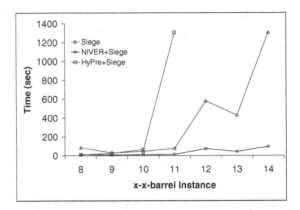

Fig. 3. Time comparison for the *x-x-barrel* family

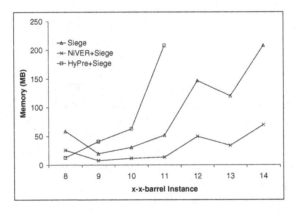

Fig. 4. Memory usage comparison for the *x-x-barrel* family

mark suite at [16]. The results obtained from experiments on *xpipe_k*, *xpipe_q0_k*, and, *xpipe_xooo_q0_T0* families are listed in Tables 5, 6, and 7, respectively. Figures for corresponding time and memory comparison follow them.

In the case of the *xpipe_k* family, without using the NiVER preprocessor, all the instances cannot be solved. Memory values listed in the tables just show the amount of memory used by the Siege SAT solver. Even when preprocessors are used, the memory values listed are just those used by Siege for solving the corresponding preprocessed instance. The amount of memory used by preprocessors is not listed. The aborts (*MA-H*) in the HyPre column of Table 5 are due to HyPre. In a machine with 2GB of memory, the HyPre preprocessor was not able to handle instances larger than *10pipe_k*. Even for the instances smaller than *10pipe_k*, HyPre took a lot of time. In case of *9pipe_k*, HyPre took 47 times the time taken by Siege. Siege SAT solver was not able to handle the *14pipe_k*

Table 5. Results for the *xpipe_k* family. Experiments were done on an Intel Xeon 2.8 GHz machine with 2 GB of memory. *MA-S*: Memory Abort by Siege SAT solver. *MA-H*: Memory Abort by HyPre preprocessor

Benchmark	N	%N↓	NiVER Time (sec)	Time (sec)			Memory (MB)		
				Siege	NiVER +Siege	HyPre +Siege	Siege	NiVER +Siege	HyPre +Siege
7pipe_k	23909	4	1	**64**	69	361	93	94	**61**
8pipe_k	35065	5	1	144	**127**	947	152	115	**87**
9pipe_k	49112	3	1	109	**106**	4709	**121**	123	**121**
10pipe_k	67300	5	2	565	**544**	5695	308	212	**196**
11pipe_k	89315	5	2	1183	**915**	MA-H	443	**296**	MA-H
12pipe_k	115915	5	3	3325	**2170**	MA-H	670	**418**	MA-H
13pipe_k	147626	5	4	5276	**3639**	MA-H	842	**579**	MA-H
14pipe_k	184980	5	5	MA-S	**8559**	MA-H	MA-S	**730**	MA-H

Table 6. Results for the *xpipe_q0_k* family. Experiments were done on an Intel Xeon 2.8 GHz machine with 2 GB of memory. *MA-H*: Memory Abort by HyPre preprocessor

Benchmark	N	%N↓	NiVER Time (sec)	Time (sec)			Memory (MB)		
				Siege	NiVER +Siege	HyPre +Siege	Siege	NiVER +Siege	HyPre +Siege
8pipe_q0_k	39434	27	1	90	**68**	304	81	62	**56**
9pipe_q0_k	55996	28	2	71	**61**	1337	77	**66**	76
10pipe_q0_k	77639	29	2	295	**280**	1116	170	120	**117**
11pipe_q0_k	104244	31	3	520	**478**	1913	233	164	**154**
12pipe_q0_k	136800	32	4	1060	**873**	3368	351	227	**225**
13pipe_q0_k	176066	33	5	1656	**1472**	5747	481	**295**	315
14pipe_q0_k	222845	34	7	2797	**3751**	MA-H	616	**412**	MA-H
15pipe_q0_k	277976	35	10	5653	**4165**	MA-H	826	**494**	MA-H

Table 7. Results for the *xpipe_x_ooo_q0_T0* family. Experiments were done on an Intel Xeon 2.8 GHz machine with 2 GB of memory. In this table an entry *x* in the Benchmark column refers to an *xpipe_x_ooo_q0_T0* instance

Benchmark	N	%N↓	NiVER Time (sec)	Time (sec)			Memory (MB)		
				Siege	NiVER +Siege	HyPre +Siege	Siege	NiVER +Siege	HyPre +Siege
7	27846	24	1	97	**87**	229	60	60	**45**
8	41491	25	1	252	**226**	605	100	85	**77**
9	59024	25	1	415	**359**	1524	135	119	**113**
10	81932	27	2	**2123**	2391	3190	215	172	**171**
11	110150	28	2	9917	**8007**	8313	317	240	**234**
12	144721	29	3	56748	30392	**27071**	448	**330**	342

instance. It aborted due to insufficient memory. But after NiVER preprocessing the *14pipe_k* instance was solved by Siege using the same machine. While solving NiVER preprocessed instance, Siege consumed only one third (730MB) of the available 2GB. As the results in Table 5 and the corresponding time comparison graph (Figure 5) shows, the speed-up due to NiVER keeps increasing with the increase in size of the instance. It is also interesting to note that in case of *14pipe_k* instance, only 5% of the variables are removed by NiVER. But still it resulted in solving the instance by using just 730MB of memory. As shown in Figure 6, the memory usage ratio, between the original instance and the corresponding NiVER preprocessed instance, also increases with the increase in the instance size.

In the case of the *xpipe_q0_k* family, again HyPre was not able to handle all the instances. As in the other cases, the benefit of NiVER preprocessing increases with the increase in the size of the instances. This is shown by the graphs in

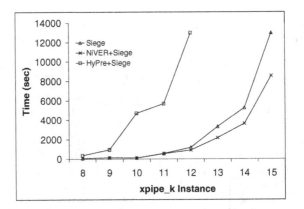

Fig. 5. Time comparison for the *xpipe_k* family

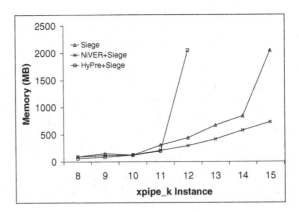

Fig. 6. Memory usage comparison for the *xpipe_k* family

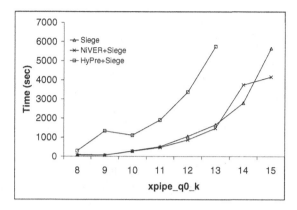

Fig. 7. Time comparison for the *xpipe_q0_k*family

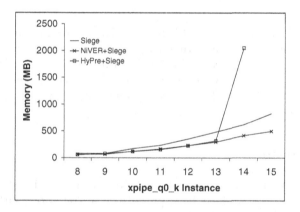

Fig. 8. Memory usage comparison for the *xpipe_q0_k* family

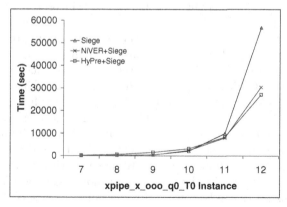

Fig. 9. Time comparison for the *xpipe_x_ooo_q0_T*0 family

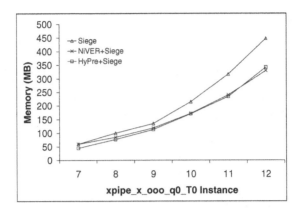

Fig. 10. Memory usage comparison for the *xpipe_x_ooo_q0_T0* family

Figures 7 and 8. For the *15pipe_q0_k* instance, NiVER resulted in decreasing the memory usage by 40%. Even in time usage, there is a significant improvement. In the case of the *xpipe_x_ooo_q0_T0* family (Table 7), both NiVER and HyPre preprocessors give similar improvement over the original instance. But still the advantages of NiVER over HyPre are its simplicity, and usefulness in a wide range of real-life instances.

5   Conclusion

We have shown that a special case of VER, NiVER, is an efficient simplifier. Although several simplifiers have been proposed, the state-of-the-art SAT-solvers do not use simplification steps other than unit propogation and pure literal elimination. We believe that efficient simplifiers will improve SAT-solvers. NiVER does the VER space efficiently by not allowing space increasing resolutions. Otherwise, the advantage of VER would be annulled by the associated space explosion. Empirical results have shown that NiVER results in improvement in most of the cases. NiVER+Berkmin outperforms Berkmin in 22 out of 32 cases (Table 2) and yields up to 33x speedup. In the other cases, mostly the difference is negligible. NiVER+Siege outperforms Siege in 29 out of 32 cases (Table 3) and gives up to 7x speedup. In the three other cases, the difference is negligible. Although, NiVER results in easier problems when some variables are removed by it, the poor performance of SAT solvers on few NiVER simplified instances is due to the decision heuristics. Experiments on four families of large SAT instances show that, the usefulness of NiVER increases with the increase in size of the problem size. NiVER also decreases the memory usage by SAT solvers. Due to that, more instances can be solved with the same machine configuration. The NiVER simplifier performs well, as most of the best runtimes in the experiments are obtained using it. Due to its usefulness and simplicity, like decision heuristics and clause learning, NiVER can also be incorporated into all general purpose DPLL SAT solvers.

Acknowledgements

Special thanks to Tom Morrisette, Lintao Zhang, Allen Van Gelder, Rune M Jensen and the anonymous reviewers for their comments on earlier versions of this paper.

References

1. M. Davis, H. Putnam. : A Computing procedure for quantification theory. J. of the ACM,**7** (1960)
2. M. Davis, et.al.,: A machine program for theorem proving. Comm. of ACM, **5(7)** (1962)
3. F. Bachhus, J. Winter. : Effective preprocessing with Hyper-Resolution and Equality Reduction, SAT 2003 341-355
4. R. I. Brafman : A simplifier for propositional formulas with many binary clauses, IJCAI 2001, 515-522.
5. E.Goldberg, Y.Novikov.: BerkMin: a Fast and Robust SAT-Solver, Proc. of DATE 2002, 142-149
6. M. Moskewicz, et.al.,: Chaff: Engineering an efficient SAT solver, Proc. of DAC 2001
7. J. Franco. : Elimination of infrequent variables improves average case performance of satisfiability algorithms. SIAM Journal on Computing **20** (1991) 1119-1127.
8. A. Van Gelder. : Combining preorder and postorder resolution in a satisfiability solver, In Kautz, H., and Selman, B., eds., Electronic Notes of SAT 2001, Elsevier.
9. H. Hoos, T. Stützle.: SATLIB: An Online Resource for Research on SAT. In: I.P.Gent, H.v.Maaren, T.Walsh, editors, SAT 2000, 283-292, www.satlib.org
10. IBM Formal Verification Benchmarks Library : http://www.haifa.il.ibm.com/projects/verification/RB_Homepage/bmcbenchmarks.html
11. UCLID : http://www-2.cs.cmu.edu/~uclid/
12. L. Ryan : Siege SAT Solver : http://www.cs.sfu.ca/~loryan/personal/
13. NiVER SAT Preprocessor : http://www.itu.dk/people/sathi/niver.html
14. E. D. Hirsch. : New Worst-Case Upper Bounds for SAT, J. of Automated Reasoning **24** (2000) 397-420
15. A. Biere: BMC, http://www-2.cs.cmu.edu/~modelcheck/bmc.html
16. M. N. Velev: Microprocessor Benchmarks, http://www.ece.cmu.edu/~mvelev/sat_benchmarks.html

Analysis of Search Based Algorithms for Satisfiability of Propositional and Quantified Boolean Formulas Arising from Circuit State Space Diameter Problems

Daijue Tang, Yinlei Yu, Darsh Ranjan, and Sharad Malik

Princeton University, NJ 08544, USA
{dtang, yyu, dranjan, malik}@Princeton.EDU

Abstract. The sequential circuit state space diameter problem is an important problem in sequential verification. Bounded model checking is complete if the state space diameter of the system is known. By unrolling the transition relation, the sequential circuit state space diameter problem can be formulated as either a series of Boolean satisfiability (SAT) problems or an evaluation for satisfiability of a Quantified Boolean Formula (QBF). Thus far neither the SAT based technique that uses sophisticated SAT solvers, nor QBF evaluations for the various QBF formulations for this have fared well in practice. The poor performance of the QBF evaluations is blamed on the relative immaturity of QBF solvers, with hope that ongoing research in QBF solvers could lead to practical success here.

Most existing QBF algorithms, such as those based on the DPLL SAT algorithm, are search based. We show that using search based QBF algorithms to calculate the state space diameter of sequential circuits with existing problem formulations is no better than using SAT to solve this problem. This result holds independent of the representation of the QBF formula. This result is important as it highlights the need to explore non-search based or hybrid of search and non-search based QBF algorithms for the sequential circuit state space diameter problem.

1   Introduction

In sequential verification, symbolic model checking is a powerful technique and has been used widely. Traditional symbolic model checking uses BDDs to represent logic functions and characteristic functions of sets. With the development of many efficient SAT solvers, bounded model checking (BMC) [1] has emerged as an alternative approach to perform model checking. Although BMC uses fast SAT solvers and may be able to quickly find counter examples, it is incomplete in the sense that it cannot determine when to stop the incremental unrolling of the transition relation. The maximum number of unrollings needed to complete BMC is the diameter of the corresponding sequential circuit state space. Therefore, determining the diameter is crucial for the completeness of BMC and thus has practical significance.

The sequential diameter can be computed as a byproduct of the sequential reachability analysis [2]. In this case, image computation is repeated from the set of initial states until the least fixed point is reached. The number of image computation steps

H.H. Hoos and D.G. Mitchell (Eds.): SAT 2004, LNCS 3542, pp. 292–305, 2005.

needed to reach a fixed point is the diameter of the sequential system. The state sets are enumerated implicitly and stored in the form of either BDDs or other representations of Boolean formulas. Images are calculated using either BDD operations or SAT evaluations or a combination of these two methods. Circuit diameter computation in [3] is purely SAT based. It does not calculate reachable states. The diameter is computed by solving a series of SAT problems derived from the unrolled next state functions. More detailed analysis of this method will be given in Section 3. In [4], search based SAT procedure is used in model checking algorithms. However, that work mainly concerns the method for preimage computation. It is not clear how to use preimage computation in the calculation of the circuit state space diameter.

The sequential circuit state space diameter problem can also be tackled by formulating it as a quantified Boolean formula (QBF) and using QBF solvers to solve them. A QBF is a Boolean formula with its variables quantified by either universal ($\forall$) or existential ($\exists$) quantifiers. The problem of deciding whether a quantified Boolean formula evaluates to true or false is also referred to as the QBF problem. Theoretically, QBF belong to the class of P-SPACE complete problems, widely considered harder than NP-complete problems like SAT. Many problems in AI planning [5] and sequential circuit verification [6] [1] can be formulated as QBF instances. In recent years, there has been an increasing interest within the verification community in exploring QBF based sequential verification as an alternative to Binary Decision Diagram (BDD) based techniques. Therefore, finding efficient QBF evaluation algorithms is gaining interest in sequential verification. Like SAT evaluation, QBF evaluation can be search based and does not suffer from the potential space explosion problem of BDDs. This makes it attractive to use QBF over BDD based algorithms since the problem of QBF evaluation is known to have polynomial space complexity. An obvious linear space algorithm to decide QBF assigns Boolean values to the variables and recursively evaluates the truth of the formula. The recursion depth is at most the number of variables. Since we need to store only one value of the variable at each level, the total space required is $O(n)$ where n is the number of variables. However, using state-of-the-art QBF solvers to solve the diameter problem lags behind other approaches. The immaturity of QBF evaluation techniques is often considered as the major reason for this. In this paper, we show that for the existing QBF formulations of the circuit diameter problem, search based QBF algorithms (this includes all DPLL based solvers) have no hope to outperform algorithms based on SAT. This result is important as it underscores the need to explore non-search based or possibly hybrids of search and non-search based techniques if we hope to do better using QBF.

2   The Sequential Circuit State Space Diameter Problem

Many problems in hardware verification concern verifying certain properties of logic circuits. To formulate such problems, one Boolean variable is introduced for each circuit node. A circuit node is either a primary input or a gate output. For a combinational circuit, the circuit consistency condition is expressed as the conjunction of the consistency conditions for each gate. This in turn is the set of consistent values at the inputs and output of this gate based on its function.

The behavior of a sequential circuit over a number of time frames can be modeled using the conventional time frame expansion approach, which creates a combinational circuit by unrolling the next state function of the sequential circuit. The sequential circuit state space diameter problem can be formulated as either a propositional formula or a QBF by unrolling the next state function. The shortest path from one state s_i of a sequential circuit to another state s_j is defined as the minimum number of steps to go from s_i to s_j in the corresponding state transition graph. Clearly, every state on a shortest path appears at most once, which means that a shortest path has no loop. A path with no loop is also called a simple path. The state space diameter of a sequential circuit is the longest shortest path from one of the starting states of this sequential circuit to any other reachable state.

3   SAT Formulation and Its Analysis

The Boolean satisfiability problem is the problem of deciding whether there exists an assignment to the Boolean variables that makes a given propositional formula true. In [3], the authors presented a purely SAT based method to compute the diameter of the state space of a sequential circuit.

They formulate the diameter problem by unrolling the next state function. Figure 1 shows the combinational circuit constructed at each step of the circuit state space diameter calculation. We have two expansions of the combinational logic, one for $n + 1$ time frames and the other for n time frames. I_i and $I'_i (i = 1, 2, \cdots)$ are sets of primary inputs, O_i and $O'_i (i = 1, 2, \cdots)$ are sets of primary outputs and S_i and $S'_i (i = 0, 1, \cdots)$ are sets of state variables. Let C_1 denote the set of variables of the $n + 1$ time frame expansion part and C_2 denote the set of variables of the n time frame expansion part. Let $F(C_1)$ and $F(C_2)$ be the Boolean functions representing the logic consistency conditions of C_1 and C_2 respectively. If $Init(S)$ is the characteristic function of the initial states, $T(I_i, S_i, S_{i+1})$ is the characteristic function of the state transition relation, then $F(C_1) = Init(S_0) \wedge \bigwedge_{i=0}^{n} T(I_i, S_i, S_{i+1})$ and $F(C_2) = Init(S'_0) \wedge \bigwedge_{i=0}^{n-1} T(I'_i, S'_i, S'_{i+1})$. Typically, n is incremented from zero until the diameter is reached. Let the state space diameter of the sequential circuit be d. If $n < d$, there must exist a simple path in the state space starting from S_0 with length $n + 1$. This condition can be checked by evaluating the following propositional formula:

$$F(C_1) \wedge (\bigwedge_{i \neq j} S_i \neq S_j) \tag{1}$$

If (1) is not satisfiable, then it must be true that $n \geq d$, which means the diameter is found. If (1) is satisfiable, DPLL based SAT solvers usually give satisfiable assignments to the Boolean variables. Since (1) is the conjunction of two terms: $F(C_1)$ and $\bigwedge_{i \neq j} S_i \neq S_j$, the satisfying assignment of (1) must satisfy every term. To satisfy every inequality of the second term, not every bit in the vectors S_i and S_j needs to be assigned. For two vectors S_i and S_j to be different, one bit difference of these two vectors is sufficient. Therefore, the satisfying assignment of (1) can be either a partial assignment, i.e. a cube, or a complete assignment, i.e. a minterm. The assigned values to the variables in S_{n+1} give us one or more states at the end of the simple paths with length $n + 1$. We use st to denote these states. However, (1) being satisfiable does not necessarily mean that $n < d$. We must check if these st are reachable in less than $n + 1$ time frames. This is checked by testing the satisfiability of the following propositional formula:

$$F(C_2) \wedge (\bigvee_{i=0 \cdots n} st = S'_i) \tag{2}$$

If (2) is unsatisfiable, we must have $n < d$. Then n is incremented and the above procedure is repeated starting by evaluating the satisfiability of (1). If (2) is satisfiable, we need to see if there is another simple path with a different end state. This is done by conjuncting (1) with the negations of the previously reached st and testing the satisfiability of it. When (1) becomes unsatisfiable, it means that all end states of simple paths whose length is $n + 1$ are reachable within n time frames. Thereby $n \geq d$. Otherwise, we continue to find a cube of new st and evaluate (2) again.

To simplify the above procedure, a new transition from the initial state to itself is added. Then the set of states reachable at a certain time frame must contain the set of states reachable at previous time frames. Therefore, (2) can be simplified as:

$$F(C_2) \wedge (st = S'_n) \tag{3}$$

Using SAT to solve the sequential circuit diameter problem, which is summarized above, has its drawbacks. For a fixed n, the algorithm basically enumerates cubes of end states of simple paths whose length is $n + 1$ by solving a series of SAT problems. However, the number of times to call the SAT procedure could be large, as the set of states S_{n+1}^{simple} that end in simple paths of length $n + 1$ from S_0 is enumerated in terms of its cubes.

Research efforts in using QBF for the diameter problem hope to overcome this cube enumeration of S_{n+1}^{simple}. However current implementation of QBF solvers have not been able to provide practically efficient solutions to the various QBF formulations for the diameter problem which is generally attributed to the relative immaturity of the QBF solvers. The next few sections provide greater insight into this through an analysis of the solution for the various QBF formulations.

4   QBF Formulations

A QBF is of the form $Q_1 x_1 \cdots Q_n x_n \; \varphi$, where $Q_i (i = 1 \cdots n)$ is either an existential quantifier $\exists$ or a universal quantifier $\forall$. φ is a propositional logic formula with $x_1 \cdots x_n$ as its variables. Adjacent variables quantified by the same quantifier in the prefix can be grouped together to form a quantification set. The order of the variables in the same quantification set can be exchanged without changing the QBF evaluation result (true or false). Variables in the outermost quantification set are said to have quantification level 1, and so on.

The propositional part φ in a QBF is usually expressed in the Conjunctive Normal Form (CNF). If φ of a QBF is in CNF, the innermost quantifier of this QBF is existential because the innermost universal quantifier can always be dropped by removing all the

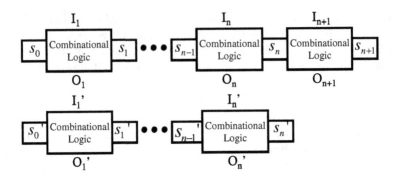

Fig. 1. Time frame expansions for state space diameter calculation

occurrences of the variables quantified by this universal quantifier in the CNF. The innermost quantifier can be a universal quantifier if φ is not in CNF. Converting φ to CNF typically introduces new variables which are quantified with existential quantifiers put inside the originally innermost quantifier. The number of quantification levels of a QBF may change if the representation of φ of this QBF changes. When φ is in CNF, the QBF having k levels of quantification is called kQBF. Most practical QBF instances are 2QBF or 3QBF. In the rest of the paper, when we talk about kQBF, k is the number of quantification levels when φ is in CNF.

In Figure 1, if for all possible input sequences $I_1 I_2 \cdots I_{n+1}$, the state S_{n+1} can be reached at one or more of the states of $S_i'(i = 0 \cdots n)$ for some input sequence $I_1' I_2' \cdots I_n'$, then $n + 1$ is greater than the state space diameter of the sequential circuit. A straightforward translation of the above sentence gives us the following 2QBF formulation of the circuit state space diameter problem:

$$\forall I_1 I_2 \cdots I_{n+1} \exists ((C_1 \setminus \bigcup_{i=1 \cdots n+1} I_i) \cup C2)\ F(C_1) \wedge F(C_2) \wedge (\bigvee_{i=0 \cdots n} (S_{n+1} = S_i')) \quad (4)$$

Let the state space diameter of the circuit be d. If $n < d$, (4) evaluates to false; if $n \geq d$, (4) evaluates to true.

A very similar formulation is:

$$\forall C_1 \exists C_2\ \neg F(C_1) \vee (F(C_2) \wedge (\bigvee_{i=0 \cdots n} (S_{n+1} = S_i'))) \quad (5)$$

Here all of the variables of C_1 are universally quantified. (5) evaluates to true iff for all possible assignments to variables in C_1, either the assignments are not consistent ($F(C_1)$ is false) or all possible states of S_{n+1} can be reached in less than $n + 1$ state transitions. Therefore, (5) evaluates to true iff $n \geq d$.

If the sequence of states $S_0 S_1 \cdots S_{n+1}$ has one or more loops in the state transition graph, then a sequence of states starting from S_0 and ending in S_{n+1} with less than $n + 1$ state transitions must exist because we can always take the path without looping. Therefore, (5) is equivalent to:

$$\forall C_1 \exists C_2\ \neg(F(C_1) \wedge (\bigwedge_{i \neq j} S_i \neq S_j)) \vee (F(C_2) \wedge (\bigvee_{i=0 \cdots n} S_{n+1} = S_i')) \quad (6)$$

The constraint $(\bigwedge_{i \neq j} S_i \neq S_j)$ in the propositional part of (6) is actually the condition to determine the recurrence diameter [1], adding this constraint does not change the QBF evaluation since the recurrence diameter is an upper bound of the state space diameter.

If an additional transition from the initial state to itself is added to the sequential transition of the circuit, as described in Section 3, (6) can be simplified to:

$$\forall C_1 \exists C_2 \neg (F(C_1) \wedge (\bigwedge_{i \neq j} S_i \neq S_j)) \vee (F(C_2) \wedge (S_{n+1} = S'_n)) \tag{7}$$

The outermost quantifiers of formula (4)-(7) are all universal quantifiers. We can take the dual of (4)-(7) to get QBFs with existential quantifiers as the outermost quantifiers. For example, the dual of (7) is:

$$\exists C_1 \forall C_2 \; F(C_1) \wedge (\bigwedge_{i \neq j} S_i \neq S_j) \wedge \neg (F(C_2) \wedge (S_{n+1} = S'_n)) \tag{8}$$

When $n < d$, (8) is true; when $n \geq d$, (8) is false.

The propositional part of (4)-(8) can be transformed to CNF by introducing new variables that are all added to the innermost existential quantification level. This makes (8) become a 3QBF while (4)-(7) are 2QBFs. However, the propositional parts of these formulas do not need to be CNF for search algorithms to solve them. In all cases, n is incrementally tested from 1 until $n = d$. d is the minimum number for n that makes (4)-(7) true and (8) false. The drawback of formulations (5)-(8) is that they have a large set of universal variables and when represented in CNF too many new variables need to be introduced. This greatly increase the search space that needs to be explored.

5   QBF Algorithms

5.1   Overview

Just as in SAT, the propositional part φ of a QBF is often in CNF. Many existing QBF solvers require φ to be CNF. Currently, most QBF algorithms are complete and can be roughly divided into two categories: resolution based and search based.

QBF algorithms based on resolution use Q-resolution to eliminate variables until there is no more variable to eliminate or an empty clause is generated [7]. Only an existential variable can be resolved out in a Q-resolution. A universal variable in a Q-resolution generated clause can be eliminated when there are no existential variable having higher quantification level than this universal variable. Like most resolution based decision methods, resolution based QBF algorithms have the potential memory blow up problem. Therefore they are seldom used in practice.

The majority of recent QBF solvers are search based. A search based algorithm tries to evaluate QBF by branching on variables to determine the value of φ at certain branches in the search tree. Note that we may not need to go all the way to the leaves of the search tree to determine the value of φ. A partial assignment to the variables may be enough for φ to be 0 or 1. Also we do not limit the search based algorithms to any particular search method like depth-first search or breadth-first search. Nor do we have any limitation on the ordering of the nodes in the search tree. The well-known Davis Logemann Loveland (DPLL) algorithm [8], which is a depth-first search procedure, is just one example of the search algorithms. Partly due to its success in SAT solvers, the DPLL algorithm has been adapted to many QBF evaluation procedures [9][10][11] [12][13] [14][15][16]. Although DPLL based QBF solvers do not blow up in space, they consume significant CPU time and are unable to handle practical sized problems as of now.

Plaisted *et al.* proposed an algorithm for evaluating QBF that belongs to neither of the above categories [17]. This algorithm iteratively eliminates a subset of variables in the innermost quantification level. This is done by partitioning the propositional formula using a set of cut variables and substituting one partition with a CNF of only the cut variables. The conflicting assignments of the cut variables are enumerated and the negations of the conflicting assignments are conjuncted to form the new CNF part. Unlike Q-resolution which can only eliminate one variable at a time, this algorithm can eliminate multiple variables simultaneously. However, enumerating conflicts may take exponential time in the number of cut variables. Therefore, using this method is very expensive for formulas without a small cut. In our experience, in practice, for many QBF instances, during the execution of this method, the variables are so much interleaved that it is impossible to find a small cut. From another point of view, the process of searching for conflicts is similar to the search in search based algorithms. Particularly, for a 2QBF instance, if the cut set is chosen to be the universal variables, then this algorithm is essentially a DPLL search algorithm. Moreover, since enumerating conflicts of cut variables achieves the effect of resolving multiple variables at one time, this algorithm still has the potential memory blow up problem.

5.2   The DPLL Algorithm

The DPLL algorithm is the most widely used search based algorithm for QBF as well as SAT evaluation. It only requires polynomial space during execution. The original DPLL algorithm is a recursive procedure for SAT and is not very efficient. Modern SAT solvers enhance the original DPLL algorithm with techniques like non-chronological backtracking and conflict-driven learning[18][19], which greatly accelerate the SAT solvers. Some of the most efficient SAT solvers today [20] [18][21] are based on the DPLL framework. Because SAT is a restricted form of QBF in the sense that it only has existential quantifiers, most existing QBF solvers incorporate variations of the DPLL procedure and many of the techniques that work well on SAT can also be used in QBF evaluation with some modifications. [9] is probably the first paper that extends the DPLL algorithm for SAT to QBF evaluation. It gives the basic rules for formula simplification like rules for monotone literals and unit propagation for existential variables. Conflict driven learning and non-chronological backtracking are adapted to later DPLL based QBF solvers [22] [16][15]. Also, the idea of satisfiability directed learning, which is a dual form of conflict driven learning and is specifically for QBF, is introduced and incorporated in these solvers. Deduction techniques such as inverting quantifiers [10] and partial implicit unfolding [11] were proposed and implemented by Rintanen. These deduction rules deduce forced assignments to existential variables by assigning truth values to universal variables having higher quantification levels.

Note that DPLL based QBF evaluation requires the branching order obey the quantification order, which corresponds to the semantics of the formula. Other decision orderings may require exponential memory to store the already searched space. They also make search hard to control and result in many fruitless searches. One exception is in 2QBF evaluation where no useless enumeration of the existential variables occurs since no existential variables precede universal variables in the prefix. Decision strategies with and without restriction to quantification order for 2QBF are described and compared in [23].

6   Analysis of Search Based QBF Algorithms

6.1   Handling Conflicts and Satisfying Assignments

Search based algorithms evaluate QBF by assigning Boolean values to variables. The propositional part φ of a QBF has three possible evaluations under a partial assignment: false, true and undetermined. If the value of φ is false, the partial assignment is called a conflict; if the value of φ is true, the partial assignment is called a satisfying partial assignment. In these two cases, the search procedure will backtrack and may do some learning. The search procedure will continue assigning unassigned variables if the value of φ is undetermined. Search based QBF algorithms often learn from the result of φ being true or false to prevent getting into the same conflicting or satisfying space again and again. Learning can be considered as choosing a subset of the current partial assignment such that this subset can still result in φ being true or false. These subsets of partial assignments are usually cached for future search space pruning.

Unlike in SAT, a satisfying assignment in QBF does not mean the end of the search. The search algorithms need to see if for all combinations of universal variables φ is satisfiable. The pruning of the satisfying space is usually done by constructing a partial assignment that is sufficient for φ to be true. For example, when φ is in CNF, this partial assignment can be constructed by choosing from every clause at least one of the value 1 literals. We call this partial assignment a *cover set* of the satisfying assignment [14]. The idea of using cover set for satisfying space pruning is incorporated in many QBF solvers. It is also called good learning in [22] and model caching in [16]. For a QBF instance of n variables, a cover set with m literals implies 2^{n-m} satisfying assignments. In fact, when a cover set is stored in the database for future pruning, existential variables belonging to the highest quantification level can be eliminated from the cover set due to the semantics of QBF. Thus the cover sets for a 2QBF instance consist only of universal variables. The conjunction of a set of literals is called a cube. A cover set is a cube.

When a QBF is derived from a circuit netlist and the value of this QBF denotes whether or not certain property of this circuit holds, a conflict in the QBF is either an

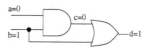

Fig. 2. Example for critical and non-critical signals

inconsistent assignment to gate inputs and outputs or a consistent assignment to the circuit nodes that does not satisfy the property. A satisfying assignment to the propositional part φ of the QBF is a consistent assignment to the circuit nodes that satisfy the property. For some circuit node variables, their values do not affect the satisfiability of the property. This is because some logic gate output does not depend on the values of all the gate inputs. In this case, some gate input is unobservable at the gate output. Any signal which fans out only to the transitive fanin cone of this input also becomes unobservable at this output. Suppose a circuit property is represented as a variable p, then p is the primary output of the property testing circuit C. Consistent assignments to other circuit nodes can either satisfy p or violate p, but some circuit nodes might become unobservable at p. For example, in Figure 2, d represents the circuit property. If $b = 1$, then $d = 1$ which makes signals a and c unobservable at d. Such unobservable signals are called *non-critical* signals. The set of *critical* signals S_c is both sufficient and necessary as the reason for the satisfiability of the property. Removing any signal from S_c results in p being undetermined. So caching the satisfiability result of p should include all the variables in S_c. Note that if φ is in CNF and is satisfied, S_c may not satisfy every clause of the CNF. Critical signals for a consistent assignment to circuit signals that unsatisfies the property are both sufficient and necessary for the result of p being violated. The selection of S_c does not require the knowledge of φ, the information of circuit structure is enough. S_c is generally much smaller than cover sets for CNF clauses. The above idea is very similar to the dynamic removal of inactive clauses in SAT proposed in [24].

6.2   QBF Evaluation with Satisfiability Driven Learning

We now demonstrate that using the search based QBF algorithms to solve the diameter problem with existing QBF formulations is no better than using SAT to solve this problem.

We first analyze the 2QBF formulation of the diameter problem which is shown in (4). We use $PROP_1$ to denote the propositional part of (4). $PROP_1$ consists of the conjunction of three terms: $F(C_1)$, $F(C_2)$ and $(\bigvee_{i=0\cdots n}(S_{n+1} = S_i'))$. Any partial or complete satisfying assignment to $PROP_1$ must make every term in this conjunction true. Therefore, the third term:

$$((S_{n+1} = S_0) \vee (S_{n+1} = S_1') \vee \cdots \vee (S_{n+1} = S_n')) \tag{9}$$

must be true if $PROP_1$ is satisfied. Note that the equality $S_{n+1} = S_i'(i = 0\cdots n)$ implies that the corresponding state bits in the two vectors S_{n+1} and S_i' are equal. Since (9) is a disjunction of $n + 1$ equalities and every equality in (9) has S_{n+1} on one side, (9) cannot be evaluated if the value of any state bit in S_{n+1} is unknown. This implies that every state variable in S_{n+1} is critical. In another words, any partial or complete assignment that is satisfying for $PROP_1$ must be a complete assignment for the state variables of S_{n+1}. Now consider the two possibilities. Case 1: (4) is false. In this case, QBF search algorithms need to find out at least one complete assignment to the universal variables for which $PROP_1$ is unsatisfiable. Case 2: (4) is true. In this case, QBF search algorithms need to prove that every universal variable assignment will make

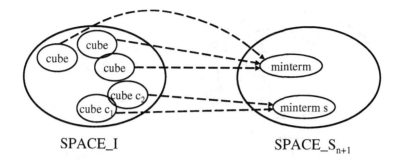

Fig. 3. Space mapping for formula (4)

$PROP_1$ satisfiable. Consider the Boolean space of the universal variables of (4), denoted as $SPACE\_I$, and the reachable state space at the $(n+1)^{th}$ time frame, denoted as $SPACE\_S_{n+1}$. $SPACE\_I$ and $SPACE\_S_{n+1}$ are shown respectively as the left circle and right circle in Figure 3. When using search based QBF algorithms to evaluate (4), we want to derive satisfying cubes, i.e. partial assignments, to cover as much of $SPACE\_I$ as possible every time a satisfying assignment of $PROP_1$ is found. The reason is that in this way, we can rule out more minterms, i.e. complete assignments, in $SPACE\_I$ when (4) is false and there is less space in $SPACE\_I$ for future searches when (4) is true. Note that any satisfying cube must contain all the critical signals. Thus any satisfying cube contains all the variables in S_{n+1} since all of them are critical. Also note that each satisfying cube that covers some part of $SPACE\_I$ corresponds to exactly one minterm, i.e. a complete assignment, in $SPACE\_S_{n+1}$. This is illustrated in Figure 3.

Moreover, we can further show that minterms in $SPACE\_S_{n+1}$ are covered by non-overlapping sets of cubes in $SPACE\_I$. This can be proved by contradiction. Suppose two distinct minterms m_1 and m_2 in $SPACE\_S_{n+1}$ correspond to sets of cubes sc_1 and sc_2 in $SPACE\_I$ respectively. If sc_1 and sc_2 overlap, then we randomly pick a minterm m_c that is contained in the intersection of sc_1 and sc_2. However, variables in m_c are primary inputs, they determine a unique minterm in $SPACE\_S_{n+1}$. Therefore, m_1 and m_2 should be the same. This is contradictory to the assumption that m_1 and m_2 are distinct.

If (4) is true, the entire $SPACE\_I$ needs to be covered by satisfying cubes. In this case, search procedures also need to cover the entire $SPACE\_S_{n+1}$ because every state at the $(n+1)^{th}$ time frame results from a primary input sequence with the length of $(n+1)$ time frames. Therefore the number of satisfying assignments of $PROP_1$ that needs to be searched is at least the number of minterms in $SPACE\_S_{n+1}$. This means when applying search based algorithms to the evaluation of (4), although we might get some pruning of $SPACE\_I$ through cubes, we cannot prune any part of $SPACE\_S_{n+1}$. In addition, when a search based QBF algorithm finds a new satisfying assignment of $PROP_1$, it does not necessarily mean that a new minterm in $SPACE\_S_{n+1}$ is found. In another words, it is possible that a single minterm in $SPACE\_S_{n+1}$ is enumerated again and again when we use search based QBF algorithms. For example, in Figure 3, two cubes c_1 and c_2 map to the same state s. However, a search procedure might need

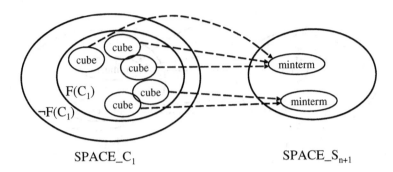

Fig. 4. Space mapping for formula (5)

to find two satisfying assignments of $PROP_1$ to derive c_1 and c_2 respectively. Note that the SAT based method described in Section 3 essentially uses a search based SAT solver to enumerate either cubes or minterms of the set of states at the $(n+1)^{th}$ time frames of simple paths. If the elements of this set are $SPACE\_S_{n+1}^{simple}$, then $SPACE\_S_{n+1}^{simple}$ is contained in $SPACE\_S_{n+1}$ because not every state in $SPACE\_S_{n+1}$ needs to end a simple path. Moreover, $SPACE\_S_{n+1}^{simple}$ may contain cubes and not just minterms. Therefore, using search based QBF solvers to solve (4) is no more efficient than the SAT method in Section 3.

The analysis of (5) is similar to the analysis of (4) although more complicated. We denote the propositional part of (5) as $PROP_2$. It is easy to see that $PROP_2$ equals to:

$$\neg F(C_1) \vee (F(C_1) \wedge F(C_2) \wedge (\bigvee_{i=0\cdots n} S_{n+1} = S_i')) \tag{10}$$

Note that (10) consists of two disjuncts, the second disjunct is precisely $PROP_1$. Also, since the conjunction of $\neg F(C_1)$ and $PROP_1$ is zero, there is no overlapping between satisfying assignments of $\neg F(C_1)$ and satisfying assignments of $PROP_1$. Thus the Boolean space of the universal variables in (5), denoted as $SPACE\_C_1$, can be divided into two parts: those that make $F(C_1)$ true and those that make $F(C_1)$ false. In Figure 4, the left circle is $SPACE\_C_1$, the inner circle of $SPACE\_C_1$ is the part that makes $F(C_1)$ true and the outer loop of $SPACE\_C_1$ is the part that makes $F(C_1)$ false. The set of assignments to $C1$ for which $PROP_1$ is satisfiable is contained in the inner circle of $SPACE\_C_1$. If (5) is false, then search procedures need to find at least one assignment to C_1 variables that makes $\neg F(C_1)$ false and $PROP_1$ unsatisfiable. This equals to finding a circuit consistent assignment to C_1 that makes $F(C_2) \wedge (\bigvee_{i=0\cdots n} S_{n+1} = S_i')$ unsatisfiable. If (5) is true, all possible assignments to C_1 make $PROP_2$ satisfiable. Thus search procedures need to cover the entire $SPACE\_C_1$. Since the variables in $SPACE\_I$ are all primary inputs and $SPACE\_C_1$ consists of all of the circuit signals including primary inputs, there is a one to one correspondence between minterms in $SPACE\_I$ and minterms in the inner circle of $SPACE\_C_1$. Using similar reasoning as in the previous analysis of (4), we can show that using search based QBF algorithms to cover the inner circle of $SPACE\_C_1$ is no more efficient than enumerating states in

$SPACE\_S_{n+1}$. Moreover, to prove (5) true, search algorithms also need to cover the part of $SPACE\_C_1$ that satisfies $\neg F(C_1)$, which makes applying search algorithms to solve (5) strictly less efficient than enumerating states in $SPACE\_S_{n+1}$.

If we change every occurrence of $F(C_1)$ to $F(C_1) \wedge (\bigwedge_{i \neq j} S_i \neq S_j)$ and let $SPACE\_S_{n+1}$ be the set of states at the $(n+1)^{th}$ time frames of simple paths in the above analysis, the same conclusion can be drawn for (6).

In (7), due to the addition of a transition from the initial state to itself, the disjunction of equalities in (6) becomes a single equality $S_{n+1} = S_n'$. If separate sets of variables are used for S_{n+1} and S_n', previous analysis can still be applied. Otherwise, we have the opportunity to replace the equality check $S_{n+1} = S_n'$ in (7) by using the same set of variables for S_{n+1} and S_n'. However, S_{n+1} and S_n' are still primary outputs in the circuits of Figure 1 and all primary outputs are fully observable, thus they must be critical. Once it is determined that S_{n+1} variables are critical, we can draw the same conclusion using the same reasoning as in the previous analysis.

The analysis of the dual formulations of (4)-(7) are basically the same as the analysis of (4)-(7). The only exceptions are that in the dual formulations, the evaluation of true becomes false and false becomes true. The outermost quantification sets in the dual formulations of (4)-(7) are existentially quantified. In these cases, instead of covering the satisfying assignments, the search procedures search for conflicts.

It is worth emphasizing that although all the existing QBF solvers that we are aware of take CNF as their inputs, the above analysis holds independent of the representation of the formula because our analysis only depends on the logic of the circuit. This is to say that if there was a QBF solver that worked directly off the circuit, the result would still hold since it depends only on the critical nature of the state variables in the formula.

7   Future Directions

From the analysis of last section, we can see that formulating the circuit diameter problem as a QBF and using search based algorithms to evaluate the QBF currently is no better than using SAT to solve this problem, which has not been efficient despite the availability of efficient SAT solvers. However, there may be some room for optimism here. One possibility of making the QBF approach more efficient is changing the formulation of the diameter problem. The formulations described in Section 4 are mainly based on unrolling the transition relations. Other formulations may require less universal variables thus greatly reduce the satisfying search space.

Another possible direction is devising more effective QBF algorithms. A partial search or a non-search based algorithm is definitely worth exploring. Current non-search based QBF algorithms include resolution based algorithms and Plaisted's algorithm proposed in [17]. Pure resolution based approach will likely blow up in space, but we could do some simplification of the resolved formula to alleviate the space explosion. It is also possible to combine the search based algorithms with resolution to get around the drawbacks of both approaches. Plaisted's algorithm in [17] is in some sense a hybrid of search and resolution. But this algorithm works well for circuits that are long and thin. However, our experience shows that the circuit constructed for calculating the sequential state space diameter in Figure 1 is not long and thin. Thus, Plaisted's algorithm is likely to be inefficient for the QBFs arising from the circuit state space diameter problem. Expanding the class of problems that Plaisted's algorithm work well on is another possibility of future research.

Overall for an algorithm to be successful, every part of the reachable state space needs to be enumerated implicitly rather than explicitly by a QBF algorithm. In addition, a clever way of implicitly storing the already explored state space is critical as well.

8   Conclusions

In this paper we describe the QBF formulations of the circuit diameter problem. We prove that using search based QBF algorithms to determine the circuit state space diameter is no more efficient than previous SAT based methods. This result will direct the future QBF approaches for the circuit state space diameter problem away from pure search based algorithms. A non-search based algorithm or a hybrid of search based and non-search based methods are possible candidates for using QBF evaluation to solve the circuit state space diameter problem.

References

1. Biere, A., Cimatti, A., Clarke, E.M., Zhu, Y.: Symbolic model checking without BDDs. In: Proceedings of Tools and Algorithms for the Analysis and Construction of Systems (TACAS'99). (1999)
2. Gupta, A., Yang, Z., Ashar, P., Gupta, A.: SAT-based image computation with application in reachability analysis. In: Proceedings of Third International Conference Formal Methods in Computer-Aided Design (FMCAD 2000). (2000)
3. Mneimneh, M., Sakallah, K.: Computing vertex eccentricity in exponentially large graphs: QBF formulation and solution. In: Sixth International Conference on Theory and Application of Satisfiability Testing. (2003)
4. McMillan, K.L.: Applying SAT methods in unbounded symbolic model checking. In: Proceedings of 14th Conference on Computer-Aided Verification (CAV 2002), Springer Verlag (2002)
5. Rintanen, J.: Constructing conditional plans by a theorem prover. Journal of Artificial Intelligence Research 10 (1999) 323–352
6. Sheeran, M., Singh, S., Stålmark, G.: Checking safety properties using induction and a SAT-solver. In: Proceedings of the Third International Conference on Formal Methods in Computer-Aided Design (FMCAD 2000). (2000)
7. Büning, H.K., Karpinski, M., Flögel, A.: Resolution for quantified Boolean formulas. Information and Computation 117 (1995) 12–18
8. Davis, M., Logemann, G., Loveland, D.: A machine program for theorem proving. Communications of the ACM 5 (1962) 394–397
9. Cadoli, M., Schaerf, M., Giovanardi, A., Giovanardi, M.: An algorithm to evaluate quantified Boolean formulae and its experimental evaluation. Journal of Automated Reasoning 28 (2002) 101–142
10. Rintanen, J.: Improvements to the evaluation of quantified Boolean formulae. In: Proceedings of International Joint Conference on Artificial Intelligence (IJCAI). (1999)
11. Rintanen, J.: Partial implicit unfolding in the Davis-Putnam procedure for quantified Boolean formulae. In: International Conf. on Logic for Programming, Artificial Intelligence and Reasoning (LPAR). (2001)

12. Giunchiglia, E., Narizzano, M., Tacchella, A.: Qube: a system for deciding quantified Boolean formulas satisfiability. In: Proceedings of International Joint Conference on Automated Reasoning (IJCAR). (2001)

13. Giunchiglia, E., Narizzano, M., Tacchella, A.: Backjumping for quantified Boolean logic satisfiability. In: Proceedings of International Joint Conference on Artificial Intelligence (IJCAI). (2001)

14. Zhang, L., Malik, S.: Towards a symmetric treatment of satisfaction and conflicts in quantified Boolean formula evaluation. In: Proceedings of 8th International Conference on Principles and Practice of Constraint Programming (CP2002). (2002)

15. Zhang, L., Malik, S.: Conflict driven learning in a quantified Boolean satisfiability solver. In: Proceedings of the IEEE/ACM International Conference on Computer-Aided Design (ICCAD). (2002)

16. Letz, R.: Lemma, model caching in decision procedures for quantified Boolean formulas. In: International Conference on Automated Reasoning with Analytic Tableaux and Related Methods (Tableaux2002). (2002)

17. Plaisted, D.A., Biere, A., Zhu, Y.: A satisfiability procedure for quantified Boolean formulae. Discrete Appl. Math. **130** (2003) 291–328

18. Marques-Silva, J.P., Sakallah, K.A.: GRASP - a search algorithm for propositional satisfiability. IEEE Transactions in Computers **48** (1999) 506–521

19. Zhang, L., Madigan, C.F., Moskewicz, M.W., Malik, S.: Efficient conflict driven learning in a Boolean satisfiability solver. In: Proceedings of the IEEE/ACM International Conference on Computer-Aided Design (ICCAD). (2001)

20. Moskewicz, M.W., Madigan, C.F., Zhao, Y., Zhang, L., Malik, S.: Chaff: Engineering an efficient SAT solver. In: Proceedings of the Design Automation Conference (DAC). (2001)

21. Goldberg, E., Novikov, Y.: BerkMin: A fast and robust SAT-solver. In: Proceedings of the IEEE/ACM Design, Automation, and Test in Europe (DATE). (2002)

22. Giunchiglia, E., Narizzano, M., Tacchella, A.: Learning for quantified Boolean logic satisfiability. In: Proceedings of the 18th National (US) Conference on Artificial Intelligence (AAAI). (2002)

23. Ranjan, D.P., Tang, D., Malik, S.: A comparative study of 2QBF algorithms. In: The Seventh International Conference on Theory and Applications of Satisfiability Testing. (2004)

24. Gupta, A., Gupta, A., Yang, Z., Ashar, P.: Dynamic detection and removal of inactive clauses in SAT with application in image computation. In: Design Automation Conference. (2001) 536–541

UBCSAT: An Implementation and Experimentation Environment for SLS Algorithms for SAT and MAX-SAT

Dave A.D. Tompkins and Holger H. Hoos

Department of Computer Science,
University of British Columbia, Vancouver BC V6T 1Z4, Canada
{davet, hoos}@cs.ubc.ca

Abstract. In this paper we introduce UBCSAT, a new implementation and experimentation environment for Stochastic Local Search (SLS) algorithms for SAT and MAX-SAT. Based on a novel triggered procedure architecture, UBCSAT provides implementations of numerous well-known and widely used SLS algorithms for SAT and MAX-SAT, including GSAT, WalkSAT, and SAPS; these implementations generally match or exceed the efficiency of the respective original reference implementations. Through numerous reporting and statistical features, including the measurement of run-time distributions, UBCSAT facilitates the advanced empirical analysis of these algorithms. New algorithm variants, SLS algorithms, and reporting features can be added to UBCSAT in a straightforward and efficient way. UBCSAT is implemented in C and runs on numerous platforms and operating systems; it is publicly and freely available at www.satlib.org/ubcsat.

1 Introduction

The propositional satisfiability problem (SAT) is an important subject of study in many areas of computer science and is the prototypical $\mathcal{NP}$-complete problem. MAX-SAT is the optimisation variant of SAT; while in unweighted MAX-SAT, the goal is to find a variable assignment that satisfies a maximal number of clauses of a given CNF formula, in weighted MAX-SAT, a weight is assigned to each clause, and the goal is to find an assignment that maximises the total weight of the satisfied clauses. MAX-SAT is a conceptually simple $\mathcal{NP}$-hard combinatorial optimisation problem of substantial theoretical and practical interest; many application-relevant problems, including scheduling problems or most probable explanation (MPE) finding in Bayes nets, can be encoded and solved as MAX-SAT.

Some of the best known methods for solving certain types of SAT instances are Stochastic Local Search (SLS) algorithms; these are typically incomplete, *i.e.*, they cannot determine with certainty that a formula is unsatisfiable, but they often find models of satisfiable formulae surprisingly effectively [8]. For MAX-SAT, SLS algorithms are by far the most effective methods for finding optimal or close-to-optimal solutions [5, 8]. Although SLS algorithms for SAT and MAX-SAT differ in their details, the basic approach is mostly the same. In the following, we mainly focus on SLS algorithms for SAT, while MAX-SAT algorithms will be discussed in more detail in Section 6.

H.H. Hoos and D.G. Mitchell (Eds.): SAT 2004, LNCS 3542, pp. 306–320, 2005.

```
procedure SLS-for-SAT(F)
    input: propositional formula F
    output: satisfying assignment of F or  no solution found'
    a := InitialiseSearch(F);
    while not Terminate(F, a) do
        if Restart(F, a) then
            a := ReInitialiseSearch(F);
        else
            X := SelectVarsToFlip(F, a);
            a := FlipVars(F, a, X);
        end
    end
    if Solved(F, a) then
        return a
    else
        return  no solution found'
    end
end SLS-for-SAT
```

Fig. 1. Pseudo-code for a typical Stochastic Local Search algorithm for SAT; a is a variable assignment, X is a set of variables in the given formula F

In Figure 1 we provide pseudo-code for a typical SLS algorithm for SAT. Each *run* of the algorithm starts by determining an initial, complete assignment of truth values to all variables in the given formula F (*search initialisation*). Then, in each *search step*, a set of variables is selected, whose truth values are then changed from true to false or vice versa. Each change of a single variable's truth value is called a *variable flip*; almost all SLS algorithms perform exactly one variable flip in each search step, but there are cases in which a given SLS algorithm may flip no variables in a given search step (a so-called *null-flip*), or several variables at once (also known as a *multi-flip*). Variable flips are typically performed with the purpose of minimising an *evaluation function* that measures *solution quality* in terms of the number of unsatisfied clauses under a given variable assignment. The search process is terminated when a *termination condition* is satisfied; this is typically the case when either a solution, *i.e.*, a satisfying assignment of F, has been found or when a given bound on the run-time, which is also referred to as *cutoff time* and which may be measured in search steps or CPU time, has been reached or exceeded. To overcome or avoid search stagnation, many SLS algorithms for SAT make use of a *restart mechanism* that re-initialises the search process whenever a *restart condition* is satisfied. For example, all GSAT and WalkSAT algorithms restart the search periodically [14, 13]. While restart mechanisms are crucial for the performance of some SLS algorithms for SAT, such as basic GSAT [14], they have been found to be ineffective in other cases [8].

Even though SLS algorithms for SAT and MAX-SAT have achieved great levels of success, we believe that there is still significant potential for further improvements. To further explore this potential, we developed UBCSAT: an implementation and experimentation framework for SLS algorithms for SAT and MAX-SAT. Our primary objective was to create a software environment that facilitates research on and development of SLS algorithms. Specifically, the development of UBCSAT was based on the following six design principles and goals:

1. include highly efficient, conceptually simple and accurate implementations of a wide range of prominent SLS algorithms for SAT and MAX-SAT;
2. facilitate the development and integration of new algorithms (and algorithm variants);
3. provide support for advanced empirical analysis of the performance and behaviour of SLS algorithms without compromising implementation efficiency;
4. provide explicit support for algorithms designed to solve the weighted and unweighted MAX-SAT problems;
5. provide an open-source software package that is publicly available to the academic community;
6. implement the project in a platform-independent way, avoiding non-standard programming language extensions.

Before discussing the design and features of UBCSAT in more detail, we briefly discuss two related software projects: OpenSAT and COMET.

The OpenSAT project [1] (www.opensat.org) was developed as a Java-based open source project for complete SAT solvers. A primary goal of OpenSAT was to make the advanced techniques and data structures used by state-of-the-art *complete* SAT solvers openly available in order to accelerate the development of new SAT solvers. Generally, the architecture and implementation of complete SAT solvers, which are based on the David-Putnam-Loveland procedure, differs considerably from that of SLS-based SAT algorithms, and traditionally there has been very little overlap between the algorithmic and implementation details used in these two types of SAT solvers. Therefore, using OpenSAT as the basis for achieving the previously stated goals, while probably not completely infeasible, appears to be problematic. In addition to the difficulty of supporting the development and implementation of SLS algorithms in a straightforward way, the current lack of support for MAX-SAT solvers, and the fact that OpenSAT currently does not provide dedicated support for the advanced empirical analysis of SAT algorithms, it is somewhat questionable whether its Java-based implementation makes it possible to achieve performance that is competitive with the native reference implementations of high-performance SLS algorithms such as WalkSAT [13] or SAPS [9].

COMET [17] is an object-oriented language that supports a constraint-based architecture for local search. The COMET language is very sophisticated and can model SLS algorithms for solving complex constraint satisfaction problems, but it neither offers explicit support for SAT/MAX-SAT nor does it provide tools for advanced empirical evaluation. While in principle, both of these issues could be addressed by realising the respective functionality within COMET, implementing UBCSAT in COMET seemed to pose the risk that in order to take full advantage of UBCSAT, users would have to understand both the idiosyncrasies of COMET as well as the architecture and interface of UBCSAT; we believe that as a consequence, UBCSAT would have been less accessible to its main target group, namely researchers interested in SAT and MAX-SAT. While there is evidence that COMET algorithm implementations are quite efficient, we do not have any insight as to how these would compare with the native reference implementations of the state-of-the-art SLS algorithms covered by UBCSAT.

To achieve our goals of a platform-independent and highly efficient implementation, UBCSAT has been developed in strict ANSI C and tested on some of the most popular operating systems (Linux, WindowsXP, SunOS). In order to provide a state-of-the-art and platform-independent source of pseudo-random numbers, we have incorporated the "Mersenne Twister" pseudo-random number generator [10]. UBCSAT is publicly available for academic (non-commercial) use without restriction to encourage free and open use throughout the SAT research community[1].

In the remainder of this paper, we will describe the UBCSAT project in greater depth. In Section 2 we give an overview of the UBCSAT architecture and illustrate the fundamental concept of *triggered procedures*, which lies at the core of UBCSAT's efficient yet highly flexible design and implementation. In Section 3, we outline the current collection of SLS algorithms for SAT that are currently implemented within UBCSAT and compare their performance against that of the respective native reference implementations. In Section 4 we demonstrate how new algorithms are implemented within UBCSAT. In Section 5 we discuss the importance of empirical analysis in SLS research, and how UBCSAT can help facilitate empirical analysis. In Section 6, we describe how UBCSAT supports SLS algorithms for weighted and unweighted MAX-SAT. Finally, in Section 7 we summarise the key features and contributions of the UBCSAT project and outline some directions for future work.

2   The UBCSAT Architecture

One of the challenges of developing the UBCSAT project was to build a flexible, feature-rich environment without compromising algorithmic efficiency. To achieve our goals, UBCSAT has been designed according to what we have named a *triggered procedure architecture*. The main ideas underlying this architecture are closely related to certain concepts from object- and event-oriented programming.

The UBCSAT software is structured around a set of *event points* that occur throughout the execution of a SLS algorithm for SAT. For each event point p, a list of procedures is maintained that are executed whenever event point p is reached; this list is called the *triggered procedure list* of p and its elements are called the *triggered procedures* of p. A *trigger* is simply a mapping of a software procedure to an event point. When a trigger is *activated*, then its associated procedure is added to the triggered procedure list of the corresponding event point.

Initially, the triggered procedure lists for all of the event points are empty; it is only when triggers are activated that procedures become executed when an event point is reached. For example, you may have a procedure for displaying the current status of

---

[1] The UBCSAT source code and x86 executables for Windows and Linux are available for download at http://www.satlib.org/ubcsat. Throughout this paper we have endeavoured to keep our descriptions and examples consistent with the UBCSAT software package version 1.0, but as development on UBCSAT continues some aspects may deviate from these descriptions.

an algorithm as it searches. You can create a trigger that maps your procedure to an appropriate event point, perhaps at the end of each search step. Whenever you would like to have the status displayed you can activate your trigger, which will ensure that at the end of each search step your procedure is executed. However, if you do not wish to have the status displayed then you do not have to do anything; your trigger will not be activated, no procedure will be added to a triggered procedure list, and your algorithm will not be slowed down by your status display procedure.

In addition to its associated procedure, event point and activation status, a trigger t can have a *dependency list* and a *deactivation list*, which are lists of other triggers that are activated or deactivated (respectively) when t is activated. The dependency list is used, for example, to ensure that when the procedure of a trigger relies on the existence of special data structures, the triggers for the procedures that create and update those data structures are also activated. The deactivation list is intended for advanced UBC-SAT users, and can be used to override default routines and to avoid conflicts between incompatible routines. In practice, deactivation lists are used in UBCSAT to improve implementation efficiency by combining the functionality of multiple procedures into one. For example, consider triggers t_a and t_b that have procedures $a()$ and $b()$, but when both triggers are activated it would significantly more efficient if the functionality of procedures $a()$ and $b()$ were combined into one procedure. In this case, a new procedure $ab()$ could be created and assigned to a trigger t_{ab} which would include t_a and t_b in its deactivation list and be available to algorithms that require the functionality of both t_a and t_b. UBCSAT detects and produces a warning if deactivated triggers are somehow reactivated, which might indicate a flaw in the design of an SLS algorithm that is being developed within the UBCSAT framework.

There is also a special type of trigger called a *container trigger* that has no associated procedure, but instead a list of secondary triggers that are activated whenever the container trigger is activated. Container triggers are used as convenient shortcuts for activating groups of triggers that are used simultaneously. Conceptually, container triggers are very similar to dependency lists; by activating one trigger several others are also activated. While dependency lists are an important part of ensuring the triggered procedure architecture works properly, container triggers simply provide shortcuts for added convenience. As we will show in a later example, many data structures in UBC-SAT require three triggers to operate properly: one to create the data structure, one to initialise it and one to update it. A container trigger can be created to activate all three of those triggers simultaneously. If a trigger corresponds to a complicated procedure that requires four different data structures to be in place, the dependency list can comprise of just four container triggers, instead of all twelve required triggers. An additional container trigger could be created to encompass those four other container triggers, but unless that container trigger would be used by other triggers, there is no added benefit in doing so.

UBCSAT has over one hundred triggers, most of which have associated procedures that fall into one of the following four categories: heuristic selection (*e.g.*, of variables), data structure maintenance, report and statistic data collection, and file I/O. Triggers are activated based on the SLS algorithm to be run, the reports/statistics requested and other system command line parameters. In the UBCSAT implementation, the triggered pro-

```
procedure UBCSAT
    SetupUBCSAT();
    ParseParameters();
    ActivateAlgorithmTriggers();
    ActivateReportTriggers();
    RunProcedures(PostParameters);
    RunProcedures(ReadInInstance);
    RunProcedures(CreateData);
    RunProcedures(CreateStateInfo);
    RunProcedures(PreStart);
    StartClock();
    while iRun < iNumRuns do
        RunProcedures(PreRun);
        while ((iStep < iCutoff) and (not bSolutionFound)) and (not bTerminateRun)) do
            RunProcedures(PreStep);
            RunProcedures(CheckRestart);
            if bRestart or (iStep = 1) then
                RunProcedures(InitData);
                RunProcedures(InitStateInfo);
            else
                RunProcedures(ChooseCandidate);
                RunProcedures(PreFlip);
                RunProcedures(FlipCandidate);
                RunProcedures(PostFlip);
            end
            RunProcedures(PostStep);
            RunProcedures(CheckTerminate);
        end
        RunProcedures(PostRun);
    end
    EndClock();
    RunProcedures(Final);
end UBCSAT
```

Fig. 2. High-level pseudo-code of UBCSAT; event points are indicated by asterisks

cedure lists are simply arrays of function pointers, so when each event point is reached, it is very efficient to call its triggered procedures.

Figure 2 shows a high-level pseudo-code representation of UBCSAT and indicates many of the most important event points. The following example further illustrates the use of event points and the concept of triggered procedures.

Let us consider WalkSAT/TABU, a well-known high-performance SLS algorithm for SAT that is based on the WalkSAT architecture [12]. As in most WalkSAT-based algorithms, WalkSAT/TABU starts each search step by uniformly selecting a clause from the set of currently unsatisfied clauses. Each variable in the clause is assigned a score, corresponding to the change in the number of unsatisfied clauses that would occur if that variable were flipped. The variable with the best score that is not *tabu* is selected as the flip variable (breaking ties randomly). A variable is tabu if it has been flipped within the last *TabuTenure* search steps, where *TabuTenure* is a parameter of the WalkSAT/TABU algorithm. If all of the variables in the selected clause are tabu, then no flip occurs at that step.

In the UBCSAT implementation of WalkSAT/TABU, the main heuristic procedure is *PickWalksatTabu()*, and a trigger of the same name exists which maps the procedure to the *ChooseCandidate* event point. Most algorithms in UBCSAT also activate the *DefaultProcedures* trigger, a container trigger that includes triggers for handling common tasks, such as keeping track of the current truth assignment and reading the formula into

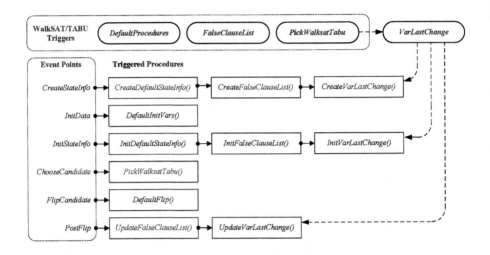

Fig. 3. The WalkSAT/TABU algorithm triggers and the triggered procedures that appear in the event point triggered procedure lists. The dashed arrows illustrate how the *VarLastChange* procedures were added to the triggered procedure lists by the activation of the *PickWalksatTabu* trigger. Note that some procedures and event points are not listed, including a few additional procedures triggered by *DefaultProcedures*

memory. Efficient implementations of WalkSAT-based algorithms require a list of the currently unsatisfied clauses, which is maintained by a set of procedures whose triggers are all included in the *FalseClauseList* container trigger.

Different from, say, WalkSAT/SKC, WalkSAT/TABU needs to know when each variable has been flipped last, in order to determine its tabu status. This requires a simple data structure (an array of values) that is maintained using three triggered procedures: *CreateVarLastChange()* allocates the memory required for the data structure, *InitVarLastChange()* initialises it at the beginning of each run and after restarts, and *UpdateVarLastChange()* updates it after each flip. Each of these procedures has a trigger that associates it with the event points *CreateStateInfo*, *InitStateInfo*, and *PostFlip*, respectively. For convenience, these three triggers are grouped into a container trigger named *VarLastChange*. When the *PickWalksatTabu* trigger is registered in UBCSAT, it lists *VarLastChange* in its dependency list, so when the Walksat/TABU algorithm is selected, the *PickWalksatTabu* trigger is activated, which will activate the *VarLastChange* trigger, and hence the three previously described triggers. (See also Figure 3.)

The primary advantage of the triggered procedure architecture lies in the fact that of the many procedures that are needed to realise the many SLS algorithms and report formats supported by UBCSAT, only those required in any given run are activated and used, while the remaining inactive or non-triggered procedures do not affect UBCSAT's performance. A secondary advantage is that different algorithms and reports can share the same data structures and procedures, saving much programming effort. Potential drawbacks stem from the implementation overhead of having to register all triggers, and from the fact that in this framework, algorithms are typically split into many rather small procedures, which can lead to decreased performance compared to more monolithic implementations. However, we have found that these disadvantages are far outweighed by the advantages of UBCSAT's triggered procedure architecture. In particular, as we will demonstrate in the following section, the performance of UBC-SAT is very competitive with native reference implementations of the respective SAT algorithms.

3   A Collection of Efficient Algorithm Implementations

UBCSAT can be seen as a collection of many different SLS algorithms. Compared to the respective reference native implementations of these algorithms, by integrating them into the UBCSAT framework several advantages can be realised: Generally, by using a single executable with a uniform interface, working with different algorithms becomes easier and more convenient. From an implementation point of view, different algorithms share common data structures and procedures, which reduces implementation effort and the likelihood of programming errors. And from an empirical algorithmics point of view, comparing two algorithms is facilitated by the fact that UBCSAT allows fairer comparisons between algorithms that share components and use the same statistical calculations, input and output formats.

The UBCSAT software package currently implements the following SLS algorithms for SAT:

- GSAT [14]
- GWSAT [13]
- GSAT/TABU [11]
- HSAT [3]
- HWSAT [4]
- WalkSAT/SKC [13]

- WalkSAT/TABU [12]
- Novelty and R-Novelty [12]
- Novelty$^+$ and R-Novelty$^+$ [6]
- Adaptive Novelty$^+$ [7]
- SAPS and RSAPS [9]
- SAPS/NR [16]

UBCSAT is designed to support weighted MAX-SAT versions (see also Section 6) as well as variants that may differ in their behaviour or implementation from the basic version of a given algorithm. Consequently, each algorithm within UBCSAT is identified as a triple (*"algorithm"*, b*Weighted*, *"variant"*), selectable on the command line as ubcsat -alg *algorithm* [-w] [-v *variant*].

For each of the previously listed algorithms, we ensured that the UBCSAT implementation behaves identically to the respective original reference implementation, taking into consideration the stochastic nature of the algorithms. This is illustrated in Figure 4, in which run-time distributions for the UBCSAT implementations of GWSAT and WalkSAT/SKC are compared with those for the original GSAT (version 41) and WalkSAT (version 43) implementations.

At the same time, the UBCSAT versions of all algorithms were optimised for efficiency, with the goal of matching or exceeding the time performance of the respective reference implementations. For many SLS algorithms, the key to an efficient implementation lies in the way crucial data structures are organised and incrementally main-

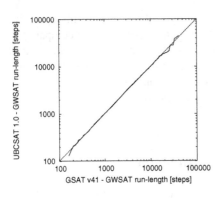
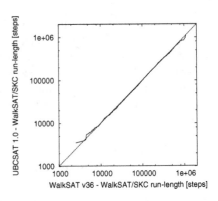

Fig. 4. Quantile-quantile plots of the run-time distributions for UBCSAT *vs.* GSAT v41 on instance `uf200-easy` (left side) and WalkSAT v43 on `bw_large.a` (right side) based on 5 000 runs per algorithm (run-time is measured in search steps)

Table 1. Total run times (in seconds) for 100 000 000 search steps on a dual-processor 1GHz Pentium III (Coppermine) machine with 256KB cache and 1GB RAM running SuSE Linux 9.1. The speedup factor *(s.f.)* shows the software speedups of the UBCSAT implementation over the original implementations (GSAT v41 and WalkSAT v43). Note by choosing *unsatisfiable* instances for this speed comparison, we ensured that in all cases exactly the same number of search steps have been performed. The `uuf-*` instances are uniform random 3-SAT, the `jnh` instance is random *P*-SAT, and the `rg` instance is a structured encoding of a graph colouring instance

Algorithm	uuf100-01			uuf400-01		
	UBCSAT	Original	*s.f.*	UBCSAT	Original	*s.f.*
WalkSAT/SKC	97.7	144.7	**1.48**	98.5	150.3	**1.53**
Novelty	117.1	151.6	**1.29**	114.5	153.4	**1.34**
GSAT	106.7	305.0	**2.86**	114.1	316.5	**2.77**
GWSAT	172.1	590.1	**3.43**	266.8	768.2	**2.88**

Algorithm	jnh202			rg-200-2000-4-11		
	UBCSAT	Original	*s.f.*	UBCSAT	Original	*s.f.*
WalkSAT/SKC	134.0	217.2	**1.62**	142.1	310.7	**2.19**
Novelty	168.4	230.8	**1.37**	159.5	323.0	**2.02**
GSAT	202.3	1541.6	**7.62**	233.0	397.8	**1.71**
GWSAT	254.3	1894.7	**7.45**	541.5	1354.5	**2.50**

tained. For example, many algorithms (such as GSAT and its variants) assign a score to each variable that is defined as the net change in the total number of satisfied clauses caused by flipping that variable. Rather than recomputing all variable scores in each step, they can be stored and incrementally updated, such that after each flip only the scores affected by that flip are recalculated [8]. However, we have found that in some situations too much time can be spent by using this scheme; in particular, using it in the implementation of WalkSAT algorithms actually decreases their performance. To further complicate matters, the optimal incremental update strategy often depends on the characteristics of the given problem instance.

In our UBCSAT implementation, we strove to use data structures and incremental updating schemes that are efficient, yet reasonably straightforward to understand and implement. The UBCSAT architecture supports functionally identical algorithm variants that are implemented using different data structures and/or incremental updating schemes in a straightforward way, which makes it easy to implement new developments in this area (such as Fukunaga's recent scheme [2]).

The performance of the UBCSAT implementations of all supported algorithms has been tested against that of the respective reference implementations in order to ensure that the former are at least as efficient (in terms of run-time) as the latter. More importantly, for GSAT- and WalkSAT-algorithms, the UBCSAT implementations have been shown to be significantly faster (see Table 1 for representative results).

4   A Framework for Developing New Algorithms

As discussed in the previous section, the UBCSAT environment includes a wide variety of algorithms and data structures. To facilitate the development and integration of new SLS algorithms, UBCSAT has been designed in such a way that new algorithms can easily re-use the existing procedures and data structures from other algorithms; the basis for this is provided by the triggered procedure architecture discussed in Section 2.

To illustrate how new algorithms are added to UBCSAT, in Figure 5 we present the pseudo-code required to add a new WalkSAT/TABU algorithm variant to UBCSAT. We have named the new variant WalkSAT/TABU-NoNull, and it differs from the regular WalkSAT/TABU algorithm in only one detail: if all of the variables in the selected clause are tabu, then a variable will be selected from the clause at random and flipped. (This variant is interesting from a practical point of view, since WalkSAT/TABU is one of the best-performing WalkSAT algorithms, but often suffers from search stagnation as a consequence of null-flips.)

Within UBCSAT, the new algorithm will be identified as a (*"walksat-tabu"*, *false*, *"nonull"*); it differs from the already supported WalkSAT/TABU only in its variable selection procedure, whose trigger we name *PickWalksatTabuNoNull*. An algorithm can explicitly specify the data structure procedures required, or it can *inherit* them from another algorithm. In this case, we will simply inherit everything from regular WalkSAT/TABU (*"walksat-tabu"*, *false*, *""*). When an algorithm requires algorithm-specific command-line parameters (such as the *tabuTenure* parameter in WalkSAT/TABU) they must be defined or optionally inherited from an existing algorithm. In addition to creating and registering the new trigger in the system, its associated procedure, here also called *PickWalksatTabuNoNull*, has to be implemented, which in this example simply calls the regular WalkSAT/TABU variable selection procedure and then handles the special case when a null-flip occurs. While this example illustrates a particularly simple variant of an existing algorithm, the process of adding implementations of new SLS algorithms to UBCSAT is typically similarly straightforward.

```
procedure AddWalksatTabuNoNull()
    CreateAlgorithm("walksat-tabu", false, "nonull",        % algorithm, bWeighted, variant
        "WalkSAT/TABU without null-flips",                  % description
        "McAllester, Selman, Kautz [AAAI 97] (modified)",  % authors
        "PickWalksatTabuNoNull",                           % heuristic trigger(s)
        ...);                                              % details omitted
    InheritDataTriggers("walksat-tabu", false, "");
    InheritParameters("walksat-tabu", false, "");
    CreateTrigger("PickWalksatTabuNoNull",                 % trigger name
        ChooseCandidate,                                   % event point
        PickWalksatTabuNoNull,                             % pointer to procedure
        ...);
end AddWalksatTabuNoNull

procedure PickWalksatTabuNoNull()
    PickWalksatTabu();
    if iFlipCandidate = NULL then
        iFlipCandidate := PickRandomVarFromClause(iWalkSATClause);
    end
end PickWalksatTabuNoNull
```

Fig. 5. Pseudo-code of the procedures required for extending UBCSAT with a new variant of WalkSAT/TABU

5   An Empirical Analysis Tool

Empirical analysis plays an important role in the development and successful application of SAT algorithms. To characterise or measure the behaviour of an SLS algorithm, typically data needs to be collected from multiple independent *runs* of the algorithm. Each run corresponds to a complete execution of the algorithm outlined in Figure 1; note that the pseudo-code of Figure 2 performs multiple runs. (Note that when restart mechanisms are used, a single run can be punctuated by one or more restarts, but this does not partition it into multiple runs.) As an example, consider the run-time data shown in Figure 4, which is based on 5 000 independent runs of each algorithm involved in the respective experiment. To facilitate the advanced empirical analysis of the SLS algorithms it implements, UBCSAT not only provides support for measuring and reporting basic descriptive statistics over multiple runs, but also strongly supports the analysis of run-time distributions (RTDs) [8]. In particular, UBCSAT can measure and report RTDs in a format that can easily be plotted (see left side of Figure 7) or further analysed with specialised statistical software.

Reports currently implemented in UBCSAT include the satisfying assignments found in each run, detailed information about the search state at each search step, flip statistics for individual variables and many others. In UBCSAT, statistics are special objects that are used to collect and summarise data for the default reports. Statistics can be shown for each individual run (column objects), or be summarised over all runs (stat objects). Additional reports and statistics can be added to UBCSAT in a straightforward manner that is conceptually closely related to the way in which new algorithms are added. Reports can be in any format and are implemented based on a list of triggered procedures that collect and print the required information.

In Figure 6, we show the creation of a column object that will calculate the average *age* of variables flipped during a run. The age of a flipped variable is calculated as

```
integer iCurVarAge;                              % global variable for statistic

procedure AddAgeStat()
    AddColumn("agemean",                         % column name
        "Mean Age of Variables when flipped",
        &iCurVarAge,                             % pointer to data variable
        "UpdateCurVarAge",                       % trigger to activate
        TypeMean,                                % type of statistic to collect on data
        ...);
    CreateTrigger("UpdateCurVarAge", PreFlip, UpdateCurVarAge,
        "VarLastChange",                         % trigger dependency
        ...);
end AddAgeStat

procedure UpdateCurVarAge()
    iCurVarAge := iStep − aVarLastFlip[iFlipCandidate];
end UpdateCurVarAge
```

Fig. 6. Pseudo-code for adding a new statistic that measures the mean age of variables when flipped

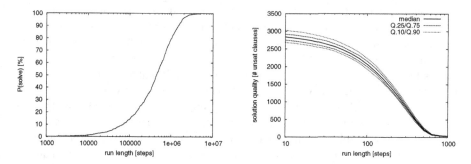

Fig. 7. An example of a run-length distribution (RLD) (left side), and a time-dependent solution quality statistics (SQT) plot (right side). The data underlying these curves can be easily generated by the UBCSAT software package and plotted using gnuplot scripts which are available on the UBCSAT website

the number of steps that have occurred since the last time the variable was flipped (the calculation is shown in *UpdateCurVarAge()*. For this statistic, the trigger *UpdateCurVarAge* is required to ensure that the correct age value is calculated at the event point *PreFlip*. The trigger *UpdateCurVarAge* depends on the trigger *VarLastChange* (see Figure 3), so if the algorithm already collects this data (as does, *e.g.*, WalkSAT/TABU) then the statistic will simply share the same data structure, but if the algorithm does not normally require this data, then the trigger will ensure that it is collected. Because this column statistic has been identified as a *TypeMean* (average over all search steps of a run), an additional trigger will be automatically activated to collect the data at the end of each search step. Like many statistics added to UBCSAT, this age statistic is now available to *all* algorithms (that use a single-flip strategy). UBCSAT facilitates comparisons between algorithms on statistics such as these, which can help further our understanding of how SLS algorithms behave.

6   Support for MAX-SAT

One area where SLS algorithms have been very successful, and have defined the state-of-the-art for more than a decade, is in solving the MAX-SAT problem, and in particular, the *weighted* MAX-SAT problem; for this reason, supporting MAX-SAT was one of our primary goals. Although there are interesting differences between the state-of-the-art SLS algorithms for SAT and MAX-SAT, at the conceptual and implementation level, there are many similarities. Unweighted MAX-SAT can be seen as a special case of weighted MAX-SAT where all clauses have uniform weight; therefore, in the following, we will focus on the weighted MAX-SAT problem. It should be noted, however, that in terms of implementation, SLS algorithms for unweighted MAX-SAT are much more closely related to SLS algorithms for SAT. In UBCSAT, unweighted MAX-SAT algorithms are therefore typically equivalent to the corresponding SAT algorithm, while weighted MAX-SAT algorithms are implemented separately, facilitating conceptually simpler and highly efficient implementations for both cases.

The main differences between SAT and MAX-SAT is that the optimal solution quality (*i.e.*, maximal total weight of satisfied clauses) for a given problem instance is often unknown. Hence, the best assignment encountered during the search, the so-called *incumbent assignment*, is memorised and returned at the end of the search. This memorisation of the incumbent assignment is accomplished in UBCSAT via a report. Typically, SLS algorithms for MAX-SAT are not guaranteed to find *optimal solutions, i.e.*, maximal weight assignments, but many state-of-the-art SLS algorithms for MAX-SAT have the property that if they search long enough, the probability of finding an optimal solution approaches one (the so-called *PAC property*, see also [6,8]), and in many practical cases assignments that are provably optimal or believed to be optimal can be found within reasonable run-times. UBCSAT supports termination criteria that end a run whenever a user-specified solution quality (*e.g.*, the known optimal solution quality for the given problem instance) is reached or exceeded; alternatively, particularly when dealing with instances whose optimal solution quality is unknown, UBCSAT can be configured with advanced criteria to determine when to terminate a run.

Currently, UBCSAT includes implementations of two dedicated algorithms for MAX-SAT, SAMD [5] and IRoTS [15], as well as weighted MAX-SAT variants for many of the SLS algorithms listed in Section 3. The mechanism for implementing new MAX-SAT algorithms within UBCSAT is exactly the same as described for the case of SAT in Section 4. While for unweighted MAX-SAT instances, the same DIMACS CNF file format as for SAT is used, for weighted MAX-SAT instances, UBCSAT currently supports a straightforward extension of the this format known as the *weighted CNF file format* (.wcnf). To support the empirical analysis of the behaviour and performance of SLS algorithms for MAX-SAT, in addition to the previously mentioned statistics and reports (see Section 5), UBCSAT supports advanced analysis methods for stochastic optimisation algorithms. In particular, the following types of empirical performance characteristics can be easily measured (see also [8]):

– qualified run-time distributions (QRTDs), *i.e.*, empirical probability distributions of the run-time required for reaching or exceeding a specific target solution quality measured over multiple runs of the algorithm;

- solution quality distributions (SQDs), *i.e.*, empirical probability distributions of the best solution quality reached within a given amount of run-time, measured in terms of search steps or CPU time over multiple runs of the algorithm;
- time-dependent solution quality statistics (SQTs), *i.e.*, the development of descriptive statistics (such as quantiles) of the SQDs as run-time increases.

QRTDs, SQDs, and SQTs are determined from so-called *solution quality traces*, which contain information on every point in time the incumbent solution was updated during a given run of the algorithm. The solution quality traces are collected by UBSAT with minimal overhead during the run of any MAX-SAT algorithm. Figure 7 (right side) shows a sample SQT measured by UBCSAT.

7   Conclusions and Future Work

In this paper we have introduced UBCSAT, a new software environment that we created with the specific goal of facilitating and supporting research on SLS algorithms for SAT and MAX-SAT. UBCSAT is built on the basis of a novel triggered procedures architecture and includes highly efficient, conceptually simple, and accurate implementations of a wide range of prominent SLS algorithms for SAT and MAX-SAT. UBCSAT facilitates the development and integration of new algorithms (and algorithm variants). It provides support for advanced empirical analysis of the performance and behaviour of SLS algorithms without compromising implementation efficiency. UBCSAT has been implemented in a platform-independent way and is publicly available to the academic community as an open-source software package.

While this paper has summarised the work on the UBCSAT project to date, UBCSAT is an ongoing effort, and we are very enthusiastic about expanding and building upon the project in the future. We plan to expand UBCSAT by incorporating existing and new SLS algorithms for SAT and MAX-SAT. While we have so far focussed on an ANSI C compliant implementation, there is some interest in adding C++ interfaces, as well as extending our implementation beyond the 32-bit boundary for counters. We will continue to add more sophisticated reports and empirical analysis tools, and we are also interested in providing more external support features, such as gnuplot scripts and better integration with the R statistical software package. It has been suggested that the UBCSAT project could benefit from support for parallel implementations, a more formalised object-based system with advanced integrity checking, and even a graphical user interface for constructing new algorithms and adding triggers. We are very interested in adding features that will make the software more accessible and useful to the research community, and welcome feedback and suggestions for further improvements.

But above all else, we hope that our UBCSAT framework will help advance state-of-the-art research in SLS algorithms, to help better understand how and why SLS algorithms behave the way they do, and to unlock some of the unexplored potential of SLS algorithms for SAT and MAX-SAT.

References

1. G. Audemard, D. Le Berre, O. Roussel, I. Lynce, and J. Marques-Silva. OpenSAT: an open source SAT software project. In *Sixth Int'l Conf. on Theory and Applications of Satisfiability Testing (SAT2003)*, pages 502–509, 2003.
2. A. Fukunaga. Efficient implementations of SAT local search. In *Seventh Int'l Conf. on Theory and Applications of Satisfiability Testing (SAT2004)*, pages 287–292, 2004.
3. I. P. Gent and T. Walsh. Towards an understanding of hillclimbing procedures for SAT. In *Proc. of the Eleventh Nat'l Conf. on Artificial Intelligence (AAAI-93)*, pages 28–33, 1993.
4. I. P. Gent and T. Walsh. Unsatisfied variables in local search. In *Hybrid Problems, Hybrid Solutions*, pages 73–85, 1995.
5. P. Hansen and B. Jaumard. Algorithms for the maximum satisfiability problem. *Computing*, 44:279–303, 1990.
6. H. H. Hoos. On the run-time behaviour of stochastic local search algorithms for SAT. In *Proc. of the Sixteenth Nat'l Conf. on Artificial Intelligence (AAAI-99)*, pages 661–666, 1999.
7. H. H. Hoos. An adaptive noise mechanism for WalkSAT. In *Proc. of the 18th Nat'l Conf. in Artificial Intelligence (AAAI-02)*, pages 655–660, 2002.
8. H. H. Hoos and T. Stützle. *Stochastic Local Search: Foundations and Applications*. Morgan Kaufmann Publishers, San Francisco, CA, USA, 2005.
9. F. Hutter, D. A. D. Tompkins, and H. H. Hoos. Scaling and probabilistic smoothing: Efficient dynamic local search for SAT. In *LNCS 2470: Proc. of the Eighth Int'l Conf. on Principles and Practice of Constraint Programming (CP-02)*, pages 233–248, 2002.
10. M. Matsumoto and T. Nishimura. Mersenne twister: a 623-dimensionally equidistributed uniform pseudo-random number generator. *ACM Trans. on Modeling & Computer Simulation*, 8(1):3–30, 1998.
11. B. Mazure, L. Saïs, and É. Grégoire. Tabu search for SAT. In *Proc. of the Fourteenth Nat'l Conf. on Artificial Intelligence (AAAI-97)*, pages 281–285, 1997.
12. D. McAllester, B. Selman, and H. Kautz. Evidence for invariants in local search. In *Proc. of the Fourteenth Nat'l Conf. on Artificial Intelligence (AAAI-97)*, pages 321–326, 1997.
13. B. Selman, H. A. Kautz, and B. Cohen. Noise strategies for improving local search. In *Proc. of the 12th Nat'l Conf. on Artificial Intelligence (AAAI-94)*, pages 337–343, 1994.
14. B. Selman, H. Levesque, and D. Mitchell. A new method for solving hard satisfiability problems. In *Proc. of the Tenth Nat'l Conf. on Artificial Intelligence (AAAI-92)*, pages 459–465, 1992.
15. K. Smyth, H. H. Hoos, and T. Stützle. Iterated robust tabu search for MAX-SAT. In *Proc. of the 16th Conf. of the Canadian Society for Computational Studies of Intelligence*, pages 129–144, 2003.
16. D. A. D. Tompkins and H. H. Hoos. Warped landscapes and random acts of SAT solving. In *Proc. of the Eighth Int'l Symposium on Artificial Intelligence and Mathematics (ISAIM-04)*, 2004.
17. P. Van Hentenryck and L. Michel. Control abstractions for local search. In *LNCS 2833: Proc. of the Ninth Int'l Conf. on Principles and Practice of Constraint Programming (CP-03)*, pages 65–80, 2003.

Fifty-Five Solvers in Vancouver: The SAT 2004 Competition

Daniel Le Berre[1] and Laurent Simon[2]

[1] CRIL-CNRS FRE 2499, Université d'Artois,
Rue Jean Souvraz SP 18 – F 62307 Lens Cedex, France
leberre@cril.univ-artois.fr
[2] LRI, Université Paris-Sud,
Bâtiment 490, U.M.R. CNRS 8623 – 91405 Orsay Cedex, France
simon@lri.fr

Abstract. For the third consecutive year, a SAT competition was organized as a joint event with the SAT conference. With 55 solvers from 25 author groups, the competition was a clear success. One of the noticeable facts from the 2004 competition is the superiority of incomplete solvers on satisfiable random k-SAT benchmarks. It can also be pointed out that the complete solvers awarded this year, namely Zchaff, jerusat1.3, satzoo-1.02, kncfsand march-eq, participated in the SAT 2003 competition (or at least former versions of those solvers). This paper is not reporting exhaustive competition results, already available in details online, but rather focuses on some remarkable results derived from the competition dataset.

The SAT 2004 competition is ending a 3-year take-off period that attracted new SAT researchers and provided many new benchmarks and solvers to the community. The good participation rate of this year (despite the addition of the anti-blackbox rule) establishes the competition as an awaited yearly event. Some new directions for a new way of thinking about the competition are discussed at the end of the paper.

1   Introduction

Building efficient SAT solvers is one of the many sides of SAT research. Whether these solvers are built as a proof of concept for a theoretical result, or are the result of a careful software engineering process for industrial purposes, they are useful for the whole community because they provide a snapshot of current algorithmic performances. Efficient SAT solvers help us estimate which SAT instances are solvable on current computers, and which methods works on which kind of problems.

As we've done in the previous competitions, we partitioned the set of benchmarks in three categories. The recent use of SAT solvers as embedded engines in

H.H. Hoos and D.G. Mitchell (Eds.): SAT 2004, LNCS 3542, pp. 321–344, 2005.
© Springer-Verlag Berlin Heidelberg 2005

successful model checkers or planning systems[1] created a huge interest in building efficient SAT solvers especially dedicated to solving SAT benchmarks with thousands (sometimes millions) of variables. These very large benchmarks are coming from an automated translation of problems from bounded model checking [2], formal verification [23], planning [11], etc. They may be found in the *industrial category* of the competition: they provide an "optimistic" bound of the kind of problems solvable by current state of the art SAT solvers. However, if many solvers have been reported to solve most of these large industrial benchmarks today, it is still possible to build a three-hundred-variable benchmark that they won't be able to solve. Most of these benchmarks arise from the theoretical result of NP-completeness of the satisfiability problem. They may be found in the *crafted category* of the contest. Such category often provides a "pessimistic" bound on the size of brute force solvable problems and may also emphasize new inference rules for strengthening SAT solvers. At last, the uniform random k-CNF – formulas containing exactly k different literals per clause – is still a widely used class of benchmarks, both useful at the theoretical and the practical level. These benchmarks represent a very particular but still hard challenge (especially the unsatisfiable random benchmarks) for the SAT community. We classified them in the *random category*.

As previously, the competition was based on a two-stage ranking. The deadline for submitting both solvers and benchmarks was February, 23rd. Solvers ran first on randomly generated random k-SAT, then on industrial benchmarks. The first stage finished early April by running the solvers on the remaining crafted instances. The authors of the SAT solvers were able to follow almost in real time the progress of their solver, ensuring the correctness of the collected information. The timeout for the first stage was only 10 minutes, because of the large number of solvers competing this year. After this first stage, the solvers were ranked for each category according to the number of solved series, then according to the total number of solved benchmarks[2]. The aim of this ranking is to focus on SAT solvers able to solve a wide range of SAT benchmarks. From an anonymized version of those rankings the judges, João Marques-Silva, Hans Kleine-Büning and Fahiem Bacchus, decided for each category which solvers were eligible to enter the second stage. Note that the solvers for which a detailed description was not available were not eligible to enter the second stage.

These "second stage solvers" were then launched on the benchmarks that remained unsolved with a longer timeout (40 minutes). Technically, the competition ran this year on two clusters of Linux boxes. One, from the "Laboratoire de Recherche en Informatique" (LRI, Orsay, France), was composed of Athlon 1800+ with 1 GB memory. It was used last year for the SAT 2003 competition.

---

[1] SATPLAN04, powered by the SAT solver Siege, got the first place in the 2004 planning competition in the optimal track http://ls5-www.cs.uni-dortmund.de/~edelkamp/ipc-4/results.html.

[2] A series of benchmarks is a collection of similar benchmarks (e.g. for the random category, benchmarks with the same number of variables and the same number of clauses). A series is solved if at least one of its benchmarks is solved.

The second one, from "the beta lab" (UBC, Vancouver, Canada), was composed of Intel Xeon 2.4 GHz with 1 GB memory.

2   The Competitors

2.1   Solvers

Due to lack of space, we'll not describe here each solver. We invite the reader to take a look at the 2-page descriptions of the solvers available on the competition results web page[3]. 25 submitters (see the online solver descriptions for the co-authors) provided 55 solvers:

- **Complete (SAT and UNSAT) solvers**:
  brchaff (R. Bruni), circush0, circush1 (H. Jin) [9], cls (W. Ruml) [6], compsat (A. Biere), cquest (I. Lynce), eqube1, eqube2 (M. Narizzano), forklift , frankensat-high frankensat-low (M. Dufourt) , funex (A. Biere), isat1, isat2, isat3 (N.S. Narayanaswamy), jerusat1.3 (A. Nadel), lsatv1.1 (R. Ostrowski) [15], march-001, march-007, march-eq-010, march-eq-100 (M. Heule), minilearning.jar (D. Le Berre), modoc (A. Van Gelder) [21], nanosat (A. Biere), oepira, oepirb, oepirc (J. Alfredsson), ofsat (O. Fourdrinoy), quantor (A. Biere) [1], sato4.2, sato4.3 (H. Zhang) [27], satzoo-1.02 (N. Een), tts-2-0 (I. Spence), werewolf (J. Roy), wllsatv1 (R. Ostrowski), zchaff, zchaff-rand (Z. Fu) [14],
- **Incomplete (SAT) solvers**:
  adaptnovelty, novely35, novely50 (H. Hoos)[8], gasat(F. Lardeux) [10], qingting (X.-Y. Li) [13], rsaps, saps, sapsnr (D. Tompkins), saprover, unitwalk (A. Kojevnikov) [7], walksatauto, walksatrnp [8] and walksatsck [17] (H. Kautz).
- **Portfolio** (contain both complete and incomplete solvers):
  satzilla, satzilla-nr, satzilla-r (E. Nudelman)

The solvers forklift (E. Goldberg and Y. Novikov) and kncfs (Gilles Dequen and Olivier Dubois [5]) were not submitted by their authors but entered the contest as last year winners.

2.2   The Benchmarks

Uniform Random k-SAT. 300 benchmarks were generated in 30 series: 15 series with 3-SAT benchmarks, 15 series with k-SAT formulas ($k > 3$). For the 3-SAT series, 3 series of most probably SAT (ratio number of clauses over number of variables equals 4) and UNSAT (ratio 4.5) benchmarks for 500, 700 and 900 variables were generated. The remaining 9 series were generated at the threshold (ratio 4.25) for 400 to 800 by 50 variables. For the 15 k-SAT benchmarks, $k = 4, 5, 6, 7, 8$, the instances were generated at the threshold for

---

[3] http://www.lri.fr/~simon/contest/results/

3 different number of variables (different for each k). Let us emphasize that the number of variables used were adapted from last year solvers performances: we needed instances not too easy but not too hard to discriminate the solvers, considering only last year solvers performances.

Industrial. During last year competition, some competitors spotted that their solvers were behaving much better on the original industrial benchmarks than on the shuffled version we were using in the competition. The order of the clauses and the numbering of the variables were supposed to have a meaning in those benchmarks and some solvers might be able to take that information into account. As a consequence, the performance of their solver during the competition did not match their observation in an industrial setting. We took this remark into account and generated 3 series from one in this category: one with the original benchmarks, and two with different shuffling of the original benchmarks.

The industrial category benchmark set was composed of 157 original benchmarks in 18 series. Two of them were dedicated to benchmarks that remained unsolved since the first competition. Another 7 formed the benchmarks not solved last year. The 10 new series were proposed by:

Hoonsang Jin 8 benchmarks from BMC using the Vis system [16].
Marijn Heule 1 benchmark from Philips
Allen Van Gelder 12 benchmarks encoding into SAT coloring problems [20].
Miroslav Velev Formal Verification problems in two series, pipe-sat-1.1 [22] (10 benchmarks) and vliw-unsat-2.0 (8 benchmarks).
Emmanuel Zarpas 5 series from IBM BMC benchmarks [25]: 01-rule, 11-rule-2, 13-rule, 22-rule and 30-rule.

Crafted. Called "Hand Made" in the previous competition, this category contains 29 series for a total of 228 benchmarks. Four series contained instances unsolved since the first competition (`hgen`, `ezfact`, `urquhart`, `others`), 9 series contained benchmarks that remained unsolved last year (`bqwh`, `chesscolor`, `factoring`, `anton SAT/UNSAT`, `hirsch`, `moore`, `markstrom UNSAT`, `station hwb`). The new benchmarks were submitted by:

Andrew Slater two series of randomly generated clustered benchmarks.
Hoonsang Jin one Ramsey benchmark.
Ke Xu 2 series from forced satisfiable CSP instances of Model RB [24].
Marijn Heule 2 series of equivalency chains.
Calin Anton 4 series of Random l-clustered k-SAT.
Harold Connamacher 5 series of benchmarks encoding the generic uniquely extendible constraint satisfaction problem [3].

Note that this year the crafted category contains the "look like random series" `hgen+`, `hirsch`, `moore` that belonged to the random category in the previous competition.

3   First Stage Results

3.1   First Stage on Random

The first obvious comment about the results of the first stage in the random category is that the incomplete solvers outperformed the complete ones this year. A nice way to see it is to take a look at Figure 1. Incomplete solvers appear on the right (from saps to saprover): they solve quickly a large number of benchmarks. Complete solvers are on the left. The complete local search solver cls lies in between. The satzilla's solvers that use a local search solver as a preprocessing step are located with the local search solvers.

Note that the best incomplete solver in number of series solved was able to solve all the series containing SAT benchmarks (3 series of 3-CNF were generated to be UNSAT). While all the incomplete solvers were able to solve at least 100 benchmarks, the best complete solver cls was only able to solve 65 of them. Note that last year winner kncfs was only able to solve 60 benchmarks. So it looks like we have a good new complete SAT solver for the random category this year? Not really. cls is a complete local search solver: compared to its local search siblings, it performs poorly on satisfiable benchmarks. Compared to its complete siblings, it perform poorly on UNSAT problems (it did not solve one!).

Another obvious fact is that none of the conflict-driven clause-learning algorithms extending zchaff (including the original) succeeded to enter the second stage. None of them was able to solve an UNSAT instance. The best of them, with respect to the number of benchmarks solved, satzoo-1.02, was only able to solve 13 instances in the series k3-r4.25-v400 (9), k3-r4-v500 (3), k4-r9.88-v155 (1). satzoo-1.02 solved the first series in 2500 s while the incomplete solvers solved it in less than a second and specialized complete solver such as kncfs and march needed respectively 38s and 18s.

The three very strong solvers saps, walksatrnp and adaptnovelty are similar to each other. Basically, walksatrnp and adaptnovelty are very similar in the sense that they are two implementations of the Rnovelty+[8] version of Walksat. The difference lies in the value of the noise (p): walksatrnp uses a fixed value of $p = 0.5$ while adaptnovelty adapts the value of p during the search. saps is using a different approach (scaling and probabilistic smoothing) but is implemented in the same framework as adaptnovelty: UBCSAT[4]. The series solved by walksatrnp but not solved by adaptnovelty and saps is k7-r88.7-v110, the collection of 10 7-CNF with 110 variables at ratio 88.7. gasat also failed to solve that series. Note that gasat solved many series but significantly fewer benchmarks than the three best solvers. This may be explained by either a greater algorithmic complexity or a less efficient implementation of the solver. This tends to be confirmed in the second stage (Table 2) since gasat solved 17 new benchmarks while most of the other incomplete solvers solved only a few (2 to 7), by the shape of gasat in Figure 1 and its *crr* value of 1.82 (Detailed explanation of this value is given later, when table 5 is introduced).

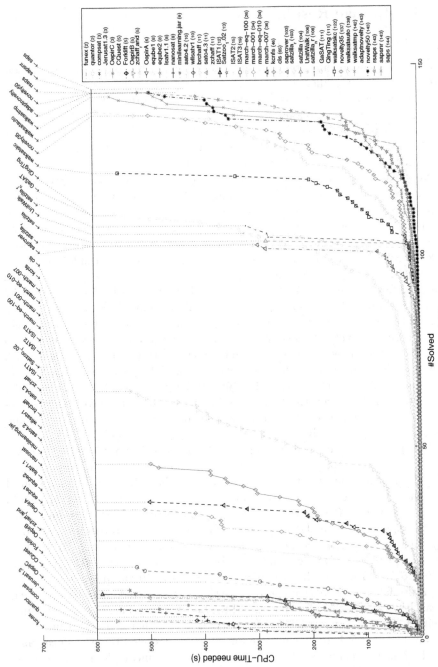

Fig. 1. # of instances solved vs. CPU time for all solvers on SAT instances

Table 1. Results of the first stage for the solvers that entered the second stage by category plus the results of `forklift` and `oepir`

ALL			SAT			UNSAT		
INDUSTRIAL								
Solver	#series	#benchs	Solver	#series	#benchs	Solver	#series	#benchs
Forklift	30	81	CQuest	10	25	Forklift	24	51
zchaff_rand	27	67	CirCUsH1	10	23	zchaff_rand	18	40
brchaff	25	58	OepirC	9	40	brchaff	17	29
CirCUsH1	25	53	Jerusat1.3	9	39	OepirC	16	55
OepirC	24	95	compsat	9	36	Satzoo_1.02	16	18
compsat	23	57	Forklift	9	30	CirCUsH1	15	30
CQuest	22	45	brchaff	9	29	quantor	15	21
minilearning.jar	22	44	zchaff_rand	9	27	Jerusat1.3	15	17
Jerusat1.3	21	56	minilearning.jar	8	25	compsat	14	21
quantor	21	37	Satzoo_1.02	7	22	minilearning.jar	14	19
Satzoo_1.02	19	40	quantor	6	16	CQuest	12	20
CRAFTED								
Solver	#series	#benchs	Solver	#series	#benchs	Solver	#series	#benchs
march-eq-100	13	63	sapsnr	11	39	Satzoo_1.02	7	15
satzilla_nr	13	48	satzilla_nr	11	38	march-eq-100	6	27
Satzoo_1.02	12	53	march-eq-100	10	36	brchaff	6	8
brchaff	12	42	Satzoo_1.02	9	38	satzilla_nr	5	10
sapsnr	11	39	Jerusat1.3	9	37	OepirA	5	10
march-007	10	59	march-007	9	36	zchaff	5	9
zchaff	10	41	brchaff	9	34	march-007	4	23
Jerusat1.3	9	45	zchaff	8	32	nanosat	4	8
OepirA	9	41	nanosat	8	31	Jerusat1.3	3	8
nanosat	9	39	OepirA	7	31	sapsnr	0	0
RANDOM								
Solver	#series	#benchs	Solver	#series	#benchs	Solver	#series	#benchs
walksatrnp	27	142	walksatrnp	27	142	kcnfs	4	14
saps	26	144	saps	26	144	march-007	3	9
adaptnovelty	26	143	adaptnovelty	26	143			
GaSAT	26	111	GaSAT	26	111			
satzilla_nr	20	108	satzilla_nr	20	108			
QingTing	19	111	QingTing	19	111			
cls	19	65	cls	19	65			
UnitWalk	17	106	UnitWalk	17	106			
kcnfs	11	60	kcnfs	10	46			
march-007	9	45	march-007	8	36			

On the UNSAT instances, only a limited number of series and benchmarks were solved. As a result, the ranking of the overall category (SAT+UNSAT) is quite close to that of the SAT category.

At the satisfiability threshold, the generated instances have the same probability to be SAT or UNSAT, thus one could expect half of the instances to be SAT and the other ones UNSAT. During the first stage, exactly 150 benchmarks were proved SAT by the incomplete solvers, which is exactly half of the benchmarks. In order to check this result, after the competition, we launched `kncfs` and `adaptnovelty` with a 15-hour timeout, on the still-unsolved random benchmarks, and they were unable to find new SAT ones (`kncfs` found 18 new UNSAT benchmarks). Even if such a result may look nice, let us notice that the number of SAT/UNSAT benchmarks was not uniformly dispatched across the series.

Table 2. Second stage results. On the random category, we also report the number of newly solved benchmarks

ALL			SAT			UNSAT	
INDUSTRIAL							
Solver	#benchs		Solver	#benchs		Solver	#benchs
zchaff-rand	**20**		**Jerusat1.3**	**3**		**zchaff-rand**	**18**
Satzoo-1.02	10		CirCUsH1	2		Satzoo-1.02	9
brchaff	8		zchaff-rand	2		brchaff	7
Jerusat1.3	6		brchaff	1		quantor	6
quantor	6		compsat	1		CQuest	5
CQuest	5		Satzoo-1.02	1		minilearning	3
CirCUsH1	3		quantor	0		Jerusat1.3	3
minilearning	3		CQuest	0		compsat	2
compsat	3		minilearning	0		CirCUsH1	1
CRAFTED							
Solver	#benchs		Solver	#benchs		Solver	#benchs
march-eq-100	**10**		**Satzoo-1.02**	**2**		**march-eq-100**	**10**
Satzoo-1.02	10		brchaff	1		Satzoo-1.02	8
satzilla_nr	3		nanosat	1		satzilla_nr	3
Jerusat1.3	2		sapsnr	0		Jerusat1.3	2
brchaff	1		satzilla_nr	0		brchaff	0
nanosat	1		march-eq-100	0		nanosat	0
sapsnr	0		Jerusat1.3	0		sapsnr	0
march-007	0		march-007	0		march-007	0
zchaff	0		zchaff	0		zchaff	0
RANDOM							
Solver	#benchs	#new	Solver	#benchs	#new	Solver	#benchs	#new
adaptnovelty	**150**	7	**adaptnovelty**	**150**	7	**kcnfs**	**26**	12
sapsnr	148	4	sapsnr	148	4	march-007	18	9
walksatrnp	145	3	walksatrnp	145	3			
GaSAT	128	17	GaSAT	128	17			
QingTing	113	2	QingTing	113	2			
UnitWalk	112	6	UnitWalk	112	6			
satzilla_nr	110	2	satzilla_nr	110	2			
kcnfs	90	30	cls	78	13			
cls	78	13	kcnfs	64	18			
march-007	63	18	march-007	45	9			

3.2   First Stage on Industrial

The industrial category is a very competitive category, mainly because of its economic aspect. Some industry-related solvers competed in the previous competitions but authors of last year winner, forklift, declined the invitation to submit a detailed description of their solver in the conference post proceedings, due to intellectual property reasons. If this can be easily understood from their point of view, from the organizers point of view, and with the agreement of the judges, it was decided not to use the SAT competition to publicize a solver without providing anything back to the SAT community.

We introduced for that reason the "Anti-Blackbox rule" that prevents SAT solvers for which the details are not known (no source code available, no publicly available technical report about it) to participate in the competition. This is a clear path on which we want the competition to stay. For those reasons, the popular SAT solver siege[4] did not enter the contest this year, but forklift

---

[4] http://www.cs.sfu.ca/~loryan/personal/

entered the contest as last year winner. The `oepir` solvers entered the competition, an overall description of the solvers being provided, but a more detailed description was borderline with a non-disclosure agreement. For that reason, the solvers became ineligible for an award. The following results do include both `forklift` and `oepirs` ones, because we think that awarding a solver and reporting its results are different things. *It is important for the community to see where those solvers stand compared to the other ones.*

Without any surprise, Conflict-Driven Clause-Learning (CDCL) solvers outperform the other ones in that category. All the solvers entering the second stage either include or extend a CDCL solver. The more robust solver is the last year winner `forklift`, which solved 30 series for a total of 81 benchmarks. The best solver with respect to the total number of benchmarks solved is a variant of Oepir, `oepirc`, which solved 24 series for a total of 95 benchmarks. The series solved by `forklift` and not by `oepirc` are Schuppan `12s` and Miroslav Velev `VLIW-UNSAT-2`, for a total of 8 benchmarks. Note that `oepirc` was able to solve them neither under their original form nor after shuffling. Both series contain huge benchmarks (15 000 to 1M variables for `12s`, 25 000 to 500 000 variables for `VLIW`).

`oepirc` outperformed `forklift` on IBM benchmarks, especially on the original series. `zchaff-rand` is also a strong solver with 27 series and 67 benchmarks solved. It was not able to solve the `12s` benchmarks. `brchaff`, `circush1`, `compsat` and `jerusat1.3` were able to solve more than 50 benchmarks. On the one hand, `oepirc`, `jerusat1.3` and `compsat` solved significantly more SAT benchmarks than the other solvers, but on the other hand, `oepirc` and `forklift` outperformed the other solvers in the UNSAT category, followed by `zchaff-rand`. Interestingly, `jerusat1.3` was the worst performer in the UNSAT category (among the second stage solvers).

Twice as many UNSAT than SAT series were solved, same ratio for the number of benchmarks solved, which emphasizes strong UNSAT solvers in the overall ranking.

3.3 First Stage on Crafted

The results in this "everything not uniform random or industrial" category are quite close. The incomplete solver `sapsnr` was the most robust solver with 11 series solved in the satisfiable category, but all the second stage solvers were quite close: there is only a difference of 8 benchmarks solved between the first and the last solver entering the second stage. In the UNSAT category, the most robust solver with 7 series solved is `satzoo-1.02`, while the strongest solvers with respect to the number of benchmarks solved are `march-007` and `march-eq-010`. On the overall ranking, `march-eq-010` and `satzilla-nr` are the most robust solvers while `march-007` and `march-eq-100` remain the best solvers with respect to the number of benchmarks solved. It is worth noting that in the crafted category, the three top solvers in the overall ranking are using completely different technologies.

4   The Second Phase: The Winners

Table 2 summarizes the result of the second stage in the three categories. The solvers `oepirc` and `forklift` do not appear since they were not awardable. During the second stage, the solvers were ranked according to the number of benchmarks solved *among the benchmarks remained unsolved during the first stage.*

Winners in the industrial category are consistent with the results of the first stage (once `forklift` and `oepirc` have been discarded): `zchaff-rand` is awarded in the industrial UNSAT category while `jerusat1.3` is awarded in the industrial SAT category. Since the number of UNSAT benchmarks solved is greater than the number of SAT benchmarks solved, the overall ranking emphasizes strong UNSAT solvers. As a results, `zchaff-rand` is also the winner in the industrial overall category.

For the random category, no solver was able to find a new satisfiable benchmark among those remaining unsolved during the first stage. As a consequence, it was decided with the agreement of the judges, to run the second stage solvers on all the benchmarks they did not solve during the first stage. It can be viewed as re-running all the solvers with an increased timeout on the initial benchmark set. As a consequence, because `adaptnovelty` was able to solve the 150 benchmarks proved satisfiable during the first stage, it was declared winner in the random SAT category. Last year's winner `kncfs` is winning again this year in the UNSAT category. Because of the unbalanced number of proved SAT/UNSAT benchmarks, `adaptnovelty` was also declared winner of the overall random category.

In the crafted category, `march-eq` was awarded in the UNSAT and overall categories while `satzoo-1.02` defended successfully his award in the satisfiable category.

Another interesting data that can be derived from the SAT competition is the smallest instance (with respect to their number of literals) in both SAT and UNSAT categories remained unsolved after the second stage. As stated earlier, those benchmarks should belong to the crafted category, since this is the aim of that category.

The smallest UNSAT benchmark still unsolved after the second stage remains last year's award winner for the smallest unsolved UNSAT benchmark `hgen8-n260-01-S1597732451` (260 variables, 391 clauses, 888 literals) produced by the hgen8 generator[5] submitted last year by Edward A. Hirsch.

Last year's smallest unsolved satisfiable benchmark `hgen2-v400-s161064952` (400 variables, 1400 clauses, 4200 literals) was solved this year by **rsaps** in 461 seconds during the first stage and **sapsnr** in 902s during the second stage.

---

[5] Available at `http://logic.pdmi.ras.ru/~hirsch/benchmarks/hgen8.html`

The new smallest unsolved satisfiable benchmark was also generated by the hgen2 generator[6] submitted by Edward A. Hirsch for the first competition in 2002 with 450 variables instead of 400:
`hgen2-v450-s41511877.shuffled-as.sat03-1682.cnf`
(1575 variables, 450 clauses, 4725 literals).

One can note that the smallest unsolved satisfiable benchmark is significantly larger than the smallest unsolved UNSAT benchmark.

5   State of the Art Contributors

It may be interesting to put all the solvers in a single entity that can decide for each SAT benchmark the best solver to solve it (a sort of perfect `satzilla`). Such entity is called the State-Of-The-Art (SOTA) solver in [19]. Not all the solvers are useful to build the SOTA solver, so we name the solvers that are needed SOTAC (State of the Art Contributor).

Any solver that uniquely solves any benchmark is obviously a SOTAC. Table 3 lists the different solvers that are the only one to solve a given benchmark in the three categories. We computed those numbers from the first stage results, with and without the black boxes `oepir` and `forklift`. The measure can only be made on the first stage results because we need exhaustive results to give all solvers a chance. Thus, some solvers not strong enough to enter the second stage can be distinguished here.

Table 3. SOTAC ranking: the number indicates the number of benchmarks that sovlers uniquely solve. Numbers in parenthesis detail the number of SAT/UNSAT benchmarks respectively

		Without black-boxes					Including black-boxes		
Solver	Total	Random	Ind.	Crafted	Solver	Total	Rand.	Ind.	Craft.
circush0	6	–	–	6 (0/6)	forklift	8	–	8 (1/7)	–
kcnfs	6	5 (0/5)	–	1 (1/0)	circush0	6	–	–	6 (0/6)
jerusat1.3	5	–	4 (2/2)	1 (1/0)	kcnfs	6	5 (0/5)	–	1 (1/0)
brchaff	4	–	3 (0/3)	1 (1/0)	jerusat1.3	4	–	3 (1/2)	1 (1/0)
zchaff	4	–	4 (4/0)	–	brchaff	3	–	2 (0/2)	1 (1/0)
quantor	3	–	3 (0/3)	–	oepira	3	–	3 (1/2)	–
satzoo-1.02	3	–	1 (0/1)	2 (0/2)	oepirb	3	–	2 (0/2)	1 (0/1)
adaptnovelty	2	–	–	2 (2/0)	oepirc	3	–	3 (0/3)	–
circush1	2	–	2 (2/0)	–	quantor	3	–	3 (0/3)	–
compsat	2	–	2 (1/1)	–	satzoo-1.02	3	–	1 (0/1)	2 (0/2)
rsaps	2	1 (1/0)	–	1 (1/0)	adaptnovelty	2	–	–	2 (2/0)
zchaff-rand	2	–	2 (0/2)	–	circush1	2	–	2 (2/0)	–
cquest	1	–	1 (0/1)	–	rsaps	2	1 (1/0)	–	1 (1/0)
march-eq-100	1	–	–	1 (1/0)	compsat	1	–	1 (0/1)	–
novelty35	1	–	1 (1/0)	–	march-eq-100	1	–	–	1 (1/0)
novelty50	1	1 (1/0)	–	–	novelty35	1	–	–	1 (1/0)
sapsnr	1	1 (1/0)	–	–	novelty50	1	1 (1/0)	–	–
satzilla	1	–	1 (1/0)	–	sapsnr	1	1 (1/0)	–	–
					satzilla	1	–	1 (1/0)	–
					zchaff	1	–	1 (1/0)	–

---

[6] Available at `http://logic.pdmi.ras.ru/~hirsch/benchmarks/hgen2.html`

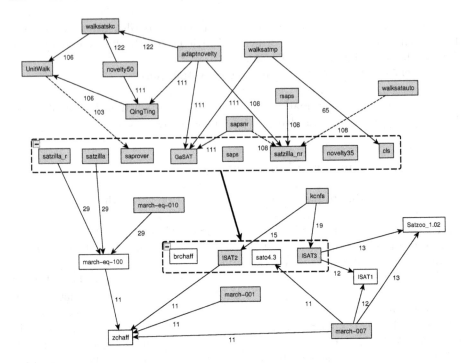

Fig. 2. Subsumptions relations on random instances

Note that adding `oepirc` and `forklift` does not change much the result: `zchaff` looses 3 benchmarks, `cquest` disappears, `jerusat1.3`, `compsat` and `brchaff` loose one benchmark. Note also that having variants of solvers minimizes the number of uniquely solved instances: with three variants, `oepir` solvers have small numbers of uniquely solved benchmarks compared to the other strong solvers.

In the random category, `kncfs`, `rsaps`, `novely50` and `sapsnr` are three SOTAC in the first stage. Note that `rsaps`, `novely50` and `sapsnr` would not be SOTAC if we take into account the second stage since `adaptnovelty` was able to solve all the SAT benchmarks. It is thus sufficient with a 2400 second timeout to keep adaptnovelty in our SOTA solver to solve all the satisfiable benchmarks solved during the competition in the random category. As a consequence, `adaptnovelty` would be a SOTAC without being the only one to solve one benchmark. Moreover, the smallest SOTA is simply composed by `kncfs` and `adaptnovelty`. Both solvers solves all the solved benchmarks in the random category.

It is more difficult to find the smallest SOTA in the other categories, because one would have to choose the smallest subset of solvers that solves all the benchmarks. If all the SOTAC are indeed in this subset, the choice has to be made for the remaining solvers.

Following the idea of SOTAC, one important issue may be to identify solvers that are subsumed by another solver. Such solvers would be useless in a SOTA

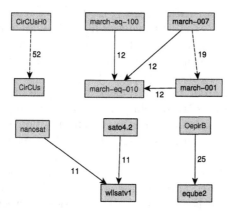

Fig. 3. Subsumptions relations on industrial instances

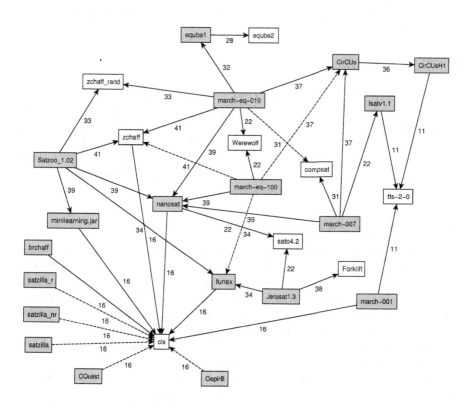

Fig. 4. Subsumptions relations on crafted instances

solver because all the problems solved by such solver would be also solved by the subsuming solver. However, in some cases, a subsumed solver may be more efficient on its subset of benchmarks. We represent, in figures 2, 3 and 4, the subsumption relations for each category. Dotted lines represent a subsumption relation but with a loss of CPU-time, and plain lines are strong subsumptions, with CPU-time savings. Each edge is labeled with the number of common solved benchmarks. For instance, figure 2 told us that `novely50` subsumes `walksatauto` and `qingting` by solving respectively all the 122 and 111 benchmarks that the solvers also solve. In order to obtain clean figures, we only considered solvers that may solve more than 15 benchmarks in each category and we deleted redundant edges. Thus, a lot of very-simple subsumption relations are missing but are not really interesting ones, especially all the relations with inefficient solvers.

One may observe that while there are many subsumption relations between solvers in the random and crafted category, it is not true in the industrial category. Especially, there is no subsumption relation between the second stage solvers in that category, which is not the case in the random category (`qingting`, `gasat` and `satzilla-nr` are subsumed by `adaptnovelty` for instance) and the crafted category (`nanosat` is subsumed by `march` solvers and `satzoo-1.02`).

6   Other Rankings

In order to demonstrate that the competition final results do not change much when the evaluation rules change, we propose here other ways to analyze the results, other methods to rank solvers and to compare the rankings. For instance, in the contest, we tried to forget about the CPU-time, by counting only the number of series and benchmarks solved. Of course the CPU-time had a direct impact on the competition due to the timeout, but it is also important to try to characterize efficient solvers, in terms of CPU-time.

6.1   Ranking *à la* SatEx

This ranking used in the SatEx web site[18] allows us to give a picture of solver performances in a very simple but meaningful way. It is not obvious to rank solvers on a basis of a penalty time when the status (SAT or UNSAT) of some benchmarks are not known. We chose to only count, on SAT and UNSAT subcategories, the subset of benchmarks that have been at least solved by a solver (including black-boxes here). Thus, the ranking is preserved even if all or none of the unknown benchmarks are SAT or UNSAT (all solvers will then have to pay the same penalty in the corresponding category).

The winners of the random categories are ranked first using the ranking, same thing for the overall ranking and UNSAT ranking in the industrial category. For the industrial SAT category `compsat` is ranked first but solves fewer benchmarks than `jerusat1.3` and `oepirc`. The same thing happens in the overall and UNSAT crafted categories, where `march-001` is ranked before `march-eq-100`. Concerning the satisfiable crafted category, the incomplete solver `sapsnr` is ranked first closely followed by `satzoo-1.02`. It is the only case for which the SatEx ranking does not guess the second stage winner. The consistency of the SatEx

Table 4. Cumulative CPU-Time Ranking (SatEx style), given with relative values. Only second-stage solvers appear here

Random benchmarks

All Benchs. (over 300)			SAT (over 150)			UNSAT (over 14)		
Solver	Time (s)	Nb.	Solver	Time (s)	Nb.	Solver	Time (s)	Nb.
adaptnovelty	97757	143	adaptnovelty	7757	143	kcnfs	4066	14
saps	+747	+1	saps	+747	+1	satzilla_r	+1689	-6
sapsnr	+1091	0	sapsnr	+1091	0	satzilla	+1704	-6
walksatmp	+1354	-1	walksatmp	+1354	-1	march-007	+1795	-5
novelty35	+4780	-6	novelty35	+4780	-6	march-001	+2064	-7
satzilla-nr	+18832	-35	satzilla-nr	+18832	-35	march-eq-010	+2937	-10
qingting	+19141	-32	qingting	+19141	-32	march-eq-100	+4027	-11
unitwalk	+20494	-37	unitwalk	+20494	-37			
gasat	+21722	-32	gasat	+21722	-32			
cls	+51259	-78	cls	+51259	-78			
kcnfs	+56281	-83	kcnfs	+60615	-97			
march-007	+60694	-98	march-007	+63233	-107			

Industrial benchmarks

All Benchs. (over 477)			SAT (over 77)			UNSAT (over 97)		
Solver	Time (s)	Nb.	Solver	Time (s)	Nb.	Solver	Time (s)	Nb.
oepirc	249839	95	compsat	28269	30	forklift	36333	52
forklift	+1751	-14	jerusat1.3	+1506	+3	oepirc	+1695	+3
zchaff-rand	+7842	-28	oepirc	+1715	+4	zchaff-rand	+6477	-12
compsat	+10315	-38	forklift	+4771	-6	brchaff	+11053	-23
brchaff	+12654	-37	zchaff-rand	+4775	-9	circush1	+11575	-22
jerusat1.3	+13161	-39	brchaff	+5011	-7	cquest	+13255	-32
circush1	+16176	-42	circush1	+8011	-13	quantor	+13430	-31
cquest	+18532	-50	cquest	+8687	-11	compsat	+13726	-31
minilearning.jar	+20066	-51	minilearning.jar	+8717	-11	satzoo-1.02	+14484	-34
quantor	+21715	-58	satzoo_1.02	+11296	-13	minilearning.jar	+14759	-33
			quantor	+11695	-20	jerusat1.3	+15064	-35

Crafted benchmarks

All Benchs. (over 228)			SAT (over 69)			UNSAT (over 41)		
Solver	Time (s)	Nb.	Solver	Time (s)	Nb.	Solver	Time (s)	Nb.
march-007	104789	59	sapsnr	20624	39	march-007	11159	23
march-eq-100	+2611	+4	satzoo-1.02	+472	-1	march-eq-100	+82	+4
satzoo-1.02	+5188	-6	jerusat1.3	+1472	-2	satzoo-1.02	+6922	-8
jerusat1.3	+8841	-14	march-007	+2206	-3	jerusat1.3	+9575	-15
brchaff	+10941	-17	brchaff	+2651	-5	oepira	+9886	-13
sapsnr	+11235	-20	zchaff	+3843	-7	zchaff	+10194	-14
zchaff	+11832	-18	satzilla-nr	+4659	-1	brchaff	+10495	-15
satzilla_nr	+13280	-11	march-eq-100	+4735	-3	satzilla-nr	+10827	-13
nanosat	+14362	-20	nanosat	+5444	-8	nanosat	+11124	-15
oepira	+14583	-18	oepira	+6903	-8			

ranking with the contest results also suggests that the time out was large enough to serve as a penalty.

6.2 Relative Efficiency of Solvers

One of the hard things to handle for ranking solvers is that only partial information is available, because of the timeout. One solution that we adopted last year

Table 5. Relative Efficiency (*re*) and CPU-Time Reduction Ration (*crr*) values for all solvers that participated the second stage. Values are computed over all the solvers of the category, for all benchmarks (SAT and UNSAT)

Random 12 Solvers			Industrial 11 Solvers			Crafted 10 Solvers		
Solver	re	crr	Solver	re	crr	Solver	re	crr
adaptnovelty	0.95	0.37	oepirc	0.86	0.74	satzoo-1.02	0.89	1.06
walksatrnp	0.95	0.49	forklift	0.77	0.68	march-eq-100	0.86	1.30
sapsnr	0.95	0.56	zchaff-rand	0.73	0.76	march-007	0.84	1.00
saps	0.95	0.67	brchaff	0.66	0.89	satzilla-nr	0.80	1.55
novelty35	0.93	0.86	compsat	0.64	0.70	nanosat	0.79	1.95
qingting	0.81	2.81	satzoo-1.02	0.64	1.94	jerusat1.3	0.77	1.16
satzilla-nr	0.80	0.93	jerusat1.3	0.63	1.55	oepira	0.74	3.13
unitwalk	0.79	2.49	circush1	0.62	1.05	brchaff	0.73	1.28
gasat	0.78	1.82	cquest	0.61	1.17	zchaff	0.72	1.80
cls	0.50	10.81	quantor	0.58	1.43	sapsnr	0.45	0.27
kcnfs	0.43	280.54	minilearning.jar	0.56	1.48			
march-007	0.33	284.76						

[12] was to compare only pairs of solvers (X,Y) on the subset of benchmarks they both solve. Let us write $s(X)$ as the set of benchmarks solved by X. Then, we can compare X and Y on their respective performances on the set $s(X) \cap s(Y)$. When doing this, we have a strong way of comparing the *relative efficiency* (RE) of X and Y : $re(X,Y) = s(X) \cap s(Y)/s(Y)$ gives the percentage of instances of Y that X solves too. Let us write now $CPU(X,b)$ the CPU time needed for X to solve all the benchmarks in b, without any timeout. Because there was a timeout in the competition, only $CPU(X,s')$, with $s' \subseteq s(X)$ are defined here for the solver X. We can compare the relative efficiency of X with respect to Y by computing $crr(X,Y) = CPU(X, s(X) \cap s(Y))/CPU(Y, s(X) \cap s(Y))$. This last measure is called here the *CPU-time reduction ratio* (CRR), and means that, on their common subset of solved benchmarks, X needs $crr(X,Y)$ percent of the time needed by Y. To summarize all the possible values, we average these two measures over all the possible values for Y, while keeping X fixed, and we thus defined $re(X)$ and $crr(X)$.

However, to have a relevant measure, one have to restrict the set of considered solvers: an inefficient solver, lucky on one instance, may have a very low re value. We thus only consider here the set of solvers that participated in the second stage, in each category. Since the first and second stage were done on the same computers for the Crafted and Industrial categories (on LRI's cluster, while the second stage for the random category ran on beta lab's cluster), we use extended results (1st stage + 2nd stage + post-competition runs) for those two categories (the timeout may be considered as 2400 s for all launches). The direct consequence of that choice is to increase the number of commonly solved benchmarks. For the random benchmarks, we restrict ourselves to the first stage results only because the second stage was done on a different cluster of computer for that category. Results are given in Table 5.

For the Random category, one can see that adaptnovelty solves 95% of the benchmarks that the other solvers solve too (corresponding to the 150 SAT bench-

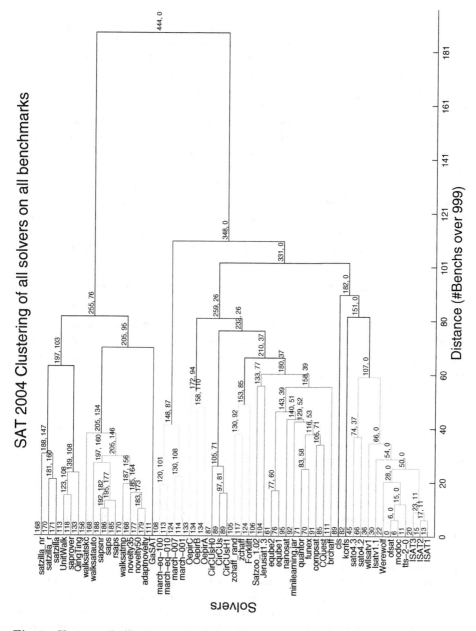

Fig. 5. Clusters of all solvers on all benchmarks. Solvers that closely solve sets of benchmarks are close together in the cluster. The height of nodes in the tree indicates the average distance of the two clusters at each considered branch of the tree. The number on the left indicates the number of instances solved for each solver, and the couple of numbers at each cluster link indicates respectively the number of common benchmarks solved by at least one member of the cluster and the number of benchmarks solved by all the members of the cluster

marks), but in only 37% of their time: `adaptnovelty` is definitively the fastest solver in this category. The *crr* then grows to 49% for `walksatrnp`. There is also a clear partition between complete/incomplete solvers. `kncfs` can solve on average 43% of the benchmarks of the other solvers, but in 280 times their CPU time.

For the Industrial category, `oepirc` exhibits the best relative efficiency, showing a very strong solver. It is interesting to see that `forklift` and `zchaff-rand` are very close from a relative efficiency point of view and that `compsat` exhibits the second best *crr* values for this table, which certainly results from its very good running time in the satisfiable category, even if its relative efficiency drops to 64%.

On the Crafted benchmarks, `satzoo-1.02` has the best *re* value, very close to `march-eq-100`. It is interesting to notice the differences between `march-001` and `march-eq-100` for the *crr*. This is likely due to the additional equivalency reasoning used in the latter. The good runtime of `sapsnr` previously observed on satisfiable crafted benchmarks is denoted here by the best *crr* of the category.

6.3   Clustering of Solvers According to Their Performances

We grouped the solver according to their ability to solve each benchmark. Each solver is represented by a vector of boolean indicating whether or not a given benchmark was solved by that particular solver. Then we use a hamming distance between those vectors to group the solvers: solvers with similar behavior (solving the same benchmarks) have a small hamming distance. The clusters are represented with a tree figure 5.

7   Discussion

7.1   The Effect of Shuffling

Shuffling the industrial benchmarks was initially introduced in the competition rules to prevent any cheating. However, two main problems occurred. First, we noticed that two runs of the same solver on the same –but shuffled differently– benchmark can lead to very different results (see *the lisa syndrome* in [12]). Second, competitors claimed that their solver were behaving much better on the original benchmarks than on the shuffled versions used for the competition. It is not a matter of cheating, but seems to be related to clauses and variables proximity observed in real-world problems encoded into SAT. To observe this phenomenon, three benchmarks were associated to each original industrial benchmark: the original one and two shuffled ones. We study here how solvers behaved on those three benchmarks.

Table 6 details the results of the solvers on the original industrial benchmarks (left column) and the same ones shuffled twice. Roughly all the solvers performed better on the original version of the benchmarks, apart `brchaff`, `jerusat1.3` and `quantor`. While the most robust solver with respect to shuffling is `zchaff-rand`, `forklift`, `circush1` and `oepirc` are the solvers that were the most sensitive to the shuffling.

Table 6. Study of the *lisa* syndrome

Solver	original #series	#benchs	Solver	first shuffling #series	#benchs	Solver	second shuffling #series	#benchs
Forklift	10	31	Forklift	10	24	Forklift	10	26
zchaff_rand	9	23	zchaff_rand	9	21	zchaff_rand	9	23
CirCUsH1	9	22	OepirC	8	28	brchaff	9	22
OepirC	8	38	brchaff	8	19	OepirC	8	29
compsat	8	21	CirCUsH1	8	17	compsat	8	17
Jerusat1.3	8	17	CQuest	8	15	CirCUsH1	8	14
brchaff	8	17	compsat	7	19	Jerusat1.3	7	20
Satzoo_1.02	8	16	quantor	7	14	minilearning.jar	7	14
minilearning.jar	8	16	minilearning.jar	7	14	CQuest	7	14
quantor	8	12	Jerusat1.3	6	19	Satzoo_1.02	6	12
CQuest	7	16	Satzoo_1.02	5	12	quantor	6	11

Table 7. The lisa syndrome, focusing on IBM benchmarks

Solver	original #series	#benchs	Solver	first shuffling #series	#benchs	Solver	second shuffling #series	#benchs
OepirC	4	22	OepirC	4	16	OepirC	4	15
zchaff_rand	4	13	brchaff	4	13	zchaff_rand	4	13
Forklift	4	13	compsat	4	12	Forklift	4	13
compsat	4	13	zchaff_rand	4	11	brchaff	4	13
CQuest	4	12	Forklift	4	11	CQuest	4	10
brchaff	4	12	CQuest	4	11	compsat	4	10
CirCUsH1	4	11	quantor	4	10	quantor	4	9
Satzoo_1.02	4	9	minilearning.jar	4	8	minilearning.jar	4	8
minilearning.jar	4	9	CirCUsH1	4	8	CirCUsH1	4	8
Jerusat1.3	4	9	Jerusat1.3	3	10	Jerusat1.3	3	10
quantor	4	8	Satzoo_1.02	2	5	Satzoo_1.02	3	6

Table 7 focuses on the five series from IBM. oepirc is again quite sensitive to shuffling while the other solvers look relatively robust (the number of benchmarks solved is too small to draw any conclusions).

The effect of shuffling the benchmarks in the industrial category is making the benchmarks harder for most of the solvers. Furthermore, the better the solvers are, the more sensitive to shuffling they are.

7.2   Progress or Not?

While the number of solvers has constantly increased since the first competition to reach 55 solvers this year, most of the solvers awarded this year are not new. The very same versions of satzoo-1.02 and kncfs were awarded last year in the same categories. jerusat1.3 was designed a bit after the first competition and participated in the the second one. zchaff was one of the first solvers to be awarded in 2002. march solvers also participated from the very beginning of the competition. The multi-strategies solvers oepir that demonstrated a great potential also participated in the 2003 competition.

These solvers have been improved since last year, for sure. However, none of the strong solvers is a result from a brand new approach.

The good news of this edition lies around local search solvers: the UBCSAT library demonstrated that it was an efficient platform for building various local search solvers. Such a common platform may help developing new efficient local search algorithms for the next competitions. The original `walksat` with `Rnovelty+` strategy also demonstrated its strength, which may tend to prove that `Rnovelty+` strategy itself may be viewed as the winner for the random category.

7.3   Is the Competition Relevant or Not?

Emmanuel Zarpas, from IBM, one of the industrial users embedding SAT solvers as engines in his tools and contributing many benchmarks to the community, published some results he obtained on his own computers on his benchmarks[7], and concluded that it was not possible to tell that Berkmin561 was better than Zchaff on those benchmarks (while berkmin561 performed better than Zchaff during the SAT 2003 competition). A recent update of the experiment showed the Zchaff II (this year winner) was not really better than Zchaff, and performed even worst when taking into account the total CPU time. The underlying question is *whether or not the result of the SAT competition means anything in an industrial setting?*

In a technical report[26], Emmanuel Zarpas proposed some guidelines for a good BMC evaluation, among them, "use relevant benchmarks, use relevant timeout".

Use Relevant Benchmarks. While the aim of IBM is to find the best solver to solve IBM BMC benchmarks, the aim of the competition is to award a robust solver, that is a solver able to solve a wide range of benchmarks across several different kinds of problems split into categories. This different point of view is alone a reason for the discrepancies observed between the results of the competition and those on the IBM benchmarks. Note that some of the IBM benchmarks were part of the SAT 2003 competition benchmarks and the results on those benchmarks were confirmed.

Use Relevant Timeout. In his experiment, Emmanuel Zarpas used a 10000 s timeout while we used 600 s for the first stage and 2400 s for the second stage. For practical reasons, we simply cannot afford spending a 10000 s timeout for the competition with 55 solvers. Figure 6 illustrates the results of the complete solvers on all the benchmarks of the industrial category. This representation allows to check the choice of the CPU timeout and to have clues about solver behavior on all the benchmarks in a given category. The 600 s timeout (dotted line) is the first stage timeout. Only the second stage solvers have an extended timeout of 2400 s. Note that this figure was obtained by launching all the second stage solvers on the benchmarks they did not solve during the first stage, which is not the current 2nd stage setting. In our opinion, the first stage results are confirmed for most solvers with that extended timeout. `satzoo-1.02` is the

---

[7] http://www.haifa.il.ibm.com/projects/verification/RB_Homepage/
bmcbenchmarks_illustrations.html

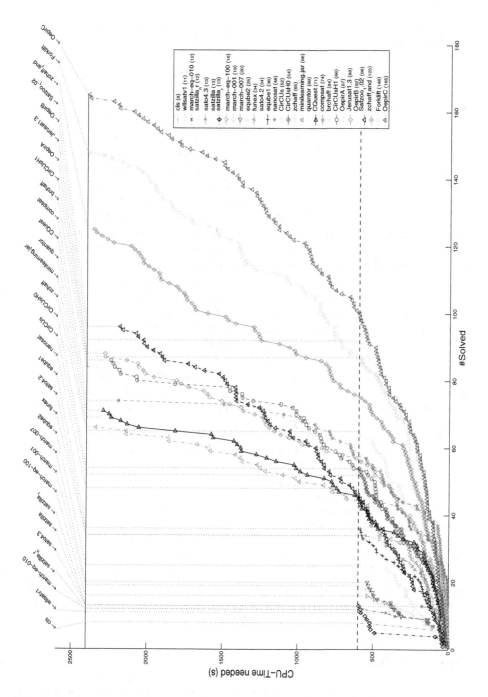

Fig. 6. # of instances solved vs. CPU time for complete solvers on all industrial benchs. Two timeouts were used here, 600s and 2400s, which explain that some curve are stopped before the 600s horizontal limit

exception: while being ranked 10th after the 600 s timeout, it would be ranked 4th after 2400 s. So the use of a "small timeout" is reasonable to isolate the best solvers in a given category.

8   The Next Competitions

Several remarks concerning the competition arose this year, and we think it is time to rethink the competition as a whole. For that reason, we already formed the board of judges for the next competition. It will be composed of Oliver Kullmann, Armin Biere and Allen Van Gelder. We already decided that a special track about certified UNSAT answers will take place during the next competition.

We designed some rules for the first competition in 2002 and tried to follow them during the second and third edition of the competition. The major changes since the beginning were the limitation of the number of variants entering the second stage and the board of judges that appeared in the second edition, and the "anti black box rule" and the solver description booklet that was added this year. Apart from that, the competitions were pretty similar.

One first step was to disallow solvers not fully described -either by a detailed report or by its source code- from the competition this year (so-called back boxes). However, it is always difficult to ask both a solver and a detailed report for the same deadline. It was already difficult to obtain a two-page description for each solver this year after the solvers were submitted. In order to fulfill both the aim to open the competition to as many solvers as possible and to award a good fully described solver, we propose to separate the competition in two steps:

- the first one requires only a 2-page description of the solver to enter the first stage, that is to see a solver running shortly on the competition benchmarks.
- a more detailed report will be required (or the source code of the solver) to participate in the second stage and being awardable.

One of the biggest issues for the next competition is to gather adequate benchmarks for the competition. Some of the benchmarks that remained unsolved during the first competition are still unsolved. It is time to get rid of those benchmarks and to find new fresh benchmarks for the competition (or to use good old ones). Furthermore, the balancing between SAT and UNSAT benchmarks of similar difficulty remains the main problem (c.f. UNSAT benchmarks in the industrial category or SAT benchmarks in the random one). Finally, the idea of a booklet dedicated to the benchmarks is appealing.

Acknowledgments

The authors would like to thank the anonymous reviewers for their help to improve that paper, the three judges, Fahiem Bacchus, Hans Kleine-Büning and João Marques-Silva for their help during the competition, and the "Laboratoire de Recherche en Informatique" (LRI, Orsay, France) and the beta lab (UBC,

Vancouver, Canada) for providing us with clusters of machines. At last, they thank all the authors of solvers and benchmarks for their participation. The first author was supported in part by the IUT de Lens, The Région Nord/Pas-de-Calais and the Université d'Artois.

References

1. Armin Biere. Resolve and expand. In *this issue*, 2004.
2. Armin Biere, Alessandro Cimatti, Edmund M. Clarke, M. Fujita, and Y. Zhu. Symbolic model checking using SAT procedures instead of bdds. In *Proceedings of Design Automation Conference (DAC'99)*, 1999.
3. Harold Connamacher. A random constraint satisfaction problem that seems hard for dpll. In *this issue*, 2004.
4. Dave A. D.Tompkins and Holger H. Hoos. Ubcsat: An implementation and experimentation environment for sls algorithms for sat and max-sat. pages 37–46, 2004.
5. Olivier Dubois and Gilles Dequen. A backbone-search heuristic for efficient solving of hard 3-sat formulae. In *Proceedings of the Seventeenth International Joint Conference on Artificial Intelligence (IJCAI'01)*, Seattle, Washington, USA, August 4th-10th 2001.
6. Hai Fang and Wheeler Ruml. Complete local search for propositional satisfiability. In *Proceedings of AAAI'04*, 2004.
7. E. A. Hirsch and A. Kojevnikov. UnitWalk: A new SAT solver that uses local search guided by unit clause elimination. PDMI preprint 9/2001, Steklov Institute of Mathematics at St.Petersburg, 2001. A journal version is submitted to this issue.
8. Holger Hoos. On the runtime behavior of stochastic local search algorithms for SAT. In *Proceedings of AAAI'99*, pages 661–666, 1999.
9. HoonSang Jin and Fabio Somenzi. CirCUs: A Hybrid Satisfiability Solver. In *this issue*, 2004.
10. Frédéric Lardeux Jin-Kao Hao and Frédéric Saubion. Evolutionary computing for the satisfiability problem. In *Applications of Evolutionary Computing*, number 2611 in LNCS, pages 259–268, University of Essex, England, UK, 14-16 April 2003.
11. Henry A. Kautz and Bart Selman. Pushing the envelope: Planning, propositional logic, and stochastic search. In *Proceedings of the Twelfth National Conference on Artificial Intelligence (AAAI'96)*, pages 1194–1201, 1996.
12. Daniel Le Berre and Laurent Simon. The essentials of the SAT 2003 competition. In *Proceedings of the Sixth International Conference on Theory and Applications of Satisfiability Testing (SAT2003)*, number 2919 in Lecture Notes in Computer Science, pages 452–467, 2003.
13. X. Y. Li, M.F. Stallmann, and F. Brglez. QingTing: A Fast SAT Solver Using Local Search and Efficient Unit Propagation. In *Sixth International Conference on Theory and Applications of Satisfiability Testing*, S. Margherita Ligure - Portofino (Italy), May 2003. See also http://www.cbl.ncsu.edu/publications/, and http://www.cbl.ncsu.edu/OpenExperiments/SAT/ .
14. M. W. Moskewicz, C. F. Madigan, Y. Zhao, L. Zhang, and S. Malik. Chaff: Engineering an efficient SAT solver. In *Proceedings of the 38th Design Automation Conference (DAC'01)*, pages 530–535, June 2001.
15. R. Ostrowski, E. Grégoire, B. Mazure, and L. Sais. Recovering and exploiting structural knowledge from cnf formulas. In *Proc. of the Eighth International Conference on Principles and Practice of Constraint Programming (CP'2002)*, LNCS, pages 185–199, Ithaca (N.Y.), September 2002. Springer.

16. R. K. Brayton, G. D. Hachtel, A. Sangiovanni-Vincentelli, F. Somenzi, A. Aziz, S. -
    T. Cheng, S. Edwards, S. Khatri, Y. Kukimoto, A. Pardo, S. Qadeer, R. K. Ranjan,
    S. Sarwary, T. R. Shiple, G. Swamy, and T. Villa. VIS: a system for verification
    and synthesis. In Rajeev Alur and Thomas A. Henzinger, editors, *Proceedings of
    the Eighth International Conference on Computer Aided Verification CAV*, volume
    1102, pages 428–432, New Brunswick, NJ, USA, / 1996. Springer Verlag.
17. B. Selman, H. A. Kautz, and B. Cohen. Noise strategies for improving local search.
    In *Proceedings of the 12th National Conference on Artificial Intelligence, AAAI'94*,
    pages 337–343, 1994.
18. Laurent Simon and Philippe Chatalic. SATEx: a web-based framework for
    SAT experimentation. In Henry Kautz and Bart Selman, editors, *Electronic
    Notes in Discrete Mathematics*, volume 9. Elsevier Science Publishers, June 2001.
    http://www.lri.fr/ simon/satex/satex.php3.
19. Geoff Sutcliff and Christian Suttner. Evaluating general purpose automated theo-
    rem proving systems. *Artificial Intelligence*, 131:39–54, 2001.
20. Allen Van Gelder. Another Look at Graph Coloring via Propositional Satisfiability.
    In *Proceedings of Computational Symposium on Graph Coloring and Generaliza-
    tions (COLOR02)*, IThaca, NY, September 2002.
21. Allen Van Gelder and Yumi K. Tsuji. Satisfiability Testing with More Reasoning
    and Less Guessing. In D. S. Johnson and M. A. Trick, editors, *Second DIMACS
    implementation challenge : cliques, coloring and satisfiability*, volume 26 of *DI-
    MACS Series in Discrete Mathematics and Theoretical Computer Science*, pages
    559–586. American Mathematical Society, 1996.
22. M.N. Velev. Automatic abstraction of equations in a logic of equality. In M.C.
    Mayer and F. Pirri, editors, *Proceedings of Automated Reasoning with Analytic
    Tableaux and Related Methods (TABLEAUX '03)*, number 2796 in LNAI, pages
    196–213. Springer-Verlag, September 2003.
23. M.N. Velev and R.E. Bryant. Effective use of boolean satisfiability procedures in
    the formal verification of superscalar and vliw microprocessors. In *Proceedings of
    the 38th Design Automation Conference (DAC '01)*, pages 226–231, June 2001.
24. K. Xu and W. Li. Many hard examples in exact phase transitions with ap-
    plication to generating hard satisfiable instances. Technical report, CoRR Re-
    port cs.CC/0302001, 2003. http://www.nlsde.buaa.edu.cn/ kexu/benchmarks/
    benchmarks.htm.
25. Emmanuel Zarpas. Bmc benchmark illustrations. http://www.haifa.il.ibm.com/
    projects/verification/RB_Homepage/bmcbenchmarks.html.
26. Emmanuel Zarpas. Benchmarking sat solvers for bounded model checking. Tech-
    nical report, IBM Haifa Research Laboratory, 2004.
27. Hantao Zhang. SATO: an efficient propositional prover. In *Proceedings of the
    International Conference on Automated Deduction (CADE'97), volume 1249 of
    LNAI*, pages 272–275, 1997.

March_eq:
Implementing Additional Reasoning into an Efficient Look-Ahead SAT Solver

Marijn Heule[*], Mark Dufour,
Joris van Zwieten and Hans van Maaren

Department of Information Systems and Algorithms,
Faculty of Electrical Engineering,
Mathematics and Computer Sciences,
Delft University of Technology
marijn@heule.nl, m.dufour@student.tudelft.nl,
zwieten@ch.tudelft.nl, h.vanmaaren@its.tudelft.nl

Abstract. This paper discusses several techniques to make the look-ahead architecture for satisfiability (SAT) solvers more competitive. Our contribution consists of reduction of the computational costs to perform look-ahead and a cheap integration of both equivalence reasoning and local learning. Most proposed techniques are illustrated with experimental results of their implementation in our solver march_eq.

1 Introduction

Look-ahead SAT solvers usually consist of a simple DPLL algorithm [5] and a more sophisticated *look-ahead procedure* to determine an effective branch variable. The look-ahead procedure measures the effectiveness of variables by performing *look-ahead* on a set of variables and evaluating the reduction of the formula. We refer to the look-ahead on literal x as the Iterative Unit Propagation (IUP) on the union of a formula with the unit clause x (in short $\text{IUP}(\mathcal{F} \cup \{x\})$). The effectiveness of a variable x_i is obtained using a look-ahead evaluation function (in short DIFF), which evaluates the differences between $\mathcal{F}$ and the reduced formula after $\text{IUP}(\mathcal{F} \cup \{x_i\})$ and $\text{IUP}(\mathcal{F} \cup \{\neg x_i\})$. A widely used DIFF counts the newly created binary clauses.

Besides the selection of a branch variable, the look-ahead procedure may detect *failed literals*: if the look-ahead on $\neg x$ results in a conflict, x is forced to true. Detection of failed literals can result in a substantial reduction of the DPLL-tree.

During the last decade, several enhancements have been proposed to make look-ahead SAT solvers more powerful. In satz by Li [9] pre-selection heuristics PROP_z are used, which restrict the number of variables that enter the look-ahead

---

[*] Supported by the Dutch Organization for Scientific Research (NWO) under grant 617.023.306.

H.H. Hoos and D.G. Mitchell (Eds.): SAT 2004, LNCS 3542, pp. 345–359, 2005.

procedure. Especially on random instances the application of these heuristics results in a clear performance gain. However, the use of these heuristics is not clear from a general viewpoint. Experiments with our pre-selection heuristics show that different benchmark families require different numbers of variables entering the look-ahead phase to perform optimally.

Since much reasoning is already performed at each node of the DPLL-tree, it is relatively cheap to extend the look-ahead with (some) additional reasoning. For instance: integration of equivalence reasoning in satz - implemented in eqsatz [10] - made it possible to solve various crafted and real-world problems which were beyond the reach of existing techniques. However, the performance may drop significantly on some problems, due to the integrated equivalence reasoning. Our variant of equivalence reasoning extends the set of problems which benefit from its integration and aims to remove the disadvantages.

Another form of additional reasoning is implemented in the OKsolver [1] [8]: *local learning*. When performing look-ahead on x, any unit clause y_i that is found means that the binary clause $\neg x \vee y_i$ is implied by the formula, and can be "learned", i.e. added to the current formula. As with equivalence reasoning, addition of these local learned resolvents could both increase and decrease the performance (depending on the formula). We propose a partial addition of these resolvents which results in a speed-up practically everywhere.

Generally, look-ahead SAT solvers are effective on relatively small, hard formulas. Le Berre proposes [2] a wide range of enhancements of the look-ahead procedure. Most of them are implemented in march_eq. Due to the high computational costs of the an enhanced look-ahead procedure, elaborate problems are often solved more efficiently by other techniques. Reducing these costs is essential for making look-ahead techniques more competitive on a wider range of benchmarks problems. In this paper, we suggest (1) several techniques to reduce these costs and (2) a cheap integration of additional reasoning. Due to the latter, benchmarks that do not profit from additional reasoning will not take significantly more time to solve.

Most topics discussed in this paper are illustrated with experimental results showing the performance gains by our proposed techniques. The benchmarks range from uniform random 3-SAT near the threshold [1], to bounded model checking (longmult [4], zarpas [3]), factoring problems (pyhala braun [12]) and crafted problems (stanion/hwb [3], quasigroup [14]). Only unsatisfiable instances were selected to provide a more stable overview. Comparison of the performance of march_eq with performances of state-of-the-art solvers is presented in [7], which appears in the same volume.

All techniques have been implemented into a reference variant of march_eq, which is essentially a slightly optimised version of march_eq_100, the solver that won two categories of the SAT 2004 competition [11]. This variant uses exactly the same techniques as the winning variant: full (100%) look-ahead, addition of all constraint resolvents, tree-based look-ahead, equivalence reasoning, and removal of inactive clauses. All these techniques are discussed below.

---

[1] Version 1.2 at http://cs-svr1.swan.ac.uk/~csoliver/OKsolver.html

2   Translation to 3-SAT

The translation of the input formula to 3-SAT stems from an early version of march_eq, in which it was essential to allow fast computation of the pre-selection heuristics. Translation is not required for the current pre-selection heuristics, yet it is still used, because it enables significant optimisation of the internal data structures.

The formula is pre-processed to reduce the amount of redundancy introduced by a straightforward 3-SAT translation. Each pair of literals that occurs more than once together in a clause in the formula is substituted by a single *dummy* variable, starting with the most frequently occurring pair. Three clauses are added for each dummy variable to make it logically equivalent to the disjunction of the pair of literals it substitutes. In the following example $\neg x_2 \vee x_4$ is the most occurring literal pair and is therefore replaced with the dummy variable d_1.

$$
\begin{array}{lll}
\begin{array}{l}
x_1 \vee \neg x_2 \vee \neg x_3 \vee x_4 \vee \neg x_5 \\
x_1 \vee \neg x_2 \vee \neg x_3 \vee x_4 \vee x_6 \\
\neg x_1 \vee \neg x_2 \vee \neg x_3 \vee x_4 \vee \neg x_6 \\
\neg x_1 \vee \neg x_2 \vee x_4 \vee x_5 \vee x_6
\end{array}
\Leftrightarrow
\begin{array}{l}
x_1 \vee d_1 \vee \neg x_3 \vee \neg x_5 \\
x_1 \vee d_1 \vee \neg x_3 \vee x_6 \\
\neg x_1 \vee d_1 \vee \neg x_3 \vee \neg x_6 \\
\neg x_1 \vee d_1 \vee x_5 \vee x_6
\end{array}
\wedge
\begin{array}{l}
d_1 \vee x_2 \\
d_1 \vee \neg x_4 \\
\neg d_1 \vee \neg x_2 \vee x_4
\end{array}
\end{array}
$$

It appears that to achieve good performance, binary clauses obtained from the *original* ternary clauses should be given more weight than binary clauses obtained from ternary clauses which were *generated by translation*. This is accomplished by an appropriate look-ahead evaluation function, such as the variant of DIFF proposed by Dubois *et al.* [6], which weighs all newly created binary clauses.

3   Time Stamps

March_eq uses a time stamp data structure, *TimeAssignments* (TA), which reduces backtracking during the look-ahead phase to a single integer addition: increasing the *CurrentTimeStamp* (CTS).

All the variables that are assigned during look-ahead on a literal x are stamped: if a variable is assigned the value true, it is stamped with the CTS; if it is assigned the value false, it is stamped with $CTS + 1$. Therefore, simply adding 2 to the CTS unassigns all assigned variables.

The actual truth value that is assigned to a variable is not stored in the data structure, but can be derived from the time stamp of the variable:

$$
\mathrm{TA}[x] = \begin{cases}
stamp < CTS & \text{unfixed} \\
stamp \geq CTS \text{ and } stamp \equiv 0 \pmod 2 & \text{true} \\
stamp \geq CTS \text{ and } stamp \equiv 1 \pmod 2 & \text{false}
\end{cases}
$$

Variables that have already been assigned before the start of the look-ahead phase, i.e. during the solving phase, have been stamped with the *Maximum-TimeStamp* (MTS) or with $MTS + 1$. These variables can be unassigned by stamping them with the value *zero*, which happens while backtracking during the solving phase (i.e. *not* during the look-ahead phase). The MTS equals the maximal even value of an (32-bit) integer. One has to ensure that the CTS is always smaller than the MTS. This will usually be the case and it can easily be checked at the start of each look-ahead.

4   Constraint Resolvents

As mentioned in the introduction, a binary resolvent could be added for every unary clause that is created during the propagation of a look-ahead literal - provided that the binary clause does not already exist. A special type of resolvent is created from a unary clause that was a ternary clause prior to the look-ahead. In this case we speak of *constraint resolvents*.

Constraint resolvents have the property that they cannot be found by a look-ahead on the complement of the unary clause. Adding these constraint resolvents results in a more vigorous detection of failed literals. An example:

First, consider only the original clauses of an example formula (figure 1 (a)). A look-ahead on $\neg r$, IUP($\mathcal{F} \cup \{\neg r\}$), results in the unary clause x. Therefore, one could add the resolvent $r \lor x$ to the formula. Since the unary clause x was originally a ternary clause (before the look-ahead on $\neg r$), this is a constraint resolvent. The unique property of constraints resolvents is that when they are added to the formula, look-ahead on the complement of the unary clause results in the complement of the look-ahead-literal. Without this addition this would not be the case. Applying this to the example: after addition of $r \lor x$ to the formula, IUP($\mathcal{F} \cup \{\neg x\}$) will result in unary clause r, while without this addition it will not.

IUP($\mathcal{F} \cup \{\neg r\}$) also results in unary clause $\neg t$. Therefore, resolvent $r \lor \neg t$ could be added to the formula. Since unary clause $\neg t$ was originally a binary clause, $r \lor \neg t$ is not a constraint resolvent. IUP($\mathcal{F} \cup \{t\}$) would result in unary clause r.

$$
\left.
\begin{array}{c}
r \lor \neg s \\
s \lor \neg t \\
s \lor t \lor x
\end{array}
\right\} \qquad r \lor x
$$

$$
\left.
\begin{array}{c}
u \lor \neg v \\
u \lor \neg w \\
v \lor w \lor x
\end{array}
\right\} \qquad u \lor x
$$

$$
\neg r \lor \neg u \lor x
$$

$$
\left.
\begin{array}{c}
\\
\\
\\
\\
\\
\\
\\
\\
\end{array}
\right\} \qquad x
$$

(a)   (b)   (c)

Fig. 1. Detection of a failed literal by adding constraint resolvents. (a) The original clauses, (b) constraint resolvents and (c) a forced literal

Table 1. Performance of march_eq on several benchmarks with three different settings of addition of resolvents during the look-ahead phase

Benchmarks	no resolvents		all binary resolvents		all constraint resolvents	
	time(s)	treesize	time(s)	treesize	time(s)	treesize
random_unsat_250 (100)	1.61	4059.1	1.51	3389.2	**1.45**	3391.7
random_unsat_350 (100)	55.41	89709.4	51.28	72721.1	**48.78**	73357.2
stanion/hwb-n20-01	31.52	282882	24.76	180408	**23.65**	183553
stanion/hwb-n20-02	41.32	345703	33.94	219915	**30.91**	222251
stanion/hwb-n20-03	30.54	280561	23.48	161687	**21.7**	163984
longmult8	139.13	15905	341.46	8054	**90.8**	8149
longmult10	504.92	330094	915.84	11877	**226.31**	11597
longmult12	836.78	41522	847.95	5273	**176.85**	5426
pyhala-unsat-35-4-03	781.19	29591	1379.33	19100	**662.93**	19517
pyhala-unsat-35-4-04	733.44	28312	1366.19	18901	**659.04**	19364
quasigroup3-9	11.67	2139	11.09	1543	**7.97**	1495
quasigroup6-12	117.49	3177	66.13	1362	**58.05**	1311
quasigroup7-12	14.47	346	11.06	248	**10.03**	256
zarpas/rule14_1_15dat	> 2000	-	46.59	0	**20.7**	0
zarpas/rule14_1_30dat	> 2000	-	> 2000	-	**186.27**	0

Constraint resolvent $u \vee x$ is detected during IUP$(\mathcal{F} \cup \{\neg u\})$. After the addition of both constraint resolvents (figure 1 (b)), the look-ahead IUP$(\mathcal{F} \cup \{\neg x\})$ results in a conflict, making $\neg x$ a failed literal and thus forces x. Obviously, IUP$(\mathcal{F} \cup \{\neg x\})$ will not result in a conflict if the constraint resolvents $r \vee x$ and $u \vee x$ were not added both.

Table 1 shows the usefulness of the concept of constraint resolvents: in all our experiments, the addition of mere constraint resolvents outperformed a variant with full local learning (adding all binary resolvents). This could be explained by the above example: adding other resolvents than constraint resolvents will not increase the number of detected failed literals. These resolvents merely increase the computational costs. This explanation is supported by the data in the table: the tree-size of both variants is comparable.

When we look at zarpas/rule_14_1_30dat, it appears that only adding constraint resolvents is essential to solve this benchmark within 2000 seconds. The node-count of zero means that the instance is found unsatisfiable during the first execution of the look-ahead procedure.

5   Implication Arrays

Due to the 3-SAT translation the data structure of march_eq only needs to accommodate binary and ternary clauses. We will use the following formula as an example:

$$\mathcal{F}_{example} = (a \vee c) \wedge (\neg b \vee \neg d) \wedge (b \vee d) \wedge (a \vee \neg b \vee d) \wedge (\neg a \vee b \vee \neg d) \wedge (\neg a \vee b \vee c)$$

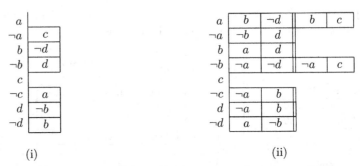

Fig. 2. The binary (i) and ternary (ii) implication arrays that represent the example formula $\mathcal{F}_{\text{example}}$

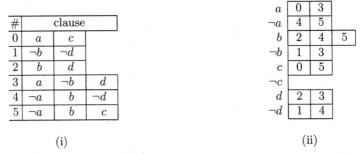

Fig. 3. A common clause database / variable index data structure. All clauses are stored in a clause database (i), and for each literal the variable index lists the clauses in which it occurs (ii)

Binary and ternary clauses are stored separately in two *implication arrays*. A binary clause $a \vee c$ is stored as two implications: c is stored in the binary implication array of $\neg a$ and a is stored in the binary implication array of $\neg c$. A ternary clause $(a \vee \neg b \vee d)$ is stored as three implications: $\neg b \vee d$ is stored in the ternary implication array of $\neg a$ and the similar is done for b and $\neg d$. Figure 2 shows the implication arrays that represent the example formula $\mathcal{F}_{\text{example}}$.

Storing binary clauses in implication arrays requires only half the memory that would be needed to store them in an ordinary clause database / variable index data structure. (See figure 3.) Since march_eq adds many binary resolvents during the solving phase, the binary clauses on average outnumber the ternary clauses. Therefore, storing these binary clauses in implication arrays significantly reduces the total amount of memory used by march_eq. Furthermore, the implication arrays improve data locality. This often leads to a speed-up due to better usage of the cache.

March_eq uses a variant of iterative unit propagation (IFIUP) that propagates binary implications before ternary implications. The first step of this procedure is to assign as many variables as possible using only the binary implication arrays. Then, if no conflict is found, the ternary implication array of each variable that was assigned in the first step is evaluated. We will illustrate this second step with an example.

Suppose look-ahead is performed on $\neg c$. The ternary implication array of $\neg c$ contains $(\neg a \lor b)$. Now there are five possibilities:

1. If the clause is already satisfied, i.e. a has already been assigned the value false or b has already been assigned the value true, then nothing needs to be done.
2. If a has already been assigned the value true, then b is implied and so b is assigned the value true. The first step of the procedure is called to assign as many variables implied by b as possible. Also, the constraint resolvent $(c \lor b)$ is added as two binary implications.
3. If b has already been assigned the value false, then $\neg a$ is implied and so a is assigned the value false. The first step of the procedure is called to assign as many variables implied by $\neg a$ as possible. Also, the constraint resolvent $(c \lor \neg a)$ is added as two binary implications.
4. If a and b are unassigned, then we have found a new binary clause.
5. If a has already been assigned the value true and b has already been assigned the value false, then $\neg c$ is a failed literal. Thus c is implied.

The variant of DIFF used in march_eq weighs new binary clauses that are produced during the look-ahead phase. A ternary clause that is reduced to a binary clause that gets satisfied in the same iteration of IFIUP, should *not* be included in this computation. However, in the current implementation these clauses are in fact included, which causes noise in the DIFF heuristics. The first step of the IFIUP procedure, combined with the addition of constraint resolvents, ensures that the highest possible amount of variables are assigned *before* the second step of the IFIUP procedure. This reduces the noise significantly.

An advantage of IFIUP over general IUP is that it will detect conflicts faster. Due to the addition of constraint resolvents, most conflicts will be detected in the first call of the first step of IFIUP. In such a case, the second step of IFIUP is never executed. Since the second step of IFIUP is considerably slower than the first, an overall speed-up is expected.

Storage of ternary clauses in implication arrays requires an equal amount of memory as the common alternative. However, ternary implication arrays allow optimisation of the second step of the IFIUP procedure. On the other hand, ternary clauses are no longer stored as such: it is not possible to efficiently verify if they have already been satisfied and early detection of a solution is neglected. One knows only that a solution exists if all variables have been assigned and no conflict has occurred.

6 Equivalence Reasoning

During the pre-processing phase, march_eq extracts the so-called equivalence clauses $(l_1 \leftrightarrow l_2 \leftrightarrow \cdots \leftrightarrow l_i)$ from the formula and places them into a separate data-structure called the *Conjunction of Equivalences* (CoE). After extraction, a solution for the CoE is computed as described in [7, 13].

In [7] - appearing in the same volume - we propose a new look-ahead evaluation function for benchmarks containing equivalence clauses: let eq_n be a weight
for a reduced equivalence clause of new length n, $\mathcal{C}(x)$ the set of all reduced
equivalence clauses ($\mathcal{Q}_i$) during a look-ahead on x, and $\mathcal{B}(x)$ the set of all newly
created binary clauses during the look-ahead on x. Using both sets, the look-
ahead evaluation can be calculated as in equation (2). Variable x_i with the
highest $\text{DIFF}_{eq}(x_i) \times \text{DIFF}_{eq}(\neg x_i)$ is selected for branching.

$$eq_n = 5.5 \times 0.85^n \tag{1}$$

$$\text{DIFF}_{eq} = |\mathcal{B}| + \sum_{\mathcal{Q}_i \in \mathcal{C}} eq_{|\mathcal{Q}_i|} \tag{2}$$

Besides the look-ahead evaluation and the pre-selection heuristics (discussed
in section 7), the intensity of communication between the CoE- and CNF-part
of the formula is kept rather low (see figure 4). Naturally, all unary clauses in all
phases of the solver are exchanged between both parts. However, during the solving phase, all binary equivalences are removed from the CoE and transformed to
the four equivalent binary implications which in turn are added to the implication arrays. The reason for this is twofold: (1) the binary implication structure is
faster during the look-ahead phase than the CoE-structure, and (2) for all unary
clauses y_i that are created in the CoE during $\text{IUP}(\mathcal{F} \cup \{x\})$, constraint resolvent
$\neg x \vee y_i$ can be added to the formula without having to check the original length.

We examined other forms of communication, but only small gains were noticed on only some problems. Mostly, performance decreased due to higher com-

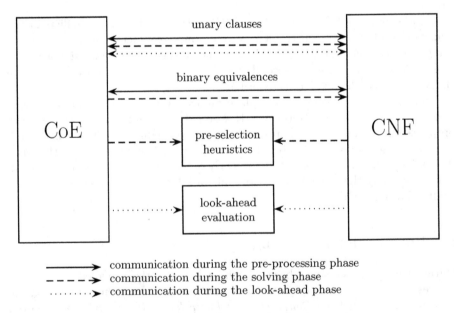

unary clauses

binary equivalences

CoE

pre-selection
heuristics

look-ahead
evaluation

CNF

$\longrightarrow$ communication during the pre-processing phase
$- - - \to$ communication during the solving phase
$\cdots\cdots\cdots\to$ communication during the look-ahead phase

Fig. 4. Various forms of communication in march_eq

Table 2. Performance of march_eq on several benchmarks with and without equivalence reasoning

Benchmarks	without equivalence reasoning		with equivalence reasoning		speed-up
	time(s)	treesize	time(s)	treesize	
random_unsat_250 (100)	1.45	3391.7	1.45	3391.7	-
random_unsat_350 (100)	48.78	73357.2	48.78	73357.2	-
stanion/hwb-n20-01	42.88	182575	23.65	183553	44.85 %
stanion/hwb-n20-02	55.34	222487	30.91	222251	44.15 %
stanion/hwb-n20-03	42.08	164131	21.70	163984	48.43 %
longmult8	76.69	8091	90.80	8149	-18.40 %
longmult10	171.66	11597	226.31	11597	-31.84 %
longmult12	126.36	6038	176.85	5426	-39.96 %
pyhala-unsat-35-4-03	737.15	19513	662.93	19517	10.07 %
pyhala-unsat-35-4-04	691.04	19378	659.04	19364	4.63 %
quasigroup3-9	7.97	1495	7.97	1495	-
quasigroup6-12	58.05	1311	58.05	1311	-
quasigroup7-12	10.03	256	10.03	256	-
zarpas/rule14_1_15dat	21.68	0	20.70	0	4.52 %
zarpas/rule14_1_30dat	219.61	0	186.27	0	15.18 %

munication costs. For instance: communication of binary equivalences from the CNF- to the CoE-part makes it possible to substitute those binary equivalences in order to reduce the total length of the equivalence clauses. This rarely resulted in an overall speed-up.

We tried to integrate the equivalence reasoning in such a manner that it would only be applied when the performance would benefit from it. Therefore, march_eq does not perform any equivalence reasoning if no equivalence clauses are detected during the pre-processing phase (if no CoE exists), making march_eq equivalent to its older brother march.

Table 2 shows that the integration of equivalence reasoning in march rarely results in a loss of performance: on some benchmarks like the random_unsat and the quasigroup family no performance difference is noticed, since no equivalence clauses were detected. Most families containing equivalence clauses are solved faster due to the integration. However, there are some exceptions, like the longmult family in the table.

If we compare the integration of equivalence reasoning in march (which resulted in march_eq) with the integration in satz (which resulted in eqsatz), we note that eqsatz is much slower than satz on benchmarks that contain no equivalence clauses. While satz[2] solves 100 random_unsat_350 benchmarks near the treshold on average in 22.14 seconds using 105798 nodes, eqsatz[3] requires on

---

[2] Version 2.15.2 at http://www.laria.u-picardie.fr/~cli/EnglishPage.html
[3] Version 2.0 at http://www.laria.u-picardie.fr/~cli/EnglishPage.html

average 795.85 seconds and 43308 nodes to solve the same set. Note that no slowdown occurs for march_eq.

7   Pre-selection Heuristics

Overall performance can be gained or lost by performing look-ahead on a subset of the free variables in a node: gains are achieved by the reduction of computational costs, while losses are the result of either the inability of the *pre-selection heuristics* (heuristics that determine the set of variables to enter the look-ahead phase) to select effective branching variables or the lack of detected failed literals. When look-ahead is performed on only a subset of the variables, only a subset of the failed literals is most likely detected. Depending on the formula, this could increase the size of the DPLL-tree.

During our experiments, we used pre-selection heuristics which are an approximation of our combined look-ahead evaluation function (ACE) [7]. These pre-selection heuristics are costly, but because they provide a clear discrimination between the variables, a small subset of variables could be selected. Experiments with a *fixed number* of variables entering the look-ahead procedure is shown in

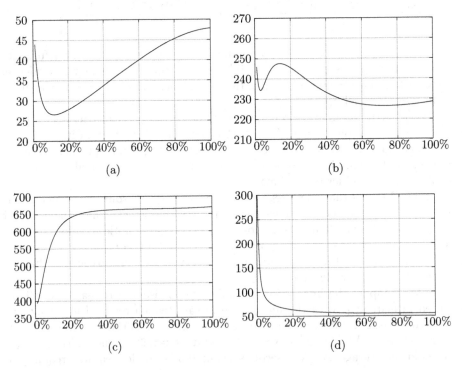

Fig. 5. Runtime(s) vs. percentage look-ahead variables on single instances: (a) random_unsat_350; (b) longmult10; (c) pyhala-braun-unsat-35-4-04; and (d) quasigroup6-12

figure 5. The fixed number is based on a percentage of the original number of variables and the "best" variables (with the highest pre-selection ranking) are selected.

The plots in this figure do not offer any indication of which percentage is required to achieve optimal general performance: while for some instances 100% look-ahead appears optimal, others are solved faster using a much smaller percentage.

Two variants of march_eq have been submitted to the SAT 2004 competition [11]: one which selects in every node the "best" 10 % variables (march_eq_010) and one with full (100%) look-ahead (march_eq_100). Although during our experiments the first variant solved the most benchmarks, at the competition both variants solved the same number of benchmarks, albeit different ones. Figure 5 illustrates the influence of the number of variables entering the look-ahead procedure on the overall performance.

8   Tree-Based Look-Ahead

The structure of our look-ahead procedure is based on the observation that different literals, often entail certain shared implications, and that we can form 'sharing' trees from these relations, which in turn may be used to reduce the number of times these implications have to be propagated during look-ahead.

Suppose that two look-ahead literals share a certain implication. In this simple case, we could propagate the shared implication first, followed by a propagation of one of the look-ahead literals, backtrack the latter, then propagate the other look-ahead literal and only in the end backtrack to the initial state. This way, the shared implication has been propagated only once.

Figure 6 shows this example graphically. The implications among a, b and c form a small tree. Some thought reveals that this process, when applied recursively, could work for arbitrary trees. Based on this idea, our solver extracts - prior to look-ahead - trees from the implications among the literals selected for look-ahead, in such a way that each literal occurs in exactly one tree. The look-

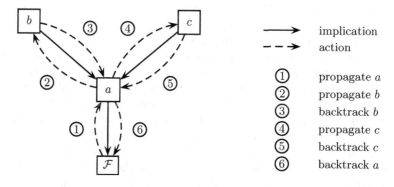

Fig. 6. Graphical form of an implication tree with corresponding actions.

Table 3. Performance of march_eq on several benchmarks with and without the use of tree-based look-ahead

Benchmarks	normal look-ahead		tree-based look-ahead		
	time(s)	treesize	time(s)	treesize	speed-up
random_unsat_250 (100)	1.24	3428.5	1.45	3391.7	-16.94 %
random_unsat_350 (100)	40.57	74501.7	48.78	73357.2	-20.24 %
stanion/hwb-n20-01	29.55	184363	23.65	183553	19.97 %
stanion/hwb-n20-02	40.93	227237	30.91	222251	24.48 %
stanion/hwb-n20-03	25.88	155702	21.70	163984	16.15 %
longmult8	332.64	7918	90.80	8149	72.70 %
longmult10	1014.09	10861	226.31	11597	77.68 %
longmult12	727.01	4654	176.85	5426	75.67 %
pyhala-unsat-35-4-03	1084.08	19093	662.93	19517	38.85 %
pyhala-unsat-35-4-04	1098.50	19493	659.04	19364	40.01 %
quasigroup3-9	8.85	1508	7.97	1495	9.94 %
quasigroup6-12	78.75	1339	58.05	1311	26.29 %
quasigroup7-12	13.03	268	10.03	256	23.02 %
zarpas/rule14_1_15dat	25.62	0	20.70	0	19.20 %
zarpas/rule14_1_30dat	192.30	0	186.27	0	3.14 %

ahead procedure is improved by recursively visiting these trees. Of course, the more dense the implication graph, the more possibilities are available for forming trees, so local learning will in many cases be an important catalyst for the effectiveness of this method.

Unfortunately, there are many ways of extracting trees from a graph, so that each vertex occurs in exactly one tree. Large trees are obviously desirable, as they imply more sharing, as does having literals with the most impact on the formula near the root of a tree. To this end, we have developed a simple heuristic. More involved methods would probably produce better results, although optimality in this area could easily mean solving NP-complete problems again. We consider this an interesting direction for future research.

Our heuristic requires a list of predictions to be available, of the relative amount of propagations that each look-ahead literal implies, to be able to construct trees that share as much of these as possible. In the case of march_eq, the pre-selection heuristic provides us with such a list.

The heuristic now travels this list once, in order of decreasing prediction, while constructing trees out of the corresponding literals. It does this by determining for each literal, if available, one other look-ahead literal that will become its parent in some tree. When a literal is assigned a parent, this relationship remains fixed. On the outset, as much trees are created as there are look-ahead literals, each consisting of just the corresponding literal.

More specifically, for each literal that it encounters, the heuristic checks whether this literal is implied by any other look-ahead literals that are the root of some tree. If so, these are labelled child nodes of the node corresponding to the implied literal. If not already encountered, these child nodes are now recursively checked in the same manner. At the same time, we remove the corresponding elements from the list, so that each literal will be checked exactly once, and will receive a position within exactly one tree.

As an example, we show the process for a small set of look-ahead literals. A gray box denotes the current position:

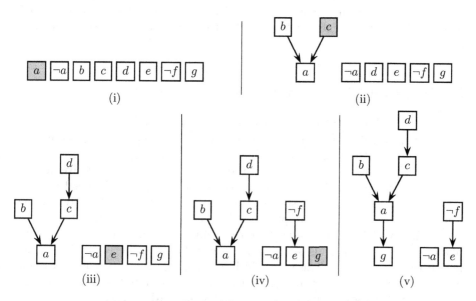

Fig. 7. Five steps of building implication trees

Because of the order in which the list is travelled, literals which have received higher predictions are labelled as parent nodes as early as possible. This is important, because it is often possible to extract many different trees from an implication graph, and because every literal should occur in exactly one tree.

Availability of implication trees opens up several possibilities of going beyond resolution. One such possibility is to detect implied literals. Whenever a node has descendants that are complementary, clearly the corresponding literal is implied. By approximation, we detect this for the most important literals, as these should have ended up near the roots of larger trees by the above heuristic. For solvers unable to deduce such implications by themselves, we suggest a simple, linear-time algorithm that scans the trees.

Some intriguing ideas for further research have occurred to us during the development our tree-based look-ahead procedure, but which, we have not been able to pursue due to time constraints. One possible extension would be to add variables that both positively and negatively imply some look-ahead literal as full-fledged look-ahead variables. This way we may discover important, but previously undetected variables to perform look-ahead on and possibly branch upon. Because of the inherent sharing, the overhead will be smaller than without a tree-based look-ahead.

Also, once trees have been created, we could include non-look-ahead literals in the sharing, as well as in the checking of implied literals. As for the first, suppose that literals a and b imply some literal c. In this case we could share not just the propagation of c, but also that of any other shared implications of a and b. Sharing among tree roots could be exploited in the same manner, with the difference that in the case of many shared implications, we would have to determine which trees could best share implications with each other. In general, it might be a good idea to focus in detail on possibilities of sharing.

9   Removal of Inactive Clauses

The presence of inactive clauses increases the computational costs of the procedures performed during the look-ahead phase. Two important causes can be appointed: first, the larger the number of clauses considered during the look-ahead, the poorer the performance of the cache. Second, if both active and inactive clauses occur in the active data-structure during the look-ahead, a check is necessary to determine the status of every clause. Removal of inactive clauses from the active data-structure prevents these unfavourable effects.

When a variable x is assigned to a certain truth value during the solving phase, all the ternary clauses in which it occurs become inactive in the ternary implication arrays: the clauses in which x occurs positively become satisfied, while those clauses in wich it occurs negatively are reduced to binary clauses. These binary clauses are moved to the binary implication arrays.

Table 4. Performance of march_eq on several benchmarks with and without the removal of inactive clauses on the chosen path

Benchmarks	without removal		with removal		
	time(s)	treesize	time(s)	treesize	speed-up
random_unsat_250 (100)	1.70	3393.7	1.45	3391.7	14.71 %
random_unsat_350 (100)	63.38	73371.9	48.78	73357.2	23.04 %
stanion/hwb-n20-01	24.92	182575	23.65	183553	5.10 %
stanion/hwb-n20-02	33.78	222487	30.91	222251	8.50 %
stanion/hwb-n20-03	23.68	164131	21.70	163984	8.36 %
longmult8	114.71	8091	90.80	8149	20.84 %
longmult10	287.37	11597	226.31	11597	21.25 %
longmult12	254.51	6038	176.85	5426	30.51 %
pyhala-unsat-35-4-03	783.52	19513	662.93	19517	15.39 %
pyhala-unsat-35-4-04	772.59	19378	659.04	19364	14.70 %
quasigroup3-9	11.73	1497	7.97	1495	32.05 %
quasigroup6-12	136.70	1335	58.05	1311	57.53 %
quasigroup7-12	22.53	256	10.03	256	55.48 %
zarpas/rule14_1_15dat	29.80	0	20.70	0	30.54 %
zarpas/rule14_1_30dat	254.81	0	186.27	0	26.90 %

Table 4 shows that the removal of inactive clauses during the solving phase is useful on all kinds of benchmarks. Although the speed-up is only small on uniform random benchmarks, larger gains are achieved on more structured instances.

10   Conclusion

Several techniques have been discussed to increase the solving capabilities of a look-ahead SAT solver. Some are essential for solving various specific benchmarks: a range of families can only be solved using equivalence reasoning, and as we have seen, march_eq is able to solve a large **zarpas** benchmark by adding only constraint resolvents.

Other proposed techniques generally result in a performance boost. However, the usefulness of our pre-selection heuristics is as yet undoubtedly subject to improvement and will be subject of future research.

References

1. D. Mitchel, B. Selmon and H. Levesque, *Hard and easy distributions of SAT problems.* Proceedings of AIII-1992 (1992), 459–465.
2. D. Le Berre, *Exploiting the Real Power of Unit Propagation Lookahead.* In LICS Workshop on Theory and Applications of Satisfiability Testing (2001).
3. D. Le Berre and L. Simon, *The essentials of the SAT'03 Competition.* Springer-Verlag, Lecture Notes in Comput. Sci. **2919** (2004), 452–467.
4. A. Biere, A. Cimatti, E.M. Clarke, Y. Zhu, *Symbolic model checking without BDDs.* in Proc. Int. Conf. Tools and Algorithms for the Construction and Analysis of Systems, Springer-Verlag, Lecture Notes in Comput. Sci. **1579** (1999), 193–207.
5. M. Davis, G. Logemann, and D. Loveland, *A machine program for theorem proving.* Communications of the ACM **5** (1962), 394–397.
6. O. Dubois and G. Dequen, *A backbone-search heuristic for efficient solving of hard3-sat formulae.* International Joint Conference on Artificial Intelligence 2001 **1** (2001), 248–253.
7. M.J.H. Heule and H. van Maaren, *Aligning CNF- and Equivalence-Reasoning.* Appearing in the same volume.
8. O. Kullmann, *Investigating the behaviour of a SAT solver on random formulas.* Submitted to Annals of Mathematics and Artificial Intelligence (2002).
9. C.M. Li and Anbulagan, *Look-Ahead versus Look-Back for Satisfiability Problems.* Springer-Verlag, Lecture Notes in Comput. Sci. **1330** (1997), 342–356.
10. C.M. Li, *Equivalent literal propagation in the DLL procedure.* The Renesse issue on satisfiability (2000). Discrete Appl. Math. **130** (2003), no. 2, 251–276.
11. L. Simon, *Sat'04 competition homepage.* http://www.lri.fr/~simon/contest/results/
12. L. Simon, D. Le Berre, and E. Hirsch, *The SAT 2002 competition.* Accepted for publication in Annals of Mathematics and Artificial Intelligence (AMAI) **43** (2005), 343–378.
13. J.P. Warners, H. van Maaren, *A two phase algorithm for solving a class of hard satisfiability problems.* Oper. Res. Lett. **23** (1998), no. 3-5, 81–88.
14. H. Zhang and M.E. Stickel, *Implementing the Davis-Putnam Method.* SAT2000 (2000), 309–326.

Zchaff2004: An Efficient SAT Solver

Yogesh S. Mahajan, Zhaohui Fu, and Sharad Malik

Princeton University, Princeton, NJ 08544, USA
{yogism, zfu, malik}@Princeton.EDU

Abstract. The Boolean Satisfiability Problem (SAT) is a well known NP-Complete problem. While its complexity remains a source of many interesting questions for theoretical computer scientists, the problem has found many practical applications in recent years. The emergence of efficient SAT solvers which can handle large structured SAT instances has enabled the use of SAT solvers in diverse domains such as electronic design automation and artificial intelligence. These applications continue to motivate the development of faster and more robust SAT solvers. In this paper, we describe the popular SAT solver zchaff with a focus on recent developments.

1 Introduction

Given a propositional logic formula, determining whether there exists a variable assignment that makes the formula evaluate to true is called the Boolean Satisfiability Problem (SAT). SAT was the first problem proven to be NP-Complete[1] and has seen much theoretical interest on this account. Most people believe that it is unlikely that a polynomial time algorithm exists for SAT. However, many large instances of SAT generated from real life problems can be successfully solved by heuristic SAT solvers. For example, SAT solvers find application in AI planning[2], circuit testing[3], software verification[4], microprocessor verification[5], model checking[6], etc. This has motivated research in efficient heuristic SAT solvers.

Consequently, there are many practical algorithms based on various principles such as Resolution[7], Systematic Search[8], Stochastic Local Search[9], Binary Decision Diagrams[10], Stålmarck's[11] algorithm, and others. Gu *et al.*[12] provide a review of many of the algorithms.

Given a SAT instance, SAT algorithms which are complete either find a satisfying variable assignment, or prove that no such solution exists. Stochastic methods, on the other hand, are geared toward finding a satisfiable solution quickly but do not prove unsatisfiability. Stochastic methods are likely to be adopted in AI planning[2] and FPGA routing[13], where instances are likely to be satisfiable and proving unsatisfiability is not required. However, for many other domains, particularly some verification problems[4, 6], the primary task is to prove unsatisfiability of the instances. Hence, complete SAT solvers are required in these cases. The zchaff SAT solver is a complete solver.

H.H. Hoos and D.G. Mitchell (Eds.): SAT 2004, LNCS 3542, pp. 360–375, 2005.

The well known Davis-Logemann-Loveland (DLL)[8] algorithm forms the framework for many successful complete SAT solvers. DLL is sometimes referred to as DPLL for historical reasons. Researchers have been working on DPLL-based SAT solvers since the 1960s. The last ten years have seen tremendous growth and success in DPLL-based SAT solver research. In the mid 1990's, techniques like conflict driven clause learning and non-chronological backtracking were integrated into the DPLL framework[14, 15]. These techniques have greatly improved the efficiency of the DPLL algorithm for structured (as opposed to random) SAT instances. Improvements to the memory efficiency of the Boolean constraint propagation procedure[16, 17] [17] have helped modern SAT solvers cope with large problem sizes. A lot of research has gone into developing new decision strategies. Chaff[17] introduced an innovative conflict clause driven decision strategy and BerkMin[18] introduced yet another decision strategy making use of recent conflict clauses. Today, the latest generation of SAT solvers like zchaff, BerkMin, siege[19], and others[20, 21] are able to handle structured instances with tens of thousands of variables and hundreds of thousands of clauses.

The performance of SAT solvers varies significantly according to the domain from which the problem instance is drawn. The SAT 2004 competition[21] broadly categorizes the instances into instances derived from industrial problems, handmade instances, and randomly generated instances. Solvers that perform well in one category rarely perform well in another category as the techniques that are successful differ from category to category.

Zchaff is a solver that targets the industrial category and hopes to be reasonably successful in the handmade category. It implements the well known Chaff algorithm[17] which includes the innovative VSIDS decision strategy and the very efficient two literal watching scheme for Boolean constraint propagation. Zchaff is a popular solver whose source code is available to the public. It is possible to compile zchaff into a linkable library for easy integration with other applications. Successful integration examples include the BlackBox AI planner[22], NuSMV model checker[23], GrAnDe theorem prover[24], and others. Zchaff compares well with other SAT solvers based on solving runtime performance – versions of zchaff have emerged as the Best Complete Solver in the 'industrial' and 'handmade' instances categories in the SAT 2002 Competition[25] and as the Best Complete Solver in the 'industrial' category in the 2004 SAT Competition[21].

This paper provides an overview of the zchaff solver with a focus on recent developments. Section 2 gives an overview of the DPLL framework on which zchaff is based. Section 3 gives an overview of the main features of the 2003 version of the zchaff solver. Section 4 presents the new features in the SAT 2004 versions of zchaff. Section 5 lists some additional features recently integrated with zchaff after SAT 2004. Section 6 gives some experimental results. Section 7 concludes the paper.

2   The DPLL Algorithm with Learning

In 1960, Davis and Putnam[7] proposed an algorithm for solving SAT which was based on resolution. Their method used resolution for the existential abstraction of variables from the original instance and produced a series of equivalent SAT decision problems with fewer variables. However, their proposed algorithm had impractically large memory requirements. Davis, Logemann and Loveland[8] proposed an algorithm that used search instead of resolution. This algorithm is often referred to as the DPLL algorithm. It can be argued that these two algorithms are closely related because the DPLL search tree can be used to derive a corresponding resolution proof, but we note that the types of proofs of unsatisfiability that the two methods discover can be different.

In most implementations, the propositional formula is usually presented in a Product of Sums form, which is usually called Conjunctive Normal Form (CNF). There exist polynomial algorithms[26] to transform any propositional formula into a CNF formula that has the same satisfiability as the original one. Henceforth, we will assume that the problem is presented in CNF. A formula in CNF is a conjunction of one or more clauses, where each clause is a constraint formed as the disjunction of one or more literals. A literal, in turn, is a Boolean variable or its negation. A propositional formula in CNF has some nice properties that can help prune the search space and speed up the search process. To satisfy a CNF formula, each clause must be satisfied individually. If a variable assignment causes any clause in the formula to have all its literals evaluate to 0 (false), then that current variable assignment or any extension of it will never satisfy the formula. A clause that has all its literals assigned to value 0 is called a *conflicting clause* and directly indicates to the solver that some of the currently assigned variables must be unassigned first before continuing the search for a satisfying assignment.

DPLL is a depth-first backtracking framework. At each step, the algorithm picks a variable v and assigns a value to v. The formula is simplified by removing the satisfied clauses and eliminating the false literals. An inference rule may then be applied to assign values to some more variables which are implied by the current assignments. If an empty clause results after simplification, the procedure *backtracks* and tries the other value for v. Modern DPLL algorithms have an additional feature – they can learn and remember new clause constraints via a procedure called conflict analysis. The worst case time complexity remains exponential in terms of the total number of variables. However, in the case of some classes of real-life applications, a good implementation shows a manageable time complexity when combined with appropriate heuristics.

An outline of 'DPLL with learning' as it is used in zchaff is given in Fig. 1. Initially, none of the variables of the CNF are assigned a value. The unassigned variable are called free variables. The function decideNextBranch() uses some heuristics to choose a free variable v to branch upon and assigns it a value. The assignment operation is said to be a decision made on variable v. The heuristics used here constitute the Decision Strategy of the solver. Each assigned v also has a decision level associated with it which equals the solver decision level at the time the decision was made. The decision level starts at 1 for the first decision and is incremented by 1 for subsequent decisions until a backtrack occurs. After each decision, the function deduce() determines some variable assignments which are implied by the current set of decisions. This inference is referred to as

```
while(there exists a free variable)
  decideNextBranch();               // pick & assign free variable
  status = deduce();                // propagate assigned values
  if(status == CONFLICT)
    blevel = analyzeConflict();     // & learn conflict clause
    if(blevel > 0)
      backTrack(blevel);           // resolve the conflict
    else if(blevel == 0)
      return UNSATISFIABLE;        // conflict cannot be resolved
  runPeriodicFunctions();
}
return SATISFIABLE
```

Fig. 1. Algorithm DPLL with Learning

Boolean Constraint Propagation (BCP). Variables that are assigned during BCP will assume the same decision level as the current decision variable. If deduce() detects a conflicting clause during BCP, then the current partial variable assignment cannot be extended to a satisfying assignment, and the solver will have to backtrack. The solver calls the conflict analysis procedure analyzeConflict() which finds a reason for the discovered conflict and returns the decision level to backtrack to. The reason for the conflict is obtained as a set of variable assignments which imply the current conflict and gets recorded by the solver as a clause.[1] The solver decision level is updated appropriately after the backtrack. The reader is referred to [27, 28] for details of conflict analysis. The solver enters decision level 0 only when making an assignment that is implied by the CNF formula and a backtrack to level 0 indicates that some variable is implied by the CNF formula to be both true and false i.e. the instance is unsatisfiable. (The function runPeriodicFunctions() in the main loop is used to schedule some periodic jobs like clause deletion, restarts, etc.)

The outline in Fig. 1 can be extended to include some simplification procedures - like applying the Pure Literal Rule or identifying equivalence classes[29]. Since these can be expensive to implement dynamically during the search, they may be used in a separate preprocessing phase.

3   Overview of the Zchaff Solver Till 2003

In this section, we will present a quick overview of the main features of the zchaff solver. The overall structure of zchaff is as in Fig. 1.

3.1   Decision Strategy - VSIDS

During the search, making a good choice for which free variable is to be assigned and to what value is very important because even for the same basic algorithm

---

[1] It is well known that the DPLL algorithm without clause recording can discover tree-like resolution proofs. With the ability to record clauses resulting from conflict analysis, the solver can discover more general proofs of unsatisfiability.

framework, different choices may produce search trees with drastically different sizes. Early branching heuristics like Maximum Occurrences in Minimum Sized clauses (MOMS)[30] used some statistics of the clause database to estimate the effect of branching on a particular variable. In [31], the author proposed literal count heuristics which count the number of *unsatisfied* clauses in which a given variable appears (in either phase). These counts are state-dependent because different variable assignments will give different counts and need to be updated every time a variable is assigned or unassigned.

The Chaff[17] solver proposed the use of a heuristic called Variable State Independent Decaying Sum (VSIDS). VSIDS keeps a score for each literal of a variable. Initially, the literal scores equal the number of occurrences of the literal in the input CNF. The literal counts are updated every time the conflict analysis procedure learns a conflict clause by incrementing the scores of each literal in the learned clause by 1. Periodically, after a fixed large number of decisions, all literal scores are divided by 2. The VSIDS literal scores are effectively weighted occurrence counts with higher weights given to occurrences in recently learned clauses. The score of a variable is considered to be the larger of the two associated literal scores. An ordering of the variables is induced by these scores and when a decision is to be made, VSIDS chooses the free variable highest in the variable order and assigns the variable to true if the score of the positive literal exceeds the score of the negative literal and false otherwise. VSIDS provides a quasi-static variable ordering which focuses the search on the recently derived conflict clause. The statistics required for VSIDS are relatively inexpensive to maintain and this makes it a low overhead decision strategy.

3.2   Boolean Constraint Propagation - Two Literal Watching

During the search for a satisfying assignment, the application of an inference rule can detect some variables whose values are implied by the current set of assignments and simplify the problem remaining to be solved. The Unit Clause Rule is a commonly used inference rule. A Unit Clause is a clause which has exactly one unassigned literal and all other literals assigned to false. The unit clause rule says that the unassigned literal in a unit clause must be assigned to true. This implied assignment is called an implication and the unit clause causing the implication is referred to as the antecedent for that variable assignment.

BCP needs to operate on very large clause databases and the pattern of accesses to the clause database often lacks locality. This leads to a large number of cache misses. BCP often contributes as much as 50-90% to the total run-time of modern solvers[32] and it is imperative to optimize the cache/memory usage of the BCP procedure. Early implementations for BCP like [33] maintained counts for the number of assigned literals in each clause in order to identify unit/conflicting clauses. This was costly to implement. The authors of SATO[16] proposed a mechanism for BCP using head/tail lists[34] which significantly improved the efficiency of BCP. In both the counting-based schemes and the head/tail lists methods, unassigning a variable is a costly operation and its complexity may be comparable to that of assigning a variable.

Zchaff uses the Two Literal Watching scheme[17] for BCP. Initially, two of the non-false literals in each clause are marked as watched literals. Each literal maintains a list of the clauses in which it is watched. Whenever a clause becomes a unit/conflicting clause, at least one of the watched literals in that clause must be assigned to false. Hence, when a literal gets assigned to false, it is sufficient to check for unit/conflicting clauses only among those clauses in which that literal is watched . The details of the mechanism for identifying unit/conflicting clauses can be found in [17]. A key benefit of the two literal watching scheme is that at the time of backtracking, there is no need to modify the watched literals in the clause database. Unassigning a variable can be done very simply by doing nothing more than just setting the variable value to "unknown".

3.3 Conflict Driven Clause Learning and Non-chronological Backtracking - Learning the FirstUIP Conflict Clause

Conflict driven clause learning along with non-chronological backtracking were first incorporated into a SAT solver in GRASP[27] and relsat[15]. These techniques are essential for efficient solving of structured problems.

Conflict Driven Clause Learning: When the BCP procedure detects a conflicting clause that results from the current variable assignments, the solver needs to backtrack. The function `analyzeConflict()` finds a subset of the current variable assignments which is also sufficient to make the analyzed clause a conflicting clause. The solver records this information as a clause which evaluates to true exactly when this subset of variable assignments occurs. This prevents the same conflict from occurring again. New clauses are learned using an operation called resolution.[2] The clauses derived using resolution are implied by the resolvents and thus such clauses are logically redundant and adding these clauses does not affect any of the satisfying assignments. However, these added clauses directly help the BCP procedure to prune some of the search space.

The question of which clauses should be selected for resolution can have many answers. In conflict driven clause learning, the solver's search process discovers sequences of clauses which are suitable to be resolved. As mentioned earlier, each assigned non-decision variable, i.e. implied variable, has an antecedent clause associated with it. The antecedent clause for setting the variable v to 1 will contain the positive literal v and all other literals will be assigned false. The conflicting clause C_f comprises of only false literals. Thus, C_f can be resolved with the antecedent of any of its variables to derive a clause C_l which will also have all false literals. The process can be continued to derive other clauses treating C_l as the conflicting clause. A lot of flexibility remains, e.g. in choosing which variable's antecedent is to be used for resolution, which of the learned clauses are to be actually added to the clause database, and when to stop learning. Zchaff

---

[2] Conflict driven conflict clause learning can be looked at in two equivalent ways as resulting from successive resolutions and as a cut in the implication graph. We refer the reader to the description in [28].

answers these questions with the FirstUIP[28] clause learning scheme. A single variable assignment at the conflict decision level, which along with all the variable assignments at previous decision levels is sufficient to cause the conflict is a Unique Implication Point (UIP) at the conflict decision level. This provides a *single* reason at the conflict decision level for the current conflict. The most recent UIP variable assignment at the conflict decision level is called the FirstUIP and can always be found since the decision at the conflict decision level is itself an UIP. In the FirstUIP scheme, all the antecedent clauses that appear in the sequence of resolved clauses are antecedents of variables at the conflict decision level and the FirstUIP clause is found when the only literal remaining at the conflict decision level corresponds to the FirstUIP assignment. Details of the procedure may be found in [27] and [28]. Such a conflict clause is called an asserting clause and it will become a unit clause after backtracking.

Non-chronological Backtracking: In order to resolve a conflict, the solver must backtrack to a prior state which does not directly entail the identified conflict i.e. none of the clauses in the database must be conflicting clauses after the backtrack. To do this, the solver finds the second highest decision level involved in the derived conflict clause (decision level 0 if a single literal clause) and unassigns all the variables assigned at decision levels greater than this level. The solver decision level is reset to be the backtrack level. The newly added conflict clause becomes a unit clause because it was a FirstUIP conflict clause and causes an implication via the BCP procedure. The original conflicting clause that was identified (and analyzed) before the backtrack will certainly be non-conflicting after the backtrack as this clause became a conflicting clause only at the very last decision level prior to backtracking.

4   The SAT 2004 Versions of Zchaff

Two new versions of the zchaff solver participated in the SAT 2004 Competition. These two versions are zchaff.2004.5.13 (submitted as zchaff) and zchaff_rand. Both zchaff.2004.5.13 and zchaff_rand can be downloaded from http://www.princeton.edu/~chaff/SAT2004_versions.html. The two versions are closely related and we will use the term zchaff2004 to refer to both of them. During its development, many features from zchaff_rand were integrated into zchaff.2004.5.13. Some features like the "shrinking" decision heuristic are implemented differently in zchaff.2004.5.13 and zchaff_rand. While there are many differences between them, the solvers are comparable in performance. In the SAT 2004 competition, zchaff.2004.5.13 was more successful on satisfiable instances while zchaff_rand appeared to be more successful on unsatisfiable instances. We have found that the performance of zchaff2004 compared to the 2003 version is slightly worse for bounded model checking, but better for microprocessor verification problems.

Many of the new features have a common theme of increased search locality and the derivation of short conflict clauses. Other researchers, e.g. the author of

Siege[19], have noted the interaction between the search heuristic and the length of the learned clauses. Like BerkMin, zchaff2004 also uses frequent restarts and an aggressive clause deletion policy. Zchaff2004 also features some heuristics whose parameters are dynamically adjusted during the search. Techniques like the VSIDS decision strategy, Two Literal Watching based BCP, FirstUIP based conflict clause learning and non-chronological backtracking which have proved to be useful in earlier versions of zchaff are retained in zchaff2004.

4.1   Increased Search Locality

When VSIDS was first proposed, it turned out to be very successful in increasing the locality of the search by focusing on the recent conflicts. This was observed to lead to faster solving times. Though VSIDS scores are biased toward recent regions of the search by the decaying of the scores, the decisions made are still *global* in nature, due to the slow decay of variable scores[18]. However, recent experiments show that branching within greater locality helps dramatically to prune the search space. SAT solvers BerkMin and siege have both exhibited great speedups from such decision heuristics. Zchaff2004 has three decision heuristics. The first one to be tried is a "shrinking" heuristic. If this is not currently active and does not make a decision, then a modified BerkMin like decision heuristic is tried. The more global VSIDS decision strategy is used last.

Variable Ordering Scheme for VSIDS: This is the default decision heuristic for zchaff2004. One way of trying to make VSIDS more local is to increase the frequency of score decay. The variable ordering scheme also differs from the previous version by incrementing the scores of the literals which get resolved out during conflict analysis. Zchaff2004 increments the scores of involved literals by 10000 instead of by 1. As a result, the decaying scores remain non-zero for longer. Due to the details of the implementation of variable ordering in zchaff, incrementing scores by 10000 also has the side effect that the variable order is no longer the same as given by the variable scores, and the active variables move closer to the top of the variable order. In zchaff_rand, the VSIDS scores are reset to new initial values determined by the literal occurrence statistics in the current clause database after every clause deletion phase.

BerkMin Type Decision Heuristic: The use of the most recent unsatisfied conflict clauses as is done by BerkMin also turns out to be a good cost-effective approach to estimate the locality. The main ideas of this approach are described by the authors of BerkMin in [18]. In zchaff2004, we maintain a chronological list of derived conflict clauses. An unassigned variable with the highest VSIDS score in a recent unsatisfied conflict clause is chosen to be branched upon. As in VSIDS, the variable is assigned to true if the score of the positive literal exceeds the score of the negative literal and false otherwise. In zchaff.2004.5.13, the most recent unsatisfied clause is identified exactly. In zchaff_rand, after searching through a certain threshold number (set to 10000 or randomly) of as yet unexamined conflict clauses, the solver defaults to the VSIDS decision heuristic in case it fails to find an unsatisfied clause. Also, zchaff_rand skips conflict clauses which have all unassigned literals during the search for a recent unsatisfied conflict clause.

Conflict Clause Based Assignment Stack Shrinking: This is related to one of the techniques used by the Jerusat solver[35]. We use our modification of the general idea as presented in [35]. When the newly learned FirstUIP clause exceeds a certain length L, we use it to drive the decision strategy as follows. We sort the decision levels of the literals of the FirstUIP clause and then examine the sorted sequence of decision levels to find the lowest decision level that is less than the next higher decision level by at least 2. (If no such decision level is found, then shrinking is not performed.) We then backtrack to this decision level, and the decision strategy starts *re-assigning* to false the unassigned literals of the conflict clause till a conflict is encountered again. We found that reassigning the variables in the reverse order, i.e. in descending order of decision levels (used in `zchaff_rand`), performed slightly better than reassigning the variables in the same order as they were assigned in previously (used in `zchaff.2004.5.13`). Since some of the variables that were unassigned during the backtrack may not get reassigned, the size of the assignment stack is likely to reduce after this operation. As the variables on the assignment stack are precisely those that can appear in derived conflict clauses, new conflict clauses are expected to be shorter and more likely to share common literals. In our experiments, no fixed value for L performed well for the range of benchmarks we tested. Instead, we set L dynamically using some measured statistics. Zchaff2004 has two such metrics. The first metric, used in `zchaff.2004.5.13`, is the averaged difference between lengths of the clause being used for shrinking and the immediate new clause we get after the shrinking. If this average is less than some threshold, L is increased to reduce the amount of shrinking and if L exceeds some threshold, L is decreased to encourage more shrinking. `zchaff_rand` measures the mean and the standard deviation of the lengths of the recent learned conflict clauses and tries to adjust L to keep it at a value greater than the mean. This dynamic decision heuristic of conflict clause based assignment stack shrinking is observed to often reduce the average length of learned conflict clauses and leads to faster solving times, especially for the microprocessor verification benchmarks.

4.2   Learning Shorter Clauses

Short clauses potentially prune large spaces from the search. They lead to faster BCP and quicker conflict detection. Conflict driven learning derives new (conflict) clauses by successively resolving the clauses involved in the current conflict. The newly derived clause is small in size when the number of resolvents is small, when the resolvents are short clauses themselves, or when the resolvents share many literals in common. Zchaff2004 has the following strategies to try to derive short conflict clauses.

Short Antecedent Clauses Are Preferred: When the clauses do not share many common literals, the sum of the lengths of all the involved clauses will determine the length of the learned conflict clause. We can directly influence the choice of clauses for the resolution by preferring shorter antecedent clauses. One way to do this is to update a variable's antecedent clause with a shorter one whenever possible. As implemented in zchaff2004, BCP queues the implied variable values along with their antecedents but does not perform the assignment immediately. The assignment occurs only when the implied value is dequeued and propagated. Thus, it sometimes happens that the same variable is enqueued multiple times with the same value but different antecedent clauses. When BCP encounters a new antecedent clause for an already assigned variable, the previous antecedent can be replaced with the new one if the new antecedent is shorter. zchaff_rand maintains a separate database for binary clauses [36] and processes the binary clauses before the non-binary clauses during BCP.

Multiple Conflict Analysis: This is a more costly technique than replacing antecedents. It is observed that BCP often discovers more than one conflicting clauses (most of which are derived from some common resolvents). For each conflicting clause, zchaff2004 finds the length of the FirstUIP clause to be learned, and only records the one with the shortest length. Variables that are assigned at decision level zero are excluded from all the learned conflict clauses.

Interaction with Decision Strategy: When the clauses being resolved during conflict analysis share many common literals, the resulting conflict clause is likely to be short. There is a strong interaction between the learned clauses and a "locality centric" decision strategy. For example, the shrinking strategy reduces the size of the set of literals that can appear in new conflict clauses. This in turn increases the likelihood that the new clauses that are learned during the search are shorter and share more literals. Decision strategies like VSIDS and BerkMin which focus on recent conflict clauses can then discover which of these new clauses are suitable for resolution, and the resulting clause is again likely to be short. The observation that the decision strategy influences the length of the derived conflict clauses has been made by the author of siege [19] who considers conflict driven clause learning to be primarily a resolution strategy.

Learning Intermediate Resolvents: While performing conflict analysis, the solver remembers the result of the first 5 resolutions. If this intermediate resolution result is shorter than the recorded FirstUIP clause, then the intermediate resolvent is also recorded after the FirstUIP clause is recorded. This is implemented in zchaff_rand.

4.3   Aggressive Clause Deletion

Learned conflict clauses slow down the BCP procedure and use up memory. Clauses which are not useful must be deleted periodically in the interest of keeping the clause database small. Clauses satisfied at decision level 0 can be deleted as they no longer prune any search space. As in BerkMin, some learned conflict clauses can be deleted periodically without affecting the correctness of the solver. Zchaff2004 periodically deletes learned clauses using usage statistics and clause lengths to estimate the usefulness of a clause. Each clause has an activity counter which is incremented every time the clause is involved in the derivation of a new learned clause. This counter is used by Zchaff2004 to calculate an approximation to the clause's activity to age ratio. Any clause with this ratio less than a certain threshold is considered for deletion. The final decision to delete the clause is then made based on the irrelevance of the clause which is estimated by the number of unassigned literals in the clause. The clause is deleted only if its irrelevance exceeds a certain irrelevance threshold. The irrelevance threshold may be a constant or may be set dynamically based on the measurements of observed clause length statistics. zchaff_rand uses max$\{L,45\}$ for the irrelevance parameter where L is the length parameter used by the dynamic shrinking decision strategy. In zchaff_rand, the clause activities are also decremented periodically by a very small amount.

4.4   Frequent Restarts

Luck plays an important role in determining the solving time of a SAT solver even for the case of unsatisfiable instances. The order in which the BCP procedure queues implications and the order in which variables get watched are determined more or less arbitrarily via the order in the CNF input file. Consequently, the same CNF formula can take widely different run times after shuffling the clauses and variables. When a VSIDS decision is made with all unassigned variables having score 0, zchaff arbitrarily picks the first variable in the list. The wide distribution of run times for slightly different algorithms running on the same instance has been noted in [37] and the authors point out that a rapid restart policy of a randomized solver can help reduce the variance of run times and thereby contribute to increasing the robustness of the solver. Zchaff2004 also uses a rapid fixed interval restart policy. The frequent restarts are observed to make the solver more robust. With restarts disabled, zchaff_rand with a timeout of 300 seconds and random seed 0 takes 688 seconds on the beijing benchmark suite (16 instances) and leaves two instances unsolved. With restarts enabled, all the 16 instances get solved within 65 seconds.

5   Recent Developments

In this section, we briefly mention some of the new features have been added to zchaff2004 after it was submitted to the SAT 2004 competition. One of the motivations was to make BCP more efficient.

5.1   Early Conflict Detection

Early conflict detection is a technique used by solvers like Limmat[38]. During BCP, the variable assignment is completed at the same time that the implied value is queued. This has the advantage that conflicting values in the implications queue can be identified early - as soon as they occur. Another advantage of this is that the implied values still in the queue are already known to the Boolean constraint propagation procedure and this could help BCP by not watching literals which are set false according to the implication queue. This technique has mixed effects on the solver run times. It may be noted that replacing the antecedent clauses becomes more complicated when early conflict detection is enabled, since extra checks have to be performed to ensure that no cycle is introduced into the current implication graph. In particular, we check that all the false literals were assigned before the single true literal got assigned.

5.2   Reorganized Variable Data Structure

During the addition of the new features, the variable object had grown in size to about a hundred bytes. All the variable objects are stored in a `STL::vector<>` as a result of which the actual variable values were widely separated in memory. Since BCP mainly needs just the values, all the variable values were put into a `vector<char>` by themselves. Other fields like the watched literal lists, variable scores, implication related data, etc. were put into `vector<>`'s of their own. This reorganized variable data structure brings small but consistent speedups.

5.3   Miscellaneous Features

The features listed here are considered to be experimental in status. The first one is a modification to the BerkMin heuristic which uses short satisfied clauses on the conflict clause stack which have less than 4 true literals and length less than 10 to make decisions. An unassigned literal from such a clause is selected and set to false. The motivation is to recreate the assignments at the time the short clause was derived. With this strategy, the performance on satisfiable benchmarks improved for the tested benchmarks and no serious disadvantages were noticed for unsatisfiable instances. The second modification is to increment the literal scores by the number of conflicting clauses analyzed for the current conflict. When multiple clauses are analyzed, they share many common resolvents and have similar literals. Hence, incrementing the score by the number of discovered conflicting clauses gives more importance to literals which are frequently involved in deriving conflict clauses. Secondly, incrementing by more than 1 will also move such literals to the top of the variable ordering.

6   Impact of New Features

In this section, we try to evaluate the impact of the various modifications made to the zchaff solver. To do this, we have created a series of versions of zchaff starting with a version similar to the 2003 version and then adding features eventually ending with a recent development version of zchaff2004. While we do not explore all the possible combinations of the the features, we hope these comparisons will yield some insight into the usefulness of the features. All experiments were run on identical machines having Pentium 4 2.80 GHz processors with 1 MB L2 cache and 1 GB RAM using a random seed equal to 0.

6.1   Experimental Results

First, we present some details about the various versions that appear in the comparisons in Tables 1 and 2. The version 'base' is similar to the 2003 version of

Table 1. Comparisons of various configurations of zchaff

Benchmark	base	s	sMR	sMRB	SMRBK	SMrBKEV
01_rule(20)	11477(11)	8817(7)	6970(6)	8810(9)	9641(9)	9467(7)
07_rule(20)	11755(12)	7656(8)	14481(11)	200(0)	103(0)	99(0)
barrel(8)	99(0)	89(0)	84(0)	227(0)	235(0)	256(0)
beijing(16)	2042(2)	2813(2)	1934(2)	79(0)	56(0)	57(0)
ferry(10)	17(0)	10(0)	8(0)	4(0)	3(0)	7(0)
fifo(4)	1978(1)	1269(1)	1229(1)	1586(1)	2367(2)	2378(2)
fpga_routing(32)	1091(1)	993(1)	967(1)	254(0)	1228(1)	1315(1)
fvp-sat.3.0(20)	9216(7)	7109(4)	8500(6)	9988(9)	10149(7)	8770(1)
fvp-unsat.2.0(24)	3253(2)	3017(2)	3859(3)	1179(0)	456(0)	402(0)
hanoi(5)	1678(1)	1037(1)	1088(1)	75(0)	215(0)	143(0)
hard_eq_check(16)	12958(14)	12855(14)	12833(13)	11307(10)	9813(9)	9798(9)
ip(4)	2087(2)	1818(0)	601(0)	236(0)	279(0)	730(0)
total(179)	57653(53)	47485(40)	52554(44)	33945(29)	35066(28)	33422(20)

zchaff. The version 's' is obtained by modifying 'base' to also score the literals which get resolved out during conflict analysis. This scoring is similar to what is done in the BerkMin solver. Literals which are involved in the recorded conflict clause and which get resolved out during conflict analysis have their score incremented by 1. The version 'sMR' adds multiple conflict analysis (Sect. 4.2) to 's' and replaces antecedent clauses (Sect. 4.2). The version 'sMRB' adds BerkMin like heuristics like the decision strategy (Sect. 4.1), aggressive clause deletion (Sect. 4.3) and frequent restarts (Sect. 4.4). In this version, scores are decayed every 20 backtracks. The solver maintains activities for the clauses (Sect. 4.3) and clauses which survive deletion for many iterations have their activities increased by a small amount. This version also includes an experimental decision strategy modification (Sect. 5.3). The version 'SMRBK' adds the dynamic shrinking decision strategy as in zchaff_rand (Sect. 4.1) and clause deletion using the shrinking length parameter L to estimate irrelevance (Sect. 4.3). SMRBK also increments the scores by the number of conflicting clauses analyzed (Sect. 5.3) and resets the VSIDS scores after each restart based on the current literal occurrence counts. The version 'SMrBKEV' adds early conflict detection (Sect. 5.1) to SMRBK and has a cleaned up variable data structure (Sect. 5.2). Due to the overhead of the extra checks required before replacing antecedents when using early conflict detection, only the antecedents of binary clauses are replaced when early conflict detection is used. The version SBKEV is derived from SMrBKEV by disabling multiple conflict analysis and antecedent-replacement. The version SMrBEV is derived by disabling dynamic shrinking from SMrBKEV.

Table 1 reports the total solving time in seconds for the various versions on benchmarks spanning microprocessor verification, bounded model checking, fpga routing, etc. The random seed was 0 for all runs and the timeout was 900 seconds per instance. The numbers in parentheses give the number of instances in the benchmark suite and also the number that remained unsolved. We see that

Table 2. More comparisons on the same benchmarks

Benchmark	SMrBKEV	SBKEV	SMrBEV	zchaff.2004.5.13	zchaff_rand
01_rule(20)	9467(7)	8776(8)	9448(8)	8759(9)	9915(9)
07_rule(20)	99(0)	99(0)	174(0)	111(0)	121(0)
barrel(8)	256(0)	303(0)	349(0)	162(0)	68(0)
beijing(16)	57(0)	63(0)	61(0)	52(0)	65(0)
ferry(10)	7(0)	3(0)	3(0)	7(0)	2(0)
fifo(4)	2378(2)	2184(2)	2069(2)	1669(1)	1765(1)
fpga_routing(32)	1315(1)	1356(1)	765(0)	1102(1)	516(0)
fvp-sat.3.0(20)	8770(1)	9361(5)	9737(10)	6432(4)	7171(1)
fvp-unsat.2.0(24)	402(0)	365(0)	1753(0)	853(0)	702(0)
hanoi(5)	143(0)	516(0)	174(0)	1180(1)	764(0)
hard_eq_check(16)	9798(9)	10302(10)	12218(12)	12095(12)	9877(9)
ip(4)	730(0)	606(0)	588(0)	1146(0)	214(0)
total(179)	33422(20)	33935(26)	37340(32)	33568(28)	31180(20)

Table 3. Effect of dynamic shrinking on microprocessor verification benchmarks

Benchmark	SMrBEV	SMrBKEV
fvp-unsat.1.0(4)	65(0)	29(0)
fvp-unsat.2.0(24)	1753(0)	402(0)
engine-unsat.1.0(10)	8182(4)	8087(4)
pipe-unsat.1.1(14)	17854(12)	10168(4)
fvp-sat.3.0(20)	17550(7)	9010(0)
total(68)	45404(23)	27696(8)

multiple conflict analysis and replacing the antecedents do not have a significant effect on the solver run times by themselves. The BerkMin like decision heuristics produce a definite improvement. Adding dynamic shrinking has a mixed effect for these benchmarks. Adding early conflict detection and reorganizing the variable data structure gives a small improvement.

Table 2 shows the effect of disabling multiple conflict analysis and the effect of disabling shrinking from the final version. Performance is degraded in both cases. Table 2 also compares zchaff.2004.5.13 with zchaff_rand.

As remarked earlier, the dynamic shrinking strategy has the most significant effect for the microprocessor verification benchmarks. We present the results without and with dynamic shrinking for some microprocessor verification benchmarks in Table 3. The timeout for these experiments was 1800 seconds.

7   Summary

We have presented some details of the 2004 versions of zchaff including the versions that participated in the SAT 2004 Competition. The new features include decision strategies that increase the "locality" of the search and focus more on recent conflicts, strategies that directly focus on deriving short conflict clauses, an aggressive clause deletion heuristic that keeps only the clauses most likely to be useful, a rapid restart policy that adds robustness and some techniques that are intended to improve the efficiency of the Boolean constraint propagation procedure. We have also presented some data that might help in evaluating the usefulness of the various features.

References

1. Cook, S.A.: The complexity of theorem-proving procedures. In: Third Annual ACM Symposium on Theory of Computing. (1971)
2. Kautz, H., Selman, B.: Planning as Satisfiability. In: European Conference on Artificial Intelligence. (1992)
3. Stephan, P., Brayton, R., Sangiovanni-Vencentelli, A.: Combinational test generation using satisfiability. IEEE Transactions on Computer-Aided Design of Integrated Circuits and Systems **15** (1996) 1167–1176
4. Jackson, D., Vaziri, M.: Finding bugs with a constraint solver. In: International Symposium on Software Testing and Analysis, Portland, OR. (2000)
5. Velev, M.N., Bryant, R.E.: Effective use of boolean satisfiability procedures in the formal verification of superscalar and VLIW. In: 38th DAC, New York, NY, USA, ACM Press (2001) 226–231
6. Biere, A., Cimatti, A., Clarke, E.M., Zhu, Y.: Symbolic Model Checking without BDDs. In: Proc. of TACAS. (1999)
7. Davis, M., Putnam, H.: A computing procedure for quantification theory. Journal of ACM **7** (1960) 201–215
8. Davis, M., Logemann, G., Loveland, D.: A machine program for theorem proving. Communications of the ACM **5** (1962) 394–397
9. Selman, B., Kautz, H., Cohen, B.: Local search strategies for satisfiability testing. In: Proceedings of the Second DIMACS Challange on Cliques, Coloring, and Satisfiability. (1993)
10. Bryant, R.E.: Graph-based algorithms for boolean function manipulation. IEEE Transactions on Computers **C-35** (1962) 394–397
11. Gunnar Stålmarck: System for Determining Propositional Logic Theorems by Applying Values and Rules to Triplets that are Generated from Boolean Formula (1994) United States Patent. Patent Number 5,276,897.
12. Gu, J., Purdom, P.W., Franco, J., Wah, B.W.: Algorithms for the Satisfiability (SAT) Problem: A Survey. DIMACS Series in Discrete Mathematics and Theoretical Computer Science (1997)
13. Nam, G.J., Sakallah, K.A., Rutenbar, R.A.: Satisfiability-Based Layout Revisited: Detailed Routing of Complex FPGAs Via Search-Based Boolean SAT. In: ACM/SIGDA International Symposium on FPGAs. (1999)
14. Marques-Silva, J.P., Sakallah, K.A.: Conflict Analysis in Search Algorithms for Propositional Satisfiability. In: IEEE International Conference on Tools with Artificial Intelligence. (1996)
15. Bayardo, R., Schrag, R.: Using CSP look-back techniques to solve real-world SAT instances. In: National Conference on Artificial Intelligence (AAAI). (1997)
16. Zhang, H.: SATO: An efficient propositional prover. In: International Conference on Automated Deduction. (1997)

17. Moskewicz, M.W., Madigan, C.F., Zhao, Y., Zhang, L., Malik, S.: Chaff: Engineering an Efficient SAT Solver. In: 38th DAC. (2001)
18. Goldberg, E., Novikov, Y.: BerkMin: A Fast and Robust SAT Solver. In: DATE. (2002)
19. Siege Satisfiability Solver, http://www.cs.sfu.ca/~loryan/personal/ (2004).
20. SAT Competition 2003, http://www.satlive.org/SATCompetition/2003/ (2004).
21. SAT Competition 2004, http://www.satlive.org/SATCompetition/2004/ (2004).
22. http://www.cs.washington.edu/homes/kautz/satplan/blackbox/ (2004).
23. NuSMW Home Page, http://nusmv.irst.itc.it/ (2004).
24. GrAnDe, http://www.cs.miami.edu/~tptp/ATPSystems/GrAnDe/ (2004).
25. SAT Competition 2002, http://www.satlive.org/SATCompetition/2002/ (2004).
26. Plaisted, D.A., Greenbaum, S.: A stucture-preserving clause form translation. Journal of Symbolic Computation **2** (1986) 293–304
27. Marques-Silva, J.P., Sakallah, K.A.: GRASP - A New Search Algorithm for Satisfiability. In: IEEE International Conf. on Tools with Artificial Intelligence. (1996)
28. Zhang, L., Madigan, C.F., Moskewicz, M.W., Malik, S.: Efficient conflict driven learning in boolean satisfiability solver. In: ICCAD. (2001) 279–285
29. Li, C.M.: Integrating Equivalency reasoning into Davis-Putnam procedure. In: AAAI. (2000)
30. Freeman, J.W.: Improvements to propositional satisfiability search algorithms. PhD thesis, University of Pennsylvania (1995)
31. Marques-Silva, J.P.: The impact of branching heuristics in propositional satisfiability algorithms. In: 9th Portuguese Conf. on Artificial Intelligence. (1999)
32. Zhang, L.: Searching for Truth: Techniques for Satisfiability of Boolean Formulas. PhD thesis, Princeton University (2003)
33. Crawford, J., Auton, L.: Experimental results on the cross-over point in satisfiability problems. In: National Conf. on Artificial Intelligence (AAAI). (1993)
34. Zhang, H., Stickel, M.: An efficient algorithm for unit-propagation. In: Fourth International Symposium on Artificial Intelligence and Mathematics, Florida. (1996)
35. Nadel, A.: The Jerusat SAT Solver. Master's thesis, Hebrew University of Jerusalem (2002)
36. Pilarski, S., Hu, G.: Speeding up SAT for EDA. In: DATE. (2002)
37. Kautz, H., Horvitz, E., Ruan, Y., Gomes, C., Selman, B.: Dynamic restart policies. In: The 18th National Conf. on Artificial Intelligence. (2002)
38. http://www2.inf.ethz.ch/personal/biere/projects/limmat/ (2004).

The Second QBF Solvers Comparative Evaluation[*]

Daniel Le Berre[1], Massimo Narizzano[3],
Laurent Simon[2], and Armando Tacchella[3]

[1] CRIL, Université d'Artois,
Rue Jean Souvraz SP 18 – F 62307 Lens Cedex, France
leberre@cril.univ-artois.fr
[2] LRI, Université Paris-Sud
Bâtiment 490, U.M.R. CNRS 8623 – 91405 Orsay Cedex, France
simon@lri.fr
[3] DIST, Università di Genova,
Viale Causa, 13 – 16145 Genova, Italy
{mox, tac}@.dist.unige.it

Abstract. This paper reports about the 2004 comparative evaluation of solvers for quantified Boolean formulas (QBFs), the second in a series of non-competitive events established with the aim of assessing the advancements in the field of QBF reasoning and related research. We evaluated sixteen solvers on a test set of about one thousand benchmarks selected from instances submitted to the evaluation and from those available at www.qbflib.org. In the paper we present the evaluation infrastructure, from the criteria used to select the benchmarks to the hardware set up, and we show different views about the results obtained, highlighting the strength of different solvers and the relative hardness of the benchmarks included in the test set.

1   Introduction

The 2004 comparative evaluation of solvers for quantified Boolean formulas (QBFs) is the second in a series of non-competitive events established with the aim of assessing the advancements in the field of QBF reasoning and related research. The non-competitive nature of the evaluation is meant to encourage the developers of QBF reasoning tools and the users of QBF technology to submit their work. Indeed, our evaluation does not award one particular solver, but instead it draws a picture of the current state of the art in QBF solvers and benchmarks. Running the evaluation enables us to collect data regarding the

---
[*] The work of the first author was partially supported by Université d'Artois, the IUT-Lens and FEDER. The work of the second and the last author was partially supported by ASI, MIUR and a grant from the Intel Corporation. The authors would like to thank all the participants to the QBF evaluation for submitting their work.

H.H. Hoos and D.G. Mitchell (Eds.): SAT 2004, LNCS 3542, pp. 376–392, 2005.
© Springer-Verlag Berlin Heidelberg 2005

strength of different solvers and methods, the relative hardness of the benchmarks, and to shed some light on the open issues for the researchers in the QBF community.

With respect to last year's evaluation [1] we have witnessed to an almost 50% increase in the number of submitted solvers (from eleven to sixteen). While most of the participants are still complete solvers extending the well-known Davis, Putnam, Logemann, Loveland procedure (DPLL) [2, 3] for propositional satisfiability (SAT), the evaluation also hosted one incomplete solver (WALKQSAT [4]), a solver based on Q-resolution and expansion of quantifiers (QUANTOR [5]), and a solver using ZBDDs (Zero-suppressed Binary Decision Diagrams) to obtain a symbolic implementation of the original DP algorithm extended to QBFs (QMRES [6]). The number of the participants and the variety of the technologies confirm the vitality of the research on QBF reasoning tools. Regarding applications of QBF reasoning, three families of benchmarks obtained by encoding formal verification problems have been submitted, for a total of 88 instances. To these we must add 22 families and 814 instances that have been independently submitted to www.qbflib.org and that have been evaluated this year for the first time. Finally, the submission of a generator for model A and model B random instances [7] allowed us to run the participating solvers on a wide selection of instances generated according to these probabilistic models.

The evaluation consisted of two steps: (i) running the solvers on a selection of benchmarks, and (ii) analyzing the results. The first step is subject to the stringent requirements of getting meaningful results and completing the evaluation in a reasonable time. In order to mate these two requirements, we have extracted the non-random part of the evaluation test set by sampling the pool of available benchmarks – more than 5000 benchmarks and 68 families – to extract a much smaller, yet representative, test set of about five hundred instances. To these, we added representatives from model A and model B families of random benchmarks to obtain the final test set. Using the selected test set, we have completed the first step by running the solvers on a farm of PCs, each solver being restricted to the same amount of time and memory. The second step consisted of two stages. In the first stage we considered all the solvers and the benchmarks of the test set to give a rough, but complete, picture of the state of the art in QBF. By analyzing the results for problems and discrepancies among the solvers results, we were able to isolate solvers and instances that turned out to be problematic, and we removed them from the subsequent analysis. The second stage is an in-depth account of the results, where we tried to extract a narrow, but crisp, picture of the current state of the art.

The paper is structured as follows. In Section 2 we briefly describe all the solvers and benchmarks that participated in the evaluation. In Section 3 we present the the choice of the test set and a description of the computing infrastructure. In Section 4 we present the results of the evaluation first stage, while in Section 5 we restrict our attention to the solvers and the benchmarks that passed the first stage: here we present the result of the evaluation arranged solver-wise (Sub. 5.1), and benchmark-wise (Sub. 5.2). We conclude the paper in Section 6 with a balance about the evaluation.

2   Solvers and Benchmarks

Sixteen solvers from eleven different authors participated to the evaluation this year. The requirements for the solvers were to be able to read the input instance in a common format (the Q-DIMACS format [8]), and to provide the answer (sat/unsat) in a given output format. Noticeably, all the solvers complied on the input requirements, which was not the case on the previous evaluation [1], while a few solvers required wrapper scripts to adapt the output format (SSOLVE, QSAT, and SEMPROP) or to load additional applications required to run the solver (OPENQBF, requiring the JVM). A short description of the solvers submitted to the evaluation follows:

CLEARN by Andrew G.D. Rowley, is a search-based solver written in C++, featuring lazy data structures and conflict learning; the heuristic is a simple and efficient lexicographic ordering based on prefix level (outermost to innermost) and variable identifier (smallest first).

GRL by Andrew G.D. Rowley, is a sibling of CLEARN, but with a different learning method described in [9].

OPENQBF by Gilles Audemard, Daniel Le Berre and Olivier Roussel, is a search-based solver written in Java, featuring basic unit propagation and pure literal lookahead, plus a conflict backjumping lookback engine; the heuristic is derived from Böhm and Speckenmeyer's heuristic for SAT.

ORSAT by Olivier Roussel, is search-based solver written in C++, featuring an algorithm based on relaxations to SAT, plus special purpose techniques to deal with universal quantifiers; the solver is currently in its early stage of development, so most of the features found in other, more mature, solvers are missing.

QBFL by Florian Letombe, is a search-based solver written in C, packed with a number of features: trivial-truth, trivial-falsity, Horn and reverse-Horn formulas detection; QBFL is implemented on top of Limmat (version 1.3), and comes in two flavors: QBFL-JW uses an extension of the Jeroslow-Wang heuristic for SAT, while QBFL-BS uses an extension of Böhm and Speckenmeyer's heuristic for SAT.

QSAT by Jussi Rintanen, is a search-based solver written in C, featuring a lookahead heuristic with failed literal rule, sampling, partial unfolding and quantifier inversion.

QMRES by Guoqiang Pan and Moshe Y. Vardi, written in C and based on a symbolic implementation of the original DP algorithm, achieved using ZBDDs. The algorithm features multi-resolution, a simple form of unit propagation, and heuristics to choose the variables to eliminate.

QUANTOR by Armin Biere, is a solver written in C based on Q-resolution (to eliminate existential variables) and Shannon expansion (to eliminate universal variables), plus a number of features, such as equivalence reasoning, subsumption checking, pure literal detection, unit propagation, and also a scheduler for the elimination step.

QUBE by Enrico Giunchiglia, Massimo Narizzano and Armando Tacchella, is a search-based solver written in C++ featuring lazy data structures for unit and pure literal propagation; QUBE comes in two flavors: QUBE-BJ, featuring conflict- and solution-directed backjumping, and QUBE-LRN, featuring conflict and solution learning; the heuristic is an extension to QBF of zCHAFF heuristic for SAT.

SSOLVE by Rainer Feldmann and Stefan Schamberger, is a search-based algorithm written in C, featuring trivial truth and a modified version of Rintanen's method of inverting quantifiers. The data structures used are extensions of the data structures of Max Böhm's SAT-solver.

SEMPROP by Reinhold Letz, is a search-based solver written in Bigloo (a dialect of Scheme), featuring dependency directed backtracking and lemma/model caching for false/true subproblems.

WALKQSAT by Ian Gent, Holger Hoos, Andrew G. D. Rowley, and Kevin Smyth, is the first incomplete QBF solver based on stochastic search methods. It is a sibling of CSBJwith WalkSAT as a SAT oracle and guidance heuristic.

YQUAFFLE by Yinlei Yu and Sharad Malik, is a search-based solver written in C++, featuring multiple conflict driven learning, solution based backtracking, and inversion of quantifiers.

Most of the solvers mentioned above are described in a booklet [10] prepared for the evaluation with the contributions of the solvers authors, with the exception of QSAT, SSOLVE, and SEMPROP, which are described, respectively, in [11], [12], and [13].

The evaluation received 88 benchmarks divided in 4 different families and a random generator, all from four different authors:

Biere (1 family, 65 instances) QBF encodings of the following problem: given an $\langle n \rangle$-bit-counter with optional reset (r) and enable (e) signals, check whether it is possible to reach the state where all $\langle n \rangle$ bits are set to 1 starting from the initial state where all bits are set to 0.

Katz (2 families, 20 instances) QBF encodings of symbolic reachability problems in hardware circuits.

Lahiri/Seshia (1 family, 3 instances) QBF encodings of convergence testing instances generated in term-level model checking.

Tacchella A generator for model A and model B random QBF instances, implemented according to the guidelines described in [7].

In order to obtain the evaluation test set, we have also considered 5558 benchmarks in 64 families from www.qbflib.org:

Ayari (5 families, 72 benchmarks) A family of problems related to the formal equivalence checking of partial implementations of circuits (see [14]).

Castellini (3 families, 169 benchmarks) Various QBF-based encodings of the bomb-in-the-toilet planning problem (see [15]).

Gent/Rowley (8 families, 612 benchmarks) Various encodings of the famous "Connect4" game into QBF [16].

Letz (1 family, 14 benchmarks) Formulas proposed in [13] generated according to the pattern $\forall x_1 x_3 \ldots x_{n-1} \exists x_2 x_4 \ldots x_n (c_1 \wedge c_n)$ where $c_1 = x_1 \wedge x_2$, $c_2 = \neg x_1 \wedge \neg x_2$, $c_3 = x_3 \wedge x_4$, $c_4 = \neg x_3 \wedge \neg x_4$, and so on. The instances consists of simple variable-independent subproblems but they should be hard for standard (i.e., without non-chronological backtracking) QBF solvers.

Mneimneh/Sakallah (12 families, 202 benchmarks) QBF encodings of vertex eccentricity calculation in hardware circuits [17].

Narizzano (4 families, 4000 benchmarks) QBF-based encoding of the robot navigation problems presented in [15].

Pan (18 families, 378 benchmarks) Encodings of modal K formulas satisfiability into QBF (see [18]). The original benchmarks have been proposed during the TANCS'98 comparison of theorem provers for modal logics [19].

Rintanen (5 families, 47 benchmarks) Planning, hand-made and random problems, some of which have been presented in [20].

Scholl/Becker (8 families, 64 benchmarks) encode equivalence checking for partial implementations problems (see [21]).

3   Evaluation: Test Set and Infrastructure

As we outlined in Section 1, deciding the test set for the evaluation is complicated by two competing requirements: (*i*) obtaining meaningful data and (*ii*) completing the evaluation in reasonable time. To fulfill requirement (*i*) in the case of non-random benchmarks, we decided to extract a suitable subset from the pool of all the available benchmarks. In particular, we designed the selection process in such a way that the resulting test set is representative of the initial pool, yet it is not biased toward specific instances. This cannot be achieved by simply extracting a fixed proportion of benchmarks from all the available families, because some of them dominate others in terms of absolute numbers, e.g., the four Robots families account for more than 70% of the instances available on QBFLIB. In order to remove the bias, we have extracted a *fixed* number of instances from each available family. In this way, the extracted test set accounts for the same number of families as the initial pool (variety is preserved), but each family contains at most N representatives (bias is removed) We used the following algorithm:

- if the family in the original pool consists of $M < N$ benchmarks, then extract all M of them, while
- if the family consists of $M > N$ benchmarks, then extract only N instances by sampling the original ones uniformly at random.

Considering all the non-random benchmarks described in Section 2, we have extracted an evaluation test set of 522 instances divided into 68 families. As for random instances, the issue is further complicated by the fact that we have several parameters to choose when generating the benchmarks, namely, the number of variables, the number of clauses, the number of alternations in the prefix, the number of literals per clause, the number of existential literals per clause, and the generation model (either model A or model B). We based our selection of random benchmarks on the experimental work presented in [22], and we generated formulas with $v = \{50, 100, 150\}$ variables, $a = \{2, 3, 4, 5\}$ alternations in the prefix, and a fixed number of 5 literals per clause. For each fixed value of a and v, we generated formulas ranging over clauses-to-variables ratio of 2 to 18 with step 2. We used the above parameters both to generate model A and model B instances, and a threshold 2 for the number of existential literals in the clause, which means at least 2 existential literals per clause in model A instances, and exactly 2 existential literals per clause in model B instances. Notice that while for model A instances we were able to choose parameters based on the previous experience of [7, 22], for model B instances, the only experimental account available is that of [7], which covers only part of the space explored in the evaluation.

Overall, the evaluation test set was completed by 432 random formulas divided into 24 families of 18 samples each, bringing the total number of benchmarks to 954.

As for the computing infrastructure, the evaluation ran on a farm of 10 identical rack-mount PCs, equipped with 3.2Ghz PIV processors, 1GB of RAM and running Debian/GNU Linux (distribution `sarge`). Considering that we had 954 benchmarks to run, we split the evaluation job evenly across (9) machines, using `perl` scripts to run subsets of 106 benchmarks on all the 16 solvers on each machine. This methodology has a two points in its favor. First, testing scripts are extremely lean and simple: one server script, plus as many client scripts as there are machines running, accounting for less than 100 lines of `perl` code. This makes the whole evaluation infrastructure lightweight and easy to debug. Second, by running clusters of benchmarks on the same machine, we are guaranteed that small differences that could exist even in identical hardware, are compensated by the fact that a given benchmark is evaluated by all the participants on the very same machine. Indeed, while noise in the order of one second does not matter much when comparing benchmarks to decide their hardness, it can make a big difference when the total runtime on the benchmark is in in the order of one second or less and we are comparing solvers. Finally, all the solvers where limited to 900 seconds of CPU time and 900MB of memory: in the following, when we say that an instance has been solved, we mean that this happened without exceeding the resource bounds above.

4   Evaluation: First Stage Results

In Table 1 and Table 2, we present the raw results of the evaluation concerning, respectively, non-random (522 benchmarks) and random (432 benchmarks)

Table 1. Results of the evaluation first stage (non-random benchmarks)

Solver	Total		Sat		Unsat		Unique	
	#	Time	#	Time	#	Time	#	Time
SEMPROP	288	10303.40	133	2985.806	155	7317.55	5	814.56
QUANTOR	284	3997.10	126	2137.25	158	1859.85	10	2624.36
YQUAFFLE	256	6733.02	110	3152.37	146	3580.65	–	–
CLEARN	255	11565.30	116	3894.06	139	7671.27	–	–
SSOLVE	245	8736.64	114	3350.96	131	5385.68	–	–
GRL	240	11895.90	107	4577.22	133	7318.70	–	–
QUBE-BJ	239	9426.09	110	4538.27	129	4887.82	–	–
QUBE-LRN	237	8270.98	113	3365.83	124	4905.15	1	433.15
QMRES	224	6337.39	122	3315.42	102	3021.97	28	901.54
QSAT	218	8375.62	93	3307.13	125	5068.49	7	197.63
QBFL-JW	205	5573.55	83	2849.65	122	2723.90	–	–
CSBJ	205	6528.84	98	3407.16	107	3121.68	–	–
QBFL-BS	191	3076.62	75	1466.10	116	1610.52	–	–
OPENQBF	185	6598.94	78	3219.56	107	3379.38	–	–
WALKQSAT	163	7262.51	83	4113.37	80	3149.14	–	–
ORSAT	73	1243.83	37	1134.74	36	109.09	–	–

Table 2. Results of the evaluation first stage (random benchmarks)

Solver	Total		Sat		Unsat		Unique	
	#	Time	#	Time	#	Time	#	Time
QuBE-LRN	426	3452.67	86	93.12	340	3359.55	–	–
QuBE-BJ	418	4343.98	76	87.12	342	4256.86	–	–
SSOLVE	403	1028.30	86	346.87	317	681.43	1	0.45
SEMPROP	384	3069.28	80	1614.06	304	1455.22	–	–
CLEARN	338	5267.99	76	1939.28	262	3328.71	–	–
GRL	335	5975.14	73	1213.30	262	4761.84	–	–
QSAT	321	3491.18	60	450.01	261	3041.17	1	208.12
CSBJ	320	5956.06	74	2038.39	246	3917.67	–	–
WALKQSAT	316	5838.05	75	2262.17	241	3575.88	–	–
OPENQBF	277	5525.99	64	334.51	213	5191.48	–	–
QBFL-JW	263	6380.26	94	58.35	169	6321.91	–	–
QBFL-BS	218	3265.93	92	1.06	126	3264.87	–	–
YQUAFFLE	197	3166.62	34	184.69	163	2981.93	–	–
QMRES	142	4091.29	53	1594.07	89	2497.22	–	–
QUANTOR	120	263.75	52	0.78	68	262.97	–	–
ORSAT	60	1.24	–	–	60	1.24	–	–

instances. Each table consists of nine columns that for each solver report its name (column "Solver"), the total number of instances solved and the cumulative time to solve them (columns "#" and "Time", group "Total"), the number of instances found satisfiable and the time to solve them (columns "#" and "Time", group "Sat"), the number of instances found unsatisfiable and the time to solve them (columns "#" and "Time", group "Unsat"), and, finally, the number of instances uniquely solved and the time to solve them (columns "#" and "Time", group "Unique"); a "–" (dash) in the last two columns means that the solver did not conquer any instance uniquely. Both tables are sorted in descending order, according to the number of instances solved, and, in case of a tie, in ascending order according to the cumulative time taken to solve them.

Looking at the results on non-random instances in Table 1, we can see that all the solvers, with the only exception of ORSAT, were able to conquer at least 25% of the instances in this category. On the other hand, only two solvers, namely SEMPROP and QUANTOR, were able to conquer more than 50% of the instances. Overall, this indicates that given the current state of the art in QBF reasoning, the performance demand of the application domains is still exceeding the capabilities of most solvers. The performance of the solvers is also pretty similar: excluding ORSAT, there are only 125 instances (less than 25% of the total) separating the strongest solver (SEMPROP), from the weakest solver (WALKQSAT), and the number of instances solved by the strongest five participants are in the range [288-245] spanning only 43 instances (less than 10% of the total). Some difference arises when considering the number of instances uniquely solved by a given solver: QMRES, QUANTOR, SEMPROP, QSAT and QuBE-LRN are the only solvers able to conquer, respectively, 28, 10, 7, 5 and 1 instance. Noticeably, the strongest solvers in this respect, QMRES and QUANTOR, are not extensions of the DPLL algorithm as all the other participants, indicating that the technologies on which they are based provide an interesting alternative to the classic search-based paradigm.

Looking at the results on random instances in Table 2, we can see that all the solvers, again with the only exception of ORSAT, were able to conquer at least 25% of the instances in this category, and six solvers were able to conquer more than 75% of the instances. Overall, this indicates that the choice of parameters for the generation of random instances resulted in a performance demand well within the capabilities of most solvers. The performance of the solvers is however rather different: even excluding ORSAT, there are 306 instances (about 70% of the total) separating the strongest QUBE-LRN, from the weakest QUANTOR, and the number of instances solved by the strongest five participants are spread over 88 instances (about 20% of the total). There is no relevant change in the picture above when considering the number of instances uniquely solved by a given solver, since only SSOLVE and QSAT are able to uniquely conquer one instance each. Noticeably, some of the strongest solvers on random instances, namely SEMPROP, SSOLVE and CLEARN, are also among the strongest solvers on non-random benchmarks, indicating that these search-based engines feature relatively robust algorithms.

As we have anticipated in Section 1, a few discrepancies in the results of the solvers were detected during the analysis of the first stage results. A total of 32 discrepancies were detected, of which 9 regarding non-random instances, and the remaining regarding random instances. For each of the discrepancies we reran the solvers reporting a result different from the majority of the other solvers and/or the expected result of the benchmark. We also inspected the instances, looking for weird syntax and other pitfalls that may lead a correct solver to report an incorrect result. At the end of this analysis we excluded from the second stage the following solvers:

- QUBE-BJ and QUBE-LRN, responsible for all the discrepancies detected on random instances; although the satisfiability status of the random benchmarks is not known in advance, the two solvers do not agree with each other in 10/23 cases, and in 7/23 cases they do not agree with the majority of solvers.
- QSAT, reporting as unsatisfiable the benchmark k_ph_n-21 of the k_ph_n family in the Pan series: these benchmarks ought to be satisfiable by construction (in modal K), and the correctness of the translations is not taunted by any other evidence in our data.
- CLEARN and GRL, reporting as unsatisfiable the benchmark s27_d2_s of the s_27 family in the Mneimneh/Sakallah series; the benchmark is both declared satisfiable by its authors and by all the other solvers.

We have also excluded the following benchmarks:

- the Connect2 family (Gent/Rowley), since on some of its instances QBFL-BS, QBFL-JW, QMRESand WALKQSAT, reported apparently incorrect results, if compared to the majority of the other solvers: examining the instances, it turns out that they contain existentially quantified sets declared as separate but adjacent lists in the Q-DIMACS prefix. Although the Q-DIMACS standard does not disallow this syntax, we believe that this might be the cause of the problems, for some solvers are not prepared to handle this kind of input correctly.

- the Logn family (Rintanen), since two of its instances are pure SAT with unbound variables containing an empty input clause: their correct satisfiability status is thus "false", but some of the solvers (namely QMRES, SEMPROPand YQUAFFLE) report them as satisfiable.

The data obtained by disregarding the above solvers and benchmarks is free of any discrepancy. Clearly, for instances that were conquered by just one solver, and for which we do not know the satisfiability status in advance, the possibility that the solver is wrong still exists, but we consider this as unavoidable given the current state of the art.[4]

5   Evaluation: Second Stage Results

5.1   Solver-Centric View

In Table 3 we report second stage results about non-random benchmarks (510 benchmarks, 11 solvers), divided into three categories:

Formal Verification 29 families and 220 benchmarks, including Ayari, Biere, Katz, Mneimneh/Sakallah, and Scholl/Becker instances.

Planning 16 families and 122 benchmarks, including Castellini, Gent/Rowley, Narizzano, and part of Rintanen instances.

Miscellaneous 21 families and 168 benchmarks, including Letz, Pan and the remaining Rintanen instances.

Table 3 is arranged analogously to Tables 1 and 2, except an additional column that indicates the category. The solvers are classified independently for each category, and in descending order according to the number of instances solved: in case of ties, the solvers are prioritized according the time taken to solve the benchmarks, shortest time first.

Looking at Table 3, the first observation is that the solvers performed slightly better on the planning category: the strongest one in the category (YQUAFFLE) solves 82% of the instances, the weakest one (ORSAT) solves about 30% of the instances, and 7 out of the 11 solvers admitted to the second stage are able to solve more than 50% of the category. On the other hand, the strongest solver in the miscellaneous category (QMRES) solves 78% of the instances, but most of the solvers (8 out of 11) do not go beyond the 50% threshold; in the formal verification arena, the strongest solver in the category (QUANTOR) does not get beyond a mere 33% of the total instances. Also significant is the fact that in the planning category the strongest solver is DPLL-based (YQUAFFLE), while both in the formal verification category and in the miscellaneous category the strongest solvers (respectively QUANTOR and QMRES) express alternative paradigms.

---

[4] Notice that the same problem exists in the SAT competition when a solver is the only one to report about an instance and the answer is "unsatisfiable".

Table 3. Results of the evaluation second stage (non-random benchmarks)

Category	Solver	Total		Sat		Unsat		Unique	
		#	Time	#	Time	#	Time	#	Time
Formal Verification	QUANTOR	74	2854.87	34	1424.66	40	1430.21	10	2624.36
	SEMPROP	71	2064.38	29	294.58	42	1769.8	2	545.51
	YQUAFFLE	68	1239.56	28	319.56	40	920	–	–
	QBFL-JW	65	967.25	26	311.44	39	655.81	–	–
	CSBJ	59	1247.82	27	371.96	32	875.86	–	–
	SSOLVE	59	1814.76	26	231.47	33	1583.29	–	–
	QBFL-BS	56	262.61	25	244.81	31	17.8	–	–
	QMRES	51	1838.81	28	1166.75	23	672.06	10	171.24
	OPENQBF	49	1459.92	20	170.26	29	1289.66	–	–
	WALKQSAT	40	1599.04	21	771.14	19	827.9	–	–
	ORSAT	27	774.68	20	718.77	7	55.91	–	–
Planning	YQUAFFLE	100	3261.00	37	1763.32	63	1497.68	4	1843.74
	SSOLVE	93	2786.63	35	353.63	58	2433.00	–	–
	QBFL-JW	92	2243.19	30	706.42	62	1536.77	–	–
	SEMPROP	88	3935.57	27	348.03	61	3587.54	2	563.87
	QUANTOR	85	479.50	28	311.18	57	168.32	1	37.35
	QBFL-BS	84	569.73	21	0.79	63	568.94	–	–
	CSBJ	84	1052.90	30	872.49	54	180.41	–	–
	OPENQBF	80	3404.27	27	1867.38	53	1536.89	–	–
	WALKQSAT	55	143.52	16	0.69	39	142.83	–	–
	QMRES	37	2513.36	18	490.71	19	2022.65	–	–
	ORSAT	38	65.04	11	13.47	27	51.57	–	–
Miscellaneous	QMRES	132	1980.92	74	1657.88	58	323.04	20	749.35
	QUANTOR	117	635.82	63	401.31	54	234.51	–	–
	SEMPROP	117	4293.44	70	2339.32	47	1954.12	6	281.64
	SSOLVE	81	4115.79	48	2762.40	33	1353.39	–	–
	YQUAFFLE	76	2212.21	38	1051.83	38	1160.38	1	179.42
	WALKQSAT	62	5517.68	44	3340.22	18	2177.46	–	–
	CSBJ	54	4222.49	36	2158.17	18	2064.32	–	–
	OPENQBF	46	1710.80	26	1160.66	20	550.14	–	–
	QBFL-BS	41	2243.40	29	1220.50	12	1022.90	–	–
	QBFL-JW	38	2362.21	27	1831.79	11	530.42	–	–
	ORSAT	7	403.59	6	402.50	1	1.09	–	–

Focusing on formal verification category, we can see that all the solvers are pretty much in the same capability ballpark. Considering the three strongest solvers, namely QUANTOR, SEMPROP and YQUAFFLE, we can see that both QUANTOR and SEMPROP are able to uniquely conquer 10 and 2 instances, respectively, while YQUAFFLE is subsumed by the portfolio constituted by all the other solvers. At the same time, QUANTOR, with 38.58s average solution time, and SEMPROP, with 29.07s average solution time, seem to be slightly less optimized than YQUAFFLE, which fares a respectable 18.22s average solution time. Among the other solvers, it is worth noting that QBFL-JW and QBFL-BS perform quite nicely in terms of average solution time (14.88s and 4.68s, respectively), and QMRES stands out for its ability to conquer 10 instances that defied all the other participants.

As for the planning category, we can see that given the relative easiness of the benchmarks selected for the evaluation, the differences between the solvers are substantially smoothed. This is also witnessed by the fact that only 4, 2 and 1 instances where uniquely conquered by, respectively, YQUAFFLE (which is also the strongest in this category), SEMPROP and QUANTOR. One possible explanation of these results is that most of the benchmarks in this category,

Table 4. Results of the evaluation second stage (random benchmarks)

Category	Solver	Total		Sat		Unsat		Unique	
		#	Time	#	Time	#	Time	#	Time
Model A	SSOLVE	187	1025.92	64	346.64	123	679.28	16	498.75
	SEMPROP	168	3055.93	58	1602.71	110	1453.22	–	–
	CSBJ	104	5629.68	52	1722.02	52	3907.66	–	–
	WALKQSAT	100	5461.75	53	1896.45	47	3565.3	–	–
	QBFL-JW	94	1926.84	72	0.64	22	1926.2	–	–
	QBFL-BS	91	378.23	72	0.74	19	377.49	–	–
	OPENQBF	69	3092.8	42	183.21	27	2909.59	–	–
	YQUAFFLE	53	1375.92	27	139.79	26	1236.13	–	–
	QMRES	31	1591.19	31	1591.19	0	0	–	–
	QUANTOR	30	0.61	30	0.61	0	0	–	–
	ORSAT	0	0	0	0	0	0	–	–
Model B	SSOLVE	216	2.38	22	0.23	194	2.15	–	–
	SEMPROP	216	13.35	22	11.35	194	2	–	–
	CSBJ	216	326.38	22	316.37	194	10.01	–	–
	WALKQSAT	216	376.3	22	365.72	194	10.58	–	–
	OPENQBF	208	2433.19	22	151.3	186	2281.89	–	–
	QBFL-JW	169	4453.42	22	57.71	147	4395.71	–	–
	YQUAFFLE	144	1790.7	7	44.9	137	1745.8	–	–
	QBFL-BS	127	2887.7	20	0.32	107	2887.38	–	–
	QMRES	111	2500.1	22	2.88	89	2497.22	–	–
	QUANTOR	90	263.14	22	0.17	68	262.97	–	–
	ORSAT	60	1.24	0	0	60	1.24	–	–

with the only exception of Gent and Rowley's connect[3-9] families, have been around for quite some time, so developers had access to them for tuning their solvers before the evaluation.

Finally, considering the miscellaneous category, the first thing to be observed is that most of these benchmarks come from the Pan families. Since such benchmarks are derived from translations of structured modal K instances [19], and the translation algorithm applied is the same for all the benchmarks, it is reasonable to assume that the original structure, although obfuscated by the translation, carries over to the QBF instances. In conclusion, the best solvers in this category are probably those that can discover and take advantage of such a hidden structure. Looking at the results it seems that QMRES, both the strongest solver and the only one able to conquer 20 instances (more than 10% of the total), is clearly the one taking the most advantage of the benchmark structure. Also QUANTOR and SEMPROP perform quite nicely by conquering 117 instances: SEMPROP, although slightly slower than QUANTOR on average, is also able to uniquely conquer 6 instances. Noticeably, the fourth strongest solver (SSOLVE) trails the path of the strongest three at a substantial distance (36 instances, about 20% of the total instances).

In Table 4 we report second stage results about random benchmarks (432 benchmarks, 11 solvers), divided into two categories:

Model A 12 families and 216 benchmarks, generated according to the guidelines presented in Section 3 to cover the space $a = \{2, 3, 4, 5\}$, $v = \{50, 100, 150\}$, where a is the number of alternations in the prefix and v is the number of variables: each of the 12 families corresponds to a given setting of a, v and contains formulas with a ratio r clauses/variables in the range 2 to 18 (step 2), and 2 instances per each value of a, v, and r.

Model B 12 families and 216 benchmarks, generated according to the same parameters as model A families.

Table 4 is arranged analogously to Table 3.

Looking at Table 4, the first observation is that the solvers performed very well on model B instances: the strongest one in the category (SSOLVE) solves 100% of the instances, the weakest one (ORSAT) solves about 27%, and 9 out of the 11 solvers admitted to the second stage are able to solve more than 50% of the instances in this category. Model A instances turned out to be slightly more difficult to solve since the strongest solver (again SSOLVE) conquered about 86% of the instances, and with the only exception of SSOLVE and SEMPROP, all the other solvers do not go beyond the 50% threshold. Noticeably, most solvers that do very well on non-random instances have troubles with the random ones: this is the case of QMRES, QUANTOR and YQUAFFLE, and the phenomenon is particularly evident on model A instances. This seems to validate analogous results in SAT, where solvers that are extremely good on non-random instances, fail to be effective on random ones. On the other hand SSOLVE and SEMPROP partially contradict this result, in that they are the most effective on random instances, and still reasonably effective on non-random instances as we have seen before. While QMRES and QUANTOR abandon the top positions on the random benchmarks, WALKQSAT performs much better on this kind of instances, possibly indicating that its incomplete algorithm is much more suited to random, rather than structured instances. On the other hand, QUANTOR and QMRES results, are a clear indication that their non-DPLL based algorithms have been tuned heavily on structured instances, and are possibly less adequate for randomly generated ones.

5.2   Benchmark-Centric View

In Table 5 we show the classification of the non-random benchmarks included in the evaluation test set according to the solvers admitted to the second stage. Table 5 consists of nine columns where for each family of instances we report the name of the family (column Family), the number of instances included in the family, the number of instances solved, the number of such instances found SAT and the number found UNSAT (group "Overall", columns "N", "#", "S", "U", respectively), the time taken to solve the instances (column "Time"), the number of easy, medium and medium-hard instances (group "Hardness", columns "EA", "ME", "MH"). The number of instances solved and the cumulative time taken for each family is computed considering the "SOTA solver", i.e., all the second stage solvers running in parallel. A benchmark is thus solved if at least one of the solvers conquers it, and the time taken is the best among the times of the solvers that conquered the instance. The benchmarks are classified according to their hardness with the following criteria: easy benchmarks are those solved by all the solvers, medium benchmarks are those solved by at least two solvers, medium-hard benchmarks are those solved by one reasoner only, and hard benchmarks

Table 5. Classification of non-random benchmarks (second stage data)

Family	Overall				Time	Hardness		
	N	#	S	U		EA	ME	MH
Adder	8	6	2	4	132.54	0	2	4
C432	8	8	3	5	1787.99	0	5	3
C499	8	5	3	2	19.97	0	4	1
C5315	8	4	2	2	437.76	0	3	1
C6288	8	1	1	0	3.93	0	0	1
C880	8	2	2	0	1.66	0	2	0
comp	8	8	4	4	0.62	0	8	0
counter	8	4	4	0	0.05	2	2	0
DFlipFlop	8	8	0	8	3.16	1	7	0
jmc_quant1	8	3	2	1	18.68	0	1	2
jmc_quant2	8	3	2	1	11.45	0	0	3
MutexP	7	7	7	0	7.30	2	5	0
s1196	6	0	0	0	–	0	0	0
s1269	8	0	0	0	–	0	0	0
s27	4	4	1	3	4.43	1	3	0
s298	8	1	1	0	452.96	0	0	1
s3271	8	8	0	8	1.90	0	8	0
s3330	8	1	1	0	154.87	0	0	1
s386	8	0	0	0	–	0	0	0
s499	8	3	3	0	70.41	0	1	2
s510	8	0	0	0	–	0	0	0
s641	8	1	1	0	350.81	0	0	1
s713	8	1	1	0	287.14	0	0	1
s820	8	0	0	0	–	0	0	0
SzymanskiP	8	8	0	8	211.20	0	8	0
term1	8	8	4	4	164.71	0	7	1
uclid	3	0	0	0	–	0	0	0
VonNeumann	8	8	0	8	16.40	0	8	0
z4ml	8	8	4	4	0.03	5	3	0
Blocks	8	8	2	6	215.02	0	7	1
Connect3	8	8	0	8	6.92	0	8	0
Connect4	8	8	0	8	3.31	3	5	0
Connect5	8	8	0	8	6.72	1	7	0

Family	Overall				Time	Hardness		
	N	#	S	U		EA	ME	MH
Connect6	8	8	0	8	2.99	1	7	0
Connect7	8	8	0	8	5.22	0	8	0
Connect8	8	8	0	8	6.29	0	8	0
Connect9	3	3	0	3	2.81	0	3	0
RobotsD2	8	8	6	2	256.12	0	8	0
RobotsD3	8	7	6	1	2411.13	0	4	3
RobotsD4	8	7	5	2	580.83	0	5	2
RobotsD5	8	8	4	4	871.52	0	7	1
Toilet	8	8	5	3	7.75	2	6	0
ToiletA	8	8	1	7	6.99	3	5	0
ToiletC	8	8	3	5	0.95	4	4	0
ToiletG	7	7	7	0	0.07	6	1	0
Chain	8	8	8	0	9.25	0	8	0
Impl	8	8	8	0	0.06	4	4	0
k_branch_n	8	4	4	0	269.07	0	1	3
k_branch_p	8	3	0	3	5.86	0	1	2
k_d4_n	8	8	8	0	32.64	0	1	7
k_d4_p	8	8	0	8	12.44	0	7	1
k_dum_n	8	8	8	0	5.34	0	8	0
k_dum_p	8	8	0	8	6.88	0	8	0
k_grz_n	8	8	8	0	1069.34	0	7	1
k_grz_p	8	8	0	8	9.19	0	7	1
k_lin_n	8	8	8	0	21.54	1	6	1
k_lin_p	8	8	0	8	4.80	0	8	0
k_path_n	8	8	8	0	3.55	0	8	0
k_path_p	8	8	0	8	11.84	0	8	0
k_ph_n	8	7	7	0	184.35	1	5	1
k_ph_p	8	3	0	3	1.46	3	0	0
k_poly_n	8	8	8	0	5.59	0	8	0
k_poly_p	8	8	0	8	0.15	0	8	0
k_t4p_n	8	8	8	0	94.10	0	2	6
k_t4p_p	8	8	0	8	22.92	0	4	4
Tree	8	8	2	6	4.82	1	7	0

are those that remained unsolved. Finally, Table 5 is divided into three sections grouping respectively, families of formal verification, planning and miscellaneous benchmarks.

According to the data summarized in Table 5, the non-random part of the evaluation second stage consisted of 510 instances, of which 383 have been solved, 172 declared satisfiable and 211 declared unsatisfiable, resulting in 38 easy, 289 medium, 56 medium-hard, and 127 hard instances (respectively, the 7%, 56%, 10%, and 24% of the test set). These results indicate that the selected non-random benchmarks are not trivial for current state-of-the-art QBF solvers, since there is a little number of easy instances, and a substantial percentage of medium-to-hard instances. At the same time, the test set is not overwhelming, since most of the non-easy instances could be considered of medium difficulty, i.e., they are solved by at least two solvers.

The cumulative results about Table 5 are not balanced across each single category: formal verification families contributed 110 hard and 22 medium-hard benchmarks, planning families contributed only 2 hard and 7 medium-hard benchmarks, and the miscellaneous families (essentially the Pan families) contributed 15 hard and 27 medium-hard benchmarks. The families submitted for

Table 6. Classification of random benchmarks (second stage data)

Family	Overall				Time	Hardness		
	N	#	S	U		EA	ME	MH
2qbf-5cnf-50var	18	18	3	15	17.63	0	15	3
2qbf-5cnf-100var	18	17	2	15	453.77	0	13	4
2qbf-5cnf-150var	18	14	2	12	46.41	0	11	3
3qbf-5cnf-50var	18	18	9	9	17.83	0	18	0
3qbf-5cnf-100var	18	17	8	9	6.7	0	15	2
3qbf-5cnf-150var	18	16	8	8	23.16	0	16	0
4qbf-5cnf-50var	18	18	6	12	10.18	0	17	1
4qbf-5cnf-100var	18	16	4	12	2.49	0	15	1
4qbf-5cnf-150var	18	17	5	12	6.38	0	15	2
5qbf-5cnf-50var	18	17	10	7	9.07	0	17	0
5qbf-5cnf-100var	18	15	9	6	21.7	0	15	0
5qbf-5cnf-150var	18	16	10	6	48.98	0	16	0

Family	Overall				Time	Hardness		
	N	#	S	U		EA	ME	MH
2qbf-5cnf-50var	18	18	1	17	0.45	0	18	0
2qbf-5cnf-100var	18	18	1	17	7.49	0	18	0
2qbf-5cnf-150var	18	18	0	18	0.58	0	18	0
3qbf-5cnf-50var	18	18	2	16	0.01	11	7	0
3qbf-5cnf-100var	18	18	2	16	0.1	2	16	0
3qbf-5cnf-150var	18	18	2	16	0.15	0	18	0
4qbf-5cnf-50var	18	18	2	16	0.15	0	18	0
4qbf-5cnf-100var	18	18	2	16	0.19	0	18	0
4qbf-5cnf-150var	18	18	2	16	0.28	0	18	0
5qbf-5cnf-50var	18	18	4	14	0.1	1	17	0
5qbf-5cnf-100var	18	18	2	16	0.22	0	18	0
5qbf-5cnf-150var	18	18	2	16	0.53	0	18	0

the evaluation resulted pretty hard for the solvers: only 4 out of 8 instances in the counter (Biere) benchmarks and 3 out of 8 in the jmc (Katz) benchmarks were solved, while none of the uclid (Lahiri/Seshia) benchmarks was solved. Quite interesting are also the results for the Mneimneh/Sakallah's and Gent/Rowley's families, which have never been extensively tested before: the benchmarks in the former resulted quite hard in accordance to what reported in [17] (only 20% of the instances solved), while the latter resulted well within the capabilities of the current state-of-the-art solvers, but not trivial (100% of the instances solved, only 5 easy instances). Among the "older" benchmarks, the Adder family (Ayari), and the C499, C5315, C6288, and C880 families (Scholl/Becker) are still quite challenging as they resulted in the last year's evaluation [1].

In Table 6 we show the classification of the random benchmarks included in the evaluation test set according to the solvers admitted to the second stage. Table 6 is arranged similarly to Table 5: Table 6 (left) is about model A instances, Table 6 (right) is about model B instances. According to the data summarized in Table 6, the random part of the evaluation second stage consisted of 432 instances, of which 415 have been solved, 98 declared satisfiable and 317 declared unsatisfiable, resulting in 14 easy, 385 medium, 16 medium-hard, and 17 hard instances. These results indicate that the selected non-random benchmarks are not trivial for current state-of-the-art QBF solvers, although less challenging than the non-random ones. All other things being equal, model A instances provide a much more challenging test set than model B instances. Using model A instances, an increase in the number of variables determines an increase in the number of hard instances in the case of 2- and 3-qbfs, but this is not confirmed in the case of 4- and 5-qbfs. The number of alternations does not seem to be correlated with the hardness: considering 150 variables benchmarks, there are 4 hard 2-qbfs, 2 hard 3-qbfs, 1 hard 4-qbf, and 2 hard 5-qbfs. Although the number of samples is too small for each single point to draw definitive conclusions, the variety of solvers used for the evaluation supports the conclusions drawn in [22], and restated also in [1], that model A instances seem to show a counter-intuitive relationship between hardness and the number of alternations in their prefix.

6   Conclusions

The final balance of the second QBF comparative evaluation can be summarized as follows:

- 16 solvers participated, 15 complete and 1 incomplete: 13 search-based algorithms, 1 (QUANTOR) based on Q-resolution and expansion, and 1 (QMRES) based on a symbolic implementation.
- 88 formal verification benchmarks, plus a random generator were submitted.
- State-of-the-art solvers, both for random and non-random benchmarks, have been identified; also, a total of 144 challenging benchmarks that cannot be solved by any of the participants have been identified to set the reference point for future developments in the field.
- Some of the challenges outlined last year in [1] have been tackled by the participants this year: Challenge 1, about the 2003 hard benchmarks, was attempted by most solvers with noticeable progress made; Challenge 8, about alternative paradigms to search-based QBF solvers, was undertaken quite successfully by QUANTOR and QMRES; finally, all the other challenges have been at least surfaced by most of the participants, a good indicator of the stimulus that the QBF evaluation is providing to the researchers.

The evaluation also evidenced some critical points:

- The QBF evaluation is still a niche event if compared to the SAT competition: 55 SAT solvers from 27 authors and 999 benchmarks were submitted this year to the SAT competition.
- QBF encoding of real-world applications (e.g., Ayari's hardware verification problems, Sakallah's vertex eccentricity problems, etc.) contributed a lot to the pool of 144 challenging benchmarks. This shows that QBF developers must improve the performance of their solvers before these can be practical for industrial-sized benchmarks.
- The question of how to check the answer of the QBF solvers in an effective way is still unanswered. Specifically, the questions of what is a good certificate of satisfiability/unsatisfiability for QBF, and, if this proves too huge to be practical, what is a good approximation of such certificate, remain open.

The last point is not only an issue for the QBF evaluation, but also for the implementation of QBF solvers: indeed, while only two versions of one solver were found incorrect in the SAT competition, we had problems with 4 solvers in the QBF evaluation. Overall, the evaluation showed the vitality of QBF as a research area. Despite some technological limitations and some maturity issues, it is our opinion that the development of effective QBF solvers and the use of QBF-based automation techniques can be regarded as promising research directions.

References

1. D. Le Berre, L. Simon, and A. Tacchella. Challenges in the QBF arena: the SAT'03 evaluation of QBF solvers. In *Sixth International Conference on Theory and Applications of Satisfiability Testing (SAT 2003)*, volume 2919 of *Lecture Notes in Computer Science*. Springer Verlag, 2003.
2. M. Davis and H. Putnam. A computing procedure for quantification theory. *Journal of the ACM*, 7(3):201–215, 1960.
3. M. Davis, G. Logemann, and D. Loveland. A machine program for theorem proving. *Communications of the ACM*, 5(7):394–397, 1962.
4. A. G. D. Rowley I. P. Gent, H. H. Hoos and K. Smyth. Using stochastic local search to solve quantified boolean formulae. In *9th Conference on Principles and Practice of Constraint Programming (CP 2003)*, volume 2833 of *Lecture Notes in Computer Science*. Springer Verlag, 2003.
5. A. Biere. Resolve and Expand. In *Seventh Intl. Conference on Theory and Applications of Satisfiability Testing*, 2004. Extended Abstract.
6. Guoqiang Pan and Moshe Y. Vardi. Symbolic decision procedures for qbf. In *10th Conference on Principles and Practice of Constraint Programming (CP 2004)*, 2004.
7. Ian Gent and Toby Walsh. Beyond NP: the QSAT phase transition. In *Proc. of AAAI*, pages 648–653, 1999.
8. E. Giunchiglia, M. Narizzano, and A. Tacchella. Quantified Boolean Formulas satisfiability library (QBFLIB), 2001. www.qbflib.org.
9. Andrew G D Rowley Ian P Gent. Solution learning and solution directed backjumping revisited. Technical Report APES-80-2004, APES Research Group, February 2004.
10. D. Le Berre, M. Narizzano, L. Simon, and A. Tacchella, editors. *Second QBF solvers evaluation*. Pacific Institute of Mathematics, 2004. Available on-line at www.qbflib.org.
11. Jussi Rintanen. Partial implicit unfolding in the Davis-Putnam procedure for quantified Boolean formulae. In *Logic for Programming, Artificial Intelligence and Reasoning. 8th International Conference*, number 2250 in LNAI, pages 362–376. Springer, 2001.
12. R. Feldmann, B. Monien, and S. Schamberger. A distributed algorithm to evaluate quantified boolean formula. In *Proceedings of the Seventeenth National Conference in Artificial Intelligence (AAAI'00)*, pages 285–290, 2000.
13. R. Letz. Lemma and model caching in decision procedures for quantified boolean formulas. In *Proceedings of Tableaux 2002*, LNAI 2381, pages 160–175. Springer, 2002.
14. Abdelwaheb Ayari and David Basin. Bounded model construction for monadic second-order logics. In *12th International Conference on Computer-Aided Verification (CAV'00)*, number 1855 in LNCS, pages 99–113. Springer-Verlag, 2000.
15. C. Castellini, E. Giunchiglia, and A. Tacchella. Sat-based planning in complex domains: Concurrency, constraints and nondeterminism. *Artificial Intelligence*, 147(1):85–117, 2003.
16. Andrew G D Rowley Ian P Gent. Encoding connect 4 using quantified boolean formulae. Technical Report APES-68-2003, APES Research Group, July 2003.
17. M. Mneimneh and K. Sakallah. Computing Vertex Eccentricity in Exponentially Large Graphs: QBF Formulation and Solution. In *Sixth International Conference on Theory and Applications of Satisfiability Testing (SAT 2003)*, volume 2919 of *Lecture Notes in Computer Science*. Springer Verlag, 2003.

18. Guoqiang Pan and Moshe Y. Vardi. Optimizing a BDD-based modal solver. In *Proceedings of the 19th International Conference on Automated Deduction*, 2003.
19. P. Balsiger, A. Heuerding, and S. Schwendimann. A benchmark method for the propositional modal logics k, kt, s4. *Journal of Automated Reasoning*, 24(3):297–317, 2000.
20. Jussi Rintanen. Improvements to the evaluation of quantified boolean formulae. In *Proceedings of the Sixteenth International Joint Conferences on Artificial Intelligence (IJCAI'99)*, pages 1192–1197, Stockholm, Sweden, July 31-August 6 1999. Morgan Kaufmann.
21. C. Scholl and B. Becker. Checking equivalence for partial implementations. In *38th Design Automation Conference (DAC'01)*, 2001.
22. E. Giunchiglia, M. Narizzano, and A. Tacchella. An Analysis of Backjumping and Trivial Truth in Quantified Boolean Formulas Satisfiability. In *Seventh Congress of the Italian Association for Artificial Intelligence (AI*IA 2001)*, volume 2175 of *Lecture Notes in Artificial Intelligence*. Springer Verlag, 2001.

Author Index

Lecture Notes in Computer Science

For information about Vols. 1–3482

please contact your bookseller or Springer